Soils

Their formation, classification and distribution

SOILS

Their formation, classification and distribution

E. A. FitzPatrick
Senior Lecturer in Soil Science
University of Aberdeen

Longman
London and New York

Longman Group Limited London

*Associated companies, branches and representatives
throughout the world*

*Published in the United States of America
by Longman Inc., New York*

© E. A. FitzPatrick 1980

First published 1980. This is a completely rewritten new edition of
PEDOLOGY A Systematic Approach to Soil Science, first published by Oliver & Boyd 1971

British Library Cataloguing in Publication Data
FitzPatrick, Ewart Adsil
  Soils. − New ed.
  1. Soils
  I. Title II. Pedology
  631.4   S591   78−41165

  ISBN 0−582−44188−9

Printed in Great Britain at The Pitman Press, Bath

Contents

Preface

List of figures

Over the past few years the need for a comprehensive text on pedology which could be used by as many people as possible, because of the large number of different systems of classification used in different parts of the world; has become more and more urgent. This need is particularly strongly felt among undergraduate students who have to depend almost entirely on lecture notes, a few general references and the very dilute treatment found in some general books on soil science.

Although it is best to use a single set of jargon and classification this would please only a small percentage of the potential users. It was decided to use simultaneously three sets of terminology when necessary, but also to present the information in such a manner, that except in Chapters 6 and 7, the text could be read and understood without the reader being committed to the jargon. Certainly the first four chapters are virtually free of jargon.

Except where stated the photographs have been produced by the author who is also responsible for the line drawings, most of which are original though a small number have been adapted from the work of others.

E. A. FitzPatrick
Aberdeen
September 1978

vii

CHAPTER 3

x

CHAPTER 7

List of colour plates

List of tables

CHAPTER 7

APPENDIX II

APPENDIX III

Introduction

HISTORICAL

The development of soil studies naturally falls into two stages; the first stretches back many centuries and references to agricultural practices are found frequently in early religious literature. Extensive discussions about soils occur in the writings of the Greeks, particularly Plato who witnessed the erosion of the hills around Athens and warned of its dangers. In Roman times there was a considerable literature about agriculture, but it was merely a collection of facts about the systems in use up to that time and rather speculative interpretations about the results but surprisingly some of the conclusions have been sustained subsequently.

The second stage is relatively recent having developed only during the last two centuries; but it has a sound foundation being based on experimentation and the application of the scientific method. This stage started with the work of Théodore de Saussure (1804) who founded the quantitative experimental method and applied it to the oxygen and carbon dioxide relationships in plants. He was able to show that carbon dioxide is absorbed and oxygen released by plants. This work ushered in a completely new approach to the problem of plant nutrition and plant physiology to which the outstanding contributors included Liebig and more especially Boussingault. At first sight these studies of de Saussure on plants might appear to be unconnected with soil science but in the early days there was considerable controversy about the origin of the material in plants and the contributions made by the soil, therefore his discovery was a fundamental contribution to the study of soil–plant relationships.

In the nineteenth century Dokuchaev (1883) produced his classical work on the Chernozems of Russia and recognised for the first time that soils are made up of several layers. In addition, he and his students conducted surveys and discovered that the nature and properties of soils vary with their environmental factors, particularly climate and vegetation. These contributions initiated a major revolution in soil science for, as indicated above most of the previous work was devoted to soil fertility and plant nutrition. Thus, quite suddenly, soils became objects for academic study and the investigation of soils as a natural phenomenon emerged as a separate scientific discipline known as *Pedology* with its own set of principles, concepts and methodology. These discoveries in Russia probably took place as a result of the enormous size of the country and its span over many different types of climate and vegetation. At the same time Hilgard (1914) in the USA, Müller (1887) in Germany and others were making detailed descriptions and investigations of Podzols in a manner similar to those being conducted in Russia. Principally because of the language difficulties, knowledge of the Russian work spread very slowly to the rest of the world but wherever it reached, the impact was great and acceptance relatively quick, with the result that now there is quite a body of data about most of the major soils of the world.

Ideally, pedology is the study of soils as naturally occurring phenomena, taking into account their composition, distribution and methods of formation.

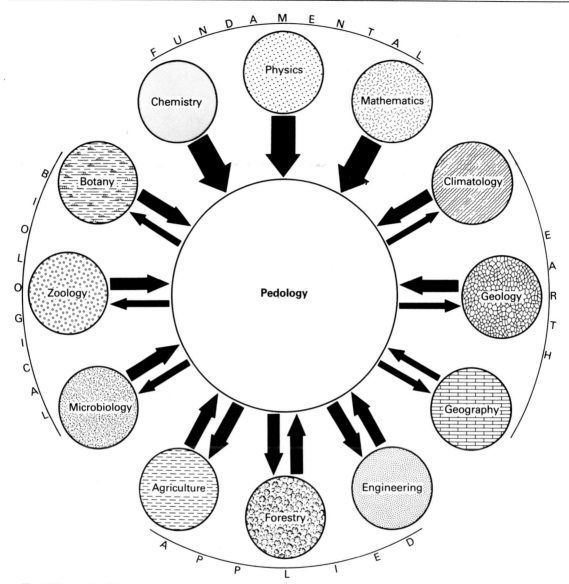

Fig. 1 The relationships between pedology and some other sciences

However, such studies cannot be separated entirely from other ramifications of soil science, including some fairly detailed aspects of soil chemistry, soil physics, soil microbiology and land·utilisation.

Unfortunately, pedology is still somewhat fragmented and lacking unification, largely because of its relationship with so many other disciplines as is shown in Fig. 1. Most sciences contribute to the study of soils, whereas the fundamental sciences make a unilateral contribution, the others enter into reciprocal relationships, the greatest and most complete being between pedology and the applied sciences of agriculture and forestry. In most disciplines unification comes through a common purpose or a generally accepted system of classification. Both are lacking in pedology, particularly the latter, but it has engaged the attention of most pedologists. Also, because of the association of pedology with so many sciences the terminology is large and diverse, which requires readers to have a wide knowledge of many disciplines or otherwise numerous definitions must be given in the text. Neither of these alternatives is realistic or satisfactory, consequently a glossary of terms is given which commences on page 334.

Purpose of the book

There are a number of textbooks covering various aspects of soil science; some include chapters on pedology, but very few are devoted entirely to this subject. There are many reasons for this. Firstly the information required to produce such a text has accumulated only during the last forty years from the significant number of purely pedological investigations throughout the world. The second reason is the lack of agreement among pedologists about the nomenclature and classification of soils. This is probably the foremost factor that has deterred many potential authors. A further reason is the subject's youth and its rapid growth, with the possibility that any textbook could quickly become out of date, irrespective of the care and effort spent over its preparation.

This book has been prepared as an aid to teaching. It is produced primarily for undergraduate students in agriculture, forestry, geography and soil science taking pedology as a part of their training in soil science. Postgraduate students having advanced training in soil science, but specialising in a branch other than pedology, should also find it useful. It must be emphasised that this is not a treatise for the specialist pedologist. The intention is to present a modest level of information about the fundamental concepts in pedology, the factors of soil formation and the characteristic soils of the world, their relationships to each other as well as to their factors of formation. Any reader requiring specific local knowledge is referred to the regional soil survey reports and memoirs.

Acknowledgements

We are grateful to the following for permission to reproduce copyright material:

The Association of American Geographers for a figure by L. Peltier from *Annals* of the Association of American Geographers, Vol. 40, 1950; John Bartholomew & Son Ltd., for free use of *Bartholomew's Regional Projection*; The British Library for two figures (originally photographs) of Silicon, by permission of the British Library; Cambridge University Collection (copyright reserved) for an aerial photograph of a salt marsh in Norfolk; CSIRO Australia for Fig. 3.4; the author, Dr. J. B. Dalrymple for a diagrammatic representation of hypothetical nine unit landsurface model, according to Dalrymple, Blong and Conacher 1968; the author, L. King for a figure from *The Morphology of the Earth: A Study and Synthesis of World Scenery* by L. King, published by Oliver and Boyd 1962; The Macaulay Institute for Soil Research for clay mineral micrographs; Macmillan Publishing Co. Inc., for 2 figures ' General Relationship Between Soil Moisture Characteristics and Soil Texture' and 'Relationships Between types of water, types of flow and suction' from *The Nature and Properties of Soils* 8th Edition by Nyle C. Brady Copyright © 1974, Macmillan Publishing Co. Inc; Rellim Technical Publications for a figure of Value/Chroma rating of Northcote; Ordnance Survey for part of the Church Stretton map; The Soil Survey of England and Wales (Rothamsted Experimental Station, Harpenden, Herts.) for a photograph showing the relationship between tone pattern and soil distribution pattern; Mr. F. M. Synge for a figure (originally a photograph) of stone polygons from Baffin Island.

I should like to thank those persons who have helped in the preparation of this book through their suggestions or by providing material in one form or another. To: Professor J. Tinsley, Drs. T. Batey, M. N. Court, N. T. Livesey, D. A. McLeod, C. E. Mullins and J. W. Parsons my grateful thanks for reading and commenting on the material at various stages. Finally, my thanks to Mr. A. M. Shanks for his help in preparing Figs. 6.1, 6.12 and 7.7.

E. A. FitzPatrick

Fundamental properties of soils

THE SOIL PROFILE

Farmers and gardeners usually regard soil as the upper few centimetres of the earth's crust that are either cultivated or permeated by plant roots. This belief or attitude, however adequate in many cases, is at the same time limited and does not allow a full appreciation of the very wide differences between various soils, nor a full comprehension of their potentialities and the problems of their amelioration. The pedologist recognises that there is not only the surface layer but many others beneath and he also considers the relationships between soils and the factors and forces of their environment. This is a much wider, more fundamental and fully comprehensive attitude.

The first step towards understanding soils is to dig a hole or pit into the surface of the earth and to carry out visual observations, which are sometimes supplemented by simple qualitative tests. The depth of the pit is determined by the nature of the soil itself, but normally varies from 1 to 3 metres below which is relatively unaltered material. After digging the pit a vertical face is prepared; this reveals a layered pattern usually characterised by differences in the colours of the various layers. Each individual layer is known as an *horizon* and the set of layers in a single pit is called a *soil profile*.

In some cases the contrast between horizons is dramatic and self-evident, while in others it is very subtle. An insight into these profile characteristics is gained by considering a Podzol profile and a Ferralsol profile described below.

A Podzol profile

This soil has been chosen as the first example because it displays sharp contrasts between the horizons and because of its widespread distribution in the humid temperate areas of the world and also of its limited but important occurrence in humid tropical and sub-tropical areas. Two illustrations are provided, the first is the idealised diagram given in Fig. 1.1 and the second is a photograph of a profile shown in Plate III d. In both cases the freshly fallen plant litter forms the uppermost layer below which is dark brown partially decomposed plant material which grades into more highly decomposed very dark brown or black organic matter. Abundant fungal mycelium and small arthropods inhabit the organic matter and are largely responsible for its decomposition. Also within these organic layers there is an abundance of plant roots which take up plant nutrients as they are released by decomposition of the organic matter.

Beneath the organic horizons there is a very dark grey mixture of black organic matter and light coloured mineral grains, mainly quartz and feldspars. This mixture is formed by two main processes. The organic matter is washed in from the horizon above, while the light coloured grains are due to weathering and the removal by percolating water of the colouring substances which are mainly compounds of iron.

This dark grey horizon is underlain by a very pale or white horizon composed mainly of quartz and feldspars and in many cases is dominantly quartz. It is also formed predominantly by weathering and the removal of colouring substances. In addition to iron, the percolating water contains many other substances including humus and aluminium. Some of these substances, particularly the iron, humus and aluminium are deposited and accumulate to form the dark brown middle horizon. This dark horizon grades gradually into the relatively unaltered underlying material which is usually a sedimentary deposit such as alluvium, glacial drift or dune sands.

A Ferralsol profile

This is an example of a soil which has developed under humid tropical conditions through progressive weathering of rock. It also illustrates those soils which exhibit distinct but subtle changes from one horizon to the other. A photograph of a profile is given in Plate Id while Fig. 1.2 is an idealised diagram showing the profile with a gradation through weathered rock into fresh rock. At the surface there is a thin loose litter of leaves and twigs containing numerous

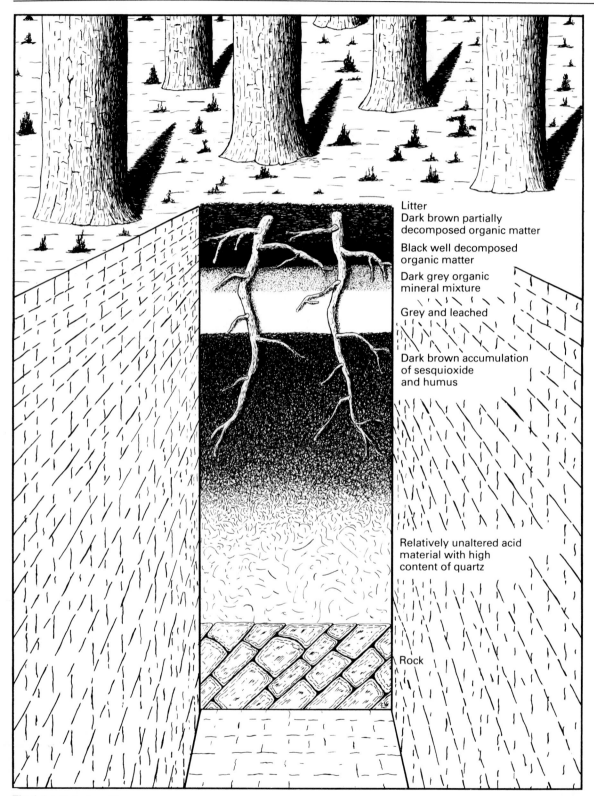

Litter
Dark brown partially
decomposed organic matter

Black well decomposed
organic matter

Dark grey organic
mineral mixture

Grey and leached

Dark brown accumulation
of sesquioxide
and humus

Relatively unaltered acid
material with high
content of quartz

Rock

Fig. 1.1 A Podzol profile

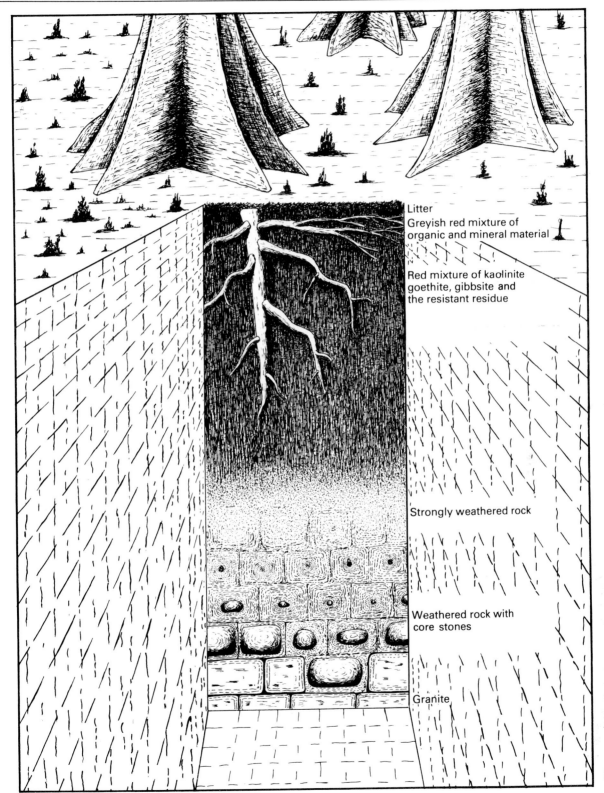

Litter
Greyish red mixture of
organic and mineral material

Red mixture of kaolinite
goethite, gibbsite and
the resistant residue

Strongly weathered rock

Weathered rock with
core stones

Granite

Fig. 1.2 A Ferralsol profile

earthworm casts. Beneath the litter there is a greyish-red mixture of organic and mineral material brought about mainly by earthworm activity. Within the litter and in the uppermost mineral horizon there is a high density of roots that take up the nutrients as they are released by decomposition of the organic matter. The organic—mineral mixture grades sharply into a bright red horizon having a high content of clay and varying amounts of iron concretions. It is formed by the complete decomposition of the primary silicates followed mainly by oxidation of the compounds of iron to give the bright red colour. This horizon grades into easily recognisable chemically weathered rock with its high content of unweathered primary silicates and some evidence of rock structure.

Below there is a gradation into the solid rock through a zone containing rounded unweathered areas of rock known as *core stones* which become progressively larger with depth. Finally, at the base of the profile there are the initial stages of weathering along the faces of the joint blocks.

These two examples serve to illustrate the wide variations that exist between soils. These two soils not only display a contrast in visible morphology they also differ in the degree of alteration of the material forming the soils. Whereas the Podzol may contain some weatherable minerals throughout the whole soil the Ferralsol virtually has none in the topsoil because it has formed as a result of profound and continuous weathering over a very long period of time.

Another difference is seen in the content of the surface organic matter there being much more in the Podzol. These two soils also demonstrate that soil formation takes place through the interaction of certain specific factors and therefore that soils are largely the product of their environment. There are five factors of soil formation, viz. parent material, climate, organisms, topography and time. Briefly, the parent material or mineral matter is altered and differentiated through physical, chemical and biological processes into horizons. Alteration takes place mainly through the interplay of moisture and temperature which are the two main components of climate.

Vegetation contributes organic matter to the soil while the soil flora and fauna aid organic matter decomposition. Development of the horizons takes place very gradually and occupies a span of time. Topography is not illustrated above but is considered fully in Chapter 2.

SOIL – A THREE-DIMENSIONAL CONTINUUM

The profile is simply a two-dimensional section through the soil. However, soil actually extends laterally in all directions over the surface of the earth forming a three-dimensional continuum. This property is of considerable and fundamental importance since false ideas can develop about soils when only profiles in isolated pits are examined. Furthermore, the horizon sequence in any one profile is not the same as at other points in the landscape. Over a short distance the sequence usually shows small changes in the degree of development of the horizons but over a long distance there is a complete change from one sequence of horizons to another sequence.

Figures 1.3 and 1.4 illustrate the three-dimensional character of a soil over a short distance, and show various types of change in the thickness and outline of the horizons. In these two diagrams the organic horizons at the surface are fairly uniform in thickness but the strongly decomposed organic matter has a wavy lower boundary. In the middle and lower parts of the soil the horizons are very irregular and appear to change in an apparently haphazard manner. Also the tops and bottoms of the horizon may have different patterns of change. Most of the horizons are continuous but the very irregular path of the thin iron pan has caused the horizon of iron accumulation to be discontinuous as best shown in Fig. 1.4. This overall pattern of change is very common, occurring in many different types of soils throughout the world.

Over a much greater distance one or more horizons gradually change laterally into other horizons with markedly different properties. These changes are often shown by variations in colour which are accompanied by changes in chemical and physical properties which may be of greater importance than the changes in colour. However, it is the colour change that is often most evident in the field. The distance through which individual horizons change is seldom the same, whereas one or more may change completely over a given distance others continue unaltered and have greater spatial distribution. The dissimilar rate of change in horizons is a further important soil characteristic.

An attempt is made in Fig. 1.5 to demonstrate how horizons change spatially. At stage one there are two horizons – solid circles and solid hexagons, while the solid triangles represent the relatively unaltered underlying material for the whole sequence. At stage two the solid circles and solid hexagons have changed into mixtures of solid and open circles overlying solid

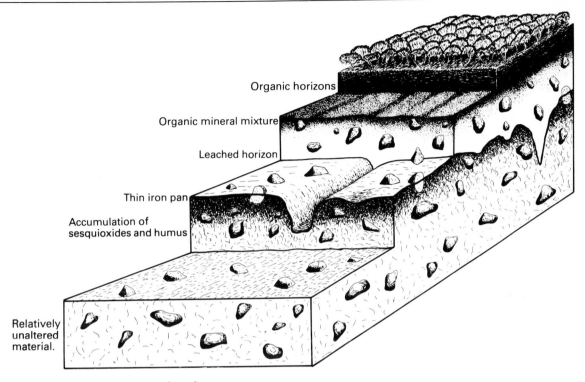

Organic horizons

Organic mineral mixture

Leached horizon

Thin iron pan

Accumulation of
sesquioxides and humus

Relatively
unaltered
material.

Fig. 1.3 Soil – a three dimensional continuum

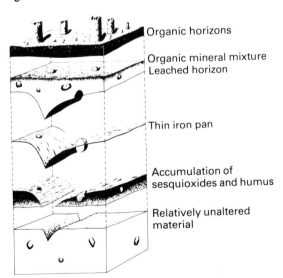

Organic horizons

Organic mineral mixture
Leached horizon

Thin iron pan

Accumulation of
sesquioxides and humus

Relatively unaltered
material

Fig. 1.4 Exploded soil showing variations in the thickness of horizons

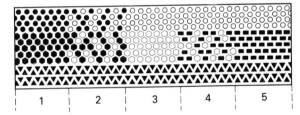

Fig. 1.5 Intergrading of horizons

gradually changes from open hexagons to solid rectangles. Thus, not only do soils change from place to place over the landscape, but the nature of the change can also vary. The nature of these lateral changes in the soil and the reasons for them vary from place to place and are determined by changes in the character of the five soil-forming factors. Probably one of the commonest types of change occurs down slopes where variations in the moisture status are chiefly responsible. On a larger scale, variations in the character of the parent material or climate often induce changes in soils. Vegetation and time also play their part in causing soil variation. Thus it is found that the different soils form a variety of patterns and have various interrelationships some of which are portrayed on maps while others can only be shown diagrammatically as given in Chapter 7.

and open hexagons and at stage three there are only open circles overlying open hexagons. These changes represent the situation in which the whole soil is changing gradually on travelling across the landscape. Stages four and five represent the situation in which the upper horizon remains uniform but the middle

5

SOIL — A PART OF THE LANDSCAPE

Soil can be considered as the thin veneer that covers a considerable part of the earth's surface and like most veneers is very vulnerable. Thus we find that the top part of the soil is constantly being disturbed and redistributed by running water. Generally, as water flows over the surface, varying amounts of soil are picked up in suspension and either deposited lower down the slope or carried away by rivers. This incessant removal of material leads to an extremely varied pattern of natural erosion and deposition. The result is that the majority of soils on slopes show some evidence of stratification, the amount varying from place to place, being greatest in the older landscapes such as those in many parts of the humid tropics where it can be considered as normal in most areas. This may cause the lower slopes to have a great thickness of stratified material. In some cases the formation of the strata is periodic so that the surface may remain static for a sufficiently long period for the formation of a soil which is buried by the deposition of the next straturn (Fig.7.15). Thus soil not only shows a vertical differentiation of horizons, but also one due to lateral surface movement. Therefore they must be considered as forming part of the landscape so that soil and landscape evolution proceed together and become more and more obvious as the age of the landscape increases. In fact, the wearing down of mountains to produce plains is really the progressive formation, removal and redistribution of the soil veneer.

SOIL — A CONTINUUM IN TIME

In addition to the redistribution of material at the surface, the internal part of the soil is either losing ions through drainage or gaining ions through lateral inwash. Thus, neither the surface of the soil nor the internal part is static. The result of these and other continuous and progressive changes is a complex dynamic system forming a continuum with three tangible dimensions as well as changing with time. These progressive changes in time may cause profound alteration in the character of the horizons, the properties of which are steadily changing. Thus horizons may change in space as well as in time. It follows therefore that soils have four dimensions and form a space—time continuum.

THE UNIT OF STUDY — THE PEDO-UNIT

Phenomena such as soils that change continuously in space and time pose a number of practical and theoretical problems with regard to their study, designation and classification.

Although the inspection of a profile is usually the first step towards categorising a soil, profiles in themselves are not units of complete study since they have only two dimensions — breadth and depth. Thus profiles can be studied only by visual examination, whereas, *it is the soil continuum that must be investigated.*

The research carried out on soils aimed at their quantitative characterisation is extremely varied and is determined primarily by the requirement of individual workers. The assessment of a tangible property such as the total amount of iron is usually conducted in the laboratory on samples collected from the continuum. On the other hand the measurement of moisture or temperature fluctuations is more satisfactorily carried out by inserting suitable probes into the soil and taking continuous or intermittent readings. The data obtained by these techniques yield information about the intrinsic properties of the continuum at any one point and by collecting data from a number of points inferences can be made about the rate and type of change from point to point. Thus the unit of study or pedo-unit can be defined as follows: the pedo-unit is a selected column of soil containing sufficient material in each horizon for adequate laboratory and field characterisation.

Figure 1.6 illustrates the pedo-unit and various other units of study. The pedo-unit occupies the volume of soil extending below ABCD and the soil profile is the exposed section ADEF. A monolith is shown extending below G and there is a core at H. The study of cores or monoliths is a useful alternative method of studying soils, having the advantage that it allows investigations to be conducted in the relative comfort of the laboratory. However, this method has certain disadvantages since the study is conducted away from the natural environment of the soil. Wherever possible, it is imperative to study soils in their normal surroundings, which play a major role in their formation. The pedological characterisation of the soil is normally performed in the laboratory on bulk or undisturbed samples obtained from the pedo-unit. Duplicate or triplicate bulk samples are collected from each horizon in a vertical sequence which ideally forms a column. The size of the individual horizon sample should always be minimal because many small samples are statistically more valid than a few large

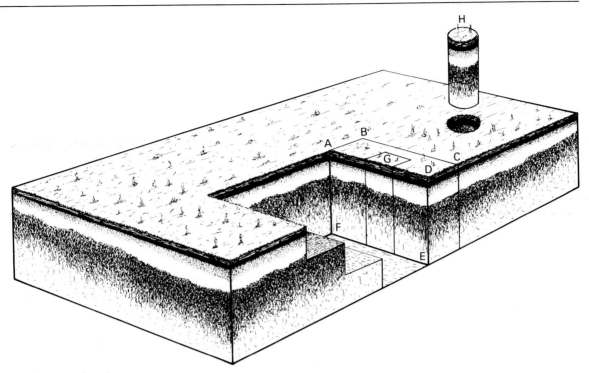

Fig. 1.6 The pedo-unit

ones. However, their precise amount is determined by the nature of the soil itself and by the requirements of the quantitative analytical techniques used to assess the various soil properties. The sample can be as small as a few grams, but if information is required about the proportion of stones, several kilograms may be needed. However, a smaller sample may have to suffice if the horizon is very thin. A rectangular column with surface dimensions of about 100 x 30 cm and extending down to include the relatively unaltered underlying material is the usual size of the pedo-unit, being large enough to give triplicate samples. This is not to be regarded as a fixed amount since the volume of soil examined will vary with the nature of the soil. Undisturbed samples can be collected for a variety of investigations, including the preparation of thin sections similar to thin sections of rock, so that the micromorphology of the soil can be studied with the microscope.

Thus there can be no fundamental unit of study or of classification since the nature of the unit is decided on the one hand by the desires of the operator and on the other by the nature of the soil itself. Sometimes the nature of the soil allows the operator to do exactly as he wishes but on many occasions the nature of the soil imposes severe limitations.

The selection of sample points should be based on some system of randomisation but it is normally made subjectively, being chosen at places to represent areas of apparent uniformity as well as to determine the type and rate of change within the continuum. The number of points examined is increased as the rate of change increases. Careful and accurate choice of sample sites is imperative since a sample of a few grams or a thin section 25 μm thick is used to categorise many hectares.

DEFINITION OF SOIL

After considering the foregoing fundamental properties, it should be possible to offer a definition of soil. However, this is not an easy task since a number of divergent attitudes towards soil are adopted by different workers. Some hold the extreme view that soils are substrates for plant growth while others regard soils as natural phenomena justifying academic study, without any consideration of the applicability and practicability of the results that are obtained. It is virtually impossible to resolve such widely dissimilar views and to formulate a single simple definition. Perhaps the attempt made below should be regarded as a short description rather than as a definition: *soil is the space—time continuum forming the upper part of the earth's crust.*

7

2

Factors of soil formation

This definition like most others may have many deficiencies. It includes as soil, the sands of the deserts and the bare rock surfaces of mountainous areas, thereby excluding the presence of plants and a stable surface as essential for soil. If an allowance is made for these latter two features then many stone walls with mossy surfaces would be included since they are stable and are covered by vegetation. Ultimately, it may be necessary to use a nonsense definition such as *soil is anything so-called by a competent authority.*

SUMMARY

Soils extend to some depth from the surface of the earth and are examined by digging a hole or pit. This reveals a number of layers or horizons which collectively are termed a soil profile.

Soils vary laterally, forming a three-dimensional continuum in which the character of the horizons changes markedly from place to place, therefore soils form the upper part or veneer of the landscape. The whole of the soil is constantly changing with time and therefore can be regarded as a space—time continuum with four dimensions.

Since soils seldom have fixed boundaries they can only be studied by delimiting arbitrary volumes as determined by the nature of the study.

Dokuchaev firmly established that soils develop as a result of the interplay of the five factors: parent material, climate, organisms, topography and time. The first four are the tangible factors interacting through time to create a number of specific processes leading to horizon differentiation and soil formation. Some workers, particularly Jenny (1941), have tried to demonstrate, quite unconvincingly, that these factors are independent variables, i.e. each of them can change and vary from place to place without the influence of any of the others. Only time can be regarded as an independent variable, the other four depend to a greater or lesser extent upon each other, upon the soil itself or upon some other factor. For example, it is now generally accepted that vegetation is a function of climate which is itself a function of wind currents, latitude, proximity to water and elevation. In addition, some parent materials such as glacial deposits, result from geological processes which are greatly influenced by climate, but some may form as a result of tectonic phenomena and are therefore independent of the other factors of soil formation.

Many attempts have been made to show that some factors are more important than others and therefore play a major role in soil formation. Such efforts are a little unrealistic since each factor is absolutely essential and none can be considered more important than any other, but locally one factor may exert a particularly strong influence.

Below are presented the more important characteristics of these factors as they are related to soil formation; this is followed in the next two chapters (3 and 4) by discussions of the principal processes and properties of the soil system. Very little is said about how each factor influences soil processes and properties; this is given and discussed in the final chapter (7). As far as possible the discussions are confined to factual statements about the factors and their influence generally.

PARENT MATERIAL

Jenny (1941) defines parent material as 'the initial state of the soil system'. The precision of this definition cannot be questioned, but most attempts to determine the initial stages of soils are fraught with difficulties, for in a number of cases the original character of the material has been changed so drastically by a long period of pedogenesis that it is possible only to speculate about the full composition of its pristine state. Sometimes, where soil formation has proceeded for a short time, the nature of the original material can be ascertained fairly accurately, but even in these cases some deductions may be necessary, particularly about soluble substances which are easily lost or redistributed in the soil system.

Usually the relatively unaltered underlying material is similar to the material in which the overlying horizons have developed, but this is not always true, especially in stratified sediments and folded metamorphic rocks where one stratum may have an entirely different chemical and mineralogical composition and structure from the material below or when there are thin superficial deposits overlying rock. In spite of these difficulties, which are concerned primarily with the evaluation of the contribution of the parent material to the soil, it is possible to state the composition of parent materials without referring them to any particular soil or set of pedogenetic processes.

Parent materials are made up of mineral material or organic matter or a mixture of both. The organic matter is usually composed predominantly of unconsolidated, dead and decaying plant remains while the mineral material which is the most widespread type of parent material contains a large number of different rock-forming minerals and can be in either a consolidated or unconsolidated state. The consolidated mineral material includes rocks like granite, basalt and conglomerate and the unconsolidated material comprises a wide range of superficial deposits of which glacial drift and loess are two important representatives. It may be of interest to know the origin of the parent material be it igneous rock or glacial drift, but the chemical and mineralogical composition are more important properties; these are responsible largely for the course of soil formation, and the resulting chemical and physical composition of the soil, including the secondary products of weathering. Parent materials also contribute to soil formation through their permeability and specific surface area.

Structure of minerals

A full discussion of the rock-forming minerals is beyond the scope of this book, nevertheless it is necessary to outline the structure of the main types in order to understand the reasons for their behaviour when subjected to various processes. Also it is of paramount importance to understand the composition and structure of the sheet silicates which include the all-important clay minerals. Minerals can be divided into non-silicates and silicates.

Non-silicates
As seen from Table 2.4 (pp. 19–20), this group contains oxides, hydroxides, sulphates, chlorides, carbonates and phosphates. All have relatively simple structures but they vary widely in their solubility and resistance to decomposition.

Silicates
Generally, these minerals have very complex structures in which the fundamental unit is the silicon–oxygen tetrahedron. This is composed of a central silicon ion surrounded by four closely-packed and equally-spaced oxygen ions. The whole forms a pyramidal structure the base of which is composed of three of the oxygen ions with the fourth forming the apex (Figs. 2.1 and 2.2). The four positive charges of Si^{4+} are balanced by four negative charges from the four oxygen ions O^{2-}, one from each ion, thus each discrete tetrahedron has four negative charges. The tetrahedra themselves are linked together in a number of different ways forming a variety of distinctive and characteristic patterns which form the basis of the classification of these minerals. Furthermore, the type of linkage determines the crystal structure as well as its resistance to weathering. A significant variation in the tetrahedral structure is the substitution of Al for Si. This is known as *isomorphous replacement* and when it occurs it produces an imbalance in the charges within the structure which is satisfied by the introduction of cations such as Na, K, Ca and Mg. It is this substitution by Al and charge imbalance that is entirely responsible for the presence of basic cations in the feldspar structure. Silicates are divided broadly into framework

9

Fig. 2.1 Models of silicon—oxygen tetrahedron (*left*) complete model; (*right*) model with apical oxygen ion removed to show the smaller central silicon ion

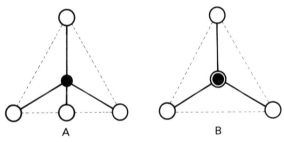

Fig. 2.2 Diagrammatic representation of a silicon-oxygen tetrahedron: (A) vertical; (B) plan

silicates, chain silicates, ortho- and ring silicates and sheet silicates.

Framework silicates

These minerals are composed of tetrahedra linked through their corners into a continuous three-dimensional structure. The simplest member of this class is quartz which is a colourless mineral composed entirely of silicon—oxygen tetrahedra, so that the resulting structure has twice as many oxygen ions as silicon ions and the formula can be written $(SiO_2)_n$. Such minerals with a continuous structure are extremely hard and resistant to weathering. The other main group of framework silicates is the colourless or white feldspars which also have a three-dimensional arrangement of tetrahedra, but in this case there is a considerable amount of isomorphous replacement and

therefore they contain a high proportion of basic cations. In many feldspars about one-quarter of the silicon ions are replaced by aluminium ions and when for example, the extra charges are balanced by sodium the new formula becomes $NaAlSi_3O_8$.

There are three principal types of feldspars, known respectively as sodium, potassium and calcium feldspars which form the ternary system, $NaAlSi_3O_8 - KAlSi_3O_8 - CaAl_2Si_2O_8$. The members of the first series, Na to K, are known as alkali feldspars and those between Na and Ca are the plagioclase feldspars. Generally, the alkali feldspars contain up to 10 per cent of calcium 'molecule', and in a similar way the plagioclase feldspars contain up to 10 per cent of the potassium 'molecule'.

Chain silicates

There are two main divisions within this group — the pyroxenes and the amphiboles. The former, of which enstatite and hypersthene are good examples, are composed of tetrahedra linked to each other by sharing two of the three basal corners to form continuous chains (Fig. 2.3). These chains have various dispositions with respect to each other and are linked laterally by various cations such as Ca, Mg, Fe, Na and Al, thus forming a variety of structures.

The amphiboles exemplified by green hornblende, have chains that are double the width of those of the pyroxenes and can be regarded as a single band of tetrahedra arranged in a hexagonal pattern. The bands

(a)

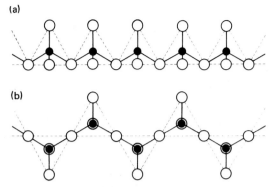

(b)

Fig. 2.3 Diagrammatic representation of a pyroxene chain (A) vertical view and (B) plan view. The points of linkage are through the unsatisfied negative charges on the oxygen ions

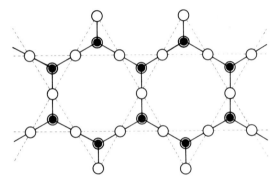

Fig. 2.4 Diagrammatic representation of an amphibole chain, plan view. The vertical view is identical to that of the pyroxene chain. The points of linkage are through the unsatisfied negative charges on the oxygen ions

have various dispositions with respect to each other and like the pyroxenes are linked by Ca, Mg, Fe, Na and Al ions (Fig. 2.4).

Ortho- and ring silicates

Probably the greatest variations in the structure of primary minerals occur within this group which includes the olivines, zircon, titanite and garnet. It is most convenient to start with the olivines, because of their relatively simple structure with separate silicon–oxygen tetrahedra arranged in sheets and linked by Mg and/or Fe ions. Olivines also differ from many of the other members of this group by being more common and forming a frequent constituent of many basic and ultrabasic igneous rocks. The other important mineral in which the linkages are provided by a cation is zircon in which each zirconium ion is surrounded by eight oxygen ions resulting in a very strong structure. Contrasted with these two are

garnet, sillimanite and kyanite which have various complex linkages provided by aluminium octahedra as described below. Thus the structure of the members of this group varies from relatively simple to highly complex, the latter being found in the majority.

Sheet silicates

This group includes minerals such as muscovite, biotite, pyrophyllite, talc and the secondary clay minerals. It is better to consider these two groups separately but they have a number of features in common and both can be regarded as being composed of various combinations of three basic sheets namely the silicon tetrahedral sheet, the aluminium hydroxide sheet and the magnesium hydroxide sheet.

Silicon tetrahedral sheet. This is composed of silicon–oxygen tetrahedra linked together in a hexagonal arrangement with the three basal oxygen ions of each tetrahedron in the same plane and all the apical oxygen ions in a second plane. Thus the silicon tetrahedral sheet is a hexagonal planar pattern of silicon–oxygen tetrahedra (Figs. 2.5 and 2.6).

Aluminium hydroxide sheet. The basic unit of this sheet is the aluminium-hydroxyl octahedron in which each aluminium ion is surrounded by six closely packed hydroxyl groups (Fig. 2.7), in such a way that there are two planes of hydroxyl ions, with a third plane containing aluminium ions sandwiched between the two hydroxyl planes. In order that all the valencies of the structure be satisfied only two out of every three positions in the aluminium hydroxide sheet are occuped by aluminium ions forming what is known as a dioctahedral structure.

Magnesium hydroxide sheet. This has a similar structure to the aluminium hydroxide sheet but the aluminium is replaced by magnesium, and because magnesium is divalent all the sites in the middle plane are occupied, forming a trioctahedral structure (Fig. 2.8).

Pyrophyllite and talc. Although these minerals are infrequent and of little importance in soil studies it is convenient to start with them since the structure of the primary and clay micas, montmorillonite and vermiculite can be thought of as being derived from their structure by relatively simple substitutions. Pyrophyllite is composed of one aluminium hydroxide sheet lying between two silicon tetrahedral sheets and is known as a 2 : 1 type mineral. In this structure each **11**

Fig. 2.5 Model of silicon–oxygen tetrahedral sheet with some of the oxygen ions removed to show the smaller silicon ions

silicon tetrahedral sheet has its apical oxygen ions facing towards the other and replacing hydroxyl ions on either side of the aluminium hydroxide sheet. Therefore, the top and bottom surfaces of each composite sheet are made up of oxygen ions in the silicon tetrahedral sheets (Fig. 2.9). Talc has a similar structure but the aluminium hydroxide sheet is replaced by a magnesium hydroxide sheet. Thus pyrophyllite is a dioctahedral mineral and talc is the trioctahedral equivalent.

Muscovite.   The structure of this mineral can be derived from the dioctahedral pyrophyllite structure by substituting one-quarter of the silicon ions by aluminium ions in the silicon tetrahedral layers. This causes an imbalance in the charges that is satisfied by

potassium which bonds the composite sheets together (Fig. 2.10).

Biotite.   The structure of this mineral can be derived from the trioctahedral talc structure by about one-third substitution of iron — Fe^{2+}, for magnesium in the magnesium hydroxide sheet. In a similar manner to muscovite the tetrahedral sheets have negative charges that are satisfied by potassium which bonds the composite sheets together. There is also a small amount of substitution of aluminium in both the octahedral and tetrahedral sheets.

Clay minerals
There are seven types of clay minerals important in soils, namely: kaolinite, halloysite, montmorillonite, hydrous mica, vermiculite, chlorite and allophane, the first six are crystalline and composed of silicon

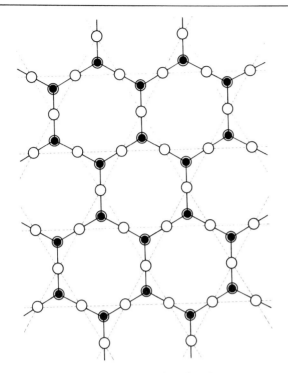

Fig. 2.6 Diagrammatic representation of a silicon−oxygen tetrahedral sheet

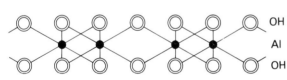

Fig. 2.7 Diagrammatic representation of an aluminium hydroxide sheet

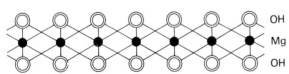

Fig. 2.8 Diagrammatic representation of a magnesium hydroxide sheet

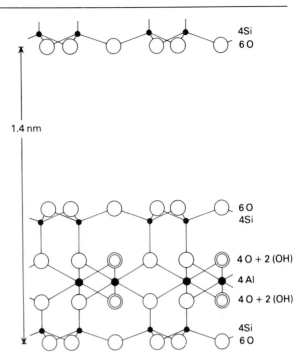

Fig. 2.9 Diagrammatic representation of the structure of pyrophyllite

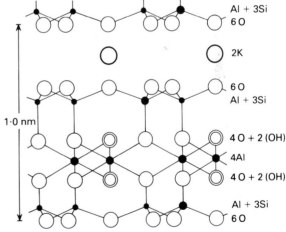

Fig. 2.10 Diagrammatic representation of the structure of muscovite

tetrahedral, aluminium hydroxide and magnesium hydroxide sheets in various combinations. Allophane is not crystalline but it is convenient to consider it here. Clay minerals can be presented in terms of their structure, properties and methods of formation but before starting these discussions it must be emphasised that a full comprehension of their structure can be achieved only by a careful study of models.

Kaolinite group

Kaolinite is the commonest member of this group and is the simplest type of clay mineral, being composed of one aluminium hydroxide sheet and one silicon tetrahedral sheet in which each apical oxygen of the silicon tetrahedral sheet replaces one hydroxyl group of the aluminium hydroxide sheet and forms

13

what is known as a 1 : 1 type of structure. Kaolinite has a well developed pseudo-hexagonal crystal structure in which the individual crystals range from 0.2–2 μm and form by growth about all the axes. Development along the *c*-axis takes place by a build up of the paired sheets in such a way that the oxygen ions of the basal plane of the silicon tetrahedral sheet are aligned opposite to hydroxyl groups of the aluminium hydroxide sheet, to which they become firmly attached by hydrogen bonding to produce a rigid structure, which cannot expand (Figs. 2.11 and 2.12).

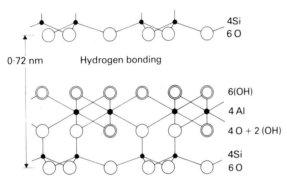

Fig. 2.11 Diagrammatic representation of the structure of kaolinite

Fig. 2.12 Photomicrograph of kaolinite sheets

Halloysite, the second most common member of the group has a similar composition to kaolinite but contains interlayer water and has a tubular form.

The water is firmly attached by hydrogen bonding resulting in a rigid structure. However, dehydration can take place very easily causing collapse of the structure forming a mineral similar to kaolinite. In a number of situations halloysite is regarded as the precursor of kaolinite (Fig. 2.13).

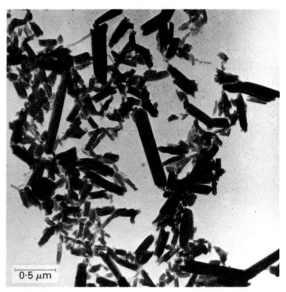

Fig. 2.13 Photomicrograph of halloysite

Montmorillonite or smectite group
These minerals have a 2 : 1 type of structure and can be derived from the pyrophyllite structure by substituting one-sixth of the aluminium ions by magnesium, which causes an imbalance of charge within the structure that is usually satisfied by basic cations (Figs. 2.14 and 2.15). Some substitution of iron can also take place. These minerals usually have a small particle size and a great affinity for water. This imparts one of the most important properties of these minerals namely their ability to expand and contract in response to the addition and loss of moisture.

Hydrous mica group (Illite)
This group can be regarded as clay size particles of muscovite and biotite (Fig. 2.16). Those similar to muscovite have the pyrophyllite structure and are of the dioctahedral variety. The trioctahedral variety is similar to biotite, which in turn is similar to talc. Both of these minerals are hydrated but neither is capable of expanding due to the additional linkages supplied by potassium but these linkages are fewer than in primary micas.

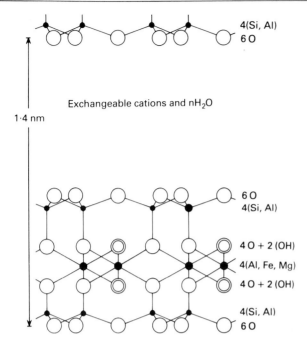

Fig. 2.14 Diagrammatic representation of the structure of montmorillonite

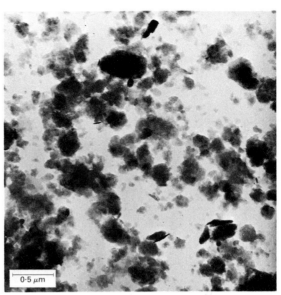

Fig. 2.16 Photomicrograph of hydrous mica

by calcium and magnesium, and in which there is increased substitution. As a result it is capable of expanding because the linkages which are broken when potassium is removed are not reformed by calcium and magnesium (Fig. 2.17).

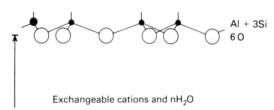

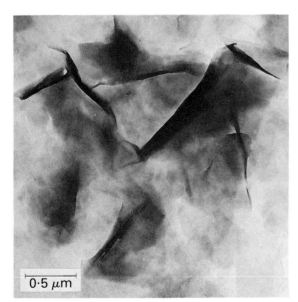

Fig. 2.15 Photomicrograph of montmorillonite

Vermiculite group

This mineral exists in both dioctahedral and triocta-hedral forms and can be regarded as hydrated mica from which potassium has been removed and replaced

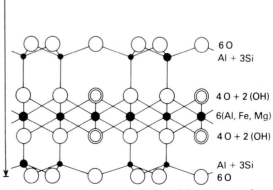

Fig. 2.17 Diagrammatic representation of the structure of trioctahedral vermiculite

15

Chlorite group

Primary chlorite is a 2 : 1 type of layered silicate plus a magnesium hydroxide sheet sandwiched between two mica layers, and replacing the potassium in the mica structure (Fig. 2.18). The chemical composition of chlorite is very variable since Mg and Fe and other cations replace aluminium to various extents. Chlorite formed in soils differs from the above by having inter-layered aluminium hydroxide.

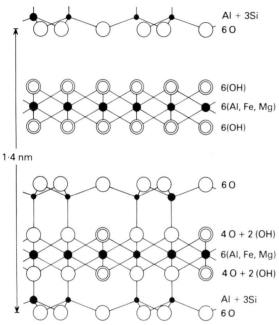

Fig. 2.18 Diagrammatic representation of the structure of trioctahedral chlorite

Allophane

This is a poorly categorised substance which is some-times regarded as a clay mineral and at other times considered among the hydroxides. Some workers regard it as a mixture of silica gel and aluminium hydroxide while others consider it to be a hydrous aluminium silicate. Because of the difficulties in determining definite values for many of its properties it is probable that it is a variable mixture of silica gel and aluminium hydroxide. The molecular ratio of silica to alumina is known to vary from 100 : 112 to 100 : 180 but the precise composition depends upon the nature of the decomposing materials as well as upon the environment and age, since it appears that it slowly changes to halloysite and then to kaolinite. Allophane was at one time thought to be associated mainly with soils derived from volcanic ash but it now appears to be much more widespread.

Mixed layers

Clay minerals scarcely ever occur in pure form in soils, often a single particle is composed of inter-stratified layers of two or more different clay minerals. Also, two or more may be intimately mixed or they may be enmeshed in large quantities of oxides hence the many difficulties encountered in their identification. Within recent years it has been dis-covered that mixed layered clays are extremely common and occur in a wide variety of soils including many soils of the humid tropics where it was thought that the soils would contain only kaolinite. Mixtures such as kaolinite—vermiculite, kaolinite—montmorillonite occur frequently. The reasons for this are not fully understood, it seems that some of the 2 : 1 minerals formed in the early stages of soil formation or inherited from the parent material become trapped and shielded by kaolinite and perhaps iron oxide and cannot undergo alteration.

A simple classification of clay minerals and related silicates is given in Table 2.1.

Properties of clay minerals. The three most import-ant properties of clay minerals are their cation-exchange capacity, i.e. their ability to absorb cations, their capacity to absorb water and their ionic double layer. Because of variations in their structure, clay minerals do not have a fixed exchange capacity but a range which is influenced by a number of factors of which the most important is pH as mentioned on page 113. Exchange reactions result mainly from:
1. Unsatisfied bonds at the edges of crystals
2. Increased number of charges due to isomorphous replacement.

Minerals such as kaolinite with their rigid structure and absence of isomorphous replacement have their exchange sites confined to the broken edges of crystals where the charges of the relatively few broken bonds are satisfied by hydrogen or basic cations, therefore the exchange capacity of kaolinite is very low. On the other hand montmorillonite and vermiculite have a relatively large amount of isomorphous replacement of silicon by aluminium giving a large number of exchange sites and a high exchange capacity. The indeterminate structure of allophane gives a very variable exchange capacity which is influenced very strongly by relatively small variations in the techniques used for its assessment. In addition to the above properties, the hydrous micas are capable of adsorbing large amounts of potassium which is relatively un-available to plants but it may be released slowly. When moist or wet, most of the individual particles are regarded as having a surface covered with negative

Table 2.1 Classification scheme for clay minerals and related silicates

Type	Interlayer material	Group	Subgroup	Species
1 : 1	nil nil	Kaolinite −serpentine Pyrophyllite−talc	Kaolinites Serpentines Pyrophyllites Talcs	Kaolinite, halloysite Chrysotile, antigorite Pyrophyllite Talc
2 : 1	individual cations or hydrated cations	Smectite or Montmorillonite Vermiculite Mica *	Dioctahedral smectites or montmorillonites Trioctahedral smectites Dioctahedral vermiculites Trioctahedral vermiculites Dioctahedral micas Trioctahedral micas	Montmorillonite Dioctahedral vermiculite Trioctahedral vermiculite Muscovite Biotite
	hydroxide sheet	Chlorite	Dioctahedral chlorites Di-trioctahedral chlorites Trioctahedral chlorites	

* Many minerals allocated to this group (e.g. illite, hydromuscovite, sericite) are alteration products of micas and may be interstratified (mixed) species.

charges which are satisfied by positive charges of the ions in the surrounding solution, the ions closest to the clay surface being most tightly held. This layer of charges and the swarm of surrounding ions is known as the *electric double layer*. As implied this layer has a higher ion concentration than the remainder of the solution and in addition the innermost ions are so firmly attached that when the particle moves part of the surrounding layer of ions also moves. The clay particle surrounded by its ionic hydrated shell is called a *micelle*.

Flocculation and dispersion are two more properties of clays. Flocculation is the process whereby the individual particles of clay are coagulated to form floccular aggregates. Although this is clearly seen in the laboratory when a small amount of calcium hydroxide is added to a suspension of clay, a similar reaction takes place in the soil. The degree and permanence of flocculation depend upon the nature of the ions present; calcium and hydrogen are very effective in this role. At the other extreme is the state of dispersion in which the individual particles are kept separate one from the other. This is accomplished by potassium and more particularly by sodium. Thus, depending upon the nature of the cations present in the soil, it may either be in a floccular or aggregated state or in a dispersed and often massive condition. Flocculation and dispersion are related to the thickness of the electric double layer, if it is thick the clays remain in suspension, if it is thin then they flocculate. Thus sodium saturated clays have a thick double layer

whereas divalent ions such as calcium suppress the double layer and cause flocculation while tri- and tetravalent ions are more efficient in causing flocculation.

A further distinctive property of clay minerals is that from any point in one layer to the same point in the adjacent layer there is a fixed distance known as the basal spacing which is utilised in X-ray studies as a differentiating criterion. Set out in Table 2.2 is the range in the exchange capacities and spacings for the commonly occurring clay minerals.

Table 2.2 Cation-exchange capacities and spacings of clay minerals

Clay mineral	CEC me % clay, pH 7	Basal spacing (nm)
Kaolinite	3−15	0.72
Halloysite (fully hydrated)	40−50	1.025
Mica	10−40	1.0
Montmorillonite	80−150	1.4
Vermiculite	100−150	1.4
Chlorite	10−40	1.4

nm = nanometre = 10^{-9} m

Fuller discussions about the structure and properties of minerals can be found in Deer *et al.* (1966), Jackson (1964) and Grim (1969).

17

Table 2.3 Chemical composition of some common rocks

Rock type	Composition (%)											
	SiO_2	Al_2O_3	Fe_2O_3	FeO	MgO	CaO	Na_2O	K_2O	TiO_2	P_2O_5	MnO	H_2O
Basalt Labradorite, augite and pseudo-morphs in chlorite and goethite after olivine; and opaque minerals. (Guppy and Sabine, 1956)	53	18	5	2	4	7	4	2	1	<1	<1	2
Diorite Hornblende, biotite, titanite, oligoclase, quartz and opaque minerals. (Guppy and Sabine, 1956)	58	17	3	4	3	5	4	3	2	1	<1	1
Granite Fine-grained grey granite, composed of quartz, microcline, oligoclase, biotite, muscovite, accessory minerals, apatite and opaque minerals. (Guppy and Sabine, 1956)	76	13	<1	1	<1	<1	3	5	<1	<1	<1	1
Garnetiferous mica-schist Muscovite, biotite, chlorite, garnet, quartz and oligoclase, iron ore and apaite. (Guppy and Sabine, 1956)	58	19	1	7	2	2	2	4	1	<1	<1	2
Limestone Calcite. (Clark, 1924)	1	<.1	Nil	.1	<1	55[+]	<.1	<1	Nil	<.1	Nil	<.1
Slate Quartz, chlorite, muscovite, apatite, tourmaline, albite and opaque minerals. (Guppy and Sabine, 1956)	57	20	10*		3	1	3	3	1	<1	Nil	Nil

* Total Fe calculated as Fe_2O_3; [+]CO_2 = 43%

Chemical and mineralogical composition of parent materials

Set out in Table 2.3 is a list of a few of the common types of rocks, their mineralogy and chemical composition. This is followed in Table 2.4 by a list of the important and widespread minerals found in parent materials, their formulae and approximate frequency. In this context it seems more appropriate to give the empirical formulae for most minerals rather than the unit cell formulae which can be found in Deer *et al.* (1966). In addition, the chemical composition of the principal minerals is given in Table 2.5. These three sets of data give the significant properties of the wide range of mineral materials forming parent materials, of which quartz, feldspars and the sheet silicates are the principal constituents. Also shown are the relatively few elements that are found in abundance in parent materials. As Clark (1924) has pointed out, the upper 10 km of the surface of the earth is made up predominantly of O, Si, Al, Fe, Ca, Na, K, Mg, Ti, P, Mn, S, Cl and C, but only the first eight elements

exceed 1 per cent. Oxygen occupies over 90 per cent by volume but usually occurs in combination with other elements, particularly with silicon the second most frequent element, forming quartz and occurring also within the more complex silicate structures. Thus the majority of soils are composed predominantly of silica, either as quartz or in a combined form. The third most frequent element is aluminium which is found mainly in feldspars and in the sheet silicates, but which occurs also in varying proportions in the other silicates. Iron is also of fairly widespread distribution and it should be noted that it is mainly in the ferrous state occurring in large amounts in relatively few minerals of which augite, biotite, hornblende, hypersthene and some olivines are the principal contributors. Sodium and potassium are found chiefly in the feldspars, but significant amounts of potassium also occur in the micas. Calcium and magnesium have a wide distribution among the silicates as well as being the principal cations in many non-silicates such as calcite and dolomite which are the dominant minerals in limestones. Phosphorus is very restricted in its

Table 2.4 Soil-forming minerals

Crystal structure	Mineral	Formula	Frequency
Ortho- and Ring silicates	Andalusite	$Al_2O_3 \cdot SiO_2$	Rare
	+ Kyanite	$Al_2O_3 \cdot SiO_2$	Rare
	Sillimanite	$Al_2O_3 \cdot SiO_2$	Rare
	Epidote group		
	Zoisite	$4CaO \cdot 3Al_2O_3 \cdot 6SiO_2 \cdot H_2O$	Rare
	Clinozoisite	$4CaO \cdot 3Al_2O_3 \cdot 6SiO_2 \cdot H_2O$	Rare
	Epidote	$4CaO \cdot 3(Al, Fe)_2O_3 \cdot 6SiO_2 \cdot H_2O$	Rare–frequent
	Garnet group		
	+ Almandite	$Fe_3Al_2Si_3O_{12}$	Rare–frequent
	Olivine group		
	Olivine	$2(Mg, Fe)O \cdot SiO_2$	Absent–common
	+ Titanite	$CaO \cdot SiO_2 \cdot TiO_2$	Rare*
	+ Tourmaline	$Na_2O \cdot 8FeO \cdot 8Al_2O_3 \cdot 4B_2O_3 \cdot 16SiO_2 \cdot 5H_2O$	Rare
	+ Zircon	$ZrO_2 \cdot SiO_2$	Rare*
Chain silicates	*Amphibole group*		
	Hornblende	$Ca_3Na_2(Mg, Fe)_8(Al, Fe)_4Si_{14}O_{44}(OH)_4$	Rare–frequent
	Tremolite – Ferroactinolite	$2CaO \cdot 5(Mg, Fe)O \cdot 8SiO_2 \cdot H_2O$	Rare
	Pyroxene group		
	Enstatite	$MgO \cdot SiO_2$	Absent–common
	Hypersthene	$(Mg, Fe)O \cdot SiO_2$	Absent–common
	Diopside	$CaO \cdot MgO \cdot 2SiO_2$	Absent–rare
	Augite	$CaO \cdot 2(Mg, Fe)O \cdot (Al, Fe)_2O_3 \cdot 3SiO_2$	Absent–common
Sheet silicates	*Serpentine group*	$Mg_3Si_2O_5(OH)_4$	Absent–rare
	Mica group		
	Muscovite (dioctahedral)	$K_2Al_2Si_6Al_4O_{20}(OH)_4$	Rare–common
	Biotite (trioctahedral)	$K_2Al_2Si_6(Fe^{2+}, Mg)_6O_{20}(OH)_4$	Rare–common
	Clay minerals		
	Kaolinite	$Si_4Al_4O_{10}(OH)_8$	Absent–dominant
	Mica	$K_1Al_4(Si_7Al_1)O_{20}(OH)_4 \, nH_2O$	Absent–dominant
	Montmorillonite	$Ca_{0.4}(Al_{0.3}Si_{7.7})Al_{2.6}(Fe^{3+}_{0.9}Mg_{0.5})O_{20}(OH)_4 \, nH_2O$	Absent–dominant
	Vermiculite (trioctahedral)	$Mg_{0.55}(Al_{2.3}Si_{5.7})(Al_{0.5}Fe^{3+}_{0.7}Mg_{4.8})O_{20}(OH)_4 \, nH_2O$	Absent–dominant
	Chlorite (trioctahedral)	$AlMg_5(OH)_{12}(Al_2Si_6)AlMg_5O_{20}(OH)_4$	Absent–rare
Framework silicates	*Feldspar group*		
	Alkali feldspar	$(Na, K)_2O \cdot Al_2O_3 \cdot 6SiO_2$	Rare–common
	Plagioclase	$Na_2O \cdot Al_2O_3 \cdot 6SiO_2 - \cdot - CaO \cdot Al_2O_3 \cdot 2SiO_2$	Rare–common
	Quartz	SiO_2	Frequent–dominant
Non-silicates	*Oxides*		
	+ Haematite	Fe_2O_3	Absent–frequent
	+ Ilmenite	$FeO \cdot TiO_2$	Absent–rare*
	+ Rutile	TiO_2	Rare*
	+ Anatase	TiO_2	Absent–rare*
	+ Brookite	TiO_2	Rare
	+ Magnetite	Fe_3O_4	Rare*

Table 2.4 – (*continued*)

Non-silicates	*Hydroxides*		
	Gibbsite	Al(OH)$_3$	Absent–dominant
	Boehmite	γAlO(OH)	Absent–dominant
	Goethite	αFeO–OH	Rare–common
	Lepidocrocite	γFeO–OH	Absent–rare
	Sulphates		
	Gypsum	CaSO$_4$ · 2H$_2$O	Absent–common
	Jarosite	KFe$_3$(OH)$_6$(SO$_4$)$_2$	Absent–rare
	Chlorides		
	Halite	NaCl	Absent–common
	Carbonates		
	Calcite	CaCO$_3$	Absent–common
	Dolomite	(Ca, Mg)CO$_3$	Absent–common
	Phosphates		
	+ Apatite	Ca$_4$(CaF or CaCl)(PO$_4$)$_3$	Rare*

* usually present   + accessory minerals

Table 2.5 Chemical composition of some common minerals

Mineral type	Composition (%)													
	SiO$_2$	Al$_2$O$_3$	Fe$_2$O$_3$	FeO	MgO	CaO	Na$_2$O	K$_2$O	H$_2$O	TiO$_2$	P$_2$O$_5$	MnO	F	CO$_2$
Apatite	–	–	<1	–	<1	56	–	–	<1	–	42	<1	4	–
Augite	47	3	1	20	7	17	1	<1	<1	1	–	1	–	–
Biotite	36	18	1	22	7	<1	<1	9	4	2	–	<1	<1	–
Calcite	–	–	–	–	<1	56	–	–	–	–	–	–	–	44
Dolomite	–	–	–	<1	21	31	–	–	–	–	–	–	–	47
Hornblende	45	11	3	13	10	12	1	1	2	1	–	<1	–	–
Hypersthene	50	1	2	28	16	1	<1	<1	–	–	–	1	–	–
Muscovite	45	37	<1	<1	<1	–	1	10	5	<1	–	<1	1	–
Olivine														
Fayalite	35	<1	<1	61	3	1	–	–	–	1	–	3	–	–
Forsterite	41	1	<1	4	54	–	–	–	<1	<1	–	<1	–	–
Serpentine														
(chrysotile)	43	1	1	1	41	<1	<1	–	12	tr	–	<1	<1	–
Alkali feldspar														
Orthoclase	65	20	<1	tr	tr	1	5	8	1	–	–	–	–	–
Plagioclase														
Labradorite	53	30	1	–	–	12	4	<1	<1	tr	–	–	–	–
Clay minerals														
Kaolinite	44	36	1	<1	<1	<1	<1	1	16	1	<1	–	–	–
Montmorillonite	51	20	1	–	3	2	<1	–	23	–	–	–	–	–

tr = trace

distribution occurring only in significant proportions in apatite which is fairly widespread in small amounts in parent materials but the total content of this mineral in soils seldom exceeds 0.2 per cent. The microelements needed by plants occur in a number of minerals, some having relatively high proportions of one or more elements. For example, tourmaline which is usually of low frequency in soils contains a large amount of boron but as this mineral is very resistant to weathering it is doubtful whether the boron is easily available to plants.

Most of the silicates are primary minerals and originate in igneous or metamorphic rocks, but the clay minerals are formed within soils or are inherited from parent materials such as lacustrine deposits and shale. Also included in Table 2.4 are a large number of minerals such as zircon, magnetite, rutile and titanite that are usually referred to as accessory minerals. They are present only in very small amounts in soils but are important for their contribution of microelements for plant nutrition and in detailed studies of soil formation as mentioned on page 68.

An extremely important characteristic of parent

materials is their wide variability in composition within short distances. Some consolidated rocks, but more especially sediments can show a three- or four-fold variation in one or more constituents within a few centimetres, this is more common among the accessory minerals such as zircon but can be very important since these are often used as index minerals in weathering studies.

Surface area of parent materials

The specific surface area of the constituent particles in the parent material determines the amount of interaction that is possible with the environment, particularly with water. Consolidated rocks have an extremely small surface area when compared with alluvial sands which in turn have a smaller surface area than lacustrine clays, thus the surface area increases as the particle size decreases. Variations in surface area and particle size distribution have a profound effect on the speed of soil formation, for it is found that a soil such as a Podzol will develop in sediments much quicker than from consolidated rock of the same mineralogical composition.

Permeability of parent materials

The permeability of the parent material influences the rate of moisture movement which in turn influences the speed of soil formation. The most permeable materials and therefore those that allow free movement of moisture usually have a high content of sand, but as the particle size decreases the general tendency is for the material to become more impermeable so that clays allow only a very slow rate of moisture movement. This can cause water to accumulate at the surface or within the body of the soil resulting in a number of important phenomena due to waterlogging, however, some sandy materials can be quite impermeable when compacted. The permeability of rocks varies with their structure; those with well developed jointing are very permeable whereas others without cracks and fissures are relatively impermeable.

Classification of parent materials

Whiteside (1953) and Brewer (1964) have suggested methods for classifying parent materials. In the system of Whiteside a somewhat general statement is made about the mineralogy and state of the material while that of Brewer is really an attempt to classify the potential of the material rather than a statement about the material itself. Thus, neither of these proposals makes a clear and precise statement about parent materials. Presented below is a classification based upon the significant and easily recognisable intrinsic characteristics employing a system of letter symbols for designating the parent material in the field and in written descriptions. Broadly, parent materials can be divided initially into nine classes according to their chemical and mineralogical composition; there are five classes established on the proportions of ferromagnesian minerals, one class on the amount of carbonate, the seventh and eighth are established on the content of salts and the ninth on the amount of organic matter. The presence of ferromagnesian minerals rather than the amount of silica as employed by geologists is used to create the first five classes, since it is the content of these minerals that is most often the important factor determining the chemical properties of the parent material. Set out in Fig. 2.19 are the nine classes, their percentage content of ferromagnesian minerals and letter symbols. There are a few exceptions to this rather simple rule, particularly rocks having a high content of easily hydrolysable minerals, such as anorthite, which are regarded as basic. Many volcanic rocks, because of their poor crystallinity do not fit easily into this scheme and have to be classified on the basis of their content of silica.

Each of these classes can exist in a number of forms and may be divided into consolidated and unconsolidated materials.

Consolidated materials
As shown in Fig. 2.19, these are subdivided into four types on the basis of the size distribution of the mineral grains as follows:

Non-crystalline: composed of a continuous phase without crystals, e.g. volcanic glass.

Fine grained: material containing minerals that are just visible with the unaided eye but there may be a few larger crystals or phenocrysts.

Medium grained: these rocks are composed mainly of minerals up to 5 mm in diameter.

Coarse grained: rocks with minerals greater than 5 mm.

Unconsolidate materials
There are two principal subdivisions of these materials in order to accommodate particle sizes greater than, as well as less than, 2 mm, these are shown also in Fig. 2.19 bottom right.

Type	Composition			Symbol
Ultrabasic	>90%	ferromags		U
Basic	40–90%	,,	45–55% SiO$_2$	B
Intermediate	20–40%	,,	55–65% ,,	I
Acid	5–20	,,	65–85% ,,	A
Extremely acid	<5%	,,	–	E
Carbonate				
slightly carbonate	1–5%	(Ca + Mg)CO$_3$		1C
moderately carbonate	5–20%	,,		2C
strongly carbonate	20–50%	,,		3C
dominantly carbonate	>50%	,,		4C

Type	Composition		Symbol
Sulphate			S
slightly sulphate	1–5%	CaSO$_4$	1S
moderately sulphate	5–20%	,,	2S
strongly sulphate	20–50%	,,	3S
dominantly sulphate	>50%	,,	4S
Saline			
slightly saline	1–5%	soluble salts	1H
moderately saline	5–20%	,,	2H
strongly saline	20–50%	,,	3H
dominantly saline	>50%	,,	4H
Organic			P
slightly organic	1–5%	organic matter	1P
moderately organic	5–20%	,,	2P
strongly organic	20–50%	,,	3P
dominantly organic	>50%	,,	4P

CONSOLIDATED

Mineral size	Symbol
Non-crystalline	N
Fine grained < 1 mm	F
Medium grained 1–5 mm	M
Coarse grained > 5 mm	K

UNCONSOLIDATED

*Particles < 2 mm

Name	Symbol	Name	Symbol
Clay	c —	Sandy clay	sc
Silt clay	zc	Sandy clay loam	sc l
Silty clay loam	zcl	Loam	l —
Silt	z —	Sandy loam	sl
Silt loam	z l	Loamy sand	l s
Clay loam	c l	Sand	s —

*Particles > 2 mm

Give modal size range and state frequency

* Same as for texture

Fig. 2.19 Classification of parent materials

Particles <2 mm

These are the twelve classes given for particle size distribution on page 88.

Particles >2 mm

These are described according to the modal size range and frequency.

By combining the symbols given in Fig. 2.19 any parent material can be given a compound symbol to indicate its mineralogy, degree of consolidation, particle size distribution and degree of stoniness.

The principal rock types, their symbols and mineralogy are given in Table 2.6, and below is the list of the predominant types of unconsolidated deposits:

Alluvial deposits
Dune sands
Glacial drifts
Lacustrine clays
Loesses
Marine clays
Pedi-sediments
Raised beaches
Solifluction deposits
Volcanic dust

It is not possible to specify these deposits as accurately as the individual rocks because of much wider variations in their properties. However, a few examples are given below to show how the system operates and definitions are given in the glossary.

AK:	acid coarse grained crystalline rock such as granite
BF:	basic fine grained rock such as basalt
2Cz-:	moderately calcareous silt
Ascl:	acid sandy clay loam
Es-5cmA:	extremely acid sand with 5 cm stones abundant

It is seen that the symbols attempt to produce an objective categorisation of parent material and not to classify them on interpretations of their methods of formation since it is often difficult to be sure about their origin. For example, it is not always easy to differentiate between alluvium and certain forms of glacial drift, also there is considerable controversy about the origin of loess.

Distribution of parent materials

Superficial deposits including glacial drift, alluvium and aeolian deposits are the predominant parent materials, having resulted from the widespread climatic changes and surface disturbances of the Pleistocene Period (see p. 304 *et seq.*). It is in tropical and subtropical countries that there are large contiguous areas where soils are developed from the *in situ* decomposition of rocks most of which are acid in composition being mainly granite, gneiss, schist, slate, sandstone and shale. Outside these areas it is mainly

Table 2.6 Major rock types, their symbols and petrography

Rock type	Symbol	Petrography
Ultrabasic Peridotite	UK	Coarse grained, dominant olivine often with enstatite, augite, hornblende sometimes, plagioclase, and biotite; accessories of magnetite and spinel.
Serpentine	UK UM	Coarse to medium grain, dominant antigorite or crysotile with accessory magnetite.
Basic Basalt	BF	Fine grained often porphyritic, dominant plagioclase with augite and often olivine; accessories of magnetite, ilmenite and apatite.
Dolerite	BM	Medium texture, dominant labradorite with much augite. Sometimes olivine, hypersthene, enstatite, hornblende and biotite; accessories of apatite, quartz, ilmenite and magnetite.
Gabbro	BM	Medium to coarse, abundant plagioclase, enstatite, hypersthene and augite, sometimes olivine with accessories of hornblende, biotite, ilmenite and magnetite.
Intermediate Andesite	IF	Glassy to fine grain, dominant plagioclase ground mass with phenocrysts of feldspars and lesser amounts of augite, enstatite, hornblende or biotite.
Amphibolite	IK	Coarse grained, crystalline, dominant hornblende with much plagioclase and often garnet; accessories of magnetite and sphene.
Diorite	IM	Medium to coarse grained dominant plagioclase with hornblende sometimes biotite, augite and olivine; accessories of magnetite, ilmenite, sphene and apatite.
Syenite	IK	Coarse to medium grained, dominant alkali feldspars with abundant hornblende, rarely biotite or augite, occasionally small amounts of quartz, accessories of sphene, zircon, apatite, magnetite and ilmenite.

Table 2.6 – *(continued)*

Rock type	Symbol	Petrography
Trachytes	IF	Fine grained ground mass of alkali feldspars with varying amounts of phenocrysts of biotite, augite, hornblende, olivine; accessories of sphene, apatite, magnetite and zircon.
Acid Conglomerate (one type)	AK	Dominantly quartz and feldspar-containing rock fragments.
Granite	AK	Coarse grained igneous, mainly alkali feldspar with much quartz, lesser amounts of plagioclase, biotite or muscovite; sometimes small amounts of hornblende, tourmaline, epidote and augite, accessories of apatite, zircon, magnetite and sphene.
Aplite	AF	Fine grained variety of granite.
Arkose	AK	Coarse grained highly feldspathic sandstone derived from granite.
Gneiss	AK	Coarse grained metamorphic banded, dominantly alkali feldspars and quartz with biotite and muscovite sometimes hornblende.
Obsidian	AN	Volcanic glass of granitic composition generally black >65% SiO_2.
Pegmatite	AK	Very coarse grained igneous dominant microcline, orthoclase, quartz and muscovite, some garnet.
Rhyolite	AN AF	Glassy or microcrystalline groundmass of alkali feldspar and quartz, generally small phenocrysts of orthoclase accessories of biotite, hornblende, magnetite, apatite and zircon.
Sandstone (siliceous)	AK	Coarse grained compact sediment, abundant quartz and feldspars with lesser amounts of biotite, muscovite, hornblende, accessories of sphene, ilmenite, magnetite, zircon.
Quartz porphyry	AF	Microcrystalline ground mass of quartz and feldspar with phenocrysts of feldspars or quartz and lesser amounts of micas and augite.

23

Table 2.6 – *(continued)*

Rock type	Symbol	Petrography
Extremely acid		
Quartzite	EK	Granulose coarse or medium grained, metamorphic dominant quartz and sometimes small amounts of muscovite, tourmaline, magnetite.
Sandstone	EK EM	Coarse or medium grained compact sedimentary dominant quartz with lesser amounts of biotite, muscovite, hornblende, accessories of sphene, ilmenite, magnetite zircon.
Schist	EM	Medium to fine grained, foliated metamorphic dominant quartz and feldspar with biotite, muscovite, sometimes garnet.
Slate	EF	Fine grained compact sedimentary dominant quartz with muscovite and chlorite.
Schist mica-schist	EM EF	Medium to fine grained, foliated metamorphic, abundant biotite, quartz and muscovite, lesser amounts of garnet, zoisite, epidote, hornblende, sphene, staurolite, sillimanite, kyanite and tourmaline.
Carbonate		
Limestone	CF CM CK	Crystalline, oolitic or earthy, sedimentary, dominant calcium carbonate with or without magnesium carbonate.
Marble	CM	Crystalline metamorphic.
Chalk	CN	Soft white limestone consisting largely of remains of foraminifera.
Dolomite	CK	Medium to coarse crystalline with nearly equal amounts of $CaCO_3$ and $MgCO_3$.
Organic		
Coal	PN	More or less altered plant remains.

in mountain ranges that soils have developed from the underlying rock but these are usually shallow. Intermediate and basic rocks occur in many places and occasionally become important. Limestone and other calcareous materials also are of restricted distribution but there are a few places where they are widespread,

the area bordering the Mediterranean Sea is probably the best example. Where limestone, basic and intermediate rocks are present they often lead to the formation of soils which contrast with the more common ones formed in acid materials.

CLIMATE

Climate is the principal factor governing the rate and type of soil formation as well as being the main agent determining the distribution of vegetation and the type of geomorphological processes, therefore it forms the basis of many classifications of natural phenomena including soils.

The climate of a place is a description of the prevailing atmospheric conditions and for simplicity it is defined in terms of the averages of its components, the two most important being temperature and precipitation. The data on which the averages are based are usually accumulated for at least 35 years so as to take account of the differences — sometimes wide — in annual rainfall and temperature patterns. Although averages are most commonly used, diurnal and annual patterns and extremes are not ignored, since they give character and sometimes are important factors. The occurrence of occasional high winds can determine the development policy of an area. However, it must be stressed that the atmospheric climatic data do not always give a true picture of the soil climate. For example, the amount of water in the soil may vary considerably within a distance of a few metres from permanently saturated to dry and quite freely draining; whereas there is virtually no difference between the amount of total precipitation of the two sites, one site may be in a depression where moisture can accumulate and the other on an adjacent slightly elevated situation. These differences in the moisture regime at the two sites regularly lead to the development of different soils and contrasting plant communities.

Data on soil climate are meagre because of the difficulties encountered in making the necessary measurements actually within the soil itself, and frequently the lack of adequate equipment; thus many inferences are made based either on occasional observations without the support of adequate measurements or upon extrapolations using atmospheric climatic data. In spite of these difficulties and deficiencies it is possible to make many general statements about soil climate and its effects using the available atmospheric and soil climatic data, coupled with reasoning based on chemical and physical principles.

Soil climate has the same two major components as atmospheric climate, namely temperature and moisture.

Temperature

Atmospheric and soil temperature variations are the most important manifestations of the solar energy reaching the surface of the earth. The atmosphere transmits most of the visible and heat rays through clean dry air but absorbs nearly all of the short wave radiation, and of the radiation reaching the soil surface part is absorbed and converted into heat while the remainder is reflected back. The percentage of the incident global radiation reflected by a surface is known as its albedo, some typical examples of which are given in Table 2.7. Of the incoming radiation 50 per cent reaches the ground or ocean, 30 per cent is lost by scatter or reflected back and 20 per cent is absorbed by water vapour, dust and clouds. About 0.1 per cent of the energy is fixed by photosynthesis but 50 per cent of that energy viz. 0.05 per cent is used in respiration. The remainder enters one of two food chains — the grazing food chain and the decay food chain. A proportion of the heat produced is maintained in the soil but some is lost to the atmosphere by convection of hot air from the soil and by back radiation. Also some may be used for evaporation of moisture into the atmosphere and is an additional

Table 2.7 Albedos for some natural surfaces

Surface	Albedo
Snow (fresh)	75–90
Clouds	
Cumuliform	70–90
Cirrostratus	40–50
Ice	30–40
Sand dunes	
dry	35–45
wet	20–30
Soil	
pale	25–45
dark	5–15
Plant communities	
Desert	25–30
Savanna	
dry season	25–30
wet season	15–20
Forest	
deciduous	10–20
coniferous	5–15
Meadow	10–20

loss. The utilisation of the greater part of the long wave solar radiation is given diagrammatically in Fig. 2.20.

The main effect of temperature on soils is to influence the rate of reactions, for every 10 °C rise in temperature the speed of a chemical reaction increases by a factor of two or three. The principal reaction in soil to which this applies is the hydrolysis of primary

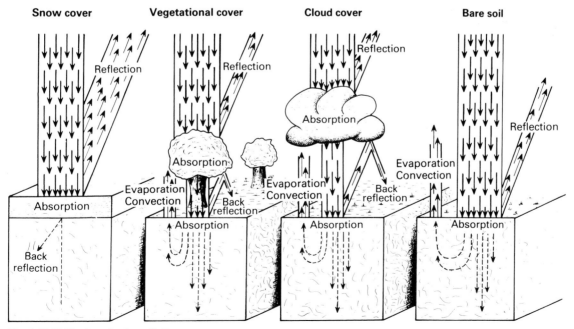

Fig. 2.20 Utilisation of solar radiation

silicates. The rate of biological breakdown of organic matter and the amount of moisture evaporating from the soil are also increased by a rise in temperature. Since most climates are seasonal the rate of chemical and biological activity will vary during the year. In the warm season, chemical weathering and biological activity are greater providing there is an adequate supply of water. In the cool or dry season the speed of reactions is reduced. Thus soil formation may have seasonality.

Development of vegetation can also be affected by soil temperature, in cool climates plants become active at 5 °C and their rate of maturation is approximately doubled for each 10 °C rise in temperature up to a maximum of about 20 °C. The amounts of radiation reaching the surface and the soil temperatures are determined by:

Hourly, diurnal and seasonal variations
Latitude and aspect
Altitude
Cloudiness and humidity
Prevailing winds
Vegetational cover
Snow cover
Soil colour
Soil moisture content
Heat conduction and porosity

Hourly, diurnal and seasonal variations
Soils have well marked hourly, diurnal and annual cycles. The hourly cycle takes place in the top 1—2 cm where there may be marked hourly variations in temperature as determined by variations in cloud cover and precipitation. The diurnal variations or cycles extend to a depth of about 50 cm while the annual cycle extends to as deep as 10—20 m where the temperature is uniform. During the diurnal cycle in tropical and subtropical areas it is normal for heat to move downwards in the soil during the day from the surface warmed by incoming radiation and upwards during the night as the surface cools. This takes place also during the summer period in the middle and polar latitudes but in these areas, during the winter, the atmosphere is generally cooler than the soil and the incoming radiation is not sufficient to heat the soil which steadily cools and eventually may freeze from the surface downwards. In tropical areas the total incoming radiation varies little from one season to another but in the middle latitudes there is ten times more radiation in summer than in winter. This difference increases towards the poles where there are long periods during the winter without radiation. During

the annual cycle in countries with contrasting seasonal climates the soil becomes warm during the summer and cool during the winter but the rhythm of these cycles, particularly the daily cycle, is regularly interrupted by wind and rain and occasionally by snow.

Heat moves very slowly down through the soil so that the diurnal maximum in the lower horizons at about 30—50 cm occurs up to 12 hours after the surface maximum as shown in Fig. 2.21. This lag is greater in the annual cycle when the lower horizons attain their mean maximum even after the surface begins to cool in response to a seasonal change. The temperature fluctuations within the soil between seasons are greater at the surface than in the lower horizons. Thus, during the summer, the diurnal mean surface temperature is higher than that in underlying layers but in winter the reverse is true as shown in Fig. 2.22.

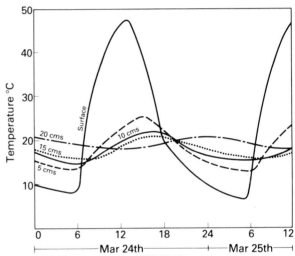

Fig. 2.21 Diurnal soil temperature variations at Cairo

Each soil has its characteristic temperature regime which for most practical purposes can be described by the mean annual soil temperature, average seasonal fluctuations from that mean and the mean cold and warm season temperatures.

A distinctive feature is that in each soil the mean annual soil temperature is almost the same at all depths, the differences being so small that they can be ignored, so that a single value can be used to represent the mean annual temperature of the soil. Since the soil temperature is uniform at 10—20 m the mean annual soil temperature can thus be determined by taking a single reading at about 15 m. If such depths are not possible then the mean of four equally spaced readings in a year at 50 cm will give similar results or more often at shallower depths.

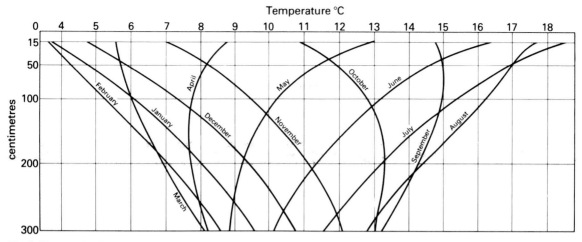

Fig. 2.22 Annual soil temperature variations in southern England

In addition, the mean annual soil temperature is about 1 °C higher than the mean annual air temperature.

Latitude and aspect

These are the two most important factors which influence soil temperature. Land surfaces normal to the rays of the sun are warmer than those at smaller angles and it is in tropical areas where the sun's rays travel their shortest distance through the atmosphere that this is most marked, particularly in the deserts where the radiation is not intercepted to any extent by water vapour, clouds or vegetation. To the north and south of the tropics, surfaces normal to the sun's rays are also warmer but as the distance from the Equator increases the mean annual temperature of such surfaces steadily decreases. This is because of the greater distance that the sun's rays have to travel through the atmosphere as enhanced by the curvature of the earth. Differences in warmth between surfaces are particularly important in the middle latitudes where sloping land surfaces facing towards the Equator usually carry a plant community indicative of a warmer and drier climate than land facing towards the poles. Therefore, slopes facing the Equator may be ready for spring cultivation days and even weeks before land facing towards the poles.

Altitude

A characteristic of the relationship between climate and altitude is the decrease of about 1 °C for each 170 m rise in elevation but this can vary within fairly wide limits, particularly in the tropics where the effect is not seen to any great extent below 500 m. In addition, the intensity of incident solar radiation increases with altitude, in middle latitudes this is about 5 to 15 per cent for each 1 000 m increase in elevation.

With increasing altitude there is initially a steady increase in precipitation, which reaches a maximum and then decreases again. In very mountainous areas these factors combine to produce a vertical climatic, vegetational and soil zonation which often terminates in glacial phenomena. An additional characteristic feature of mountainous areas is that cold air being more dense than warm air flows from the higher cooler areas into the valleys where it can cause harmful frosts.

Cloudiness and humidity

Both of these factors when of high intensity absorb radiation and reduce the amount reaching the soil. On the other hand they reduce heat losses at night by reflection back to the earth's surface. Dust particles and pollution of the atmosphere play a similar role.

Prevailing winds

Soil temperatures can rise or fall depending upon the character of the prevailing wind, a dry wind increases evaporation which in turn causes a loss of heat and a drop in soil temperature. On the other hand the soil can either gain or lose heat through the presence of warm or cold winds. Ocean currents and winds are particularly important in transferring heat to extra-tropical areas from the tropics where the incoming solar energy exceeds the outgoing.

Vegetational cover

This has a strong buffering effect on the soil temperature for during the day it absorbs or reflects a high

27

proportion of the radiant energy which would otherwise reach the soil surface and at night it reflects to the surface some of the heat radiating from the warm soil to the cooler atmosphere. Thus, soils beneath vegetation are cooler during the day and warmer during the night than adjacent bare sites. The cooling effect of vegetation upon the soil is utilised in tropical agriculture where plants such as coffee and cocoa, which are very susceptible to high light intensities and high soil temperatures, are interplanted with shade-trees. The buffering effect of vegetation is also important in the middle latitudes where it prevents rapid heat losses during the winter and reduces the penetration of frost into the ground.

Snow cover

A cover of snow protects the surface from prevailing winds and reduces heat losses by reflecting the heat back thereby preventing rapid temperature fluctuations. A snow cover may maintain the soil temperature at the surface 7 to 8 °C higher than in an adjacent bare soil but the bare soil will heat up quicker in the spring. Sometimes an early shower of snow can reduce the penetration of frost during the winter months.

Soil colour

Dark coloured soils absorb the greatest amount of heat and may have temperatures a few degrees higher than neighbouring soils of lighter colour; as a consequence they lose more moisture by evaporation but they gain more by condensation at night. Because dark objects emit more heat than light coloured objects they become cooler at night hence the greater amount of condensation. Black soil absorbs about 92 per cent of the solar energy whereas snow absorbs only 10–15 per cent; the remainder of the radiation is reflected back into the atmosphere.

Soil moisture content

The moisture content of the soil can have a marked influence upon the absolute soil temperature as well as upon the rate of temperature change. Dry soils have a specific heat of about 0.84 J g^{-1} whereas that of a saturated soil is about 3.56 J g^{-1}, this means that moist soils warm up more slowly because of the considerably greater amount of heat required to raise the temperature. Some heat is also used up in evaporation and lost to the atmosphere, further preventing a rise in temperature. Therefore, wet soils never attain the same maximum temperature as adjacent dry ones. Similarly, wet soils cool much more slowly since they require to lose more heat to effect a change in temperature.

Heat conduction and porosity

The subsurface horizons receive most of their heat by conduction from the surface. The rate of heat transfer depends upon a number of factors of which porosity and moisture are the two most important variables, more especially the latter. Therefore, it is necessary to consider the movement of heat in soils in the dry and the wet state. When the soil is dry the conduction of heat is very slow because the conductivity of air in the pore space is very low, since it is only at the points of contact between mineral grains that heat can be conducted rapidly from one to the other, thus the higher the porosity the lower the conductivity. Very often porosity increases with decrease in particle size, consequently soils containing a high percentage of fine materials tend to heat up slowly; dry peat with its high content of air heats up even more slowly.

In a moist soil, water forms bridges between the particles thus increasing the areas of contact between them. Since water transfers heat both by conduction and convection better than air, heat will flow more rapidly through the bridges and therefore through the soil as a whole. However, if the soil is very wet heat conduction will be at its slowest because of the large amount of heat required to heat the water.

A limited amount of heat transfer takes place when warm water flows from the warmer upper part of the soil into the lower horizons.

Moisture

The moisture in soils includes all forms of water that enter the soil system, and is derived mainly from precipitation as rain and snow or may be supplied by the lateral movement of water over the surface, or within the body of the soil itself. Underground water may also contribute. In a number of cases flooding by rivers provides the greater part of the soil moisture, perhaps the best examples are to be found in the Nile valley.

The moisture entering soils contains appreciable amounts of dissolved CO_2 thus it is probably more correct to think of it as a dilute weak acid solution which is much more reactive than pure water. However, underground waters may be alkaline, weakly acid or even saline.

Precipitation varies considerably from place to place but the total volume is no guide to the actual amount entering the soil. This is determined by:

Intensity of rainfall
Vegetational cover
Infiltration capacity

Permeability and slope
Speed of snow melting
Original moisture content of the soil.

Intensity of rainfall

In areas of bare soil it is rainfall of moderate intensity that is most effective in entering the soil particularly when the soil is porous. Light showers of rain that hardly enter the soil are quickly lost by evaporation. Heavy showers may fall at a rate greater than the capacity of the soil to absorb the moisture, therefore it accumulates on the surface or runs off, creating an erosion hazard. Further, the impact of large raindrops on the soil during a heavy shower may cause dispersion and the formation of a slurry which seals the surface and prevents the entry of moisture into the soil. This can cause severe erosion if the site is sloping as the slurry of dispersed soil runs downhill over the surface. The impact of raindrops causes material to be splashed upward and outwards. On slopes, some of the material will fall below the original position thereby bringing about a certain amount of erosion down the slope. Drops from leaves may have the same effect since they may attain their terminal velocity between the canopy and the ground in high forest areas.

Vegetational cover

This has a profound effect on the amount of precipitation reaching the soil surface. In many cases light showers are held on the foliage and returned to the atmosphere by evaporation. Only heavy showers provide sufficient moisture to saturate the foliage and produce an excess which falls on to the soil surface or runs down the stem causing greater leaching immediately around the base of the tree. As the density of the vegetation increases the amount of moisture held on the foliage increases. This attains its maximum in tropical rain forests which, with their layered structure can intercept and retain large volumes of rainfall.

In areas of high rainfall it is vital to maintain a vegetational cover during the rainy season to avoid the impact of raindrops and the resultant erosion. In dry areas the foliage of trees intercepts nearly all of the rain, only the soil between the areas covered by the canopy receives any rainfall.

Infiltration capacity

The capacity of water to enter the soil varies considerably; soils with a well developed structure or coarse texture allow free entry of moisture whereas massive clays are virtually impermeable. Variations in the capacity of the soil to absorb moisture are particularly important in areas of high rainfall where low infiltration rates cause water to collect at the surface creating a serious erosion hazard. In areas of low rainfall, maximum infiltration is of prime importance so that the plants can utilise fully the annual precipitation.

Permeability and slope

The amount of moisture capable of percolating completely through the soil is largely dependent upon the permeability of the middle and lower horizons. When either of these is impermeable the upper layers can quickly become saturated with water, resulting in lateral movement through the soil and run-off over the surface. On sloping sites, erosion is often the normal consequence while on flat sites temporary flooding takes place.

Speed of snow melting

When snow melts the moisture can enter the soil or flow over the surface as determined by the infiltration capacity of the soil, permeability and slope as described above. If the melting takes place slowly a great proportion of the moisture will enter the soil, fast melting leads to run-off and a possible moisture deficiency for plants during the growing season.

Another factor affecting the water released by melting snow is the presence of frozen soil. If the soil is frozen the water will run off over the surface and may be lost. This can be of utmost importance in certain continental areas with low precipitation where early showers of snow prevent the soil from freezing so that a considerable proportion of the water from the melting of the snow in spring can enter and be stored in the soil.

Original moisture content of the soil

This also influences the entry of precipitation, obviously no further additions can take place if the soil is already saturated. Such conditions occur at the lower ends of slopes and in depressions where run-off usually accumulates and the water-table comes to the surface. On these sites precipitation accumulates with the formation of temporary or permanent ponds.

Water characteristics in soils

The state and movement of water in soils is very complex, therefore the overall picture has to be presented in a number of stages followed by a few examples to illustrate the interrelationships.

The state of water in soils varies from that which is free to flow through the soil to that which is adsorbed firmly on to the surfaces of particles. Thus water is held with varying degrees of *tension* or *suction* within the soil. The tension with which water is held is expressed in many ways but a common method is in bars. The tension needed to support a column of water 1 000 cm in height is about 1 000 millibars or 1 bar.

Water characteristics

Two important moisture characteristics of soils are the *field capacity* and the *wilting point*. After the soil has been saturated and the excess water drained away the soil is said to be at field capacity and the water is held at a tension of about 50 millibars. If plants are growing on the soil they will extract moisture until they cannot extract any more, then they will wilt and eventually they will die if the soil is not rewetted. The point at which permanent wilting starts is known as the wilting point where water is held at about 15 bars, but this value does differ slightly from soil to soil and plant to plant. Thus water held between field capacity and wilting point is the water available to plants. The amount varies from soil to soil being greatest in silty soils and least in sands as shown in Fig. 2.23. As the water content diminishes from field capacity to the wilting point the soil water suction increases and so the plant has greater and greater difficulties in obtaining water thus experiencing ever increasing water stress.

The absolute amount of water held at various suctions varies from soil to soil and is determined mainly by the particle size distribution and organic matter content. The relationship between the tension and moisture content for three soils is shown in Fig. 2.24. For a given tension the clay soil contains the greatest amount of water. Similarly, for a given moisture content the clay has the greatest tension. An interesting feature is that much of the water held by the clay soil beneath field capacity is also beneath wilting point and therefore unavailable to plants. In Fig. 2.23 the moisture content is given as a weight percentage but it is often more advantageous to express the results as a volume percentage which gives a better picture of the relative amounts of solid, air and moisture in the system.

Water movement in soil

Water is constantly moving in most soils but there are different types of movement depending upon the amount of water present and the nature of the soil. There are three types of water movement: saturated flow, unsaturated flow and vapour flow.

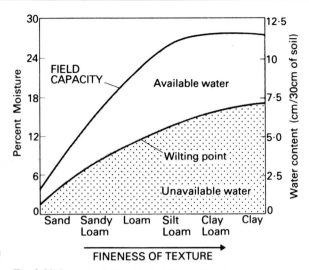

Fig. 2.23 General relationship between soil moisture characteristics and soil texture. Note that the wilting point increases as the texture becomes finer. The field capacity increases up to silt loams then levels off

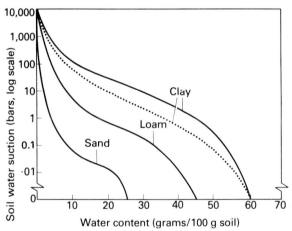

Fig. 2.24 Soil moisture suction curves for three representative mineral soils. The solid lines show the relationship obtained by slowly drying completely saturated soils. The dotted line for the clay soil is the relationship expected when a dry soil is wetted. The difference between the two lines for the clay is due to hysteresis

Saturated flow

Generally this takes place when water is moving through soil in which all of the pores are filled with water. This occurs most commonly beneath the water-table but under some conditions a part of the soil may become saturated even in soils that are usually aerobic. Thus, saturated flow may also take place in individual parts of the soil even though the large pores may be filled with air. Movement takes place in any direction, it may be vertically downwards or laterally over a pan. The rate of flow is determined by the hydraulic driving force and hydraulic conductivity, i.e. the ease with which water may pass through the soil, this in turn is determined by the size and disposition of the pores.

Unsaturated flow

Under these conditions water moves from pore to pore by flowing over the surfaces of peds and/or particles while there is a considerable amount of air in the larger pore spaces. Movement can be in any direction. In soils that have recently been wetted by rain the flow is downwards in response to gravity but after field capacity has been attained movement is usually either lateral or upwards in response to a moisture gradient due to drying at the surface or uptake by plant roots. Unsaturated flow is the main process responsible for leaching soils and for attempting to equilibrate the moisture tension in soils. Thus it takes place in response to gravity as well as in response to a moisture tension gradient.

Vapour flow

This is the movement of water in the vapour phase from one place to another in the soil or from the soil to the atmosphere. The rate of movement is determined by the relative humidity and temperature gradient, the size and discontinuity of the pores and the amount of water present. The humidity of the above ground atmosphere very largely determines the amount of water that is lost from the soil. The temperature gradient is most important at the surface which as shown on page 26 may have very high temperatures causing a large amount of vaporisation of water. The size and continuity of the pores will determine whether there is free movement from place to place within the soil. The amount of water will determine the amount of pore space through which the vapour can move and as the water content decreases the air filled porosity increases. In order to understand more fully the movement of water in the soil system it is necessary to consider the following six situations:

    Water movement in a coarse sand
    Water movement in a well structured loam
    Water movement in a clay
    Water movement in response to freezing
    Water movement in stratified materials
    Comparative water movement in contrasting soils

Water movement in a coarse sand

Consider the situation in which water is continuously added to a column of sand. It will flow freely downwards by saturated flow in response to gravity thus wetting more and more sand. If the addition of water is stopped, downward saturated flow will continue but it will soon change to unsaturated flow because there will not be enough water to fill all the pores. There will only be enough to flow over the surfaces of the grains. Eventually, downward flow will almost cease when the upper part of the sand reaches field capacity and when there is only enough water to form thin films on the pore surfaces and bridges between the particles and is held in these positions by forces greater than gravity. However, there will still be a small amount of downward flow because of the moisture gradient with the dry sand. At the surface evaporation results in the loss of water thus creating a moisture gradient which causes water to be drawn upwards to the surface by unsaturated flow.

In such a system any fine particles and ions in solution will move downwards in the percolating water and some dissolved ions will move to the surface when there is unsaturated flow upwards. If plants are growing on the soil they will extract water and ultimately the wilting point will be attained. At the wilting point there is still some water in the soil which will continue to dry until it equilibrates with the atmosphere. The water held at this stage is known as hygroscopic water. Some of the water below wilting point can move by unsaturated flow even though it cannot be taken up by plants. Figure 2.25 shows the relationship between the soil surface and the thickness of water films at various water states and suctions.

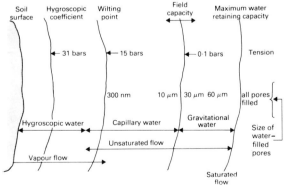

Fig. 2.25 Relationships between the types of water, types of flow and suction

Water movement in a well structured loam

Consider a similar situation to the above but in this case there is a well structured soil in place of the coarse sand. After the water has drained away and field capacity has been attained the peds will be surrounded by thin films of water and there will also be bridges between the peds but most important is that the interiors of the peds will be completely saturated with water. As the soil dries water moves to the outer surfaces of the peds where it is lost by evaporation or taken up by roots so that unsaturated flow tends to be within the individual peds rather than in the whole soil mass.

Water movement in a clay

In a saturated clay water movement is extremely slow and through drainage scarcely ever takes place even when the soil is completely saturated. Thus the concept of field capacity is not really applicable to clay soils so that the available water will range from the wilting point to the point of anaerobism that will inhibit root growth. Clay soils lose their water mainly by evapotranspiration so that water is lost from the surface as well as from the body of the soil. As water is lost, the clays lose their films of water and are drawn closer and closer together. This causes the soil to shrink and ultimately soils develop a system of cracks, the amount and extent of which will depend upon the intensity of drying and the nature of the clay. As drying continues water moves from the interior of the soil to the surface of the cracks and in so doing transports soluble materials which may form an efflorescence on the crack surfaces. An interesting feature of this system is that very little air is drawn into the mass of the soil; it only enters the main crack system.

With the onset of the wet season water flows rapidly down the cracks so that rewetting of the soil takes place from the bottom upwards. Much of the water then moves laterally into the soil by unsaturated flow thereby rehydrating the clay micelles causing the whole soil to expand and the cracks to close.

Water movement in response to freezing

When the water in soil freezes it has the same effect as drying so a moisture gradient is produced. In a soil that is freezing from the surface downwards there is also a temperature gradient, the result is that water moves towards the freezing front. The rate and amount can be quite large since stones in the surface soil can develop a coating of ice several millimetres thick on their undersurface after a single very cold night.

Water movement in stratified materials

Any type of stratification or horizonation with marked differences in clay content can severely influence the movement of water in soils. When water percolates through the soil and meets a horizon or layer with coarser texture the water accumulates above the junction until there is a sufficient build up which is always well above field capacity; then it drains rapidly through at one or more points. This leads to a distinctive type of leaching and translocation. If the underlying layer or horizon has more clay the water accumulates above and only very slowly enters the underlying material. This is a fairly common situation in soils, causing temporary or sometimes permanent waterlogging near the surface.

Comparative water movement in soils with different textures

Consider a well structured clay loam and a sand in an environment with marked seasonality and where precipitation exceeds evapotranspiration. Conditions are such that during the rainy period water flows vertically downwards by saturated flow. Since the clay loam has a much higher water retaining capacity and therefore more water at field capacity more rainfall will be required to wet the soil to the same depth as the sand. Thus, through percolation, will start in the sandy soil before the clay soil. This situation may be repeated several times during a rainy season.

During short dry periods in the rainy season the loss of water by evapotranspiration may cause the water content to fall below field capacity but it may be attained again after the next rain.

Thus a sandy soil is leached to a greater extent than a clayey soil where the other variables are constant. There might also be the situation in areas of fairly low rainfall where there might be sufficient precipitation to bring the sand to field capacity and even cause some leaching but be insufficient to bring the clay loam to field capacity.

Water state in soils

The classical method is to classify soil water into gravitational, capillary and hygroscopic (Fig. 2.25). The gravitational water flows freely downwards through the soil and is held at a tension of about <0.1 bar. Capillary water is held in the pores and on the surfaces of particles at 0.1 to 31 bars and moves in any direction in response to a moisture gradient. Hygroscopic water is tightly held by the soil at suction values in excess of 31 bars and moves in the vapour phase.

These concepts have been superseded by concepts of flow and potential. Now the state of water in soils is described physically in terms of its *potential energy*, i.e. its potential to do work and to move from place to place within the soil. The potential energy of soil water depends on its height above sea level, its pressure or its suction, and on the surfaces of particles, and the presence of dissolved salts. These are described as gravitational, pressure, matric and osmotic potential energies but by convention the word energy is omitted. These ideas are beyond the scope of this book but are fully described by Taylor and Ashcroft (1972).

Moisture losses

The moisture entering the soil system is lost through drainage and the combined processes of evaporation and transpiration generally referred to as *evapotranspiration*. Even in very wet and permanently saturated sites there is usually continuous percolation for seldom is the moisture truly stagnant. Losses by evapotranspiration are influenced by the nature of the vegetation and a number of atmospheric conditions including solar radiation, air temperature, the character of the prevailing winds, humidity and cloudiness. Humidity is possibly the most important factor determining the rate of moisture loss by transpiration and evaporation. In areas of particularly high humidity, losses are reduced to a minimum, sometimes resulting in waterlogging of the surface horizons as in certain maritime areas of Western Europe and at high elevation in the wet tropics.

The plants growing on the soil surface can be thought of as water extractors, removing the water from the soil by their roots and losing it to the atmosphere. Thus we find that soil devoid of vegetation remains wetter longer. This has led to the practice of dry farming which is explained on page 337.

In presenting climatic data it is often customary to give the *potential evapotranspiration*, i.e. the amount of moisture that might be lost by these two processes if there were an unlimited supply of water present and a complete cover of vegetation.

In most situations the potential evapotranspiration exceeds the actual because the rainfall may be very low but in tropical rain forests the actual evapotranspiration may exceed the estimated potential because of the luxuriance of the vegetation. The potential evapotranspiration shows a marked seasonal variation in the middle latitudes because of a marked summer growing season but there is only a small seasonal variation in the tropics.

Some typical evapotranspiration losses during the

Table 2.8 Some typical evapotranspiration losses during the growing season

Location	losses (mm)
Cool mountain valley	300
Irrigated desert	2 000
Humid areas	350
Semiarid areas	750
Hot, dry with irrigation	500 to 1 250

growing season are given in Table 2.8. Another useful method of expressing soil water in relation to plants is by the use of the *transpiration ratio* which is the transpiration in grams per gram of plant tissue produced. This ranges from 200 to 500 in humid regions to >1 000 in arid regions. Thus a crop of wheat containing 4 000 kg of dry matter per hectare and having a transpiration ratio of 250 (England) or 500 (Colorado) will use about 100–200 mm of rain during the growing season plus evaporation from the soil surface, thus moisture may be a limiting factor.

There are a number of practical difficulties in measuring evapotranspiration so that it is usually estimated using a formula based on meteorological data. Among the various formulae suggested, the Penman (1956) formula is the best method for estimating potential evapotranspiration from a simple set of meteorological data and gives very reliable predictions. However, the main value of this information lies outside the field of pedology.

Since climate plays the major role in determining the nature of the soil, vegetation and landscape, it would seem appropriate to include the classification of Strahler (1970) given below. It is a relatively simple and straightforward system and is one of the many systems currently in use and seems to be useful in pedology.

A number of workers have introduced climatic indices based on atmospheric climatic data. Most are of little value, since atmospheric climatic data often have little pedological significance. For example, the precipitation : evaporation ratio, (P/E index) of Transeau (1905) shows the amount of precipitation relative to evaporation so that as the value increases above unity, the amount of moisture available for percolation through the soil increases. This may be true for easily permeable soils on flat sites, with complete through drainage but it is of little value for soils with compact impermeable layers or those at the lower end of slopes and in depressions that receive a considerably larger volume of moisture by run-off.

Figure 2.26 is a generalised attempt to show the moisture cycle under humid conditions and Fig. 2.27

33

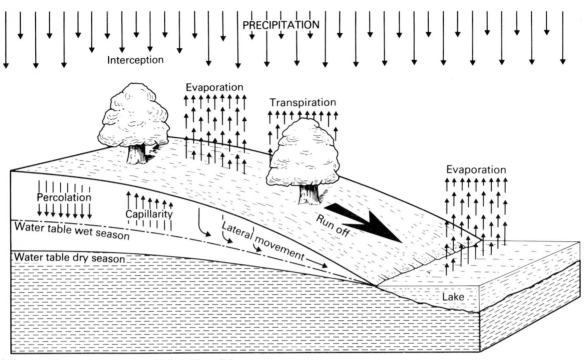

Fig. 2.26 The moisture cycle under humid conditions

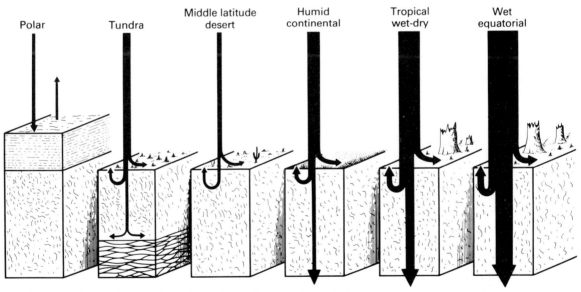

Fig. 2.27 The fate of moisture falling on the surface under a number of different climatic conditions

shows the fate of moisture falling on the surface under a number of different climatic conditions.

Since soil climate is an important property of soils and with an ever increasing amount of soil climate data being collected, attempts are being made to produce classifications of soil climate but, as with soil classification generally, opinions differ about the number and nature of the categories. The system suggested by USDA (1975) is fairly simple and is given below following the classification of atmospheric climates. This takes into account the effect of slope, aspect and seasonal variations in the atmospheric climate.

Classification of atmospheric climates (Strahler, 1970)

Group I   Low latitude climates

1. Wet equatorial climate
Precipitation.   Continuously high with 1 500 to 2 500 mm p.a. and monthly variation of 75 to 25 mm, no dry season or short dry season.

Temperature.   Mean annual temperature of 24 to 27 °C with 1 to 2 °C variation, the warmest month has a mean of 27 to 29 °C with diurnal range of 8 to 11 °C and maximum of 29 to 32 °C, the coldest month has a mean of 21 to 27 °C with diurnal range of 8 to 11 °C and minimum of 21 to 24 °C – no frosts.

Distinctive features.   Uniform high temperatures and high precipitation with thunderstorms and hurricanes in places.

Occurrence.   10 °N–10 °S (Asia 10–20 °N). The Amazon Basin, Windward coast of Central America, the Congo Basin, New Guinea, the Philippines and the east coast of Madagascar.

2. Trade wind littoral climate
Precipitation.   1 500 to 3 000 mm p.a. with a monthly variation of 25 to 700 mm and marked summer maximum.

Temperature.   Mean annual of 24 to 27 °C with 1 to 2 °C variation, the warmest month has a mean of 21 to 27 °C with diurnal range of 11 to 14 °C and maximum of 29 to 32 °C; the coldest month has a mean of 21 to 27 °C with diurnal range of 10 to 14 °C and minimum of 18 to 21 °C and no frosts.

Distinctive features.   Strong winds and fairly cool winter dry season.

Occurrence.   10–25° N and S latitude. East coast of India, Burma, Vietnam, Indonesia, northern Philippines, Western Guinea coast of Africa, north-eastern coast of South America, northern coast of Haiti and Puerto Rico.

3. Tropical desert and Steppe climates
Precipitation.   Steppe: 100 to 500 mm p.a., desert: 10 to 100 mm p.a.

Temperature.   Mean annual of 21 to 27 °C with 14 to 22 °C variation, the warmest month has a mean of 35 to 38 °C with diurnal range of 14 to 17 °C and maximum of 49 to 54 °C; the coldest month has a mean of 12 to 16 °C with a diurnal range of 14 to 17 °C and minimum of −1 to 4 °C; no days with frost.

Distinctive features.   Very low rainfall, often in heavy sudden showers.

Occurrence.   15–35° N and S latitude. North-western Mexico and south-western USA, west coast of Peru, northern Chile, the Sahara, Somalia, Arabia to West Pakistan, west coast of Southern Africa and central Australia.

4. West coast desert climate
Precipitation.   Less than 250 mm p.a., generally virtually nil.

Temperature.   Mean annual of 18 to 23 °C with 5 to 10 °C variation, the warmest month has a mean of 20 to 25 °C, with diurnal range of 15 to 20 °C and maximum of 45 to 55 °C. The coldest month has a mean of 12 to 18 °C with diurnal range of 15 to 20 °C and minimum of −1 to 4 °C, 0 to 25 days with frost.

Distinctive features.   Extremely dry, but relatively cool with small annual temperature range, fog in the coastal belt.

Occurrence.   15- 30 °N and S latitude. Coast of Chile, southern USA, western Sahara and South Africa, north-western Australia.

5. Tropical wet-dry climate
Precipitation.   1 000 to 1 700 mm p.a. with a monthly variation of 0 to 350 mm and a marked summer maximum.

Temperature.   Mean annual of 24 to 27 °C with 4 to 8 °C variation; the warmest month has a mean of 27 to 32 °C with a diurnal range of 8 to 10 °C; and maximum of 35 to 38 °C; the coldest month has a mean of 21 to 24 °C with diurnal range of 14 to 17 °C and minimum of 16 to 18 °C; no frosts.

Distinctive features.   Marked seasonal contrasts,

unreliable precipitation, dry winter season of two to four months.

Occurrence. 5–25 °N and S latitude. Western central America, north-western South America, southern interior of Brazil, Bolivia, Paraguay, south-central, west and eastern Africa, western Madagascar, parts of India, and south-eastern Asia and northern Australia.

Group II Middle-latitude climates

6. Humid subtropical climate

Precipitation. 750 to 1 600 mm p.a. with a monthly variation of 50 to 175 mm and distinct summer maximum.

Temperature. Mean annual of 16 to 21 °C with 11 to 22 °C variation, the warmest month has a mean of 24 to 29 °C with diurnal range of 5 to 8 °C and maximum of 29 to 38 °C; the coldest month has a mean of 4 to 13 °C with diurnal range of 8 to 11 °C and minimum of −4 to 2 °C; 15 to 65 days with frost.

Distinctive features. Moderate annual precipitation with fairly uniform distribution, occasional frosts, hurricanes and typhoons.

Occurrence. 20–35 °N and S latitude. South-eastern USA, most of China and Formosa, southern Japan, north-eastern Argentina, Natal coast, eastern Australia, north-eastern Italy, south coast of the Black Sea.

7. Marine west coast climate

Precipitation. 500 to 2 500 mm p.a. with a monthly variation of 25 to 100 mm.

Temperature. Mean annual of 7 to 13 °C with 8 to 14 °C variation; the warmest month has a mean of 16 to 18 °C with diurnal range of 8 to 11 °C and maximum of 21 to 24 °C; the coldest month has a mean of 2 to 7 °C with diurnal range of 8 to 11 °C and minimum of −4 to −7 °C; 100–150 days with frost.

Distinctive features. Dull, drizzly weather with cool wet summers and mild wet winters and cyclonic storms.

Occurrence. 40–60 °N and S latitude. California to Alaska, British Isles, north Portugal to southern Scandinavia, southern Chile, south-eastern Australia, Tasmania and New Zealand.

8. Mediterranean climate

Precipitation. 400 to 800 mm p.a. with summer minimum or winter maximum.

Temperature. Mean annual of 12 to 18 °C with 14 to 19 °C variation, the warmest month has a mean of 16 to 24 °C with diurnal range of 15 to 19 °C and maximum of 35 to 38 °C; the coldest month has a mean of 7 to 10 °C with diurnal range of 14 to 17 °C and minimum of −1 to 4 °C; 15 to 65 days with frost.

Distinctive features. Hot dry summers and mild rainy winters with some snow in many places.

Occurrence. 30–45 °N and S latitude. The Mediterranean coast, central California, central Chile, southern tip of Africa, south-western Australia, south Australia, eastern Turkey and northern Iran.

9. Middle latitude desert and Steppe climate

Precipitation. Steppe: 100 to 500 mm p.a.; desert: 10 to 100 mm p.a. both with erratic monthly variation.

Temperature. Mean annual of 4 to 16 °C with 11 to 33 °C variation, the warmest month has a mean of 18 to 27 °C with diurnal range of 14 to 17 °C and maximum of 38 to 43 °C; the coldest month has a mean of −1 to 4 °C with diurnal range of 11 to 14 °C and minimum of −35 to −45 °C; 180 to 220 days with frost.

Distinctive features. Very wide variation in temperature between winter and summer and low unreliable precipitation.

Occurrence. 35–50 °N and S latitude. Inter-montane basins and Great Plains of western USA and Canada, southern USSR, Manchuria, east and central Argentina.

10. Humid Continental Climate

Precipitation. 400 to 700 mm p.a. with a monthly variation of 75 to 125 mm and weak summer maximum.

Temperature. Mean annual of 2 to 7 °C with a variation of 28 to 42 °C; the warmest month has a mean of 18 to 24 °C with diurnal range of 11 to 14 °C and maximum of 27 to 29 °C; the mean of the coldest month is −4 to −18 °C with diurnal range of

14 to 17 °C and minimum of −35 to −40 °C; 120 to 140 frost-free days.

Distinctive features. Cool moist summers, heavy winter snowfall and wide annual temperature range.

Occurrence. 35–60 °N latitude. In North America it roughly straddles the Canadian–USA border from the east coast to central Alberta. In Europe and the USSR it takes the shape of a long isosceles triangle with southern Scandinavia, Poland and Czechoslovakia at the base and the vertex deep in Siberia; in Asia it extends from northern Manchuria out to Hokkaido and Sakhalin.

Group III High latitude climates

11. Continental Subarctic Climate
Precipitation. 250 to 500 mm p.a. with a monthly variation of 25 to 75 mm and summer maximum.

Temperature. Mean annual of −9 to 0 °C with 18 to 33 °C variation, the warmest month has a mean of 7 to 21 °C with diurnal range of 11 to 17 °C and maximum of 27 to 35 °C; the coldest month has a mean of −5 to −40 °C with diurnal range of 5 to 8 °C and minimum of −50 °C; 50 to 90 frost-free days.

Distinctive features. Precipitation falls mainly as snow and six to eight months with mean temperatures less than 0 °C.

Occurrence. 50–70 °N latitude. Broad belts from Alaska to Newfoundland and Norway to Kamchatka.

12. Marine Subarctic Climate
Precipitation. 500 to 700 mm p.a. with a monthly variation of 35 to 75 mm and winter maximum.

Temperature. Mean annual of −15 to −1 °C with variation of 10 to 20 °C; the warmest month has a mean of 2 to 5 °C with diurnal range of 5 to 8 °C and maximum of 10 to 15 °C; the coldest month has a mean of −2 to 15 °C with diurnal range of 3 to 5 °C and minimum of −25 °C; 5 to 20 frost-free days.

Distinctive features. High precipitation, cloudiness, high fog and humidity.

Occurrence. 50–60 °N and 45–60 °S latitude. West coast of Alaska, south coast of Greenland, northern Iceland and northern Norway.

13. Tundra Climate
Precipitation. 250 to 500 mm p.a. with a monthly variation of 20 to 40 mm with autumn maximum.

Temperature. Mean annual of −15 to −1 °C with a variation of 14 to 33 °C; the warmest month has a mean of 4 to 10 °C with diurnal range of 5 to 8 °C and a maximum of 16 to 18 °C; the coldest month has a mean of −5 to −30 °C with diurnal range of 5 to 8 °C and minimum of −45 °C; 0 to 20 frost-free days.

Distinctive features. Precipitation falls almost exclusively as snow; six to ten months with mean temperature less than 0 °C.

Occurrence. North of 55 °N and south of 50 °S. Arctic coast of North America and Eurasia, coastal Greenland, many islands in the Arctic Ocean, southern tip of South America.

14. Ice-cap Climates
Precipitation. 50 to 500 mm p.a. almost exclusively as snow.

Temperature. Mean annual of −45 to −18 °C with 14 to 22 °C variation; the warmest month has a mean of −5 to −18 °C with diurnal range of 5 to 8 °C and maximum of −18 to 4 °C; the coldest month has a mean of −25 to −60 °C with diurnal range of 5 to 8 °C and minimum of −75 °C; no frost-free days.

Distinctive features. Twelve months with mean temperature <0 °C and glaciers.

Occurrence. Greenland, Arctic Ocean and Antarctica.

15. Highland Climates
Generalisations are not possible, each set of highlands causes a modification of the local climate. This is usually in the form of a decrease of 1 °C per 170 m elevation and an increase in rainfall, then a decrease at very high elevations, except in the tropics where the precipitation continues to increase. A distinctive feature is the large diurnal and annual temperature range and the presence of glaciers.

Occurrence. Cascade–Sierra Nevada, Rockies, Andes, Alps, Himalayas, eastern highlands of Africa, and the mountains of Borneo and New Guinea.

Classification of soil climates (USDA, 1975)

The classification of soil climates is based on the

37

moisture and temperature regimes which are defined separately.

Classes of soil moisture regimes

The soil moisture regimes are defined on the basis of the ground water and the presence or absence of water held at a tension <15 bars in the middle of the pedo-unit.

Aquic moisture regime.   The soil is generally saturated with water and more or less free of oxygen so that reduction can take place.

Aridic and torric moisture regimes.   The soil is dry for more than 50 per cent of the growing season or never moist for more than 90 consecutive days during the growing season. There is little leaching in these regimes and soluble salts usually accumulate.

Udic moisture regime.   The soil is not dry for as long as 90 consecutive days. These soils occur in humid climates with well distributed rainfall with enough during the growing season and leaching in most years.

Ustic moisture regime.   The soil has a limited amount of moisture but it is present in sufficient quantity during the growing season. In the tropics this moisture regime occurs in monsoon climates.

Xeric moisture regime.   The soil is dry for 45 or more consecutive days after the summer solstice and moist after the winter solstice. This soil moisture regime is found in Mediterranean climates.

Classes of soil temperature regimes

Pergelic temperature regime.   The soil has a mean annual temperature <0 °C and has permafrost.

Cryic temperature regime.   The soil has a mean annual temperature from 0 to 8 °C and are commonly frozen in their upper part during winter.

Frigid temperature regime.   The soil has an annual temperature that is <8 °C but it is warmer in summer than the cryic regime so that the difference between the mean summer and mean winter temperature is more than 5 °C.

Mesic temperature regime.   The mean annual soil temperature is 8 to 15 °C and the difference between mean summer and mean winter temperature is >5 °C.

Thermic temperature regime.   The mean annual soil temperature is 15 to 22 °C and the difference between mean summer and mean winter soil temperature is >5 °C.

Hyperthermic temperature regime.   The mean annual soil temperature is 22 °C or higher and the difference between the mean summer and mean winter soil temperature is >5 °C.

The prefix iso- can be added to the name of the soil temperature regime to indicate that the difference between the mean summer and mean winter temperatures is <5 °C.

These climatic divisions are used at various category levels in the US taxonomy (USDA 1975).

Climatic changes

One of the more fascinating aspects of climate is its constant and continuous change with time as revealed by an examination of the geological column which shows that the climate at any one point on the surface of the earth has changed many times, being at one time hot and dry, at other times hot and humid, or cold and wet. During the last two and a half million years some of the most dramatic changes have taken place and in many cases these fluctuations have led to the change from one set of soil-forming process to another (cf. Table 2.12).

ORGANISMS

The organisms influencing the development of soils range from microscopic bacteria to large mammals including man. In fact, nearly every organism that lives on the surface of the earth or in the soil affects the development of soils in one way or another. The variation is so wide that any grouping in order of their taxonomy really has little meaning from the pedological standpoint. An arbitrary and somewhat crude classification of the more important soil organisms is as follows:

Higher plants
Vertebrates
    Mammals
    Other vertebrates
Microorganisms
Mesofauna
Man

Higher plants

Under natural conditions higher plants are normally found arranged in communities containing many different species, but occasionally there may be a very small number present, particularly in forest communities which may be dominated by a single tree species. When afforestation is practised it is customary to plant only one or sometimes two different tree species in a given area but there may be many species in the ground flora. Under advanced agricultural conditions there is almost invariably a single species planted but occasionally a mixture of two or three different species may be grown together. Set out below is a list of the common naturally occurring plant communities (Eyre, 1968; Riley and Young, 1966) some are briefly defined in the glossary:

Boreal forests
Coniferous forests
Deciduous forests
Desert communities
Grassland communities
Heaths
Hydroseres
Mangrove swamps
Sand dune communities
Savannas
Savanna woodlands
Sclerophyllous woodlands
Temperate rain forests
Thorn woodlands
Tropical rain forests
Tundra and mountain communities

The higher plants influence the soil in many ways. By extending their roots into the soil they act as binders and so prevent erosion from taking place; grasses are particularly effective in this role and it is suggested that the development and spread of flowering plants especially during the Tertiary Period (see p. 52) is one reason for the formation and preservation of many of the very deep soils from that period. Roots also bind together small groups of particles and help to develop a crumb or granular structure. This is also a characteristic feature of some grasses which are regularly planted purely for their improvement of the soil structure. As the large roots of trees grow and expand they bring about a certain amount of redistribution and compaction of the mineral soil. They also act as agents of physical weathering by opening cracks in stones and rocks. When plants die their root systems contribute organic matter which upon decay leave a network of pore space through which water and air can circulate more freely causing some soil processes to be accelerated. One of the best examples is seen in many poorly drained soils where the mineral soil that surrounds a decomposed root becomes oxidised to give yellow or reddish-brown colours. If

this process is intense, pipes of soil material can develop around roots through the cementing action of ferric hydroxide that gradually accumulates. A somewhat similar situation is found in arid and semi-arid areas where the freer movement of water down the old root channels leads to a greater deposition of calcium carbonate on their sides, eventually forming pipes.

The amount of organic matter contributed by plant roots varies widely; whereas trees contribute a very large amount when they die, the total averaged over the lifespan of a single tree really amounts to a very small addition. On the other hand the contribution of annuals or biennials may be quite considerable.

Plant roots exude various substances on which many microorganisms thrive so that the soil in immediate proximity to the root or *rhizosphere* is an area of prodigious microbiological activity; frequently the concentration of iron immediately next to the live root is reduced imparting a bleached appearance to the soil.

The addition of litter to the surface of the soil is probably the greatest contribution of the higher plants except in certain agricultural practices when it may be removed. The total amount of organic matter added by the different plant communities is variable but it is no guide to the amount in the soil which depends more upon the rate and type of breakdown than on the amount added. Table 2.9 gives the weight of litter supplied by some plant communities to the surface. These data demonstrate that tropical vegetation contributes the greatest amount of organic matter but its associated soils usually contain the least amounts since biological activity, particularly that of microorganisms and termites, is much greater under warmer conditions. Thus, although the organic matter production is greater so also is its decomposition by the soil fauna and flora. On the other hand under cool humid conditions, the supply of organic matter is small but the decomposition rate is slower so that the normal tendency is for organic matter to accumulate at the surface especially in moist habitats.

By intercepting rain, higher plants shelter the soil from the impact of raindrops which puddle the surface, reducing the permeability and creating an erosion hazard. Higher plants tend also to maintain more equable conditions by shading the soil from the direct rays of the sun thereby reducing losses by evaporation. It follows, therefore, that the soil beneath dense vegetation neither becomes excessively wet when the rainfall is high nor very dry during periods of low rainfall.

Plants extract nutrients from the body of the soil

39

Table 2.9   Organic matter production by vegetation

Type of vegetation	Annual production Metric tons per hectare (air dry)	Location	Author
Alpine meadow	<0.5–0.9	Poland	Swederski (1931)
Short-grass prairie	1.6	Colorado USA	Clements and Weaver (1924)
Mixed tall-grass prairie	5.0	Colorado USA	Clements and Weaver (1924)
Quercus robur	3.7	Netherlands	Witkamp and van der Drift (1961)
Pinus sylvestris	2.5	Norway (Eidseberg)	Bonnevie-Svendsen and Gjems (1957)
Fagus sylvatica	3.5–5.0	Germany	Danckelmann (1887)
Tropical forest	25.0	Java	Hardon (1936)

and under natural conditions return most of them to the surface in their litter which decomposes and releases the nutrients, rendering them available for re-absorption by the plants. This cyclic process is fundamental and under natural conditions there is a delicate balance between the demands of the plant community, the quantity of nutrients in the cycle and the amount released by decomposition of the soil minerals. In tropical areas there is a number of high forest communities sustained solely by this process, the soils being completely devoid of minerals that yield plant nutrients. This paradoxical situation of virgin high forest on an inherently infertile soil leads to disastrous consequences when the forest is used as an indicator of high fertility and agriculture is attempted. An interesting example of the differing capacity of two plant communities to recycle nutrients is reported from Wisconsin (Jackson and Sherman, 1953). Under a hardwood community the surface soil is maintained at pH 6.8 because the roots of the trees are able to penetrate to the relatively unaltered underlying calcareous material and extract calcium which is returned to the surface in the litter to maintain the high pH value. In an adjacent situation there is a tall grass community where the surface soil is at pH 5.5 because the calcareous material is well below the reach of the grass roots. This is also an excellent example of the variability in the composition of the litter supplied by different species to the soil surface.

Vertebrates

Mammals

Most mammals roam freely over the surface of the ground but there is a relatively small number that inhabit the soil and contribute directly towards its formation, they either live on other soil organisms or forage on the surface eating plant material. Included in this group are rabbits, moles, susliks and prairie-dogs which burrow deeply into the soil and cause

considerable mixing, often bringing subsoil to the surface and leaving a burrow down which topsoil can fall and accumulate within the subsoil. Perhaps the most conspicuous and classical examples of this phenomenon are the crotovinas found in many Chernozems and Kastanozems especially those of Europe (see Plate Ic) where blind mole rats are principally responsible.

Moles are of common occurrence in soils beneath deciduous forests or in cultivated fields, normally they do not burrow below the upper horizons so that there is not the same dramatic manifestation of their activity as with the blind mole rats. Their presence in grass fields is clearly demonstrated, however, by the small mounds of soil or 'mole hills' that they form (Fig. 2.28). Moles burrow in search of food particularly earthworms; consequently a high frequency of mole hills usually indicates a high earthworm population.

Fig. 2.28 Mole hills in a field of cultivated grass in Britain. Each mole hill is about 30 cm in diameter and 15 cm high.

Other burrowing organisms are the African mole rats, chipmunks, gophers, ground squirrels, marmots, mice, mountain beavers, snow shoe rabbits, shrews, voles and woodchucks. Generally, this group plays a minor role in soil formation but locally some of them can be important. The behaviour of beavers is unique in that they build dams which can cause water to accumulate, resulting in the development of very marshy conditions and waterlogging of the soil particularly when the dams are abandoned or poorly maintained.

In certain areas, where there is uncontrolled grazing and mammals such as goats are allowed to multiply freely they devour most of the vegetation leaving the soil surface bare for erosion to take place, both by wind and water. This is conspicuous in many countries bordering the Mediterranean Sea where much topsoil has been lost through continuous overgrazing and progressive erosion. Even domesticated animals may be deleterious to the soil which can become compacted at the surface due to overgrazing by large animals such as cows and sheep, thereby reducing the permeability and increasing run-off which can be an erosion hazard.

Other vertebrates

These include various birds, snakes, lizards and tortoises that make their nest in the soil; however, their importance is minimal.

Microorganisms

This group of organisms includes a large number of bacteria, fungi, actinomycetes, algae and protozoa, all of which enter into umpteen poorly understood processes in the soil. An interesting feature is the wide distribution of many species which occur in tropical as well as in polar areas.

The predominant microorganisms are the bacteria and fungi. Bacteria are the smallest and most numerous of the free-living microorganisms in the soil where they number several million per gram with a live weight variation of between 1 000 and 6 000 kg ha^{-1} in the top 15 cm. This weight is slightly less than that of the fungi but greater than that of all the other microorganisms combined (Richards, 1974).

Unique among the microorganisms are the algae which are unicellular organisms containing chlorophyll and therefore are capable of photosynthesis. However, their requirement of light for photosynthesis makes it imperative that they occur predominantly at or near to the surface of the soil. This is probably not an ideal habitat rendering them vulnerable to rapid fluctuations in moisture and temperature.

Protozoa are also found in large numbers, mainly in the organic horizons. Their preferred habitat is in moist and wet situations but they can survive dry periods by encysting. The rhizopods and flagellates are the principal forms but also present are a few ciliates and altogether may number 10^3 to 10^5 g^{-1}, with a biomass of 5 to 20 g m^{-2}. Other protozoa live in the alimentary system of termites and play a vital role in the digestion of cellulose.

The distribution of microorganisms in the soil is determined largely by the presence of food supply. Generally they need a substrate with adequate and balanced nutrition and occur in the greatest numbers in the surface organic matter, on faeces or in the rhizosphere. Most members have an optimum temperature range of 25 to 30 °C, but since the soil temperature is seldom as high as this they operate below their optimum. The thermophilic bacteria which are important in the desert require much higher temperatures. In addition, most microorganisms require an aerobic environment but some important bacteria require anaerobic conditions. However, a small number of others are active under both aerobic and anaerobic conditions. Bacteria and actinomycetes thrive in soils at pH 7 or slightly higher but are not as frequent as fungi in acid soilds, while in alkaline soils some actinomycetes find favourable conditions for their growth. As a result of their restricted mobility microorganisms can only travel short distances or are immobile as in the case of the fungi and actinomycetes which can penetrate and exploit new areas only through the growth of their mycelia. In the absence of food all microorganisms enter a resting stage until the arrival of a fresh supply. Thus the general picture of microorganisms in soils is one of the organisms consisting mainly of resting stages in a mosaic of microhabitats, bursting into activity when some event brings a fresh food supply to their immediate environment (Warcup, 1967).

Originally, microorganisms were divided into heterotrophic and autotrophic organisms on the basis of their nutritional requirements. Heterotrophic organisms were considered to get their energy and nutritional requirements entirely from other organic materials while autotrophic organisms derived their carbon from carbon dioxide of the atmosphere and their energy from the sun or oxidative processes and were thus independent of any organic materials. However, it is now known that many heterotrophic organisms get their energy from the sun while some autotrophic organisms require specific organic growth factors such as vitamins. Thus organisms are now classified on the basis of their sources of carbon and energy into the following four categories:

Photoautotrophs
Photoheterotrophs
Chemoautotrophs
Chemoheterotrophs.

Photoautotrophs

These organisms contain chlorophyll and utilise light as their energy source and CO_2 as their principal source of carbon, they include blue-green algae, certain photosynthetic bacteria (purple and green sulphur bacteria); also included are the higher plants which are the main suppliers of organic matter to the soil. Some algae may also be important contributors particularly in arid and semiarid areas. In addition, some algae are capable of fixing atmospheric nitrogen and therefore supplement the nitrogen content of the soil. This is of considerable importance in rice soils where the amount of N fixed by algae is often sufficient to dispense with the application of nitro-genous fertilisers and has led to algal inoculations in some places in India and Japan.

In sharp contrast are the photoautotrophic bacteria which are unable to fix atmospheric nitrogen nor can they easily utilise amino acids, therefore they have to obtain their nitrogen supply from a mineral source such as ammonium or nitrate ions.

Photoheterotrophs

These organisms use light as a source of energy and derive much of their carbon from organic compounds. They include a specialised group of photosynthetic bacteria known as non-sulphur purple bacteria.

Chemoautotrophs

These organisms derive their energy from the oxidation of inorganic compounds and use CO_2 as the principal source of carbon. They include several groups of specialised bacteria including nitrifying bacteria and the thiobacilli.

Chemoheterotrophs

These organisms utilise organic compounds both as a source of energy and carbon. They include protozoa, fungi, actinomycetes and most bacteria. This is the largest group of microorganisms; they are of immense importance through their participation in ammonifica-tion, humification and other processes in the soil system and are responsible for the transformation of vast quantities of dead organic matter, both plant and animal.

Within the soil all of the microorganisms do not have a separate and discrete existence but interact with each other in the following systems:

Competition and antagonism
Predation and parasitism
Commensalism and mutualism

Competition and antagonism

The chief form of competition is for food and energy, taking place both between individuals of the same species and different species. With regard to antagonism it appears that some organisms excrete substances known as antibiotics which inhibit the activity of other organisms. However, the full significance of these two interactions is not fully known.

Predation and parasitism

A predator uses its prey as a source of food whereas a parasite uses its host both as a habitat and as a source of food.

The main predators among the soil microorganisms are the protozoa whose staple diet is bacteria. Since protozoa do not appear to enter into any reaction with the soil their main function seems to be the control of the bacterial population. Also included among the predators are fungi that have a loop mechanism for trapping nematodes that are then digested.

The presence of parasitism is not as clear as preda-tion but some fungi are parasites on other fungi while some bacteria parasitise both fungi and other bacteria. Microorganisms are also parasitic on higher animals and plants and can be of considerable economic importance when they attack crop plants. Important members of this group include *Poria hypobrunnea* of tea, *Fomes lignosis* of rubber and *Gaeumannomyces graminis* var. *tritici* (Take all) of wheat and barley.

Commensalism and mutualism

Commensalism

In a commensalistic relationship an organism benefits as a result of the activity of another organism. A good example is the decomposition of cellulose to sugar by one fungus that is consumed by a second organism that would otherwise die. Another example is the requirement by some organisms of B-vitamins that are produced by other organisms. A third example is that *Nitrosomonas* produces nitrate which acts as a substrate for *Nitrobacter*. A further example is the presence of obligate anaerobic organisms such as *Clostridia* in apparently aerobic soils. This is due to the reduction of the oxygen content by aerobes to the point that anaerobes can live.

Mutualism

This is the situation in which two organisms interact and confer a benefit to both. Perhaps the best example is a lichen which is the association of a fungus and an alga. The alga provides carbon compounds as energy sources and vitamins for the fungus which in turn provides mineral nutrients, water and protection for the alga.

Mycorrhizae and root nodules are also good examples of mutualism. A mycorrhiza is a composite fungus–root organ which occurs on many arborescent angiosperms and conifers especially the **Pinaceae**. Mycorrhizae are of two types, ectotrophic and endotrophic. The former develop a complete sheath of fungal tissue which encloses the terminal rootlets of the root system while the latter, which are more widespread, ramify through the cortex of the root. Their function is similar to that of root-hairs, aiding in the uptake of nutrients and water since mycorrhizae are more competitive than plant roots particularly when substances are relatively unavailable.

Another type of mutualistic relationship exists between bacteria and the root systems of many plants particularly with members of the **Leguminoseae**. In this case bacteria (*Rhizobium* sp.) form little colonies or nodules beneath the epidermis of the root where they fix atmospheric nitrogen which then passes into the conducting system of the plant and serves as a nutrient.

Mesofauna

This group of organisms comprises the large number of varied species that are in the main discernible with the naked eye and ingest organic matter. It includes earthworms, enchytraeid worms, nematodes, mites, springtails, millipedes, centipedes, a few crustacea, rotifers, some gastropods and many insects, particularly termites. Like microorganisms, their distribution is determined almost entirely by their food supply and, therefore, they are concentrated in the top 2 to 5 cm of the soil; only a few, such as earthworms and termites, penetrate below 10 to 20 cm. Generally, the mesofauna require a well aerated soil since they require atmospheric oxygen and thus cannot live in waterlogged or wet puddled soils, they also have a narrow temperature range with optimum conditions about 25 to 30 °C. While most types prefer conditions around neutrality some are tolerant of acid conditions, particularly mites and springtails.

The density of the mesofauna under favourable conditions is given in Table 2.10 and they can be

Table 2.10   Density of mesofauna under favourable conditions

Organism	Density
Earthworms	80 g m^{-2}
Enchytraeid worms	$10^5/\text{m}^{-2}$
Millipedes	$10-25/\text{m}^{-2}$
Mites	$5 \times 10^4/\text{m}^{-2}$
Nematodes	$10^6/\text{m}^{-2}$
Protozoa	$10^5/\text{m}^{-2}$
Slugs and snails	$10-25/\text{m}^{-2}$
Springtails	$10^4/\text{m}^{-2}$
Termites	$5-1\,000/\text{m}^{-2}$
Beetles	$100/\text{m}^{-2}$

classified, according to their activity, into the following six groups:

1. Organisms ingesting organic and mineral material: earthworms, millipedes, enchytraeid worms, dipterous and coleopterous larvae.
2. Organisms ingesting organic material: earthworms, millipedes, enchytraeid worms, mites, springtails, snails, slugs, ants, termites, centipedes.
3. Organisms transporting material: earthworms, millipedes, enchytraeid worms, ants, termites.
4. Organisms improving structure and aeration: earthworms, millipedes, ants, termites.
5. Predators: nematodes, centipedes, mites, snails, slugs.
6. Parasites: nematodes, a few arthropods.

Organisms ingesting organic and mineral material

Earthworms are probably the best example of this class of organism since they consume a greater volume of material than the others, however, millipedes, enchytraeid worms and larvae are important locally.

The passage together of mineral and organic material through the alimentary system of organisms is a unique process, during which some of the organic material is decomposed to provide the energy and body tissue for the organism, and simultaneously the remainder is formed into a homogeneous blend with the mineral material. The resulting mixture is rich in ammonia from the urine of the earthworm and provides a suitable habitat for microorganisms causing them to proliferate rapidly. An interesting feature about the passage of material through the earthworm is that the microorganisms are in no way affected but continue their activity completely uninterrupted, indeed it would appear that the alimentary system of earthworms does not have a specific microflora, it is the same as that of the surrounding soil. A further

43

unique behaviour of the earthworm is the secretion of calcium carbonate by the calciferous gland which gives the cast a slightly higher pH value than the surrounding soil. Some organisms have little effect on the organic matter they ingest except for causing fragmentation, the mites and enchytraeid worms being good examples. However, earthworms have both cellulase and chitinase in their alimentary systems.

Organisms ingesting organic material

This is the largest group of the mesofauna, of which the surface feeding earthworms and millipedes are the most conspicuous. However, the most numerous and widespread are the arthropods including mites whose chief activity is the comminution of the organic matter. A few arthropods are parasitic on plants or are vectors of plant and animal diseases and, therefore, are of considerable economic importance. The mites are most common in the accumulation of surface organic matter in acid soils of cool humid areas such as Podzols, from which earthworms are normally absent. They consume mainly the easily digestible plant remains. In the case of dead roots and stems they avoid the outer lignified tissues but burrow into the softer central part where their presence and progress is usually shown by the occurrence of small ovoid faecal pellets, easily seen in thin sections under the microscope (Fig. 6.36). Mites also consume large amounts of fungal mycelium, fragments of which are found in their faecal pellets. Among the insects that ingest organic matter are the ants and termites. The latter are found mainly in hot climates where they consume and digest vast quantities of dead organic matter with such efficiency that they keep the soil surface almost bare of litter and decomposing organic material.

Organisms transporting material

Within many soils large amounts of material are transported from one place to the other by earthworms, enchytraeid worms and millipedes. Certain species of earthworms also bring material to the surface to form their characteristic casts, the shape of which varies with species. Some produce a roughly conical cast whereas others produce tower shaped forms. It has been estimated that under optimal conditions in Australia where earthworms have a biomass of 80 g m^{-2} it takes about 60 years for a volume of topsoil equal to that of the top 15 cm of soil to pass through the alimentary system of the earthworm population, but this figure can only be a very rough approximation (Barley, 1959).

Ants and termites also transport considerable amounts of material from one place to the other, particularly the latter which bring large amounts of material to the surface to build their termitaria (Fig. 2.29). Termites are selective, moving only material less than 1 mm in diameter, therefore they can alter the particle size distribution of soil in which they operate. It has been suggested that up to 2 m of fine textured topsoil in East Africa and elsewhere have been formed by their activity (Webster, 1965).

Fig. 2.29 A termitarium in western Nigeria

Organisms improving structure and aeration

When organisms burrow into the soil they produce a network of passages which facilitate the movement of water and so improve aeration. In addition, organisms that ingest both mineral and organic material produce faecal pellets which are invariably more stable than other soil peds and therefore represent an improvement in structure.

These processes can be very important in soils that tend to lose their structure easily and become compacted. In certain soils most of the aggregates in the upper horizons are fragments or entire faecal pellets of earthworms, hence their better aeration and high fertility status.

Predators

Among the mesofauna in the soil, the main predators are the nematodes, which are characterised by their very restricted feeding habits. Their food seems invariably to be protoplasm, obtained by devouring entire organisms such as bacteria or by piercing the cell walls of fungi, plant cells, rotifers or other nematodes with their stylets since the dead organic matter in the soil does not seem to form part of their diet. There are over 10 000 species of nematodes of which about 2 000 inhabit the soil. The remainder occur in sea or fresh water or are parasites in plants, animals or man.

A second group of predaceous organisms are centipedes which feed on a variety of smaller arthropods, worms, snails and slugs. Other predators include some mites, slugs and insect larvae. Echytraeid worms seem to feed mainly on bacteria and fungi but they also ingest mineral material and dead organic matter but they do not possess enzymes that decompose complex plant polysaccharides.

Parasites

In addition to being the main group of predators, nematodes are also the main group of soil inhabiting parasites among the mesofauna but in this role they play a negligible part in soil formation.

Man

The activities of man are too many and too diverse to be considered adequately, without lengthy treatment. His influence has been so widespread that in many countries it is difficult to find soils that have escaped his influence of one sort or another. Generally, his main role is the cultivation of soils for the production of food or tree crops but in carrying out these operations his practices are often very poor causing impoverishment of the soil and erosion. Brief mention is made in Chapter 6 of the different ways the various soils can be utilised for agriculture and forestry, some indication being given of the main ameliorative methods and precautions that have to be adopted.

TOPOGRAPHY

Topography refers to the outline of the earth's surface and is synonymous with relief. It includes the dramatic mountain ranges and flat featureless plains both of which give the impression of considerable stability and permanence and that seem to be timeless. However, this is not the case, for it is known from many investigations that all land surfaces, even those in areas of very hard rocks such as granites, are constantly changing through weathering and erosion. Further, it is well established that ultimately all mountainous areas will be worn down to flat or undulating surfaces but this process is very slow, taking many millions of years in the case of mountains such as the Himalayas and the Andes. However, there are a few topographic features such as sand dunes and volcanoes that can change or develop quite rapidly. Thus topography is not static, but forms a dynamic system, the study of which is known as *geomorphology*.

The nature of the topography can influence soils in many ways, for example, the thickness of the pedo-unit is often determined by the nature of the relief. On flat or gently sloping sites there is always the tendency for material to remain in place and for the pedo-unit to be thick but as the angle of slope increases so does the erosion hazard, resulting in thin soils on strongly sloping ground. Where there is a thick cover of vegetation, soil will accumulate on fairly steep slopes.

It follows that soils on steep mountain slopes are shallow and often stony, containing many primary minerals, but those of the less steeply sloping old land surfaces contain a high percentage of clay and resistant minerals. In addition, it has been pointed out already that relief determines aspect and strongly influences the moisture regime of many soils. Also, in areas where the difference in elevation between the highest and the lowest point is great, then climatic changes are introduced. These differences in elevation, aspect, slope, moisture and soil characteristics lead to the formation of a number of interesting soil sequences. In fact, topography is one of the chief factors which determine the spatial distribution or pattern of soils in the landscape, and is discussed fully on pages 297–302.

Topographic features fall into three main categories, those produced by tectonic processes, those formed by erosion and those formed by deposition. Set out in Table 2.11 are the principal topographic features and their processes of formation: definitions of the common phenomena are given in the glossary.

Tectonic processes

Initially, all major relief features are produced by tectonic processes whether they are uplift, subsidence, differential lateral movement, or vulcanism; subsequently, the surfaces are acted upon by water, ice, wind, frost and mass movement which are the principal agencies of erosion and deposition.

Table 2.11 Principal topographic features and their methods of formation

Processes		Topographic features
Tectonic	Uplift	Faults, folded mountains, plateaux, raised beaches, table lands
	Subsidence	Rift valleys (graben)
	Vulcanism	Ash and agglomerates, calderas, lava flows, volcanoes
Running water	Erosion	Bad lands, cuestas, dolines, gorges, inselberge, karst, mesas and butes, pediments, peneplains, tors, valleys
	Deposition	Alluvial fans, alluvial plains, alluvial terraces, bahadas, deltas, lévées
Moving ice	Erosion	Aretes, cirques, crag-and-tail phenomena, hanging valleys, roches moutonnées, U-shaped valleys
	Deposition	Drumlins, eskers, moraines, till (boulder clay)
Frost	Erosion	Gullies, jagged rock surfaces, tors
	Deposition	Talus cones, solifluction terraces
Wind	Erosion	Desert pavements or hamadas, dreikanters (ventifacts)
	Deposition	Dunes, loesses
Mass movement	Erosion and Deposition	Landslides, soil-creep, solifluction terraces

Running water

This is the main agency of erosion and deposition but the exact processes and stages through which the landscape must pass are not well established. This is a subject surrounded by much controversy, however, there are two principal schools of thought; those of Davis (1954) and Penck (1953). The former regards the land surface as going through a series of stages during which the surface gradually becomes subdued and eventually forms a peneplain. In contrast to this Penck suggests that after the initial incision and down cutting by running water, there is parallel retreat of the slopes to form pediments and pediplains. These two ideas are shown diagrammatically in Fig. 2.30. Although there is evidence for both theories, the study of very old land surfaces in Africa and elsewhere gives stronger support to the ideas of Penck. These contrasting theories apply more specifically to

humid environments. Since the earth's surface has a variety of climates it is reasonable to assume that each set of climatic conditions will produce its own set of topographic features and this has led to the concept of morphogenetic regions (Peltier, 1950). Each morphogenetic region has its own range of temperature and precipitation which together produce specific processes forming characteristic topographic features (Fig. 2.31). Surprisingly, when most land surfaces are examined in some detail they display well developed pediments or remnants of pediments which appear to have formed under humid conditions. This is probably the result of the prolonged period of erosion and deposition that existed during the Tertiary Period when warm and humid conditions prevailed over most of the earth's surface. These conditions have left an indelible imprint and produced landscapes similar to those suggested by Penck but now, superimposed upon these Tertiary landscapes are the influences of many of the other morphogenetic processes. However, they have changed only slightly the main outline of the landscapes, but are responsible for a large number of diverse and small topographic features such as terraces, moraines, dunes and solifluction deposits.

Based on the above most workers now use the four-element model of landscapes shown in Fig. 2.32. This has recently been challenged by Dalrymple *et al.* (1968), who have suggested a nine-slope element model (Fig. 2.33).

In some cases the geological structure may control and determine the form of the landscape. When easily weathered and eroded rocks lie next to hard rocks, these latter may form promontaries in the landscape.

Disregarding the theories of landscape formation, the material removed by erosion is normally transported in rivers and later deposited to form alluvial deposits or taken to the sea to build deltas. In arid areas where rivers are few or absent the material produced by weathering either remains *in situ* or is carried only a short distance, usually to the bottom of the slopes where it accumulates to form alluvial cones or bahadas.

In areas of limestone where solution is the principal process of weathering a unique type of landscape known as karst is developed. This is characterised by stony surfaces and the occurrence of dolines, which are depressions formed by the gradual slumping of material down solution channels (Fig. 2.34).

Moving ice

Erosion by ice, or glacial erosion, is a relatively

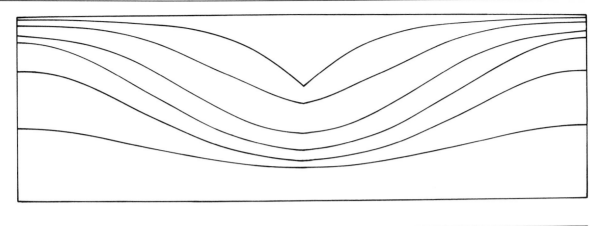

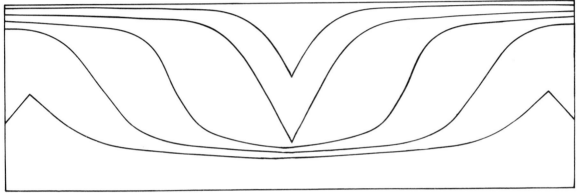

Fig. 2.30 Contrasting sequences of valley profiles from youth to old age (*above*) according to Davis; (*below*) according to Penck.

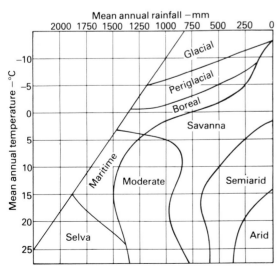

Fig. 2.31 Diagram of climatic boundaries of morphogenetic regions (adapted from Peltier 1950)

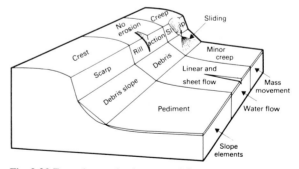

Fig. 2.32 Four element landscape model

local process at present but was more widespread in the past when it was superimposed upon land forms that developed mainly under the influence of water during the Tertiary Period. Glaciers can be powerful eroding agents particularly in very mountainous areas where their most spectacular effects are the formation   **47**

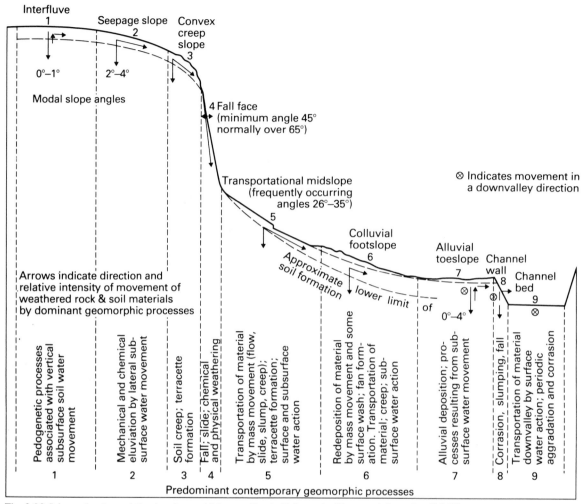

Fig. 2.33 Schematic nine unit landsurface model according to Dalrymple, Blong and Conacher (1968)

of U-shaped valleys and the removal of pre-existing soils and weathered rocks. The material carried in glaciers (Fig. 2.35) is later deposited on land or in the seas. When on the land the configuration of the surface is usually altered by characteristic depositional features, such as, drumlins, eskers, moraines and till (Fig. 2.36).

Wind

Today, vigorous wind action is confined largely to arid and semiarid areas, and to coastal positions, where the formation of sand dunes is a most characteristic and conspicuous process. However, during the Pleistocene Period, wind activity was widespread in areas peripheral to glaciers. From these areas winds

picked up much fine material from the deposits in front of melting glaciers and transported it long distances to form loess deposits such as those found in Europe and North America (Fig. 2.37). This process is still active in certain glaciated areas. In contrast, some very thick loess deposits such as those of northern China have been derived from material blown off the Gobi Desert. Loess usually forms a thick featureless blanket which entirely obliterates the pre-existing topographic features.

Frost

Repeated freezing and thawing of wet unconsolidated material on slopes causes a considerable amount of movement. This process is known as solifluction and

Fig. 2.34 Sink holes in a karst area in Yugoslavia

is probably the principal geomorphological process taking place in polar areas leading to the accumulation of great thicknesses of material on the lower part of the slopes. Often a number of topographic features such as terraces, result from this process (Fig. 7.13). Since the action of frost is intimately associated with soil formation a fuller account is given in the next chapter (see page 56 *et seq.*).

Mass movement

Mass movement, without the aid of frost is often ignored as a process of landscape formation but Sharpe (1938) has attempted to demonstrate that although the amount of soil that creeps insidiously down the slope and the frequency of landslides per annum may be small, the total amount measured over many thousands of years may bring about considerable alteration to topography. In tropical and subtropical areas the high frequency of stone lines and the arching of material down the slope is a clear indication that the top metre or so of the soil is gradually being

moved and redistributed within the landscape.

The result of the various denudation and depositional processes discussed above is to produce a wide range of geomorphological features of various stages of development. Sometimes development may be arrested or accelerated by a change of climate or by renewed tectonic activity. For example, many of the land forms in the Australian desert were produced by running water under humid conditions but development has virtually stopped since the climate became dry. Similarly, many old land surfaces with well developed pediments are found at high elevations in East Africa because of renewed tectonic activity and uplift during the Tertiary Period. A further result of geomorphological processes is that large stretches of land are now covered with glacial drift, loess or alluvium which are new parent materials, often quite unrelated to the underlying material. So that, at present the major elements of the landscape in most places reflect Tertiary erosion and sculpturing, but superimposed upon them are the various influences of the Pleistocene and Holocene Periods.

49

Fig. 2.35 A melting glacier in Svalbard containing numerous layers of rock detritus some of which may be deposited eventually to form parent material

Fig. 2.37 A loess deposit in central Poland. Note the characteristic prismatic structure in the upper part of the section and the dark coloured buried soil near the base

TIME

Soil formation is a very slow process requiring thousands and even millions of years, and since this is much greater than the lifespan of any individual human being, it is impossible to make categorical statements about the various stages in the development of soils. A further complication is introduced by the periodic changes of climate and vegetation which often deflect the paths of soil formation in one direction or another. Therefore, all that is said about time as a factor in soil development is in part speculation and in part deduction using indirect evidence based on botanical, zoological, geological or geomorphological data. The botanical data are usually in the form of pollen analyses from peat and sometimes from the soil itself. The zoological and geological data are from buried fossils that are known to

Fig. 2.36 Stratified glacial deposits in eastern Scotland. The two lower strata are unsorted compact till with angular stone and boulders. The upper stratum is a glaciofluvial deposit hence the higher concentration of stones and boulders many of which are rounded to subrounded

have existed at a given period. In a number of cases some type of absolute dating such as ^{14}C has been used to establish the age of a soil.

Not all soils have been developing for the same length of time. Most started their development at various points during the last 100 million years. In Table 2.12 are set out the principal events of the last 75 million years as related to soil formation. The divisions in the table are not fixed date boundaries for soils, indeed many soils that started their development in one period have continued to develop in a succeeding period or periods.

Some horizons differentiate before others, especially those at the surface which may take only a few decades to form in unconsolidated deposits. Middle horizons differentiate more slowly, particularly when a considerable amount of translocation of material or weathering is necessary; some taking 4 000 to 5 000 years to develop. Some other horizons require even longer periods; weathering of rock to form a Ferralsol may require more than a million years. The evidence for this last statement lies in the fact that Ferralsols are found only on very old land surfaces which have been exposed to weathering since at least the Tertiary Period.

Further evidence regarding the speed of horizon differentiation when complete weathering takes place, is gained by a study of the interglacial soils in the central USA where there are some strongly weathered horizons known as 'gumbotil'. These occur either at the surface or are sandwiched between two glacial deposits and may have formed during a single interglacial period extending over 50 000 to 70 000 years (see p. 304). The rate of development of these horizons can be regarded as maximal since they have formed in a zone of permanent saturation where hydrolysis is most vigorous. A particularly important feature of soils is that they pass through a number of stages as they develop, culminating in deep pedounits with many well differentiated horizons. The principles underlying the development of soils in relation to time are considered in the section on soil evolution which follows.

Soil evolution

Initially, pedologists tended to interpret most soil features as the result of the interaction of the prevailing environmental conditions at the time when examinations were made. However, it soon became evident that most places have experienced a succession of different climates which induced changes in the vegetation and in soil genesis. Therefore, most soils

are not developed by a single set of processes but undergo successive waves of pedogenesis. Furthermore, each wave imparts certain features that are inherited by the succeeding phase or phases. In a number of cases these properties are developed so strongly that they are evident thousands and even millions of years after their formation. The outcome is that soils are regarded as having developmental sequences which manifest not only the present factors and processes of soil formation, but also a varying number of preceding phases.

Although changes in the external environmental factors are usually responsible for progressive changes in the soil, vegetation and landscape, there are a number of cases when the environmental factors remain relatively constant but continuing soil development leads to the formation of new features within the soil. Some of these can become very prominent and may themselves influence the further course of soil formation. The progressive development of an impermeable middle horizon can cause waterlogging at the surface and changes in the soil characteristics with corresponding changes in the vegetation. These various progressive changes in soils are known as *soil evolution.* Changes produced by variation in external factors are *exogenous evolution* and those due to internal changes are *endogenous evolution.* Examples of these changes have been reported from every major land surface and, set out in Tables 2.12 and 2.13 are some of the major events that have produced evolutionary changes in soils. This is followed by an outline of their effects, while in Chapter 7 details are given of a number of the more interesting and widespread types of soil evolution.

King (1967) suggests that the oldest land surfaces, such as those in parts of Africa and Australia, developed in the mid-Tertiary Period, but one or two may date from the early part of the Tertiary or even the Cretaceous. If this is the case, then the soils which occur on these surfaces may be very old. Consequently, any consideration of soil development and change through time must commence by discussing the events that took place during the early part of the Tertiary Period but the age of the surface may not be the same as the soil which may be younger. The evidence afforded by plant remains shows clearly that the surface of the earth at that time did not have the extremes of climate that it experiences at present. The flora from the late Cretaceous to the Miocene Period in the northern latitudes grew under warmer conditions as indicated by the occurrence of palms in Canada and Britain, where conditions could be regarded as subtropical. However, full tropical condi- **51**

Table 2.12 Generalised geochronological data for North Central Europe and Central North America, Tertiary to Holocene

Period	Time (Years B.P.)	Northern Europe	Alpine region	Central North America
Holocene	Present	*Holocene*	*Holocene*	*Holocene*
	— 10 000 —			
Pleistocene (Glacial and inter-glacial stages; Glacial periods *italicised*)	125 000	*Weichsel*	*Würm*	*Wisconsinian*
	235 000	Eemian	Riss/Würm	Sangamonian
	360 000	*Saale*	*Riss*	*Illinoian*
	670 000	Holstein	Mindel/Riss	Yarmouthian
	780 000	*Elster*	*Mindel*	*Kansan*
	900 000		Gunz/Mindel	Aftonian
	1 150 000	Various deposits	*Gunz*	*Nebraskan*
	1 370 000	of 'cold' and	Donau/Gunz	
	1 600 000	'warm' phases	*Donau*	
	—2 500 000		Pre Donau	
Tertiary				
		←——————— *Pliocene* ———————→		
		(Development of man)		
	12 000 000	←——————— *Miocene* ———————→		
	25 000 000	←——————— *Oligocene* ———————→		
	32 000 000	←——————— *Eocene* ———————→		
	60 000 000	←——————— *Palaeocene* ———————→		
	70 000 000			

Table 2.13 Generalised geochronological data of North Central Europe, Late Weichsel to The Present

Period	Time (year B.P.)	Culture	Climatic phases	Vegetation	Soil
Holocene (Post Glacial)	Present	*Historical*	*Subatlantic* Cool–Dry	Beech, Spruce and Pine in the north	Cambisols Luvisols Histosols Podzols Gleysols
	2 500	*Iron age* *Bronze age* (Start of Agriculture) *Neolithic*	*Sub-boreal* Warm–Dry	Oak, Ash, Spruce and Pine in the north	Progressive leaching
	5 000		*Atlantic* Warm–Moist	Oak, Elm	Spread of Histosols
	7 500	*Mesolithic* (Hunters and Collectors)	*Boreal* Warm–Moist	Oak, Hazel	Cambisols Podzols
	9 000		Moister	Pine, Hazel	Gleysols
			Preboreal Cool	Birch, Pine	Cambisols Podzols Gleysols
Late Weichsel (Late Glacial)	10 000		*Upper dryas* Subarctic	Grasses, Sedges	Gelic Gleysols Solifluction
	11 000	*Palaeolithic*	*Allerød* Warm Subarctic	Birch	Rankers
	15 000		*Lower dryas* Subarctic	Grasses, Sedges	Gelic Gleysols, Solifluction

tions existed in many of the same areas where they are found today, as well as in many of the present subtropical and arid areas, including such places as central Australia and the southern part of the Sahara Desert. These warmer conditions caused the rocks to undergo profound weathering by hydrolysis in the cases of the silicates and by solution in the case of limestone and other soluble rocks. Considerable natural erosion also took place during this prolonged period to form the characteristic flat or gently sloping surfaces of planation associated with old landscapes. But weathering and soil formation kept pace with and even proceeded faster than erosion so that great thicknesses of soil and weathered rock were maintained. The processes of weathering were so complete in many places that rocks of all types were transformed into clay minerals, hydroxides and oxides of iron and aluminium together with their resistant residues such as quartz and zircon, but sometimes even these resistant primary minerals have been decomposed. Most of the bauxite deposits of the world are Tertiary phenomena resulting from profound chemical weathering.

The warm, humid conditions were maintained in tropical and subtropical areas throughout most of the Tertiary Period, but in the poleward areas the climate became progressively cooler from the end of the Miocene so that by the end of the Pliocene the floras of Eurasia and North America bore many similarities to those of the present, but they contained more species particularly in Western Europe, where the present flora is a poor reflection of the past. The gradual cooling of the climate culminated in the Pleistocene Period with its repeated glaciations both in the Northern and Southern Hemispheres but it is in the former with its greater circumpolar land surface that the effects on pedology are most significant and widespread. There were five distinct glacial episodes, the fourth or Saale (Illinoian) glaciation, being the most extensive and covering large parts of Eurasia and North America. These glaciations together with their associated periglacial conditions effectively removed most of the deep soils and weathered rock that formed during the Tertiary Period. However, there is still sufficient present to leave no doubt about its previous ubiquity or about its contribution to the various superficial deposits of the Pleistocene. Much of the so-called boulder clay (glacial drift) contains fine material derived from a previously weathered soil, mixed with fresh rock fragments.

Perhaps the most impressive soil evolution has taken place in Australia and various parts of central Africa particularly those bordering the Sahara where soils that developed under hot humid conditions are now partially fossilised in an arid environment.

Most other tropical and subtropical areas experienced periods of higher rainfall or *pluvial conditions* during the Pleistocene. Many workers regard these periods as coincident with the glacial episodes but the evidence for this is fragmentary and the chronology is not as firmly established as that for the glaciated areas. It is certain that erosion by water did reduce the thickness of soil and weathered mantle in many tropical and subtropical areas but the effect was small as compared with widespread and deep stripping caused by glacial and periglacial processes.

The rapid climatic amelioration at the end of the Pleistocene possibly induced the most dramatic set of changes to be experienced by the earth's surface. This caused wave after wave of vegetation to pass rapidly over many areas particularly in Western Europe and North America. These were quickly followed by the influence of man, so that soil development in many places in the middle latitudes has been subject to climatic conditions varying from full glacial to marine in the relatively short period of 10 000 years. As a result, many of the soils in this area have diverse properties inherited from the Tertiary and the Pleistocene Periods.

In historical times many evolutionary paths have been directed by man's activity: the polders of Holland, the rice soils of eastern countries and the catastrophic erosion in many of the Mediterranean countries and North America quickly spring to mind. An interesting feature to emerge from recent investigations is that many plagioclimaxes in the vegetation may be due to the activity of man. Excellent examples are afforded by the ericaceous heaths of Europe and many of the savannas of the tropics.

3

Processes in the soil system

Soils are complex, dynamic systems, in which an almost countless number of processes are taking place. Generally these processes can be classified as chemical, physical or biological, but there are no sharp divisions between these three groups; for example, oxidation and reduction are usually regarded as chemical processes but they can be accomplished by microorganisms. Similarly, the translocation of mineral particles can take place either in suspension or in the bodies of organisms such as earthworms. In this chapter the processes are discussed in general terms whereas in Chapter 6 they are related to the formation of specific soils. Some processes have been discussed in the previous chapter but they are mentioned again.

PHYSICAL PROCESSES

The main physical processes are translocation, aggregation and fragmentation of material, but the agencies responsible are very varied, thus there are many specific ways in which these three processes can be achieved. For example, freezing and thawing can bring about specific aspects of all three processes and are responsible also for certain topographic forms. The physical process can be considered under the following headings:

Aggregation
Translocation
Freezing and thawing
Wetting and drying
Expansion and contraction
Exfoliation
Unloading

Aggregation

This is the process whereby a number of particles are held or brought together to form units of varying but characteristic shapes. In many cases the details of the mechanisms are poorly understood but there appear to be certain correlations between the type and degree of aggregates and other soil properties. Therefore, for most situations it is possible only to indicate the general relationship that exists between the aggregates and other features. For example, it is easy to understand how cementing substances will bind particles together, but it is difficult to explain how the characteristic and often regular shapes are formed. The type and degree of aggregation is usually referred to as structure which is discussed on page 97. The principal agencies responsible for the formation of aggregates are:

Clay and humus
Cementing substances
Mesofauna
Plant roots
Expansion and contraction
Freezing and thawing
Microorganisms
Exchangeable cations
Cultivation

Clay and humus
Clay and humus are capable of binding particles together and are the main factors responsible for much of the aggregation in the upper horizons of soils.

Cementing substances
Many products of hydrolysis, particularly compounds of iron and aluminium, can cement small groups of particles together. In some cases progressive cementation can form very hard and massive horizons. Also, the continuous deposition of calcium carbonate can lead to the formation of massive horizons.

Mesofauna
Earthworms, enchytraeid worms and other organisms that ingest organic and mineral material produce large amounts of faecal material. Where the activity of any one of these is particularly vigorous entire horizons may be composed of faecal pellets with their characteristic granular, ovoid or vermiform shapes (Fig. 4.30).

Plant roots
The fine roots of many plants, especially those of grasses bind together small groups of particles to form a crumb or granular structure (Fig. 4.21).

Expansion and contraction
In certain soils particularly those containing a high content of clay such as montmorillonite, expansion and contraction take place in response to the gain and loss of water, resulting in extensive cracking and the formation of large prisms or wedge-shaped peds with slickensided surfaces. When only a small amount of expansible clay is present the volume change with moisture change is small but is usually sufficient to form a prismatic or blocky structure.

Freezing and thawing
These two processes can produce massive, platy and subcuboidal structures. See page 100–101.

Microorganisms
It appears that microorganisms may aid the formation of structure by secreting various mucilaginous compounds or gums. These substances bind the particles together to form crumbs or granules which are more frequent and better formed in the presence of organic matter and an active microflora.

Exchangeable cations
In some cases the nature of the exchangeable cations can have an influence, for example the presence of large amounts of calcium causes flocculation and the formation of a crumb or granular structure in the upper horizons whereas sodium causes a dispersion of the system and the formation of a massive or columnar structure.

Cultivation
Structure is a somewhat ephemeral property which can be altered easily by cultivation; usually, and through mismanagement, the change is from a naturally good crumb or granular structure to poor blocky or massive. Probably a better example is the formation of a ploughpan through repeated cultivation to a constant depth with heavy implements. However, ploughing and other tillage operations are conducted to improve the structure and aeration of the soil. In fact, it is probably true to say that any great increase in productivity in the future will come through the improvement and management of structure rather than through improved fertiliser techniques the principles of which are now fairly well understood.

Translocation
Some or all of the material in the soil system moves or is moved by a variety of methods from one place to the other. In fact, many of the processes of soil formation are concerned primarily with the reorganisation and redistribution of material in the upper 2 m or so of the earth's crust.

The principal types of translocation are:
Translocation in solution
Translocation in suspension
Translocation due to organisms
Translocation *en masse*
Translocation through freezing and thawing
Translocation through expansion and contraction

Translocation in solution
In a humid environment much of the soil solution normally migrates in response to gravity by moving vertically down through the soil to underground waters and finally to the river system. As the solution moves some of the material such as compounds of iron and aluminium may be precipitated to form horizons such as the middle horizons of Podzols, while the most soluble materials which include bicarbonates, sulphates and nitrates are lost completely from the soil system. In soils on slopes there is some lateral movement of moisture through the soil and down the slope; this attains very high proportions when the soils contain impermeable horizons over which the moisture can move. As a result ions are transferred laterally and enrich the soils in the lower part of the slope. If the volume of water is large or if the impermeable layer comes close to the surface the entire upper part of the soil can become saturated and there may be free water on the surface. Such situations are known as *flushes* and usually carry a plant community indicative of a moist habitat.

In semiarid environments there is only a partial loss of material from the soil system. The easily soluble simple salts including nitrates and chlorides are completely removed whereas carbonates and sulphates are usually precipitated in the middle or lower positions in the soil. In an arid climate there is only a short period following rainfall during which some of the most soluble materials become dissolved and are capable of movement. Since the rainfall under these conditions is small, any downward flow is quickly reversed by intense evaporation which induces upward movement of the soil solution and a deposition of salts at or near to the surface. The soil solution in humid areas will also have a period of temporary upward movement if there is a marked dry season.

55

Translocation in suspension

Fine particles and colloidal materials are often transported from one place to the other within the soil system. Perhaps the most important manifestation of this process is the removal of particles <0.5 μm from the upper horizons of some soils followed by their deposition in the middle position to form clay coatings (see p. 107). Sometimes the fine material may be redistributed within an horizon and in some cases it would appear that particles up to 50 μm in diameter can be redistributed as in the formation of fragipans (isons) (p. 257). These phenomena are considered further under coatings on page 106 *et seq.*

Translocation by organisms

Many members of the mesofauna and some mammals are responsible for the movement of much material within the soil system; these processes are discussed on page 44 *et seq.*

Translocation en masse

Mass movement in response to gravity has already been mentioned as a mechanism for the formation of landscapes involving the movement of the weathered mantle down the slope. Since soil usually constitutes the upper part of the weathered mantle, mass movement can remove the soil and expose fresh material to pedogenetic processes.

In addition, mass movement can also take place within the body of the soil on flat sites and therefore is not caused solely by gravity. There are two processes by which this can take place:

Expansion and contraction (see pp. 55, 58)
Freezing and thawing (see p. 220)

Freezing and thawing

Before discussing the effect of freezing and thawing it is necessary to discuss the variation in the properties of water with a decrease in temperature. As the temperature of water is lowered it behaves in the normal manner by decreasing in volume until it reaches 4 °C when it attains its maximum density of 1.0 g cm^{-3}. Thereafter it expands slightly until it reaches 0 °C and changes into ice whereupon there is a 10 per cent increase in volume corresponding to a reduction in density to 0.9 g cm^{-3}. With a further reduction in temperature ice behaves like a normal solid and decreases in specific volume but it has a very high coefficient of linear expansion which is 51 x 10^{-6} K^{-1}. This contraction–expansion–contraction property of water leads to the formation of a number of unique phenomena as discussed on page 220.

Freezing and thawing take place to varying degrees over a wide area of the earth's surface and in widely differing materials including mineral horizons, organic horizons and rocks, both as boulders within soils as well as large exposed rock surfaces. However, the fluctuations in temperature about 0 °C are of little importance if water is not present. When water is present many processes take place including the following:

Frost heaving
Ice crystal growth and segregations
Frost shattering
Solifluction

Frost heaving

Often when the soil cools through a loss of heat to the atmosphere the surface freezes fairly rapidly forming a massive structure, but when freezing is slow, ice crystals with a characteristic needle form or piprake grow out from the surface carrying a small capping of fine material. Alternatively, they may grow beneath stones. This process is responsible for the heaving to the surface of stones which may form a continuous layer in polar areas. Frost heaving is also responsible for breaking down large clods to form a better structure, hence the reason for ploughing before the onset of winter.

Ice crystal growth and segregations

When freezing takes place slowly beneath the surface or in controlled laboratory conditions a number of ice patterns develop, as determined by the speed of freezing, particle size distribution of the material and moisture content. In very sandy material the ice forms fillings in the large pores giving a massive structure. In loams and silts a marked lenticular structure develops with lenses of soil surrounded by continuous ramifying veins of pure clear ice (Fig. 6.23). In clays ice segregates into both vertical and horizontal layers to isolate subcuboidal blocks of frozen soil (Fig. 4.29).

The thickness of the ice segregations is a function of the speed of freezing as well as the moisture content. When freezing is very slow and prolonged and there is a plentiful supply of moisture the ice segregations are very large since water normally migrates to the freezing front and then turns into ice. Lenses >10 m thick occur in parts of Siberia and northern Canada.

The ice segregations develop as a result of the formation of ice crystals which grow and elongate about their *c*-axis in a similar manner to quartz or calcite. Thus in thin section of permafrost cut normal to the surface it is possible to see the vertically elongate ice crystals.

Ice segregations also cause a reorganisation of the soil by withdrawing water and by compacting the lenticular or subcuboidal units through expansion upon freezing and crystal growth. These units are usually so well formed that they maintain their shape when the material thaws.

Also, in many polar areas the surface soils freeze during the winter and thaw during the summer, creating an annual cycle which, coupled with the development of ice segregations, causes profound disturbance of the soil surface and the formation of a number of characteristic patterns including mud polygons and stone polygons. These are more fully discussed on page 222 *et seq.*

Since stones have a lower specific heat than the surrounding soil they cool down and heat up more quickly, so that during the cooling process they attain 0 °C before the surrounding soil, and form loci for the formation of ice which can become quite thick around the stone displacing the surrounding material. Repeated freezing and thawing resulting in the formation and disappearance of ice can cause reorientation of the stones which on flat situations steadily develop a vertical orientation of their long axes and are gradually forced towards the surface.

Frost shattering

The occurrence in polar and alpine areas of extensive surfaces strewn with angular rock fragments is generally interpreted as being due to alternate freezing and thawing. It is assumed that as water freezes in the pores in the rock, pressures of over 500 kg cm^{-2} are created and the rock splits apart leaving a number of loose fragments when it thaws. Therefore, the amount of fragmentation will increase with the number of times the rock freezes and thaws. This explanation, however plausible, does not stand up under careful analysis since many areas in the middle latitudes have more frost—thaw cycles than those of polar areas but do not have the same extensive occurrence of angular rock debris. Everett (1961) has suggested that the active agent is the steady growth of ice crystals within the fine pores in rock during prolonged periods of low temperatures. This creates high pressure which severely weakens the rocks causing them to split apart, and it is claimed that, over a long period, rocks can be reduced to accumulations of silt size material. Frost shattering of rock fragments also occurs in soils and unconsolidated sediments (Fig. 3.1).

Solifluction

On sloping situations repeated freezing and thawing causes the movement *en masse* of the material down the slope forming stratified slope deposits that are known as solifluction deposits, grezes litées or eboulis ordonnés. The exact nature of the mechanism involved is not understood. However, it appears that when the ground freezes, expansion takes place normal to the surface and when the soil thaws and contracts, it does not return to its original position but responds to gravity, thereby moving slightly down the slope. Alternatively, the soil becomes very wet and even saturated upon thawing and being in a somewhat loosened state due to freezing, flows down the slope. This process can be accelerated in the presence of permafrost which is impermeable and causes water to accumulate within the surface layers. Thus the rate of down slope movement is variable but generally tends to be rapid on slopes >10 to 15°. Where movement of material down a slope is gradual the stones become oriented with their long axis parallel to the slope. In addition, the surface of such areas is usually characterised by a number of lobes or terraces produced by differential movement.

The characteristic arcs at depths of 1 to 3 m in the

Fig. 3.1 Frost shattered rock in Svalbard. In many cases adjacent fragments can be fitted together thus indicating that they were once part of a larger unit

material on slopes in tundra areas are attributed to movements resulting from freezing and thawing and therefore are regarded as a normal type of solifluction formation (Fig. 3.2). It is probable that these arcs could be due to plastic deformation and slow flow of the frozen mass down the slope. The depth of summer thawing in these areas seldom exceeds 1 m and since the arcs occur below this depth it is difficult to explain their presence as due to freezing and thawing. However, the temperature in this layer during the summer will only be a few degrees below freezing therefore it could flow by plastic deformation during this period of the year. The warmer and less viscous upper part will flow more rapidly than the cooler and more viscous lower portion, thus accounting for the formation of the arcs.

Another feature is that these arcs are destroyed when they are incorporated into the layer above which has an annual freeze–thaw cycle. Therefore, it seems necessary to differentiate between mass movement caused by freezing and thawing and that caused by plastic flow.

Wetting and drying

These processes have been mentioned already in connection with expansion and contraction but they can be considered in a much wider context. Few, if any, reactions in soils take place in the absence of water, therefore wetting initiates soil development. Similarly, most reactions stop when the soil becomes dry. Therefore, it is possible to regard soils that are subject to periods of wetting and drying as having phases of activity followed by resting phases. Although this happens to a greater or lesser extent in most soils, it is only in extreme cases where the results are dramatic. A particularly good example is the formation of gilgai (see below). Another striking example is found in arid regions where nitrogen fixation takes place only during a wet period, thereby increasing the amount of nitrogen available to plants which begin to grow in response to the increased moisture.

Expansion and contraction

This is a very important phenomenon in soils containing a high proportion of clay such as montmorillonite and occurs principally in the soils of hot environments, with alternating wet and dry seasons. During the dry season the soil dries, contracts and eventually develops cracks which may be over 10 cm wide on the surface and extend down for over a metre. Invariably some of

the upper horizon falls to the bottom of the crack where it stays until the wet season when the soil absorbs water and expands. Since there is now a greater amount of material in the middle or lower horizons these expand to a greater volume than before and in so doing create pressures which cause a displacement of material, which can only move upwards. In addition to the displacement of the material there is a considerable amount of minor sheering at an angle of about 45°. As one portion of soil glides over the other the sliding surfaces become polished.

With the annual repetition of this process a large number of areas of soil about one metre or more in diameter are raised above the general level of the surface and the process can proceed until the subsoil is brought to the surface and exposed. Such phenomena are known as *gilgai*. The precise pattern of the gilgai on the surface is determined by a number of factors, particularly slope. On flat sites gilgai are usually subcircular mounds whereas on gentle slopes they tend to be long and narrow running normal to the contour as shown in Figs. 3.3, 3.4, 3.5, 3.6 (Hallsworth *et al.,* 1955).

There are phenomena that develop on a micro scale and are visible only in thin sections. Some of these are mentioned on page 96. Freezing and thawing are also main agencies of expansion and contraction as discussed on page 56.

Exfoliation

Boulders or outcrops of rocks at the surface experience wide temperature fluctuations especially in polar and desert areas where repeated expansion and contraction cause flaking or scaling of the rock. This is of some importance on sloping situations where the material that is released slips down forming scree or talus cones, but generally it is of little importance as a process in the soil system.

Unloading

At present some importance is attached to unloading as a mechanism of physical weathering of rocks. It is suggested that as rocks are weathered and removed the resulting reduction in the thickness of the overburden allows the underlying material to expand and spall off in large sheets parallel to the surface. The main significance of this process is that it increases the surface area of the material and the depth of moisture penetration.

Fig. 3.2 Fossil solifluction material overlain by loess in central Poland

CHEMICAL PROCESSES

The main chemical processes taking place in soils are:
  Hydration
  Hydrolysis
  Solution
  Clay mineral formation
  Oxidation
  Reduction

Hydration

This is the process whereby a substance absorbs water to form a new compound which differs only slightly from the original state. Only a few of the primary minerals undergo direct hydration, therefore very little takes place during the early stages of weathering and soil formation. The principal exception is biotite which absorbs water between its layers, expands and finally splits apart (Fig. 3.7). Hydration is more often a secondary process following the hydrolysis of primary minerals and the formation of decomposition products which are then hydrated to varying degrees. The principal compounds affected in this way are iron and aluminium oxides.

Hydrolysis

This is probably the most important process participating in the destruction of minerals and determining the course of soil formation. Because of its complexity it is not well understood, however, when defined in chemical terms it appears quite simple and straightforward. Hydrolysis of minerals is the replacement of the cations occurring in the structure of the primary silicates, by hydrogen ions from the soil solution with

59

Fig. 3.3 Profile of gilgai; the pressures produced by expansion and contraction cause contortions in the soil and heaving of the underlying material to the surface

the formation of a hydroxyl group in the mineral or a separate hydroxide, this replacement eventually leads to the complete decomposition of the minerals.

Due to variations in the structure of the primary minerals (see p. 9 *et seq.*) the course of hydrolysis does not follow the same path for each type, however, it is always initiated by the replacement of basic cations such as Na, K, Ca and Mg, by hydrogen ions. The next stage for the orthosilicates like olivine and the chain silicates (pyroxenes and amphiboles) is the removal of iron which links together the individual tetrahedra or the tetrahedral chains. This is sufficient to cause a high degree of decomposition of these minerals since these ions are responsible for holding their structure together.

During the hydrolysis of feldspars with their framework structure the removal of the basic cations does not have the same marked influence upon their structure but the subsequent removal of aluminium from the tetrahedra causes severe weakening which leads to complete decomposition of the mineral. This takes place because aluminium is in a position of 4-coordination in the feldspar structure so that its removal leaves a large number of broken bonds each of which is then individually satisfied by a hydrogen ion. However, there is only weak bonding between the hydrogen ions, and the structure eventually collapses. The rate and efficiency of this reaction is influenced by the following factors:

Surface area
pH
Volume and speed of water flowing through the soil
Temperature
Chelating agents
Removal of substances by precipitation

Surface area
Unconsolidated deposits have a greater surface area than consolidated rocks and therefore decompose

Fig. 3.4 Circular raised areas of the gilgai

much faster. The initial stages of hydrolysis of consolidated rocks destroy the 'cementing' agent between the minerals, cause disaggregation so that the constituent minerals are separated one from the other, and produce a rapid increase in the surface area. A further increase is caused by weathering along the cleavages of the minerals, thus hydrolysis effectively increases the surface area which in turn increases the rate of weathering.

pH

The lower the pH values the more vigorous is the ensuing hydrolysis but since soil acidity seldom falls below pH 3.5 the intensity of the reactions is small when compared with that in mineral acids. The lowest pH values usually occur in the surface horizons in association with acid decomposition products of organic matter.

Volume and speed of water flowing through the soil

When water flows through the soil the soluble products of hydrolysis are removed; this prevents the establishment of an equilibrium between the minerals and the soil solution, and thus perpetuates conditions for hydrolysis. At the same time the pH of the soil solution is maintained at fairly low values by the steady inflow of water charged with CO_2 and acid decomposition products from the organic matter.

Temperature

It has already been stated on page 25 that as temperatures increase so does the rate of hydrolysis.

Chelating agents

The rate of decomposition is increased greatly in the presence of chelating agents. This explains the rapid loosening of minerals beneath lichens and may be a very important factor in releasing plant nutrients by plant exudates in the rhizosphere. The precise mechanics that operate are still somewhat obscure but there seems to be a direct exchange of H^+ from the chelate for the metal ion such as Fe^{3+} from the mineral.

Removal of substances by precipitation

If substances released by hydrolysis remain in

61

Fig. 3.5 Depressions between gilgai filled with water

solution an equilibrium may be established thus preventing further hydrolysis; on the other hand if they are removed from solution by precipitation then the reaction can proceed. This applies particularly to iron and aluminium which precipitate as oxides and hydroxides.

Solution

There are only a few substances found in soils that are soluble in water or carbonic acid. Those that are very soluble in water include chlorides and nitrates but these are of restricted distribution in parent materials. However, they may accumulate in the soils of arid areas. The substances soluble in carbonic acid include calcite and dolomite which are widespread and form the major components of limestone, chalk, and some other parent materials. These substances are unique since they are almost completely soluble and therefore, when pure, supply only a very small residue after solution. Consequently, soils developed on these materials are normally quite shallow. On the other hand most other materials, particularly the silicate rocks, furnish a considerable residue of primary

minerals and secondary products. Particularly insoluble is apatite which can persist for thousands of years in some soils of humid areas developed in drift deposits.

Some minerals such as quartz, zircon, haematite and ilmenite that are usually considered to be inert and insoluble do dissolve eventually, this accounts for the small amount of primary material $<50 \mu$m found in many very old tropical soils. The solubility of quartz and some other substances found in soils is given in Fig. 3.8. The rate of solution is influenced in the same way by those factors given above that affect hydrolysis, namely, surface area, pH, volume and speed of water flow through the soils, temperature, chelating agents and removal of substances by precipitation.

Whereas some processes are confined largely to the upper horizons, solution and hydrolysis occur throughout the soil and extend to great depths, and far below what is usually regarded as soil. This is common in many tropical areas where the considerable thicknesses of decomposed rock might be regarded more correctly as geological phenomena rather than as soil but it is necessary to consider them as part of the system,

Fig. 3.6 Wheat growing well on the raised areas but only weeds grow in the wet depressions

Fig. 3.7 Splitting apart of biotite following hydration

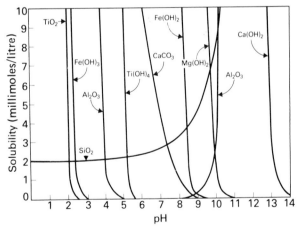

Fig. 3.8 Solubility diagram

since the soil often grades into this material or it may form parent material.

Resistance of minerals to transformations
The minerals in the soil have varying degrees of resistance to transformation; the simple salts like gypsum are soluble, and are lost comparatively quickly from the soil system in areas of high rainfall; carbonates are less soluble and are removed more slowly. The stability of the silicates is considerably greater but within this group some are more resistant than others.

As indicated above, minerals like olivine and the pyroxenes are among the easiest to hydrolyse whereas clay minerals are very resistant with the feldspars in an   63 intermediate position. The most resistant minerals include quartz, zircon, magnetite and titanite, which constitute part of the residue but they usually succumb in time. Resistance to weathering also varies with the size of the mineral particles and environment. Since the small soil particles with their high surface area are more important in hydrolysis, the modified stability sequence of Jackson (1968) for fine particles is given below in order of increasing resistance:

1. Gypsum, halite
2. Calcite, dolomite, apatite
3. Olivine, hornblende, augite, diopside
4. Biotite
5. Albite, anorthite, microcline, orthoclase, volcanic glass
6. Quartz
7. Muscovite, clay-mica
8. Vermiculite and interstratified 2 : 1 layer silicates
9. Montmorillonite
10. Pedogenic chlorite
11. Allophane
12. Kaolinite, halloysite
13. Gibbsite, boehmite
14. Haematite, goethite, lepidocrocite, magnetite
15. Anatase, zircon, titanite, rutile, ilmenite

Products of hydrolysis and solution

The end products of hydrolysis and solution are:
Weathering solution
Resistant residue
Alteration compounds

Weathering solution

This contains the basic cations together with some iron, aluminium and silicate ions. All·of these solutes are lost or redistributed in the soil system.

Resistant residue

This includes minerals such as quartz, zircon, rutile and magnetite which alter only very slowly but can decompose when present as very fine particles.

Alteration compounds

These are principally hydroxides, oxyhydroxides and oxides of iron and aluminium, silica and clay minerals; also included are manganese dioxide and titanium compounds.

Ions of iron only remain in solution at pH 2.5 or lower, while aluminium ions remain in solution at pH 4.0 or lower but since these degrees of acidity are seldom found in soils these ions precipitate out of solution very quickly as amorphous hydroxides. This

reaction can take place immediately after hydrolysis so that some hydroxides may form pseudomorphs after many minerals, then follows slow crystallisation to form goethite, haematite, gibbsite or boehmite, or there may be a recombination of aluminium hydroxide with silica to form clay minerals. Indeed it would appear that all secondary crystalline substances are derived from a pool of amorphous materials but in some cases the amorphous phase is only of short duration and gives the impression that crystalline substances are the initial products. Precipitation also takes place after translocation away from the site of release: for example on the surfaces of other mineral grains or peds, or within pores.

Iron oxides. These occur mainly as ferric hydroxide gel, goethite $FeOOH$ or haematite Fe_2O_3. Ferric hydroxide gel is amorphous with yellowish-brown colour and there is some evidence to show that it closely resembles ferric hydroxide precipitated in the cold in the laboratory. Goethite is crystalline with reddish-brown colour but changes to yellowish-brown as it becomes hydrated. It has a wide distribution ranging from the tropics to the arctic and is one of the main colouring substances in soils. Haematite is bright red and occurs chiefly in soils of tropical and sub-tropical areas or in old geological formations. Another iron oxide is lepidocrocite which is bright orange in colour but up to the present has been found only in soils subject to waterlogging where it is responsible for the bright coloration of many of the mottles. Apparently, iron is present as amorphous ferric hydroxide in young soils, becoming cryptocrystalline on ageing to form goethite or haematite. Thus, the little weathered soils of temperate areas mainly have amorphous ferric hydroxide and a small amount of goethite whereas those of older landscapes have goethite or haematite or a mixture of these latter two. Within soils ferric hydroxide and ferric oxides occur as discrete particles, as coatings on sand grains and on peds, or they form part of the soil matrix as micro-aggregates visible only with the electron microscope.

Aluminium oxides. In a crystalline form aluminium oxide occurs mainly in the soils of tropical areas as gibbsite $Al(OH_3)$ but it can occur in limited quantities in soils of temperate areas. In very old and highly weathered soils boehmite $AlO \cdot OH$, and diaspore $AlOOH$ are the principal forms. These minerals from hexagonal plates which are held together by hydrogen bonding and together or separately they can accumulate to form the main constituents of bauxite. A considerable proportion of aluminium hydroxide

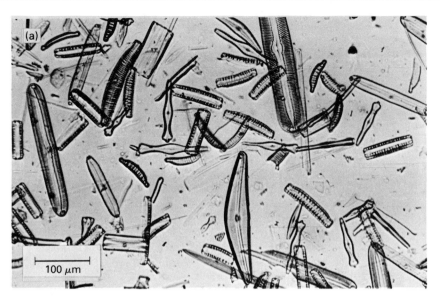

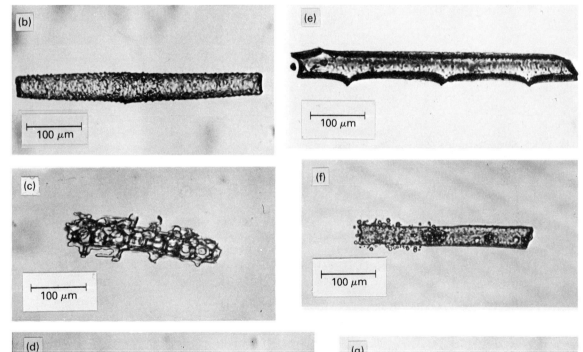

Fig. 3.9 Diatoms and Phytoliths: (a) diatoms; (b–g) phytoliths

in some soils is associated with silica gel to form allophane, particularly in young soils of temperate areas.

Silica. Amorphous or hydrous silica — $SiO_2 \cdot nH_2O$, in addition to forming part of allophane, may be lost in the drainage water or redistributed within the soil system. For example, in Australia there are certain strongly cemented layers known as silcrete that are formed principally by the accumulation of amorphous silica. Elsewhere there are a few instances of crystallisation to form secondary quartz and it has been shown that the clay fraction in the bleached layers of the so-called Kauri Podzol in New Zealand is composed largely of cristobalite — SiO_2. Many plants absorb and accumulate large quantities of silica within their structure to form masses of opaline silica. When the plants die the silica is returned to the soils as phytoliths (Fig. 3.9) which have been observed on the Island of Reamie (East Africa) to comprise the uppermost horizon with a thickness of 5—30 cm (Riquier, 1960).

Manganese dioxide. At present this is considered to be of rather restricted distribution being found as a blue-black coating on the surface of peds or within peds or associated with iron in concretionary deposits and certain massive materials. Usually these features form in soils subject to periods of waterlogging as shown in Plate IVb.

Titanium compounds. These occur in the form of needles as secondary titanite or as crystals of anatase but both are very rare. The white coating of leucoxene — TiO_2 around ilmenite and other minerals or forming pseudomorphs is more common but it is also infrequent.

All of these secondary substances are fully described in Appendix I (p. 313).

Clay mineral formation

Early workers visualised a simple transformation from primary minerals such as feldspars to clay minerals like kaolinite but as Yaalon (1959) has pointed out this requires aluminium to move from a position of 4-coordination in the feldspar to one of 6-coordination in the aluminium hydroxide sheet in kaolinite. This is not possible by simple rearrangement of the structure. For some situations many workers now favour a hypothesis that requires a complete breakdown of the feldspar structure to form amorphous materials or allophane followed by a synthesis of clay minerals. It

was thought at one time that allophane was associated mainly with soils derived from volcanic ash but now it appears to be much more widespread and represents an early stage of hydrolysis. However, there are experimental and field data to show that the decomposition of feldspars in an acid environment can lead to either kaolinite or amorphous material. Evidence for the build up of clay minerals from amorphous materials is seen in the weathering zone in many tropical soils that contain vermiform kaolinite (Fig. 3.10) which does not appear to form pseudomorphs after primary minerals but grows in pores. On the other hand in some rocks, feldspar crystals and micas are seen to be replaced entirely by an intricate intergrowth of vermiform kaolinite. In attempting to resolve this problem Fieldes and Swindale (1954) have suggested that under conditions of vigorous decomposition there is breakdown to the hydroxides and allophane followed by synthesis of clay minerals whereas, where conditions for hydrolysis are slower, there is direct formation of clay minerals.

Fig. 3.10 Macrocrystalline kaolinite in partially crossed polarised light

Even without a clear picture of the mechanism of clay formation there is sufficient field evidence to show that kaolinite is formed in an acid environment from which basic cations and iron are constantly being lost. On the other hand montmorillonite is formed in the presence of large amounts of basic cations and iron, as in mafic rocks and volcanic ash of intermediate or basic composition. The formation of montmorillonite can result from granitic rocks if basic cations accumulate in the system as happens in depressions and in arid and semiarid areas. The micas either form in an intermediate type of environment by the transformation of feldspars or some other mineral, or are merely fragments of muscovite and biotite of clay dimensions.

Although clay minerals are relatively stable they can undergo transformation. There is now a large body of evidence to show that montmorillonite can be transformed into a mica-like mineral and this into kaolinite. Under very acid conditions, clay minerals may be broken down to hydroxides which are either lost from the soil system or translocated and deposited at some point within the pedo-unit.

Hydrolysis of individual minerals

The general picture which seems to be emerging from studies in many parts of the world is that the type of breakdown and nature of the alteration compounds are determined largely by the soil climate and more particularly by the microenvironment in which the mineral is found. It is common to find similar minerals yielding different end products in contrasting environments. Similarly, contrasting minerals can produce identical substances. Another important feature is that the number of decomposition products is relatively few as compared with the wide range of primary minerals. By far the greatest amount of work on mineral transformation has been conducted on the micas. Walker (1949) has shown that during hydrolysis, biotite goes from black through brown to dull brown and at the same time the structure changes from biotite to the mixed layered vermiculite– biotite and then to vermiculite. Biotite can undergo other transformations and may be changed to chlorite and then to kaolinite or straight to gibbsite. Similarly, feldspars can be changed to mica, kaolinite, halloysite or gibbsite (Figs. 3.11 and 3.12). Hornblende goes to chlorite, vermiculite or gibbsite. Minerals having a high content of magnesium and iron such as olivine

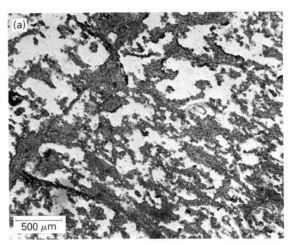

Fig. 3.12 Enlarged portion of Fig. 3.11: (*a*) shows finely crystalline gibbsite forming a network which is mainly a pseudomorph of the outline of the original feldspar crystals and cleavages; (*b*) same section in crossed polarised light

and many of the pyroxenes often form pseudomorphs of hydroxides and allophane as initial decomposition products or may have an intermediate stage of chlorite.

Weathering indices

A number of indices have been produced in an attempt to assess the rate and type of weathering under different conditions but none has been found to be of general applicability. One approach is to assume that a particular constituent of the parent material remains constant and unchanged while the others are lost, altered or redistributed in the soil system. Many workers have used either the total zirconium content or the frequency of the mineral zircon in weathering studies. The former is the less reliable of the two since

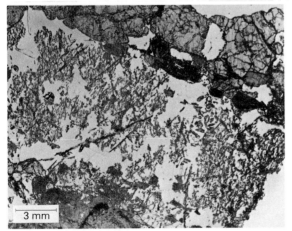

Fig. 3.11 Thin section of weathered granite in which the feldspars have been replaced by gibbsite. In some cases the outlines of the original feldspar crystals have been preserved

67

it may occur in very small zircons or within a mineral and, therefore, it can be lost from the soil by weathering. Zircons in the size range 20—200 μm are a fairly reliable guide to mineral transformation because of their relatively large size and consequent stability, but even these are partly decomposed in some situations. However, this mineral is present only in relatively small amounts, therefore it has to be used as an indicator of weathering with caution because it may be unevenly distributed in the parent material.

Quartz is extremely resistant and has an important advantage as an index mineral primarily because it is usually present in large amounts in most parent materials. Therefore, it is less liable to be affected by variations in distribution but small grains of quartz weather slowly hence the low silt content in many tropical soils.

During the initial stages of hydrolysis there appears to be little change in volume despite considerable changes in composition. Therefore, it is possible to calculate the amount of material lost and redistributed. At a later stage the structure of the rock is destroyed and some of the most resistant minerals become weathered so that it is not possible to calculate with a high degree of accuracy the relative gains and losses.

Oxidation and reduction

It is convenient to consider these two processes together since one is the reverse of the other. Iron is the principal substance affected by these processes for it is one of the few elements that is normally present in the reduced state in the primary mineral. Consequently, when it is released by hydrolysis and enters an aerobic atmosphere it is quickly oxidised to the ferric state and precipitated as ferric hydroxide. Then it is transformed slowly to crystalline goethite or haematite, this latter forming in some soils of hot countries. If, on the other hand the iron is released into an anaerobic environment it stays in the ferrous state hence the greyish-blue colour of many constantly waterlogged soils. In addition, reduction of iron from the ferric to the ferrous state is possible as in cases where soil formation is taking place under conditions of prolonged anaerobism and the parent material contains ferric iron. Dramatic examples of this are seen in areas of red Devonian or Triassic sediments which have been changed to bluish-grey and even blue. These two processes can also be accomplished by microorganisms (see p. 42).

Rock weathering

The amount of rock weathering which has taken place during the Holocene Period can be regarded as relatively insignificant as compared with the considerable amount of rock decomposition which took place during the Tertiary Period. Therefore, in order to determine the full changes that take place when rocks are weathered it is necessary to examine old soils such as some of those found in tropical areas. Probably the most outstanding contribution to rock decomposition has been made by Harrison (1933) who has described weathering sequences for a number of different types of rocks. He found that the end product of weathering as well as the course of weathering varies with the type of rock and with the specific environmental conditions. With small modifications his findings have been confirmed for many other tropical countries and the results he gives for dolerite and granite can be used to illustrate weathering in a humid tropical environment. Probably the weathering of a basic igneous rock is the best illustration of the profound changes that take place when rocks weather by hydrolysis.

Weathering of dolerite
This rock has the following composition:

Labradorite and small amounts of alkali feldspar	50%
Augite	42%
Enstatite	2%
Biotite	1%
Ilmenite and titaniferous magnetite	5%
Olivine	<1%
Quartz	<1%

This sequence of weathering starts at the base of the profile with unweathered dolerite above which there is up to 2 m of weathered rock containing boulders or core stones some more than a metre in diameter. This has been formed by weathering along joint planes which were initially the only surfaces exposed to percolating water. Each core stone has five concentric rings of progressively more weathered rock. Above this layer there is mottled clay 45 cm thick, followed by a bright red layer 3 m thick, a grey sandy subsoil 25 cm thick and finally blocks of ironstone on the surface. In thin sections the greatest alteration of the rock is seen to be taking place within the innermost ring over a distance of not more than 4 mm. The first indication of alteration is shown by the occurrence of yellow staining of iron oxide and small scales of gibbsite along the cleavages of the feldspars. This changes into highly corroded feldspars which are replaced by crystalline gibbsite. Indeed, the same feldspar crystal may vary from relatively fresh at one end to complete decomposition and be replaced by gibbsite at the other end (cf. Figs. 3.11 and 3.12). At the same time the augite is replaced by chlorite and iron oxide and there is a little secondary silica. Within the next 23 mm any remaining feldspars are changed to gibbsite and the chlorite is decomposed to iron oxides. These changes are accompanied by dramatic changes in chemical composition, the most important being the considerable reductions in the amount of combined silica and basic cations and the large relative increase in the amount of aluminium and iron as shown in Table 3.1. Within the third and fourth layers when the structure of the rock has almost disappeared equally important changes have taken place. Apparently the gibbsite is resilicated to form kaolinite which represent 18 per cent and 39 per cent respectively of these two layers; also there is a concentration of iron. The fifth layer or crust does not have as much kaolin as the underlying two but there is a further increase of iron. Thus within 66.5 mm dolerite has been transformed into a weathered product containing mainly kaolinite, iron oxides and gibbsite. The mottled clay and red layers have a similar chemical composition but with a higher content of silica indicative of further resilication and the formation of more kaolinite. Also present are iron and aluminium oxides. The data for the sandy subsoil are interesting and do not appear to be due to any chemical alteration but to differential erosion at the surface, removing the fine material. This is a common phenomenon in the humid tropics.

Weathering of granite

In the majority of cases the weathering of granite and other crystalline rocks contrasts sharply with the weathering of basic rocks. With granite the change from rock to soil proceeds gradually as shown by a slow decrease in the amount of minerals such as feldspars and slow loss of basic cations, some loss of combined silica and a marked relative increase in the amount of aluminium. The granite also loses its overall structure very slowly as illustrated in the generalised diagram in Fig. 3.17.

A good example of the weathering of granite has been described by Harrison (1933) from Guyana where he recognised six main zones with a total thickness of 439 cm.

1. Zone of unaffected granite
2. Zone of softening and disintegrating granite — 58 cm
3. Zone of active hydration and kaolinisation — 16 cm
4. Zone of gradual continuing hydration and kaolinisation — 290 cm
5. Zone of more or less complete kaolinisation — 55 cm
6. Soil — 20 cm

A brief description of each of the zones is given below while the chemical composition of the zones and their losses calculated on an aluminium constant basis is given in Tables 3.2 and 3.3.

Zone of unaffected granite
This granite is a medium textured grey granite with the following mineralogical composition:

Quartz	33%
Alkali feldspars	53%
Oligoclase	10%
Muscovite	3%
Biotite	1%

There are also inclusions of epidote, titanite and iron ore.

Zone of softening and disintegrating granite — 58 cm
In this zone there are core stones between which the granite is weathered by decomposition of the outer edges of the orthoclase and oligoclase but the microcline is quite fresh. The biotite is completely altered to chlorite but the muscovite is fresh.

Here there is only a relatively small loss of silica but there is some loss of iron oxide, a part being converted to the ferric state. Practically all of the manganese is lost along with about 10 per cent of the titanium. Over 60 per cent of the magnesium, 90 per cent of the calcium, 25 per cent of the potassium and 55 per cent of the sodium have been lost indicating the weathering of the feldspars and biotite. In contrast there are some spherulites of secondary silica.

Zone of active hydration and kaolinisation — 16 cm
The general structure of the rock is still preserved with a high proportion of the feldspars being changed to kaolinite but some of the microcline and muscovite is still fresh. There is a considerable loss of silica and much of the remaining basic cations indicating a considerable amount of weathering of the feldspars. There is also some secondary silica.

Zone of gradual continuing hydration and kaolinisation — 290 cm
This is a very thick zone in which the visible rock structure gradually decreases upwards. Most of the remaining alkali feldspars have decreased to insignificant amounts in the upper part but the microcline is still in high proportions and relatively fresh. Some

Table 3.1   Chemical composition of dolerite and weathered products (%)

A		Unaltered dolerite
B	0–4 mm	First layer of concentric weathering
C	4–27 mm	Second layer of concentric weathering
D	27–37 mm	Third layer of concentric weathering
E	37–64 mm	Fourth layer of concentric weathering
F	64–66.5 mm	Fifth layer (crust) of concentric weathering
G	330–375 cm	Mottled clay (pale colour)
H	330–375 cm	Mottled clay (dark colour)
I	375–675 cm	Red layer
J	675–765 cm	Grey sandy subsoil
K	765 cm	Ironstone on surface

	A	B	C	D	E	F	G	H	I	J	K
Quartz	<0.1	9	8	13	10	9	8	8	5	81	1
SiO_2	51	8	5	10	19	6	29	29	26	6	3
Al_2O_3	16	34	40	19	23	24	24	24	21	3	21
Fe_2O_3	2	21	18	34	30	36	20	19	29	2	65
FeO	9	4	4	1	2	1	1	1	1	2	<1
H_2O	<1	18	20	15	11	20	13	13	14	5	9
TiO_2	2	4	4	7	5	4	5	5	4	2	2
MnO	<1	Nil	Nil	Nil	Nil	Nil	<0.1	Nil	Nil	Nil	<0.1
MgO	8	1	<1	<1	<1	<1	<1	<1	<1	<1	<1
CaO	9	1	<1	<1	Nil	<0.1	Nil	Nil	Nil	<0.1	Nil
K_2O	2	<1	<1	<1	<1	<0.1	<0.1	<0.1	<0.1	<0.1	<0.1
Na_2O	2	<1	<1	<0.1	<0.1	<1	<1	<0.1	<0.1	<0.1	<0.1
P_2O_5	<0.1	<0.1	<0.1	<0.1	<0.1	<0.1	<0.1	<0.1	<0.1	<0.1	<0.1

Table 3.2   Average chemical compositions of zones of alteration of granite (%)

1 Zone of unaffected granite
2 Zone of disintegration
3 Zone of hydration and kaolinisation
4 Zone of continuing hydration and kaolinisation
5 Zone of completed kaolinisation

	1	2	3	4	5
Quartz	32	41	42	37	31
SiO_2	41	33	29	30	31
Al_2O_3	14	15	19	21	24
Fe_2O_3	1	2	2	2	2
FeO	1	<1	<1	<1	<1
H_2O	1	2	5	6	9
TiO_2	1	1	1	<1	1
MnO	<1	trace	–	–	
MgO	1	<1	<1	<1	<1
CaO	1	<1	<0.1	<1	<0.1
K_2O	5	4	3	2	1
Na_2O	3	2	<1	<1	<1

Results recalculated on basis of constant alumina and water contents (14.45 and 0.68 per cent respectively)

	1	2	3	4	5
Quartz	32	40	33	27	20
SiO_2	41	32	23	22	20
Al_2O_3	14	14	14	14	14
Fe_2O_3	1	2	1	1	1
FeO	1	<1	<1	<1	<1
H_2O	1	1	1	1	1
TiO_2	1	1	1	<1	1
MnO	<1	trace	–	–	–
MgO	1	<1	<1	<1	<1
CaO	1	<1	<0.1	<0.1	<0.1
K_2O	5	4	2	2	1
Na_2O	3	2	<1	<0.1	<0.1

Table 3.3   Losses of constituents between granite and its weathering products

I.	Changes between zones 1 and 2
II.	Changes between zones 2 and 3
III.	Changes between zones 3 and 4
IV.	Changes between zones 4 and 5
V.	Changes between zones 1 and 5 (total losses)

(a) Results expressed as percentage of original rock, assuming constant alumina (14.32 per cent) and constant water (0.50 per cent)

	I	II	III	IV	V
SiO_2	5	12	13	4	34
Fe_2O_3 and FeO	1*	1	<1	1	1
TiO_2	<0.1*	<1	<1	<1*	<0.1
MnO	<1	–	–	–	<1
MgO	1	<1	<0.1*	<1	1
CaO	1	<1	–	<0.1	1
K_2O	1	1	1	1	5
Na_2O	2	1	<0.1	<0.1	3

(b) Results expressed as percentages of constituents present in original rock

	I	II	III	IV	V
SiO_2	7	16	19	6	47
Fe_2O_3 and FeO	28*	18†	9	33	60
TiO_2	1*	23†	17	39*	<1
MnO	100	–	–	–	100
MgO	56	31	–	6	93
CaO	73	18	–	1	99
K_2O	25	30	19	19	94
Na_2O	68	27	1	1	97

* Gains   † Losses on

muscovite is also present but the chlorite has disappeared completely. There is a continuing loss of silica and much potassium which indicates the weathering of some of the microcline.

Zone of more or less complete kaolinisation – 55 cm
The rock structure has almost completely disappeared and the mass is composed almost entirely of quartz and kaolinite. There is also a small residue of microcline and muscovite. Some silica, a high proportion of the remaining iron and nearly all of the remaining potassium are lost.

The result of this weathering sequence is the almost total loss of basic cations and about 40 per cent silica which together make up about 50 per cent of the original rock. However, the formation of hydration products means that the volume loss is about 40 to 45 per cent.

These data indicate the degree of alteration and reorganisation of the constituents that can take place when rocks undergo profound hydrolysis.

Although the overall picture is the same for all rocks, i.e. the complete hydrolysis of the primary silicates and loss of some of the constituents in solution, particularly the basic cations, the details seem to vary somewhat from place to place and from rock type to rock type. Whereas basic rocks are transformed within a relatively short distance to highly altered compounds, intermediate and acid igneous rocks usually show a more gradual decomposition. The primary decomposition products also vary somewhat. Harrison has demonstrated that gibbsite does not form during the alteration of granite; instead there is a fairly direct transformation to kaolinite but in some other tropical countries one of the initial products of the weathering of granite can be gibbsite.

The moisture regime of the environment is a factor of some importance for it appears that while dolerite will form gibbsite in a humid subsurface horizon this does not happen when weathering is near to the surface. Here the weathered dolerite loses a considerable proportion of the aluminium and no gibbsite is formed. The reason for this is not clear but it has been shown that some amorphous hydroxides will not form in the presence of organic matter. This is a very important feature and should be borne in mind when examining a full set of data from a pedo-unit because differences in the chemical composition of the various layers might be due to the initial type of weathering rather than to pedogenetic reorganisation.

Granular disintegration
A number of biotite containing rocks, particularly the coarse grained crystalline rocks such as granites and gabbros, may undergo extreme physical disintegration during the early stages of weathering. The result is that the individual mineral grains are either separate one from the other or they occur as small clusters of two or three grains. In addition, a high proportion of the grains are fractured. This is the process of *granular disintegration* and is generally attributed to the exfoliation pressures of biotite resulting from hydration and the early stages of hydrolysis, since it is clearly seen that the biotite grains are seldom fractured (Fig. 3.13).

500 μm

Fig. 3.13 Granular disintegration of granite

Description and classification of rock weathering
Combining the physical and chemical weathering of rocks it is possible to give the following descriptions and classification:

Degree of preservation of rock structure
Degree of chemical alteration of individual minerals
Degree of physical alterations of minerals

Degree of preservation of rock structure
I Rock structure completely preserved – some minerals may be slightly or completely altered, e.g. replacement of olivine by iddingsite.
II Rock disaggregated into individual minerals or groups of minerals with fine pores.
   (i) Few if any minerals altered, a little secondary material may form coatings, e.g. some granites, gabbros, etc. (cf. Fig. 3.13).
   (ii) Some minerals slightly or completely altered, some secondary material forming coatings and fillings, e.g. some granites, gabbros, etc.
III Rock structure well preserved, many minerals

partly or completely altered, many pores and spaces, some secondary material forming fillings and coatings, a little translocated material may be present, e.g. granites, gneisses, gabbros.

IV Rock structure partly preserved, only the resistant minerals are unaltered and in their relative positions, they may be pitted by solution, many spaces and much secondary material, a little translocated material may be present, e.g. granites.

V Rock structure completely destroyed or not present.

VI Individual areas completely removed to form pores which may or may not have secondary material, e.g. solution of calcareous rocks.

Degree of chemical alteration of individual minerals

I Mineral partly altered:
   (i) Outside rim altered, e.g. iddingsite or olivine
   (ii) Centre altered, e.g. sericite in feldspars
   (iii) Any part altered
   (iv) Altered along cleavages:
      (*a*) without replacement, e.g. feldspars
      (*b*) with *in situ* replacement, e.g. iron oxides in amphiboles
      (*c*) with transported material, e.g. iron oxide in feldspars
   (v) Some of the mineral removed to form a pore, e.g. apatite, calcite

II Mineral strongly altered:
   (i) Partially replaced, residual fragments randomly distributed with or without pores
   (ii) Partially replaced, residual fragments distributed according to crystal structure with or without pores
   (iii) Much of the mineral removed leaving a pore, e.g. apatite, calcite

III Mineral completely altered:
   (i) Complete replacement by one or more substances, e.g. feldspar by kaolinite or gibbsite, biotite by vermiculite
   (ii) Replacement with irregular pores, e.g. feldspar by gibbsite
   (iii) Replacement along cleavages with intervening pore pattern
   (iv) Completely removed to leave a pore, e.g. calcite, apatite, olivine

Degree of physical alteration of minerals

I Continuous fracture through the rock from mineral to mineral.

II Individual random fissures within a mineral:
   (i) empty
   (ii) coatings present

III Fissures parallel to cleavages, e.g. exfoliation of micas, anastomosing in feldspars.

By combining the chemical and physical processes of weathering it is possible to produce a classification of rock weathering including a symbol that can be used in soil descriptions and soil formulae. This is given in Table 3.4.

BIOLOGICAL PROCESSES

Probably the most important biological processes taking place in soils are humification of organic matter and the translocation of material from one place to the other. Other processes included are nitrification, nitrogen fixation and those listed below that are discussed elsewhere:

Microbiological oxidation and reduction of inorganic substances (p. 42)
Water and ion uptake (p. 39)
Fragmentation (p. 39)
Aggregation (p. 54)

Translocation

Some emphasis must be given to those biological processes that bring about churning, and other soil disturbance. The most dramatic manifestation of this process is brought about by the soil inhabiting vertebrates (p. 40) but probably the greatest amount is accomplished by earthworms and termites which mix the organic and mineral material and redistribute it in the soil system. These are extremely important processes because they are continually bringing the microbiological population into contact with a fresh food supply and therefore help to maintain a constant breakdown of the organic matter and release of ions for plant nutrition. Also they cause disaggregation and reaggregation of the soil with the result that the mineral grains are being exposed to regular changes of environment resulting in greater decomposition.

Humification

The breakdown of organic matter to form humus and to release various plant nutrients is an extremely complex and little understood process involving organisms as the principal agents. Relatively little work has been completed on the stages of breakdown of the litter falling on to the surface, therefore it is

Table 3.4 Classification of rock weathering

Degree of weathering	Rock	Sediments	Symbol
Chemically weathered			
Slightly chemically weathered	Alteration along fissures and cracks, little mineral alteration	The rock fragments are easily recognisable but show signs of mineral alteration. Some alteration of fine earth	1w
Moderately chemically weathered	Core stones well developed and/or predominant dis-aggregation, clear evidence of mineral alteration	Disaggregation of most of the rock fragments. Clear evidence of mineral alteration	2w
Strongly chemically weathered	Original structure well preserved. of the minerals are weathered.	Complete disaggregation, > 20%	3w
Very strongly chemically weathered	Original structure partly preserved	Type of soil horizon Sapron	Sa
Physically weathered			
Slightly physically weathered	Occasional fracture, rock structure well preserved	Occasional shattered rock fragment, adjacent particles can be fitted together	1p
Moderately physically weathered	Frequent to abundant fractures, rock structure partially disrupted	Frequent shattered rock fragments, adjacent particles may fit together	2p
Strongly physically weathered	Dominant angular rock fragment, rock structure poorly preserved	Many rock fragments shattered, adjacent particles may not fit together	3p
Very strongly physically weathered	Type of soil horizon, Lithon		Lh
		Nearly all original rock fragment shattered into two or more fragments	Lh

not possible to make any generalisations about these processes. The example given below is taken from the work of Kendrick and Burges (1962) who made a study of the decomposition of the needles of *Pinus sylvestris* and suggested the following sequence of events.

About 40 per cent of the living needles on the tree are infected in the spring by the parasitic fungus *Lophodermium pinastri* which causes characteristic black spots and transverse black bars (Fig. 3.14). Other fungi are also present but they do not appear to attack the needles until they become old and turn brown, by which time over 90 per cent of the needles are infected. The needles fall to the ground in autumn becoming part of the litter where they spend about six months and are then infected by another fungus (*Desmazierella acicola*) which replaces the previous group. The needles now become part of the decomposing organic matter and over a period of two years *Desmazierella acicola* spreads, particularly inside the needles where it deposits areas of black pigment. The outsides of the needles are attacked by two other fungi, *Sympodiella acicola* and *Helicoma monospora* which also produce the dark coloured materials that are a characteristic feature of decomposing organic matter. After two years the needles become compressed

Fig. 3.14 Pine needles infected with the fungus *Lophodermium pinastri* as indicated by the characteristic black spots and transverse black bars

73

and fragmented and are attacked by the mesofauna. Mites, collembola and enchytraeid worms eat the conidiophores of the fungi and simultaneously instars of mites attack the soft interior of the needles (Fig. 6.36). The remaining fragments are attacked by yet another group of fungi (Basidiomycetes) and for the next seven years the remains become the amorphous, strongly decomposed organic matter where biological activity is at a very low level.

Further studies on the breakdown of organic material have been made by Saitô (1957, 1965) who has given an account of the decomposition of beech leaves. The litter consists of freshly fallen leaves and brown leaves only slightly subject to microbiological attack. In the decomposing material where the moisture content remains fairly constant, many leaves turn yellow following the attack of Basidiomycete mycelia associated with vigorous growth of bacteria. Here the infected leaves at first become much thinner without losing their structure, then mouldy from overgrowth of Basidiomycete mycelia. Later, growth of other organisms takes place and gradually the leaves are transformed into amorphous debris. Not all leaves in the same layer become infected by Basidiomycete mycelia, furthermore, decomposition does not always take place uniformly throughout a leaf. In the yellow infected leaves there is a marked disappearance of lignin followed by a rise in the quantity of water-soluble substances, finally the Basidiomycete mycelia in the decomposing leaves are broken down by bacteria.

Generally, the early stages of decay are characterised by the loss of the water-soluble materials, starches and proteins. This is followed by the decomposition of hemicelluloses and cellulose leaving a residue consisting largely of lignin and cuticularised cell walls. In the case of the mesofauna their soft proteinaceous material is rapidly decomposed leaving chitinous exoskeletons which may accumulate at the junction of the organic and mineral material.

It is clear from the above work that the breakdown of the organic matter in soil is accomplished by successive waves of organisms some of which are symbiotic chemoheterotrophs, like the fungi, which can be attacked by parasitic mites which in turn are decomposed by bacteria. This means that a given carbon atom may pass from the atmosphere to a plant leaf and then to a succession of soil organisms before it passes back to the atmosphere in CO_2; or it may become temporarily part of the soil humus.

The cycling of carbon is shown diagrammatically in Fig. 3.15.

Formation of humus accumulations at the surface

The formation or accumulation of humus at the surface seems to result from a delicate balance between the nature of the soil, viz. its ability to supply nutrients, and the nature of the vegetation on the site.

When plants are growing in a situation of stress resulting from the lack of nutrients the polyphenols in the leaves and needles form more stable condensation products with the proteins at the time of abscission so that when the leaves fall to the surface decomposition and release of the nitrogen is very slow and there is a slow but steady increase in the amount of organic matter at the surface.

In such acid situations the majority of the roots of the trees have a mycorrhizal association which serves a number of functions:

1. The invasion of the root by mycorrhizal fungi causes the root to lose its root tip. Thus the tendency to vertical growth is lost and the root grows mainly horizontally.
2. The mycorrhizae cause the roots to branch more freely, thus we find that there is a greater volume of roots near to the surface.

Thus the very large amount of roots together with their mycorrhizae cause a very efficient nutrient absorbing system to be developed. Under these conditions the nitrogenous substances in the organic matter are transformed to ammonia which can be held on the exchange complex. Only very small amounts of nitrate are formed and they are quickly absorbed by the plant. So that with regard to the nitrogen balance there is a net accumulation in the system.

Nitrification

This is the process during which nitrate is formed by soil organisms. When chemoheterotrophic organisms, including mesofauna, decompose organic residues, the principal and simplest nitrogen compound produced is ammonia; this can then be oxidised to nitrite and afterwards to nitrate, each stage being accomplished by specific autotrophic bacteria. Ammonia is oxidised by *Nitrosomonas* and *Nitrococcus* and the nitrite by *Nitrobacter*.

These organisms require fairly exacting conditions for their activity; the soils must be aerobic and must at the same time be moist with about 50 per cent of the pore space saturated. The optimum temperature is 27–32 °C but they can operate down to 4 °C.

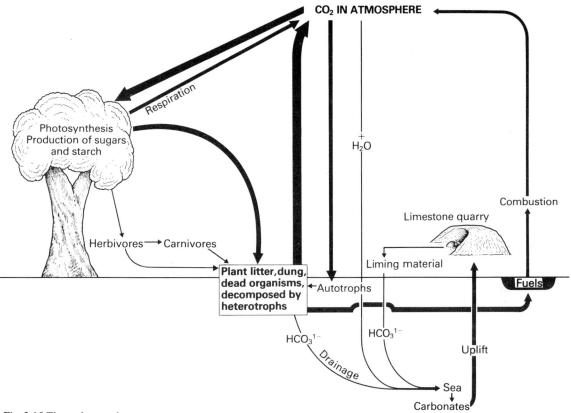

Fig. 3.15 The carbon cycle

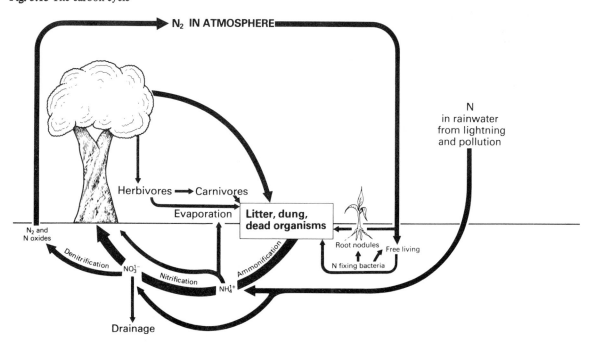

Fig. 3.16 The nitrogen cycle

Although they prefer conditions around neutrality they can also operate down to about pH 5. Under anaerobic conditions these organisms are inhibited, others take their place and reduce the nitrogenous compounds to nitrogen which is lost to the atmosphere. This process may be of relatively little significance in the direct formation of soils but it is of paramount importance in plant nutrition. Most plants in natural or semi-natural plant communities obtain the greater part of their nitrogen in the form of nitrate produced by bacterial activity from the remains of organisms including parts of plants and the soil microorganisms. Thus nitrogen is constantly being cycled from the soil to plants and then back to the soil via the litter and its decomposition products.

Nitrogen fixation

There are a number of free living chemoheterotrophic bacteria including *Azotobacter*, *Clostridium pastorianum* and *Beijerinckia* sp. that are capable of utilising atmospheric nitrogen to form their cell protein which upon the death of the organism, is decomposed to ammonia to form part of the nitrogen available to plants or to take part in nitrification. Other microorganisms capable of fixing atmospheric nitrogen include some algae.

Nitrification and nitrogen fixation can be represented in the cycle shown in Fig. 3.16.

THE WEATHERING UNIT — PRIMARY SOIL SYSTEMS

Weathering can be used as a comprehensive term to include all of the subaerial processes responsible for the transformation of mineral material, both consolidated and unconsolidated. These diverse processes combine to produce several types of

Fig. 3.17 Primary soil system produced by chemical weathering of consolidated igneous rock

weathering units or primary soil systems which are formed either from superficial drifts or from consolidated rocks. Chemical and physical processes predominate; the biological processes are significant only at the surface but are of paramount importance where they occur, being responsible largely for the formation of the pedo-unit. The following are the principal mechanisms responsible for the formation of primary soil systems.

    Hydrolysis of consolidated rock
    Hydrolysis of unconsolidated material
    Solution of consolidated rock
    Solution of unconsolidated material
    Fragmentation of consolidated rock
    Fragmentation of unconsolidated material

Of these six mechanisms the three illustrated in Figs. 3.17 to 3.19 are possibly the most common.

Hydrolysis of consolidated rock – System 1

This mechanism produces a system which starts at the top with strongly weathered material composed of clay minerals, hydrous oxides and the resistant residue, and grades with depth through progressively less weathered material into fresh rock. In the lower parts of systems developed on crystalline rocks there are usually a number of core stones which increase in number and frequency with depth but on sedimentary rock the areas of more resistant materials often occur as strata rather than as core stones. A characteristic feature of this system is the uneven surface of the weathering front which may be caused by hydrological differences or variations in the rock structure (Fig. 3.17).

Hydrolysis of unconsolidated material – System 2

This mechanism produces a system which has many similarities to that developed by the hydrolysis of hard rocks by having the zone of maximum decomposition at the surface. However, in most cases the amount of weathering throughout the material is minimal since most unconsolidated deposits such as loess or alluvium are of recent age, but when an old system is examined the material becomes progressively less weathered with depth passing through a zone where most of the stones are almost completely decomposed forming 'ghosts' (Fig. 3.18).

Solution of consolidated rock – System 3

The characteristic feature of solution is the manner in which subterranean channels develop, down which the fine soil particles can fall leaving bare rock surfaces or a shallow layer of residue. These channels gradually develop into depressions or caves (see Fig. 2.34).

Solution of unconsolidated material – System 4

Systems produced by this mechanism are characterised by extreme differential weathering to form pipes of material within the body of the relatively unweathered material.

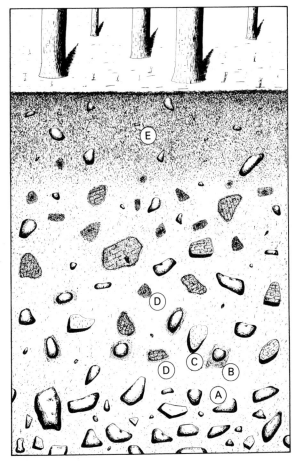

Fig. 3.18 Primary soil system produced by hydrolysis of unconsolidated material: (A) Relatively unaltered material; (B) partially weathered boulders such as granite or basalt with an unweathered core; (C) very resistant boulder such as quartzite; (D) very decomposed boulder, could be a ghost of granite or basalt; (E) soil containing very resistant boulders

Fig. 3.19 Primary soil system produced by physical disintegration of rock. This is an example of disintegration by frost on a sloping site hence the arcs indicating movement of material down the slope by solifluction.

Fragmentation of consolidated rock — System 5

The only agency of significance producing fragmentation is freezing and thawing. Systems produced by this mechanism have at the surface a matrix of fine material containing many angular rock fragments which become more frequent with depth until there is virtually no matrix. On sloping situations an additional effect of frost action is to cause the stones to become oriented parallel to the slope resulting in the development of pattern shown in Fig. 3.19.

Fragmentation of unconsolidated material — System 6

This mechanism produces systems that are similar to the one above with the exception that the amount of fine material is uniformly distributed throughout the system but again the degree of frost shattering of stones and boulders diminishes with depth and is usually absent below 2 m.

These systems are somewhat schematic and do not specify the precise type of soil present. For example a Chernozem or a Podzol may develop when basic System 2 is operating. Similarly a Vertisol or a Ferralsol may form when basic System 1 is taking place. Further, as explained in Chapter 2, it is common for one set of processes to stop and be replaced by another set. This took place on a large scale during the Pleistocene Period. Thus we find today that System 1 is best developed in tropical and subtropical areas whereas System 2 occurs mainly in the middle latitudes. Systems 5 and 6 now occur mainly in polar areas but during the Pleistocene they extended into the middle latitudes and were superimposed upon Systems 1 and 2 but now Systems 1 and 2 are superimposed upon Systems 5 and 6. For these reasons we find that in the humid areas outside the limit of the influence of Pleistocene glacial and periglacial processes the soils are deep and have developed by chemical processes whereas within that limit the soils are shallow and are mainly in System 2. This finds its maximum expression in the middle and early Pleistocene deposits of the central USA where the material, mainly loess and till, has been transformed to a clay-rich soil up to 3 m thick (see p. 304).

SUMMARY

Under natural conditions it is the rule rather than the exception to find several of the above processes operating simultaneously or superimposed one upon the other. For example, hydrolysis, oxidation and clay mineral formation usually proceed together in most freely drained soils; whereas reduction, hydrolysis and clay formation will take place together in wet soils. Therefore, it is really quite impossible to give a comprehensive picture in the form of a single diagram without making it too complex. Figures 3.20 and

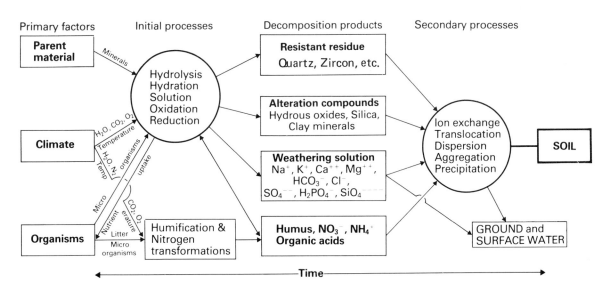

Fig. 3.20 Soil formation (adapted from Yaalon, 1960)

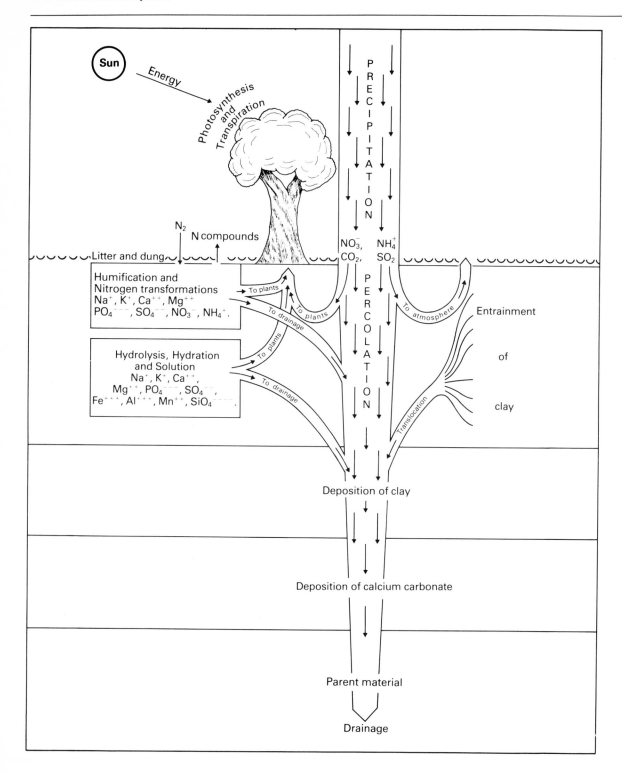

Fig. 3.21 The soil system with particular reference to those soils in which there is translocation and deposition of clay and calcium carbonate

4

3.21 are attempts to summarise much of what has already been said about the processes in the soil system in a humid climate where chemical and bio-logical processes predominate. Similar illustrations can be constructed for arid and tundra environments

All processes are time dependent but the rate of the process depends upon a number of factors and each case has probably to be considered on its own, further-more, the measurement of the rate is usually very difficult.

The processes within soils may change as the soils develop. In a soil with some carbonate, the removal of the carbonate continues until it is completely removed, then other processes speed up or change due to a drop in pH.

The rate of removal of any particular constituent will depend upon other factors. Sodium can only be removed if there is an adequate flow of water through the system so that rate of removal is determined in part by precipitation with soil permeability also play-ing its part.

Properties of soil horizons

Some soil properties are distinctive features and can be used as important differentiating criteria while others seem to have little pedological significance but are important in relation to crop production. A general account of the character of the pedologically import-ant properties used for categorising soils is given below but no attempt is made to describe methods for their assessment. Analytical and descriptive methods for use in the field and in the laboratory are covered adequately in many publications particularly the *Soil Survey Manual* (USDA, 1951), Hodgson (1974) and FitzPatrick (1977) and in standard text-books especially those by Piper (1947), Jackson (1958), Milner (1962) and Hesse (1971). Where new ideas are introduced a full characterisation is given as in the case of some aspects of structure and thin section morphology. The values for some properties vary with the analytical methods employed, therefore for some of the data it is necessary to state the tech-nique adopted.

In addition, an attempt is made to integrate the field and laboratory aspects of the various properties particularly with regard to those properties seen in hand specimens, in thin section and with electron microscopes. Hence the reason for not having a separate section on micromorphology.

Perhaps it should be stated that every attempt has been made to reduce the jargon to an absolute minimum, particularly when discussing micromorpho-logy.

Making sharp divisions between properties is some-times difficult since there is often an overlap. For example the sand fraction, 2 to 0.2 mm often contains   81 only individual minerals but it may contain small concretions, rock fragments and secondary crystalline material, all of which can be considered under their respective headings. This problem is largely obviated by studying the soil in thin sections. Similarly, cation-exchange capacity can be described independently but it does overlap with pH since it varies with pH. Therefore, the following headings are largely arbitrary but useful.

1. Surface of the soil
2. Position
3. Thickness
4. Boundaries
5. Consistence – handling properties
6. Mineral composition
   Larger separates >2 mm
      Mineralogical composition
      Shape of particles
      Size and frequency
      Orientation
      Relationship with the fine earth
   The fine earth <2 mm
      Particle size distribution
      Texture
      Shape of particles
      Mineralogical composition
      Micromorphology
7. Colour
8. Fine material
9. Detrital grains (micromorphology)
10. Structure and porosity
11. Passages – faunal and root
12. Faecal material
13. Coatings
14. Clay plugs
15. Surface residues
16. Density
17. Moisture content
18. pH
19. Cation-exchange capacity and percentage base saturation
20. Soluble salts
21. Carbonates
22. Elemental composition
23. Amorphous and microcrystalline oxides of iron, aluminium and silicon
24. Segregations and concretions
25. Organic matter
26. Secondary crystalline material

1. SURFACE OF THE SOIL

In many situations there may be a sparse or absent vegetative cover so that the surface of the soil is clearly exposed and can be examined and described. In such cases a wide variety of phenomena may form at the surface depending upon local conditions. There may be a distinctive microrelief such as gilgai (see p. 58), frost phenomena (see p. 220) or faunal mounds (see p. 40). Frost heaving and deflation lead to the accumulation of stones at the surface while desiccation particularly of clay soils leads to the formation of cracks. Some soils develop a crust while some others may have an efflorescence of salts.

2. POSITION

Horizons can be grouped into upper, middle and lower, according to the position they occupy in the pedo-unit. Those in the upper position occur at or near to the surface and are often strongly influenced by biological activity. Usually they contain the largest amount of organic matter and in a humid environment the greatest amount of water passes through them, consequently, they lose significant amounts of material, either in solution or suspension. The middle position includes those horizons that are influenced less strongly by biological processes. On the other hand they receive and sometimes retain some of the material washed in from above. Where an upper horizon is very thick it may be regarded as extending into the middle position.

The lower position is occupied regularly by relatively unaltered material as in the case of many soils developed in Pleistocene or Holocene sediments, but in a dry environment there may be accumulations of calcium carbonate or calcium sulphate.

Areas of older soils frequently have weathered materials in various stages of decomposition occupying this position. In the so-called bisequal soils a few horizons that normally occur in the upper or middle positions are found forming a second lower sequence of horizons. These situations are relatively rare but may be very important in certain local situations.

3. THICKNESS

Perhaps it is questionable whether the thickness of horizons should be considered as an important differentiating property but, in most circumstances, fully developed horizons have fairly well defined limits of thickness. Within these limits the exact thickness is used as a crude indication of the degree of development and age of the horizons. However, different fully developed horizons vary widely in their thickness, some are as thin as 1 cm whereas others may be several metres thick. In any given pedo-unit the constituent horizons are seldom uniform in thickness, sometimes they vary by not more than 1 or 2 cm from the mean thickness but others are extremely variable particularly those that form tongues into the underlying horizons. Thin horizons (Plate IVa) are often very irregular in outline and present many difficulties with regard to delimiting pedo-units.

4. BOUNDARIES

The vertical change from one horizon to the other varies in distinctness and outline. A change in colour is the principal and most easily observed property that is used to delimit horizon boundaries but in many cases other properties such as structure and texture are used. Depending upon the vertical distance over which the change takes place five classes of distinctness are used. These are described below and illustrated in Fig. 4.1.

Classes of distinctness

Abrupt   – change takes place within 2 cm
Sharp   – change takes place within 2–5 cm
Clear   – change takes place within 5–10 cm
Gradual – change takes place within 10–20 cm
Diffuse – change takes place within >20 cm

The outline of the horizon boundaries varies considerably from soil to soil. The types of outline are described below and illustrated in Fig. 4.2.

Classes of outline

Smooth   – almost straight
Wavy   – gently undulating
Lobate   – with regular lobes
Irregular – strongly undulating and mamillated
Tongued – forming tongues into the underlying
                 horizon, shallow tongues and deep
                 tongues

Usually, when an horizon is thin the change is sharp or very sharp (Plate IVa); on the other hand as horizons increase in thickness they tend to have merging boundaries (Plate Id) but there are a number of situations where there are departures from

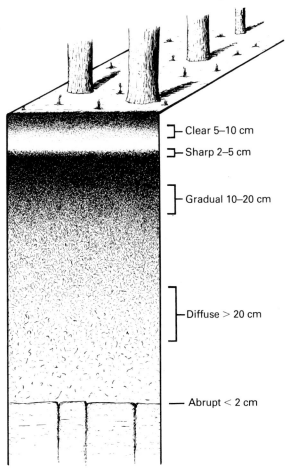

Fig. 4.1 Distinctness of soil horizon boundaries

these generalisations, for some thick horizons may have at least one sharp boundary.

5. CONSISTENCE – HANDLING PROPERTIES

Soil consistence or handling properties refers to the type and degree of cohesion and adhesion between soil particles, i.e. the resistance of the soil to deformation and rupture, or the strength and nature of the forces of cohesion and adhesion. Consistence is usually determined by pressing the soil between the first two fingers and the thumb and feeling as well as observing the changes that take place.
The terms used to describe consistence include:

Brittle	Loose
Compact	Plastic
Firm	Soapy

83

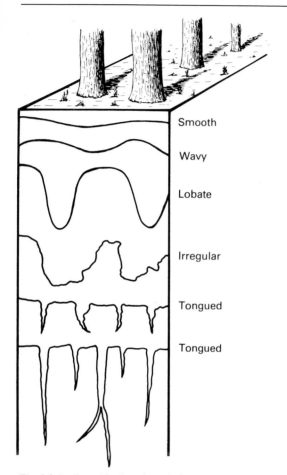

Smooth

Wavy

Lobate

Irregular

Tongued

Tongued

Fig. 4.2 Outline of horizon boundaries

Fluffy	Soft
Fluid	Sticky
Friable	Tenacious
Hard	Thixotropic

Brittle. Firm and ruptures suddenly in an explosive manner.

Compact. Firm with close packing. It can be moderately compact or very compact.

Firm. Moderately coherent but can be crushed and broken into fragments between forefinger and thumb; there is weak but noticeable resistance.

Fluffy. Friable with a low bulk density and therefore feels light in the hands.

Fluid. Some soils when wet behave like liquids and will flow. The rate at which a soil will flow depends

upon a number of factors. For example a thixotropic clay will at one time be firm and plastic while at another time it will be fluid. However, the moisture content usually determines the extent to which the soil becomes fluid but some materials will sometimes respond to gravity and flow in the dry state. Fluid materials can be divided into the following classes:

Not fluid: a 7 cm sphere will not settle when placed on a flat horizontal surface.

Very slightly fluid: a 7 cm diameter sphere will very slowly settle when placed on a flat horizontal surface.

Slightly fluid: a sphere cannot be formed but the material remains as a coherent mass.

Moderately fluid: the material cannot be kept as a coherent mass and flows through the fingers.

Very fluid: the material flows freely.

Friable. Weakly coherent and is easily crushed by weak to gentle pressure. It will again become coherent when pressed together. The moisture content can be very variable.

Hard. Dug out in coherent lumps and needs force to cause it to rupture. Hardness is usually due to compaction and/or cementation and occurs in all moisture states. The range in hardness is:

Hard: strong resistance to rupture when pressed between the finger and thumb but easily broken when held between the hands.

Very hard: can only be broken between the hands.

Extremely hard: can only be broken with an implement such as a geological hammer.

Loose. The material is loose and not coherent and is present as single grains. The moisture content can be very variable.

Plastic. The moist soil material can be moulded or pressed into specific shapes which it retains. In the field, plasticity is also determined by attempting to roll the soil material into a thread and then to determine the extent to which the thread can be formed into rings. The range is:

Non-plastic: no thread will form.

Slightly plastic: forms threads with difficulty.

Moderately plastic: forms threads which will not bend into rings.

Very plastic: forms long threads which will bend into rings.

Soapy. The material is usually sticky and plastic with a distinctly soapy feeling. This is a characteristic

84

of many soils having a high content of exchangeable sodium.

Soft. Dry or moist and very slightly coherent becoming loose with very gentle pressure.

Sticky. The soil material adheres or sticks to other objects. This property only develops in some wet soils and varies in degree from not sticky, through slightly sticky and moderately sticky to very sticky.

Non-sticky: little or no material adheres to finger and thumb.

Slightly sticky: a small amount of material adheres to finger and/or thumb but it is easily removed.

Moderately sticky: some material adheres to both finger and thumb and is drawn out when the pressure is removed.

Very sticky: the material is strongly adherent and is drawn out some distance when the pressure is removed.

Tenacious. Plastic but requires a considerable amount of pressure to mould it or to press it into specific shapes.

Thixotropic. Friable or firm and moist but becomes very moist or wet when continuously manipulated between the fingers. This is a characteristic of many soils containing allophane.

The consistence of most materials alters with a change in moisture status. For example some clay soils may be hard when dry, plastic when moist and sticky when wet. Therefore, it is necessary to describe consistence at more than one moisture level. However, the consistence in the wet state is often the most significant. Cementation is often included as part of consistence but in most cases it can only be determined in the laboratory.

Some horizons are massive and hard and offer a considerable degree of resistance to disruption as a result of cementation by substances such as iron oxides, aluminium oxide and calcium carbonate. Resistance to disruption can result also from physical compaction. Generally, the pressure needed to disrupt a dry soil increases with the content of fine material, sands are usually quite loose while clays form very hard aggregates. Moist sands have a small measure of coherence whereas moist clays are plastic and become very sticky when wet, particularly if the content of montmorillonite is high. The presence of large amounts of humified organic matter in the soil is particularly important for it increases the plasticity of sandy soils but has the reverse effect on a clay by reducing the

stickiness. The consistence of soils of medium texture does not change very much with variations in moisture content. In either the dry or moist state they are usually *friable*, i.e. firm with well formed aggregates that crumble easily when pressure is exerted on them. When wet, these soils tend to become sticky but never to the same extent as clays.

6. MINERAL COMPOSITION

Soil particles are divided initially into two sizes. The limit is normally set at 2 mm which delimits the 'fine earth' from the 'larger separates'.

Larger separates

These include most rock fragments and concretions. Rock fragments will be discussed below while concretions are treated on page 115.

A rock fragment is a unit of any size that contains two or more individual minerals with the same arrangement as in a rock mass. Rock fragments are usually larger than about 2 mm but some rocks such as shales and mud stones break down into smaller fragments.

A knowledge of the nature and properties of the larger particles in soils can lead often to important conclusions about the origin and formation of the parent material and about the soil itself.

The properties of the larger separates that are important in pedological studies are:

Mineralogical composition
Shape of particles
Size and frequency
Orientation
Relationship with the rest of the soil

Mineralogical composition

There is a substantial body of evidence to show that, in some superficial deposits the sand and silt fractions of soil have a similar mineralogical composition to the material >2 mm. Therefore, the composition of the larger separates seen in a hand specimen or in a thin section can be used to assess the chemical composition of soils and parent materials. This is a common practice in many glaciated areas.

Shape of particles

The shape of the stones often gives a clear indication of the processes which have influenced the formation of the parent material and/or the soil itself. Rounded stones are found in alluvium and beach deposits whereas in glacial drift the stones tend to be subangular to subrounded while angular stones are associated with frost action or exfoliation. Where stones or joint blocks are subjected to weathering for long periods they usually become rounded forming core stones (Fig. 1.2) but limestone tends to develop irregular shapes due to differential solution.

Size and frequency

The size of the larger separates varies from material just larger than 2 mm to boulders one or more metres in diameter. In transported materials the size increases with the vigour of the transporting medium. In the case of water the carrying capacity of a river on a constant slope varies with the 3.2 power of the velocity. Therefore, alluvial deposits associated with mountain streams normally have a high porportion of large rounded stones and boulders. Whereas those associated with meandering streams are usually free of large separates.

Glacial deposits are characterised by a large number of subangular stones and boulders of various sizes.

The size and frequency of frost shattered rock fragments are determined by the original nature of the material as well as by the intensity of frost action. Generally, the size decreases as frost action increases, but this is more the case on laminated rocks such as slates and schists than on the crystalline rocks such as granite.

An arbitrary classification of the size of rock fragments is as follows:

Boulders	>10 cm
Large stones	5–10 cm
Small stones	1–5 cm
Gravel	2 mm–1 cm

Orientation

Stones in superficial deposits are orientated in a number of specific patterns, i.e. the long axes of the stones tend to be aligned in a single compass direction and also have a specific angle of dip. In most cases the compass direction and angle of dip can be measured and the data plotted on polar coordinate paper similar to that given in Fig. 4.3. The radial lines are at 10° intervals so are the concentric circles starting with 0° on the outermost increasing to the innermost. The data for each stone are plotted as a dot on the intersection of a radius and a circle. For example, stone A has a direction of 30° and a dip of 20 °NE. Stone B has a direction of 140° and a dip of 60 °SE.

The stones in alluvium usually have an imbricate pattern with their long axes aligned in the direction of

river flow but they dip upstream. Glacial deposits often have stones aligned in the directions of ice flow, a feature that is used in glaciological studies to determine the direction of ice movement. In areas of vigorous frost action the stones become oriented by frost heaving to form a number of patterns at the surface as well as within the soil itself. On a flat site the stones within the soil are vertically oriented whereas on a sloping situation they become oriented parallel to the slope and normal to the contour. This

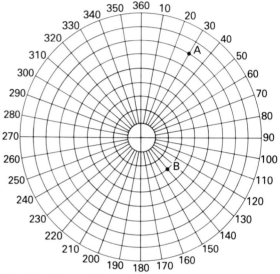

Fig. 4.3 Polar coordinate paper

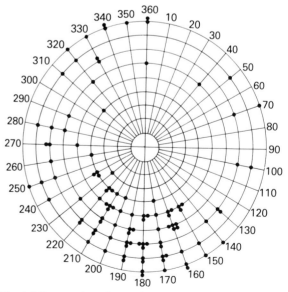

Fig. 4.4 Stones oriented normal to the contour and more or less parallel to the slope which has an angle of 19°

feature is particularly important in interpreting the origin of slope deposits (Fig. 4.4.). A similar type of orientation on slopes is associated with certain forms of soil creep found in many tropical areas.

Relationships with the rest of the soil

Rock fragments often form the nuclei for the development of other phenomena. In horizons of carbonate accumulation it is common to find some rock fragments with pendants of carbonate while others may be completely surrounded by calcite crystals. Another common feature in many lower horizons of soils of cool temperate regions is the occurrence on the top of stones of a thin capping composed mainly of silt (see Figs. 6.46 and 6.47). These are clearly seen in thin sections and are used as evidence of active or relic freezing and thawing.

The fine earth <2 mm

The most important properties of the fine earth include the following:

    Particle size distribution
    Texture
    Shape of particles
    Mineralogy
    Micromorphology

Particle size distribution

Many workers tend to regard particle size distribution and texture as synonymous. While this might be true in a number of cases it is advisable to consider them as separate properties. Confusion arises, however, since the particle size class names are usually the same as texture class names.

    The material <2 mm is divided further into various size fractions, the precise number depends upon the nature of the investigation, but normally either the International scheme or that proposed by the United States Department of Agriculture (USDA) is adopted. The former has four fractions whereas the latter is more detailed and recognises seven fractions. Recently, there has been a tendency to use the scheme introduced by the Massachusetts Institute of Technology. A comparison of the International and the USDA schemes together with that used in the USSR is shown in Fig. 4.5. The relative proportions of each size fraction vary enormously from soil to soil but the 21 classes given below are recognised by USDA.

    Sands
        Coarse sand
        Sand

        Fine sand
        Very fine sand
    Loamy sands
        Loamy coarse sand
        Loamy sand
        Loamy fine sand
        Loamy very fine sand
    Sandy loams
        Coarse sandy loam
        Sandy loam
        Fine sandy loam
        Very fine sandy loam
    Loam
        Silt loam
        Silt
        Sandy clay loam
        Clay loam
        Silty clay loam
        Sandy clay
        Silty clay
        Clay

This is a rather large number of classes having fine shades of difference between adjacent classes with the result that many workers prefer to use fewer broader classes. These 21 classes can be grouped into 12 basic classes and presented in the form of a triangular diagram by using only three size fractions as shown in Fig. 4.6. This is a very convenient method of presenting particle size classes as well as being useful in comparative studies. This scheme can be simplified even further as shown in Fig. 4.7.

    In very detailed studies, the fine earth is divided into 12 or more fractions of which the <0.5 μm (or 0.2 μm) fraction is particularly important in categorising the soil for it is composed of the clay minerals, amorphous hydroxides and microcrystalline oxides.

    The particle size distribution of soil horizons results from fairly specific processes. Soils developed on fairly recent deposits have inherited a considerable proportion of their particles from the parent materials but, as soils increase in age, more and more clay is formed and the texture gradually becomes increasingly fine. However, the translocation of clay particles from one horizon to the other or lateral washing out may cause the horizon losing clay to become coarser and the receiving horizon to become finer.

    As pointed out earlier the main influence of particle size distribution is upon permeability which gradually decreases with decreasing particle size but there are notable exceptions when there are well formed aggregates.

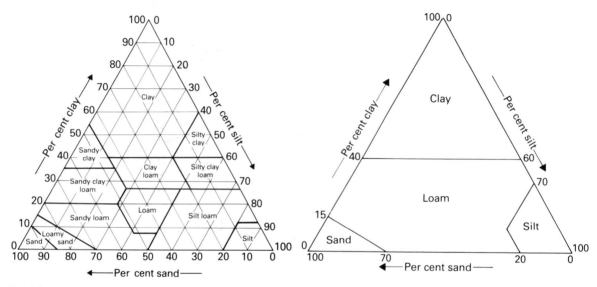

Fig. 4.5 Comparison of three major schemes of particle size distribution in millimetres

Fig. 4.6 Triangular diagram of texture classes according to the USDA (Sand 2 to 0.05 mm, silt 0.05 to 0.002 mm, clay 0.002 mm)

Fig. 4.7 Simplified triangular diagram of particle size classes

Texture

The texture of the soil refers to the 'feel' of the moist soil resulting from the mixture of the constituent mineral particles and organic matter. Therefore, it is an approximate measure of the particle size distribution or mechanical composition which is measured in the laboratory. Whereas particle size data are required for certain studies, texture as determined in the field is more closely related to the behaviour of the soil in the field and to those physical properties of significance in agriculture.

Texture is determined by rubbing the moist soil between the fingers and thumb. It is a subjective technique but can be mastered with some experience. Being a simple technique it has advantages over particle size analysis which is a lengthy laboratory process. Table 4.1 gives the various texture classes and their feel characteristics.

The sand, loamy sand, sandy loam and loam classes are divided into fine, medium and coarse according to the size of the sand fraction as given below:

fine   >50% between 50–250 μm
medium >50% between 250–500 μm
coarse   >50%   >500 μm

Certain variations in the composition of the soil can

cause disagreement between texture and particle size distribution. It is usual for the upper horizons to contain varying amounts of organic matter. When the content is small the effect is minimal but large amounts of organic matter cause the soil to be smooth and to appear to have a higher content of silt.

When the content of organic matter is sufficiently high to influence the texture the term *humose* is added to the name of the texture class.

The type of clay will also have an effect since some clays (e.g. montmorillonite) absorb more water than others (e.g. kaolinite) thus two soils may contain the same amount of clay but the one with montmorillonite will be more sticky and plastic than that with kaolinite.

Shape of particles

The shapes of the particles in the fine earth have an influence upon the consistence of the soil. In addition, the shape of the sand fraction is often a useful guide to the origin of the material. Sand varies in shape from smooth and round to very rough and angular. The smooth round shape is found in wind-blown materials and beach sand; subrounded grains occur in alluvium and beach deposits; rough, angular sand occurs in glacial deposits.

Minerals in a strongly weathering environment develop varied forms, as determined by the nature of the weathering and the crystal structure of the mineral. For example, as apatite of sand size weathers by solution a number of very irregular forms are produced, this contrasts with other minerals such as hypersthene which upon hydrolysis develops serrated edges. Minerals of silt size are either angular, having been formed by physical comminution or irregular as a result of differential decomposition and do not seem to become rounded in the same way as sand.

The shape of some particles is controlled very largely by their crystal structure. The primary micas and chlorite keep their marked platy shape in all size fractions and the amphiboles always tend to be fibrous. Some of the very hard and resistant minerals, such as zircon, which have well formed crystal outlines lose their initial form only after prolonged decomposition and even then the larger grains maintain a considerable part of their original crystalline form. Some characteristic shapes of sand particles are given in Fig. 4.8.

Finally, as has been pointed out already, most of the clay minerals have a platy shape except halloysite which is tubular.

Mineralogy

The fine earth contains both primary and secondary minerals. The former dominate the sand and silt fractions but occasionally within these fractions there may be a large amount of various types of concretions including secondary minerals.

The mineralogical examination of the sand fraction is a very long and tedious process, therefore it is customary to investigate only one size fraction, normally 50 to 100 μm as this has been found to give a fairly accurate picture of the overall sand mineralogy. This fraction usually has large but varying amounts of

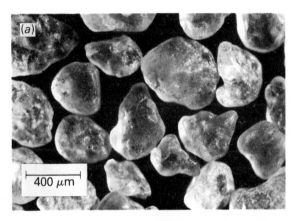

Fig. 4.8 Shape of mineral grains: (*a*) well rounded quartz grains from a sand dune: (*b*) angular and pitted, slightly weathered grains of feldspar

quartz, feldspars and muscovite and in order to obtain an accurate estimate of the minerals of low frequency the fraction is divided into two subfractions, the 'light minerals' and the 'heavy minerals' by specific gravity separation using a liquid (usually bromoform) of sp.gr. $2\,900$ kg m^{-3}. The light fraction which floats is composed mainly of quartz, feldspars and muscovite while

Table 4.1 Key for the assessment of soil texture

Grittiness	Smoothness	Stickiness and plasticity	Ball and thread formation	Texture
Non-gritty to slightly gritty	Not smooth	Extremely sticky and plastic	Extremely cohesive balls and long threads which bend into rings easily	Clay
	Moderately smooth	Very sticky and plastic	Very cohesive balls and long threads which bend into rings	Silty clay
		Moderately sticky and plastic	Moderately cohesive balls, forms threads which will not bend into rings	Silty clay loams
	Extremely smooth	Very slightly sticky and plastic	Moderately cohesive balls, forms threads with difficulty that have broken appearance	Silt
	Very smooth	Slightly sticky and plastic	Moderately cohesive balls, forms threads with great difficulty that have broken appearance	Silt loam
Slightly to moderately gritty	Slightly smooth	Moderately sticky and plastic	Very cohesive balls, forms threads which will bend into rings	Clay loam
Moderately gritty	Not smooth	Very sticky and plastic	Very cohesive balls, forms long threads which bend into rings with difficulty	Sandy clay
	Not smooth	Moderately sticky and plastic	Moderately cohesive balls, forms long threads which bend into rings with difficulty	Sandy clay loam
	Slightly smooth	Slightly sticky and plastic	Moderately cohesive balls, forms threads with great difficulty	Loam
Very gritty	Not smooth	Not sticky or plastic	Slightly cohesive balls, does not form threads	Sandy loam
Extremely gritty	Not smooth	Not sticky or plastic	Slightly cohesive balls, does not form threads	Loamy sand
		Not sticky or plastic	Non-cohesive balls which collapse easily	Sand

the heavy fraction contains a wide range of ferromagnesian and accessory minerals and is often the important fraction in categorising soils and parent materials.

The clay fraction is composed predominantly of crystalline clay minerals and amorphous materials except for a small amount of primary minerals mainly quartz that occur in the 1 to 2 μm size range. The clay fraction is studied by X-ray and D.T.A. (Differential Thermal Analysis) methods which give data that are usually more important than those of the sand mineralogy since the amount and type of the different clay minerals allow important conclusions to be made about the processes taking place in soils.

Micromorphology

The definition of the fine earth, i.e. material <2 mm is very similar to the definition of the s-matrix of Brewer (1964) but the term s-matrix has been avoided in order to integrate micromorphology into the whole of soil science and also to use the term matrix in the more conventional and accepted sense as fine material enclosing coarse material. In the case of soils, the term matrix refers generally to the clay and clay humus mixtures.

The characterisation of the fine earth in thin sections has received a considerable amount of attention and a number of classifications have been suggested particularly those of Brewer (1964) and Stoops and Jongerius (1975) but these systems tend to divorce micromorphology from the rest of soil science.

The wide range in particle size distribution coupled with the equally wide range in mineralogy and other properties have given rise to a very wide range in the type and organisation of the fine earth both within itself and with regard to the larger separates. In most cases the fine earth is aggregated to form peds of a variety of shapes as discussed below under structure.

Within the fine earth the two components to be considered are the fine material (approximately <5 μm) and detrital grains both of which, like structure, are considered below under major headings.

7. COLOUR

The colour of the soil is usually determined by the nature of the fine material. The general characteristics of colour are given below while those for the fine material specifically are given separately because they have other properties in addition to colour.

Generally, the colour of a soil is determined by the amount and state of the iron and/or the organic matter. Haematite (Fe_2O_3) is responsible for the red colour in many soils developed under strongly aerobic conditions in tropical and subtropical areas. This substance usually occurs as very small particles and needs to be present only in fairly small amounts to impart intense coloration. However, the mineral responsible for most of the inorganic coloration in aerobic soils is goethite ($FeO-OH$) which has colours that range from reddish-brown to yellow as its hydration increases; the highly hydrated yellow and yellowish-brown forms are sometimes referred to as limonite.

Many grey, olive and blue colours occur in the soils of partially or completely anaerobic situations and originate through the presence of iron in the reduced or ferrous state. Sometimes, substances with blue colours such as vivianite may form under these conditions and therefore contribute to the coloration.

The colour of the upper horizons usually changes from brown to dark brown to black as the organic matter content increases and there is a tendency for the organic matter to become darker in colour with increasing humification. Also the pH and cation status can be influenced, for acid horizons low in calcium and organic matter are often pale in colour, whereas in the presence of large amounts of calcium or sodium dark colours will form even with small amounts of organic matter. Dark colours are produced also through the presence of manganese dioxide, or by the presence of elemental carbon following burning.

Pale grey and white originate through the lack of alteration of grey or white parent materials, depositions of calcium carbonate, efflorescences of salts or as a result of the removal of iron leaving significant amounts of light coloured minerals such as quartz, feldspars and kaolinite. Each horizon usually has its own specific and fairly narrow range of colours Furthermore, most horizons exhibit some variability in colour within the same pedo-unit but this is usually ignored when small. On the other hand the marked colour pattern of some horizons are distinguishing characteristics.

Colour patterns

These may be spotted, streaked, speckled, marbled, or mottled and blotched.

Spotted: individual spots of one colour in a uniform matrix.

Streaked: elongate areas of one colour in a uniform matrix. The streaks may be in any direction.

Speckled: small particles of organic and/or mineral material fairly uniformly distributed in a self-coloured matrix.

Marbled: there are distinct differences in colour but they grade gradually from one into the other in an irregular manner.

Mottled and blotched: these are irregular patterns of two or more colours with sharp boundaries between the different areas. Mottling is used for patterns with mainly grey, olive or blue areas, and lesser yellow, brown or red areas. Such patterns are interpreted as resulting from seasonal wetting and drying of the horizon. The duration of the dry period is indicated by the yellow and brown areas which become progressively larger and more frequent as aeration increases. In these cases the yellow colour is often due to lepidocrocite formation or may be due to the oxidation of pyrite to jarosite. Blotching is used for a similar pattern but the colours are yellow, brown, red or black and are caused by processes other than wetness.

The colour patterns have contrast, sharpness of boundaries, frequency, size and shape of mottles and blotches.

Colour contrast

Faint: difference recognisable only with close examination.

Distinct: difference readily seen.

91

Prominent: differences are conspicuous and the dominant feature of the horizon.

Sharpness of boundaries

The following scale may be used:
Sharp: change in <1 mm.
Clear: change in 1–2 mm.
Diffuse: change in >2 mm.

In describing soil colour the relationships between the colour pattern and other properties should be noted. For example the outside of peds may differ from the inside.

Although the colours of most horizons are produced by pedogenic processes there are a number of instances when they are inherited from the parent material. For example, many sediments of Devonian and Triassic age are bright red owing to the presence of haematite. Unless pedogenic processes are very vigorous, soils developed from such parent materials may retain the parent material colour for thousands of years through many contrasting phases of pedogenesis.

A very high proportion of horizon and soil names are based upon colour since it is usually the most conspicuous property and sometimes the only one that is easily remembered. For example, class names such as Chernozem and Podzol are derived from the colour of a particular horizon. Such a system of nomenclature is not entirely satisfactory and now other criteria are used to establish soil names but some terms are so well entrenched in the literature that it would be impractical to introduce replacements. It should be pointed out that many inferences that are made about soils are based upon colour, since there appears in many cases to be a correlation between it and other properties.

Many soils are darker when moist or wet and in order to attain a measure of uniformity there is a tendency to report in the literature the colours of moist soils. In the past many bizarre terms were used to describe soil colours, but with the introduction of the 'Munsell Soil Color Charts' a high degree of standardisation has been achieved and has allowed colour to become a more useful criterion in soil classification. For the purpose of soil classification the value/chroma or V/C rating (Northcote, 1971) is often used according to the grouping shown in Fig. 4.9.

8. FINE MATERIAL

As stated above the fine material is usually responsible for the colour of the soil; where it ranges from coatings around grains to a complete matrix enclosing

grains and pores. Thus the term matrix is used in a general manner for the fine material enclosing or surrounding coarser material. This approach has been adopted because of the great difficulty in producing a set of terms that truly represent the heterogeneity of soils and at the same time satisfy all micromorphologists. Usually the matrix is composed of clay or is the clay–humus mixture and includes the material <5 μm but some soils may have a matrix dominated by silt or even fine sand. Since clay and clay–humus matrices are the most common and diverse they will be discussed in detail.

The clay and clay–humus matrices when present in large amounts appear in plane polarised light as a continuous phase surrounding or enclosing large grains, pores and other features. However, all of the material

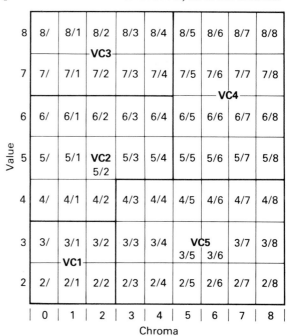

Fig. 4.9 Value/Chroma rating of Northcote

<5 μm is not always present as matrix for in some cases the finer material occurs as separate phenomena such as coatings. Except for a few cases the clay and clay–humus matrix is responsible for the colour of the soil both in the field and in thin sections. As the content and size of the detrital grains increase there is a tendency for the fine material to be arranged around the grains or to form bridges between them. However, there are some very sandy soils that are very compact with the fine material occurring in the small spaces between the grains. The fine material is very mobile and responds readily to expansion and

contraction or it may be translocated to form coatings on surfaces as discussed below. The properties of clay and clay–humus matrices include the following:

Matrix colours
Matrix anisotropism and isotropism
Matrix patterns

Matrix colours

The colour of an individual matrix is very variable and as stated on page 91 it usually differs from that of soil examined in the field. The colour pattern varies from uniform to very varied and complex.

Matrices with uniform colours

Opaque matrices
Red matrices
Reddish-brown, reddish-yellow and brownish-yellow matrices
Dark brown and brown matrices
Yellow to pale brown matrices
Grey, greyish-blue, olive-grey and olive matrices
Green matrices

Opaque matrices

These are due most commonly to the presence of large amounts of haematite or manganese dioxide, or combinations of these substances. They occur predominantly in 'laterite' and other cemented and hardened horizons in tropical and subtropical areas. These matrices are usually so dense that only sections <15 μm sometimes show transparence and in reflected light they vary from deep red to black due to haematite and manganese dioxide respectively.

Poor impregnation is a common cause for the lack of transparence and patchy opaqueness in slides, therefore one has to guard against misinterpreting such occurrences.

Red matrices

These are relatively rare in spite of the widespread occurrence of red soils which usually appear yellowish-red in thin section. Where a red matrix is found it seems to be due to finely divided haematite.

Reddish-brown, reddish-yellow and brownish-yellow matrices

The principal colouring substance occurring in soils is probably goethite whose colour depends upon its degree of hydration. Non-hydrated goethite imparts red colours to soils in the field but usually appears reddish-brown in thin section, on the other hand fully hydrated goethite is brownish-yellow, both in the field and in thin sections. Between these two extremes there is the complete range of colours.

In some cases a brownish-yellow matrix is due to lepidocrocite, particularly in the mottles in soils developed in an environment with fluctuating anaerobic and aerobic conditions.

Reddish-brown, reddish-yellow and brownish-yellow matrices are common in tropical and subtropical areas or in soils formed from sediments with these colours.

Dark brown and brown matrices

Ferric hydroxide and/or colloidal organic matter give brown colours to matrices and are common in a number of soils which have incorporated organic matter or are in a relatively early stage of weathering. Both of the substances are amorphous and cause the matrix to be isotropic when they are present in large amounts, singly or together.

Yellow to pale brown matrices

These matrices are associated with low concentrations of colouring substances especially ferric iron or they may be due to goethite in a highly hydrated state. They are characteristic of many soils that are moist for long periods of the year.

Grey, greyish-blue, olive-grey and olive matrices

Ferrous compounds are responsible for many of the matrices with this range of colours but the precise substances have not been identified. They are common in horizons that are temporarily or permanently saturated with water.

Very low contents of iron may cause the matrix to be pale in colour, this may be due to an initial low amount of iron or to its removal or segregation within the horizon.

Green matrices

These are quite exceptional but they are known to occur in certain soils in Sarawak. Their origin is uncertain but may be due to high concentrations of an element such as chromium.

Matrices with colour patterns

Often the matrix colour shows some variation which though not conspicuous in hand specimens may be seen in thin sections. These patterns may result from local concentration or differential removal of material.

The characteristic yellow and red matrix in the mottled horizon of tropical areas is possibly due to segregation and local concentration of iron. The ochreous and grey pattern in horizons with reducing conditions is often the result of local concentrations of lepidocrocite around passages and pore space.

It appears that certain red soils may change to reddish-brown if the climate becomes wetter. The matrix usually does not change uniformly but passes through a mottled or marbled phase showing both types of colour. Similarly, when soils develop from red sediments there may be a colour change to reddish-brown, reddish-yellow which in turn may change to grey or olive depending upon the degree of wetness. In these cases mottled patterns may develop during the change from one colour to the other.

Matrix anisotropism and isotropism

When transparent clayey matrices are examined in crossed polarised light they vary from those which are completely black (optically isotropic) to those which show different patterns of interference colours (optically anisotropic) throughout the whole section. Williamson (1947) has shown that if clay particles are allowed to settle on to a glass slide they form laminations, and sections cut normal to the slide show interference colours and therefore are optically anisotropic. It is considered, therefore, that the interference colours (anisotropism) of the matrix develop when the fine particles of laminar minerals become closely packed like the pages of a book. On the other hand an isotropic matrix contains a large amount of amorphous or poorly crystalline material. In some cases it appears that an optically isotropic matrix may be composed of crystalline material but the crystals are small and randomly orientated. Thus kaolinite which is an anisotropic mineral may form an isotropic matrix because the individual particles are randomly orientated. Thus optical anisotropy of the matrix refers to the organisation of the material within the matrix and not to the anisotropism of the individual particles themselves. By definition all layered substances such as soil pedo-units, mica and clay coatings are anisotropic since they are not the same in all directions. In addition, most substances such as micas and clay coatings are also optically anisotropic, i.e. they show interference colours in crossed polarised light. However, there are some materials which appear to be layered in plane transmitted light but do not show interference colours in crossed polarised light. Thus they are anisotropic in plane transmitted light but are optically isotropic in crossed polarised light. Some coatings of gibbsite are excellent examples of this phenomenon.

The areas with interference colours within the matrix vary in amount and degree of distinctness of clarity of their boundaries.

The amount of anisotropism in the fine material varies from none (isotropic) to dominant and can be described according to the following scale:

Term	Symbol	Range in %
Dominant anisotropic areas	D	>50
Very abundant anisotropic areas	VA	25−50
Abundant anisotropic areas	A	15−25
Very frequent anisotropic areas	VF	10−15
Frequent anisotropic areas	F	5−10
Occasional anisotropic areas	O	2−5
Rare anisotropic areas	R	0.5−2
Very rare anisotropic areas	VR	<0.5

The percentage of areas with interference colours in the matrix may vary when the stage is rotated, therefore it may be necessary to give some indication of the range of variation. However, this is obviated by using circular polarised light.

The distinctness of the optically anisotropic areas varies from those that have clear or marked boundaries with distinct extinction to those that are difficult to distinguish and have an undulose extinction.

Areas with distinct extinction

There are two separate phenomena with clearly visible interference colours and distinct extinction. These are:
Domains
Lines

Domains

These are discrete optically anisotropic areas that seem to be small groups of closely packed and similarly aligned clay particles. They are visible in crossed polarised light but not normally seen in plane transmitted light because they have the same refractive index as the surrounding material and therefore do not show any relief. Domains have clear to diffuse boundaries and colours of the first order − properties which serve to differentiate them from mica flakes with which they may be confused. However, differentiating between domains and small flakes of mica is often difficult, particularly if the mica is somewhat weathered and has paler interference colours. Then the relief has to be used as the differentiating criterion.

Domains vary in the following properties:
Interference colours
Size and frequency
Distribution and orientation patterns

Interference colours of domains. The exact interference colours of domains vary from grey to orange and red of the first order. This variation seems to result from variations in the type of clay minerals as well as the amount of iron present. When kaolinite is dominant the colours are grey to white becoming yellow or red when there is much iron present. Strong yellow and reds of the first order are seen when there are 2:1 clay minerals present.

Size and frequency of domains.   Domains vary considerably in length and width and can be classified on their size characteristics as set out in Table 4.2. The

Table 4.2   Classification of domains

Length		Width	
Type	Length (μm)	Type	Width (μm)
Very short	<10 μm	Very narrow	<1 μm
Short	10–50 μm	Narrow	1–5 μm
Long	50–100 μm	Broad	5–10 μm
Very long	>100 μm	Very broad	<10 μm

size and frequency of domains vary considerably from soil to soil depending upon the type of clay and the moisture status of the soil. In some strongly weathered aerobic soils (Ferralsols) the domains are usually very short and narrow and sometimes can only be seen in sections about 5 μm thick. In moist soils (Gleysols) the domains are long and broad when the clay is mainly kaolinite becoming very large when mica and vermiculite are present. In soils dominated by montmorillonite (Vertisols) they tend to be short and narrow occurring singly or in thin zones.

Distribution and orientation of domains.   Domains have a number of patterns of arrangement. They usually occur singly and randomly oriented in body of the matrix away from rock fragments, detrital grains and pores. They can also have a general overall orientation or be arranged in zones or bands. The zones may have a random orientation but sometimes they may have normal or oblique orientation to each other.

Domains sometimes have an orientation that is related to some other property of the soil. The three main related orientations are with pores, detrital grains and rock fragments. In the case of discrete pores and detrital grains domains are often aligned tangentially to their surfaces forming partial or complete aureoles. When the pores are planar the domains may be aligned parallel to one or both surfaces thus forming single or double domain surfaces.

In fine textured sediments unaffected by pedogenesis domains are horizontally orientated as a result of the gradual settling of the material but this orientation becomes deranged when pedological processes begin to operate. Some arrangements of domains are given in Figs. 4.10 to 4.15.

Fig. 4.10 Abundant domains, many surrounding pores and sand grains

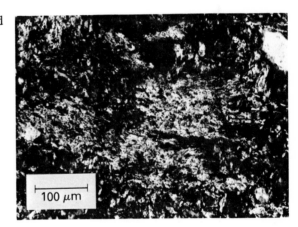

Fig. 4.11 Clusters of domains

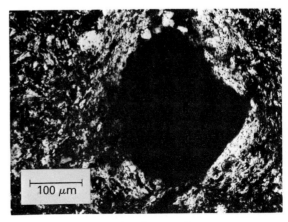

Fig. 4.12 An aureole of domains around a pore

95

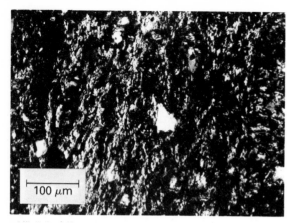

Fig. 4.13 Almost vertical zones of domains

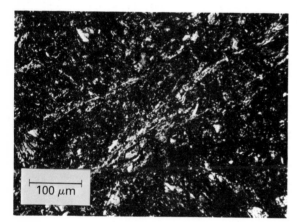

Fig. 4.14 Anisotropic lines

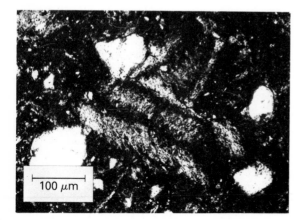

Fig. 4.15 Reorganisation of domains by soil fauna

Formation of domains. The exact mechanism for the formation of domains is obscure but most probably stresses resulting from expansion and contraction are responsible. Furthermore, it is not certain whether domains are fairly permanent features or whether they form, disintegrate and reform regularly. However, it does appear that their alignment is very easily altered when the soil is disturbed since those at the sides of specimens tend to be orientated parallel to the sides as a result of disturbance and pressure exerted during the collection of the sample. In soils with vigorous faunal activity the domains are usually smaller and less frequent in the faecal material than in the surrounding soil. In a number of cases domains can form from clay coatings (see p. 107).

Lines

Thin optically anisotropic lines occur on ped surfaces as well as within the matrix. Generally they are oblique and are regarded as a thin layer of pressure orientated clay particles. In some cases they appear also to be formed by the orientation of domains. These lines seem to be the thin section manifestation of slickensides when they occur on ped surfaces. When they occur within the matrix they are also regarded as slickensides but the two surfaces have not separated.

Thus from the above it is seen that the matrix can become organised in a number of different ways.

Areas with undulose extinction

In many cases these are more apparent than real since they may be seen in sections of normal thickness but in very thin sections areas with distinct extinction are seen. In some red soils containing much ferric oxide there is a tendency for the optical anisotropism to be masked by the ferric hydroxide but again it is visible in very thin sections.

In some highly weathered pale coloured soils in which the matrix is composed largely of kaolinite, undulose extinction is present. This would seem to be due to the low first order interference colours of kaolinite which do not allow any marked contrast to develop.

The areas with undulose extinction are regarded as having domains but either their orientation changes gradually or they have very diffuse edges and grade gradually from one into the other. Such situations sometimes give a reticulate anisotropic pattern.

Matrix patterns

The matrices of soils do not appear to be static but are constantly changing at various rates through physical, chemical or biological processes and, in addition, within them there may occur a number of phenomena which appear to have formed *in situ* by diffusion of material through the matrix itself. Good examples of this are crystals of pyrite, gypsum and dendritic manganese dioxide.

Frequently fauna ingest the matrix and redistribute it within the soil in their faecal material. Other physical disturbances such as expansion and contraction exert stresses and pressures causing deformation and realignment of domains or formation of a flow pattern or disruption and incorporation of other phenomena such as coatings and may exfoliate and even break flakes of mica. Sometimes movement and disturbance of the matrix can be inferred from the distribution pattern of the sand grains particularly biotite which may become orientated in the direction of movement.

9. DETRITAL GRAINS (MICRO-MORPHOLOGY)

A detrital grain is a unit composed of a single mineral grain of any size and derived from the parent material. They are usually <2 mm in size but they can be larger.

The nature of the organisation of the detrital grains is very variable. In most soils they occur separate one from the other and embedded in the matrix and are described in terms of their size, shape, composition, frequency and other properties. Where there is a small amount of fine material the grains may be separated one from the other and form a single grain structure or they may have bridges of fine material between them. Another type of fabric is composed of closely packed grains with a small amount of matrix. This is a characteristic feature of many soils having a high content of fine grains such as loess and is most marked in soils containing a high proportion of small calcite grains. In addition, detrital grains may have a specific distribution pattern such as occurring in bands, clusters, or related to some other feature. A random orientation pattern is striking when there is a high proportion of mica present. This is seen as a false diamond type of arrangement in cross polarised light since it is at the 45° position that mica flakes show their maximum interference colours (Fig. 4.16). The arrangement of the grains may be related also to a weathering pattern, for it is found that in some horizons formed by a strong weathering the quartz grains may retain the arrangement present in the original rock (Fig. 6.7).

The ease with which the detrital grains can be identified depends largely upon their size. Grains larger than about 100 μm in diameter (fine sand) can be identified fairly easily when they occupy the entire thickness of the specimen. Smaller grains, those <25 μm (fine silt and clay) are particularly difficult to identify with certainty because they may be completely surrounded by the matrix which may obscure some of their properties.

The shapes of detrital grains are often an indication of the processes taking place in the soil. This is discussed on page 89.

Fig. 4.16 Random orientation of mica flakes which show their maximum interference colours in the 45° position hence the apparent lattice-like arrangement

10. STRUCTURE AND POROSITY

These refer to the spatial distribution and total organisation of the soil system as expressed by the degree and type of aggregation and the nature and distribution of pores and pore spaces.

In most soils the individual particles do not exist as discrete entities but are grouped into aggregates or peds with fairly distinctive shapes and sizes. This is found most often in horizons of medium texture. Soils that are composed predominantly of coarse sand are loose and without well formed aggregates, while those composed mainly of clay tend to be coherent or massive and to form fewer large aggregates. It is found that in soils with well formed aggregates or when there is mainly sand the individual units may have only a few points of contact but generally are surrounded by a continuous pore phase. On the other hand in a coherent or massive soil it is the mineral material that forms the continuum but it may and

97

often does contain discrete pores. Thus in some instances there is a continuous phase of pore space whereas in others the soil material forms the continuum Frequently, an intermediate situation is encountered when both the pores and the soil material form continuous phases by having interlocking systems.

When soils are manipulated in the hand the shape and size of the peds clearly become evident if they are present. When they are absent the large lump of soil is referred to as a clod and can be broken into smaller and smaller fragments as more and more pressure is exerted. Therefore, it is necessary to exercise great care when examining the soil for structure in order to differentiate very exactly between clods, peds and fragments.

The degree and type of aggregation determine aeration and permeability and, therefore, the infiltration capacity and moisture movement. They often determine the volume of the pores and pore-space and therefore the volume of the soil atmosphere. This is usually saturated with moisture in a humid climate and has the following composition: 79—80 per cent nitrogen, 15—20 per cent oxygen and 0.25—5 per cent carbon dioxide. The above ground atmosphere differs by having only 0.03 per cent carbon dioxide.

Structure also influences the erosive potential since the presence of surface horizons with a massive structure reduces infiltration, which increases run-off thereby increasing the erosion hazard.

There are at present several good systems for classifying aggregates but there is no general agreement about the classification of pores. However, the system recently presented by Brewer (1964) has merit. Nevertheless, in some respects it seems superfluous to have two systems of classification — one for aggregates and the other for pores — since they are largely interdependent or the complement of each other, except when discrete pores occur within the aggregates. It seems more appropriate to have one system based on the type of aggregates since aggregation is the more common type of arrangement, and to describe discrete pores in terms of their size and shape.

Most systems of structure classification used in pedology are based mainly on field characteristics and consequently are somewhat crude.

A fuller understanding of the undisturbed structure can be obtained by impregnating the soil with a resin containing a fluorescent dye. The impregnated block is cut open to expose the undisturbed central part which is studied or photographed using ultraviolet light. This latter technique gives photographs in which the black areas are soil and the light areas are pores and pore space. Furthermore, it is possible to study

these specimens at many levels of magnification thus obtaining a detailed insight into the architecture of the soil.

By this method it has been found that much of the aggregation observed in a hand specimen in the field is caused by disturbance. Many soils that appear to have aggregates are seen to be poorly aggregated when impregnated and examined under UV light. Therefore, it is necessary to differentiate between those soils that have discrete aggregates and those in which the aggregates are poorly formed or partially joined to each other. When the aggregates are separate the structure is termed *complete* and when the aggregates are partially joined the structure is *incomplete*.

Alternatively, structure can be studied in thin sections. By using large thin sections it is possible to examine the complete range of structures for most soils, starting with the unaided eye followed by examination under the microscope at various magnifications. The structural forms seen under the microscope are similar in shape to those seen with the unaided eye and in some cases the macro-peds are the only structural units present. Therefore, the terms given below can be used for the descriptions of thin sections. However, a conspicuous additional feature of peds and massive soils is the presence of discrete pores which are described in terms of their frequency, size, shape, etc. There are a number of recurring shapes of discrete pores such as small circular or ovoid pores which are attributed to the formation of gas bubbles. Figures 4.18 and 4.31 illustrate a number of different types of intrapedal pores. By using thin sections, fluorescing blocks and conventional field methods the seventeen types of structure recognised are described below.

In a number of cases the structure of an horizon may be compound or composite. Compound structure usually occurs in soils with prismatic or columnar structure in which the large peds are made up of incomplete smaller peds. Composite structure occurs when two or more contrasting processes are operating simultaneously in the same horizon. This is common when the soil has a fairly vigorous fauna such as termites and worms resulting in areas of granular faecal material ramifying through areas with other forms of aggregates (Fig. 4.23).

Pores

Pores are that part of the soil occupied by soil atmosphere or water. In thin sections they are occupied by resin. Pores are either discrete or form a continuous phase through which movement may take place. In some cases the formation of pores can be regarded as the first stage in the formation of peds; this is seen in those cases where pores are formed by shrinkage and cracking. In this mechanism of pore formation one, two or three cracks start at the same point and radiate outwards forming an angle of 120° between each radial limb; such three-limbed pores are common in soils of tropical and subtropical areas. When shrinkage is marked each limb will bifurcate and connect with limbs from other three-limbed pores forming complete or incomplete peds and continuous pores.

In other cases pores are discrete with spherical or oval shapes and formed by gas bubbles released from the soil solution during drying or freezing. Pores may also be formed by the solution of material or the disappearance of ice. On the other hand there are a number of irregular forms of unknown origin.

Types of structure

The various types of structure set out alphabetically are described below.

Alveolar (Fig. 4.17)
Peds: Absent to rare — continuous soils phase.
Pores: Abundant circular, ovoid or irregular pores form a honeycomb type of arrangement with a continuous soil phase. This type of structure is most easily identified in thin sections and may go unrecognised in the field.
Occurrence: Upper horizons of desert soils, many bleached horizons and middle horizons of medium texture.
Genesis: Release of gasses during wetting and drying or freezing and thawing.

Angular blocky (Fig. 4.18)
Peds: Complete with flat, concave and/or convex faces, sharp angular corners and strongly accordant surfaces.
Pores: Continuous between peds, few to many discrete pores within peds.
Size:

Very fine angular blocky	<5 mm
Fine angular blocky	5–10 mm
Medium angular blocky	10–20 mm
Coarse angular blocky	20–50 mm
Very coarse angular blocky	>50 mm

Occurrence: Some middle and lower horizons of medium to fine texture.
Genesis: Wetting and drying or freezing and thawing.

Bridge (Fig. 6.40)
Peds: Absent to rare with the fine material coating and forming bridges between sand grains.
Pores: Irregular pores separated by fine material linking single grains. This type of structure is most easily identified in thin sections and may go unrecognised in the field. It may be mistakenly called single grain in the field.
Occurrence: In some very sandy horizons.
Genesis: Deposition and/or concentration of fine material around sand grains.

Columnar (Fig. 4.19)
Peds: Large, incomplete vertically elongate with domed upper surfaces, the peds are continuous with the underlying material. Flat, vertical strongly accordant surfaces. The individual peds are usually larger than even the largest thin section, therefore only a portion of each ped or two adjoining peds can be studied at one time.
Pores: Continuous at the surface and sides of peds, occasional to frequent discrete pore within peds.
Size:

Very fine columnar	<10 mm
Fine columnar	10–20 mm
Medium columnar	20–50 mm
Coarse columnar	50–100 mm
Very coarse columnar	>100 mm

Occurrence: Middle horizon of many Solonetz and Solodic Planosols.
Genesis: Expansion and contraction in many highly alkaline soils.

Composite (Fig. 4.20)
Various combinations of many of the other types.

Compound (Fig. 4.32)
Large peds such as prisms or columns that are themselves composed of smaller incomplete peds.

Crumb (Fig. 4.21)
Peds: Complete irregular shapes with rough surfaces that are not accordant — they occur separately or in clusters.
Pores: Abundant and continuous around peds, few to many small discrete pores within peds sometimes imparting a spongy appearance.
Size:

Very small crumb	<1 mm
Small crumb	<2 mm
Medium crumb	2–5 mm

Occurrence: Upper horizons beneath grass.
Genesis: Loose binding of particles by fine roots, organic matter or clay.

Granular (Fig. 4.22)

Peds: Complete subspherical peds with smooth or
re-entrant surfaces that are not accordant. They
may occur singly or in clusters.

Pores: Usually frequent to abundant and continuous
around peds, discrete pores, occasional to absent
within peds.

Size: Very small granular may not be easily recognised
in the field but is very distinctive in thin sections.

Very fine granular	<1 mm
Fine granular	1–2 mm
Medium granular	2–5 mm
Coarse granular	5–10 mm
Very coarse granular	>10 mm

Occurrence: In some upper and middle horizons.
Very well developed in the upper horizons of some
Vertisols and the middle horizons of some Podzols.

Genesis: (1) Faecal pellets of some organisms,
particularly small arthropods.
(2) Wetting and drying of upper horizons with
expanding lattice clays (self-mulching process).
(3) Precipitation and flocculation of sesquioxides.

Irregular blocky

Peds: Complete with irregular shapes and flat concave
and/or convex faces, mixed rounded and angular
corners and re-entrant surfaces. Strongly accordant
surfaces, sometimes there is a tendency to be
incomplete.

Pores: Usually frequent and continuous around peds,
few to many discrete pores within peds.

Very fine irregular blocky	<5 mm
Fine irregular blocky	5–10 mm
Medium irregular blocky	10–20 mm
Coarse irregular blocky	20–50 mm
Very coarse irregular blocky	>50 mm

Occurrence: Middle horizons of many fine textured
soils.

Genesis: Wetting and drying.

Labyrinthine (Fig. 4.23)

Peds: Abundant granular.

Pores: Abundant intertwining faunal passages containing varying amounts of granular material,
discrete pores rare to absent.

Occurrence: Upper or middle horizons particularly in
tropical and subtropical countries.

Genesis: Produced by the vigorous activity of arthropods, particularly termites.

Laminar

Peds: Horizontally elongate with flat parallel surfaces
incomplete.

Pores: Horizontally continuous and usually frequent
to occasional, discrete pores, occasional to absent.

Very fine laminar	<1 mm
Fine laminar	1–2 mm
Medium laminar	2–5 mm
Coarse laminar	5–10 mm
Very coarse laminar	>10 mm

Occurrence: Some upper horizons and in some
sedimentary materials.

Genesis: Compaction by implements in upper
horizons or sedimentary deposition.

Lenticular (Fig. 4.24)

Peds: Lens-shaped with convex surfaces and overlap-
ping arrangement. Surfaces are relatively smooth
and strongly accordant, often with finer textured
upper surfaces.

Pores: Usually occasional to frequent and continuous
around peds, discrete pores occasional to absent
within peds.

Very fine lenticular	<1 mm
Fine lenticular	1–2 mm
Medium lenticular	2–5 mm
Coarse lenticular	5–10 mm
Very coarse lenticular	>10 mm

Occurrence: Some upper, middle or lower horizons
with loamy textures.

Genesis: Freezing and thawing, compaction by
implements.

Massive or coherent (Fig. 4.25)

Peds: Absent to rare, continuous soil phase.

Pores: Continuous pores few to absent, occasional
discrete pores with various shapes, often with three
limb intersection.

Occurrence: Upper, middle and lower horizons of
some fine textured soils.

Genesis: Destruction of aggregates in the upper
horizon, usually by poor cultivation. Inhibition of
aggregate formation in lower horizons due to the
weight of the overburden.

Prismatic (Fig. 4.26)

Peds: Incomplete vertically elongate with irregular
upper surface and continuous with the underlying
material. Flat vertical strongly accordant surfaces.
The individual peds are usually larger than even the
largest thin section, therefore, only a portion of
each ped or two adjoining peds can be studied in
thin sections at one time.

Pores: Continuous at the surface and sides of peds,
occasional to frequent discrete pores within peds.

Size: Very fine prismatic   <10 mm
      Fine prismatic   10–20 mm
      Medium prismatic   20–50 mm
      Coarse prismatic   50–100 mm
      Very coarse prismatic   >100 mm

Occurrence: Middle and lower horizons of some fine textured soils sometimes extend to the surface.

Genesis: Wetting and drying.

Single grain (Fig. 4.27)

Peds: No aggregates – individual detrital grains and small rock fragments.

Pores: Continuous and usually abundant, no discrete pores.

Occurrence: Coarse sandy soils and some strongly leached horizons.

Genesis: (1) Inherited from the parent material.
      (2) Removal of fine particles by vertical leaching or lateral differential erosion.

Spongy (Fig. 4.28)

A tangled sponge-like mass of organic matter that usually occurs at the surface.

Subangular block

Peds: Complete with flat concave and/or convex faces rounded corners and strongly accordant surfaces.

Pores: Continuous between peds, few to many discrete pores within peds.

Size: Very fine subangular blocky   <5 mm
      Fine subangular blocky   5–10 mm
      Medium subangular blocky   10–20 mm
      Coarse subangular blocky   20–50 mm
      Very coarse subangular blocky >50 mm

Occurrence: Some middle and lower horizons of medium to fine texture.

Genesis: Wetting and drying.

Subcuboidal (Fig. 4.29)

Peds: Complete with square or rectangular outline, angular corners and flat strongly accordant surfaces.

Pores: Continuous between peds, occasional to frequent discrete pores within peds.

Size: Very fine subcuboidal   <5 mm
      Fine subcuboidal   5–10 mm
      Medium subcuboidal   10–20 mm
      Coarse subcuboidal   20–50 mm
      Very coarse subcuboidal   >50 mm

Occurrence: Lower horizon of some fine texture soils

Genesis: Freezing and thawing.

Vermicular (Fig. 4.30)

Peds: Vermiform (worm-like) aggregates or forms with a characteristic U in U structure.

Pores: Abundant intertwining faunal passages containing varying amounts of faecal material, some of the passages may be completely filled and recognised only by the characteristic vermiform areas or aggregates. The vermiform areas are regarded as aggregates that have coalesced with the soil. In some cases they are very old and are recognised by the orientation of domains or have become impregnated with iron compounds. It is sometimes difficult to recognise this structure in the field. (Fig. 4.15).

Occurrence: Upper middle or lower horizons with vigorous earthworm activity particularly in many Chernozems of Europe and Ferralsols of Australia.

Genesis: Produced by earthworms.

Wedge (Fig. 4.31)

Peds: Absent to rare.

Pores: Continuous pores that are mainly linear or sinuous and intersecting at 45–60° giving wedge-shaped soil areas with pores on two or three sides. The soil surfaces are strongly accordant, occasional to frequent linear or sinuous discrete pores.

Size: Very fine wedge   <5 mm
      Fine wedge   5–10 mm
      Medium wedge   10–20 mm
      Coarse wedge   20–50 mm
      Very coarse wedge   >50 mm

Occurrence: Fine textured middle horizons containing a high content of expanding lattice clays, very common in Vertisols.

Genesis: Expansion and contraction in response to wetting and drying. As the soil absorbs water and expands high pressures are created which are released as one portion of soil slips over the other. The slip planes form slickensides and upon drying the soil separates along many of the slip planes giving the characteristic pore pattern. In crossed polarised light the slickensides have a thin anisotropic layer caused by the orientation of clay particles or domains as a result of the movement of one soil mass over the other.

Depending upon the shape of the aggregates, opposing faces of adjacent aggregates are parallel to a greater or lesser extent. This property is referred to as the *degree of accordance* and can be classified as follows:

Degree of accordance

Non-accordant: <5 per cent of the opposing surfaces have a similar outline.

Weakly accordant: 5–25 per cent of the opposing surfaces have a similar outline.

Moderately accordant: 25–50 per cent of the opposing surfaces have a similar outline.

101

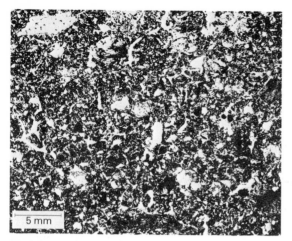

Fig. 4.17 Thin section of alvelolar structure

Fig. 4.19 Columnar structure – the width of the structure is 1 m

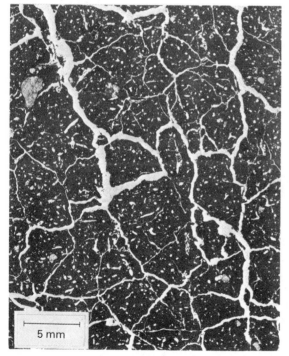

Fig. 4.18 Thin section of angular blocky structure with a high content of ovoid and circular pores within the individual peds

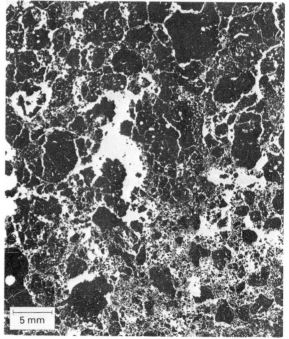

Fig. 4.20 Thin section of composite structure of complete subangular blocky and granular

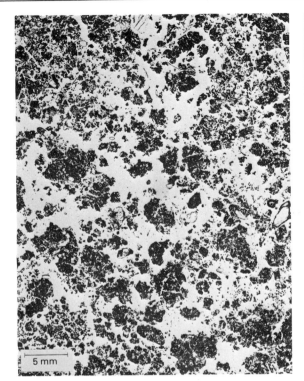

Fig. 4.21 Thin section of crumb structure − composed of numerous loose porous aggregates

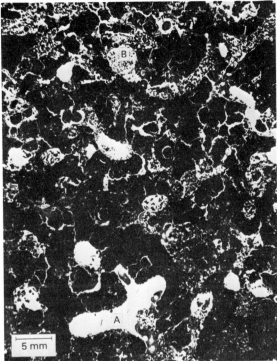

Fig. 4.23 Fluorescent block of labyrinthine structure produced by termites. A is a termite passage and B are granules produced by termites

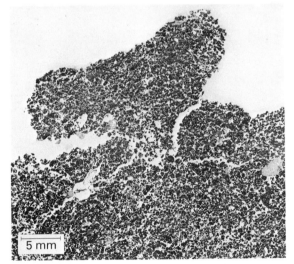

Fig. 4.22 Thin section of granular structure in the upper horizon of a Vertisol − there are abundant subrounded peds

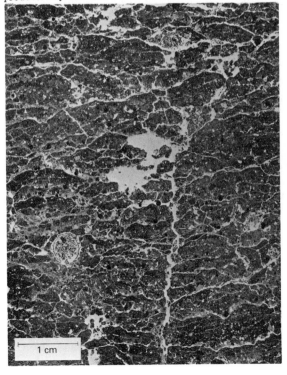

Fig. 4.24 Thin section of lenticular structure

103

Strongly accordant: 50–75 per cent of the opposing surfaces have a similar outline.

Very strongly accordant: >75 per cent of the opposing surfaces have a similar outline.

Degree of aggregation

In addition the degree of aggregation can be classified as follows:

None: single grain or massive.

Weak: (poorly developed) poorly formed peds.

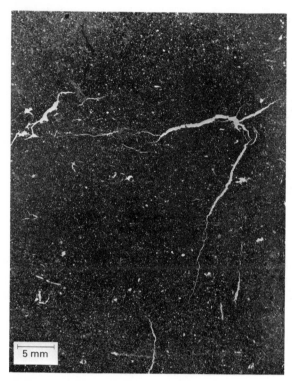

Fig. 4.25 Thin section of massive or coherent structure with contraction cracks caused by drying

Moderate: (well developed) well formed peds that may be visible in the undisturbed soil in hand specimens.

Strong: (very well developed) very well formed peds that are quite clear in the undisturbed soil.

Considering now the distribution of structure within soils, the surface horizons usually have a crumb or granular structure with peds up to about 3 mm diameter. This normally changes to blocky or prismatic in the middle horizon where the peds may vary from 1–10 cm in diameter and up to 30 cm high; finally to massive or coherent in the lower position,

Fig. 4.26 Prismatic structure in the middle horizon. Note the sharp edges and flat faces of the prisms

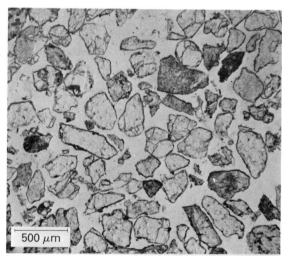

Fig. 4.27 Thin section of single grain structure, there are no aggregates and the particles are separate one from the other

but there are very wide departures from this simple outline. Some variations of structure with depth are given in Fig. 4.32.

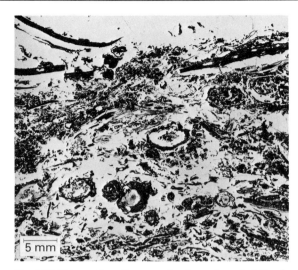

Fig. 4.28 Thin section of spongy structure, characteristic of many surface organic horizons

Fig. 4.29 Subcuboidal structure formed by freezing

11. PASSAGES — FAUNAL AND ROOT

Organisms such as earthworms, termites and beetle larvae that burrow through the soil create passages and chambers which in cross-section are usually circular or ovoid with smooth surfaces. The behaviour of many termites seems to be unique. As they burrow through the soils they produce small granules by manipulating the soil between their mandibles. These granules are either transported to their termitaria or are left in varying concentrations in their passages. Other organisms such as earthworms and beetle larvae leave in their passages varying amounts of faecal material which may be in a variety of forms. It may be as discrete faecal pellets or it may fill the passages

Fig. 4.30 Thin section of vermicular structure produced by earthworms in a Chernozem

completely or form a lining on the sides of the passages. Chambers differ from discrete pores by containing varying amounts of faecal material and also by having interconnecting pores. In most cases the faecal material has the same particle size distribution as the surrounding soil but in certain circumstances there may be marked differences when the organism has transported material over a long distance. When the passages of organisms occur in the subsoil the contrast between the material they contain and the surrounding soil is usually striking.

The intensity of burrowing organisms varies from soil to soil, therefore the frequency of faunal passages

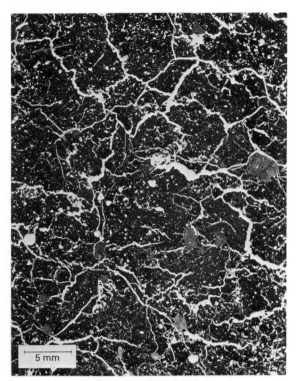

Fig. 4.31 Thin section of wedge structure characterised by pore space running and intersecting at an angle of about 45°

and faecal material is similarly variable. When there are many fauna the soil becomes a tangled mass of passages filled to a greater or lesser extent with faecal material. This may be in the form of discrete granular units or it may be continuous and worm-like hence the terms vermicular aggregate and vermicular structure as shown in Fig. 4.30. A peculiarity of some earthworms is that they transport and deposit in their burrows rounded gravel up to about 3 mm in diameter.

Plant roots also form passages of similar shape and may become filled with material when the plant dies and the roots decompose. Good examples of this are seen in semiarid and arid areas where pipes of calcium carbonate may form in the spaces left by the roots. Also, in wet soils, local oxidation around the sides of the old root passages forms tubes of material cemented by iron oxide. In many cases roots follow pre-existing pores.

The formation of root passages differs fundamentally from those formed by burrowing organisms. Whereas burrowing organisms excavate a passage and probably only disturb the surrounding soil slightly, plant roots force their way through the soil and as

their diameter increases they create pressures within the soil but as yet this effect has not been demonstrated in thin sections.

12. FAECAL MATERIAL

Up to the present sufficient detailed studies have not been conducted to be certain about the characteristics of all faecal material so that misidentification is sometimes possible but for a number of situations the position seems to be quite clear.

Faecal material is usually recognised by its characteristic morphology. Arthropods produce small individual or clustered granular or ovoid units, worms produce vermicular (worm-like) deposits. Since the distribution of soil organisms is determined by an adequate food supply the greatest amount of faecal material occurs in association with the soil organic matter near to the surface. Some organisms penetrate deeply into the soils where their distribution is often shown by their faecal material and passages. Some organisms ingest both mineral and organic material so that their faecal material will contain both materials. The concentration of mineral ingesting fauna varies considerably from soil to soil, whereas some soils show little or no evidence of such organisms, there are many soils that are composed largely or entirely of faecal material, forming characteristic types of structures (Fig. 4.30).

Faecal material is not permanent but is usually subjected to constant change by a number of processes. It usually provides an ideal habitat for microorganisms which proliferate rapidly and decompose the organic portion leading ultimately to its collapse and loss of form.

Faecal material occurring at the surface is affected by physical processes such as raindrop impact which will cause rapid fragmentation and disintegration. Conversely, mainly mineral faecal material deposited at depth may remain unchanged for a very long time. Thus, it is possible to follow the morphological changes from fresh to completely altered faecal deposits.

13. COATINGS

These are deposits of material on the surfaces of peds, pores, rock fragments and detrital grains. They include thin films of translocated clay, coatings of silt, calcite crystals and opaline silica. Thus they vary widely in their properties and generally their identification and study is best conducted on thin sections. They can be classified on the basis of their composition and mode of formation into the following main types:

- Clay coatings
- Sesquioxide coatings
- Silt coatings
- Calcite coatings
- Silica coatings
- Manganese dioxide coatings
- Organic coatings
- Whole soil coatings
- Ice coatings
- Composite coatings
- Compound coatings

Clay coatings

The clay fraction of the soil is generally regarded as comprising all of the mineral material $<2\,\mu m$, both crystalline and amorphous. Clay coatings are probably the most common type of coatings and are recognised in plane polarised light by their layered structure (anisotropism) and absence or low frequency of coarse material while in crossed polarised light they are most often anisotropic and usually show interference colours. Clay coatings are formed by the progressive deposition of clay size ($<2\,\mu m$) particles on the surfaces of pores, peds and mineral grains. The particles are deposited tangential or parallel to the surface in such a manner that they gradually build up like the layers within a mica. Clay coatings are usually fairly easily differentiated from the micas because they are generally curved and follow the outline of the surface upon which they have been deposited (Fig. 4.33). The curved nature imparts a distinctive extinction characteristic in crossed polarised light to those which show interference colours. This is a black extinction band which sweeps through the coating as the stage is rotated as shown in Fig. 4.34. Such a clear extinction band occurs only when the constituent particles in the coatings are very strongly orientated. With a decreasing measure of orientation the band becomes more and more diffuse and ultimately does not form. Most clay coatings appear to be non-pleochroic but a small number of cases of distinct pleochroism do occur.

Some coatings do not show interference colours and seem to form when the constituent material is either amorphous or does not have laminar crystals so that although the coating may appear strongly anisotropic in plane polarised light it behaves like an isotropic crystal in crossed polarised light. An excellent example of coatings without interference colours is reported from the Solomon Islands by Webb (1973)

where the entire soil is composed of intertwining coatings composed predominantly of gibbsite.

It is difficult to be certain about the mineralogy of coatings from optical examination though a fairly general interpretation can sometimes be made on the basis of their interference colours. Kaolinite and chlorite give pale colours of the first order and lower second order colours. However, using peals it is possible to make an X-ray examination of small areas of thin sections and thereby identify most features with greater certainty.

Clay coatings are very fragile and in certain cases they are transient phenomena. For example, it appears that the homogeneous upper red coating-free horizons in many tropical soils are derived from the underlying mottled clay which may have abundant coatings. The processes causing homogenisation of the mottled clay to form the red topsoil are not understood, presumably faunal activity, expansion and contraction resulting from wetting and drying and the growth and death of plant roots are largely responsible. In some cases it is possible to see coatings being fragmented by physical processes and incorporated within the matrix of the soil and in others, coatings appear to be destroyed by fauna which ingest and homogenise the coatings as they burrow through the soil. Evidence for this can be seen when faunal passages cut through coatings and in a few isolated cases fragments of coatings occur in faecal material.

Clay coatings appear to age with time by gradually losing their sharp outline and disappearing to form part of the matrix. In some very old soils there are zones or bands of clay within the soil that appear to have been coatings but the constituent particles are no longer strongly orientated so that they do not have the characteristic black extinction band. In extreme cases the clay particles have become reorganised to form domains. Thus, it appears that a given clay particle which forms part of the matrix can be translocated to form part of a coating which later disintegrates to form part of the matrix thereby going through a complete cycle.

The presence of clay coatings was used at one time as evidence for vertical clay translocation and accumulation. It seems now that clay coatings can form in a variety of ways. The presence of many coatings in many moist or wet horizons suggests that local reorganisation of the material within the horizon is a very active mechanism and is responsible for their formation. In some horizons where weathering is proceeding rapidly the clays so formed are often deposited within cracks and on the surfaces of the more resistant minerals (Fig. 6.9) so that the end product of weather- **107**

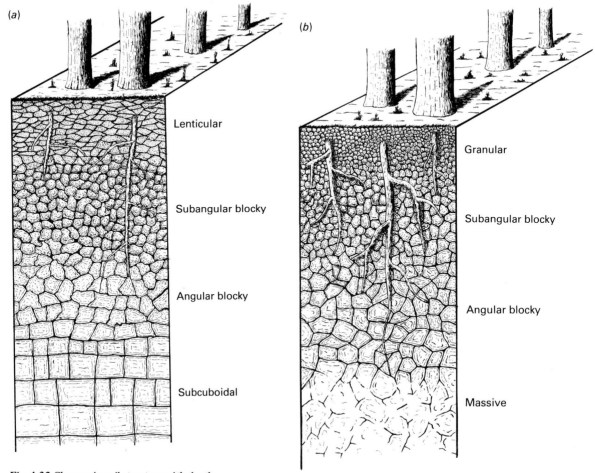

Fig. 4.32 Changes in soil structure with depth

ing contains many coatings. It is possible that coatings may form *in situ* on free surfaces, for it has been shown that certain Andosols have coatings of halloysite while the matrix contains allophane, X-ray analyses of thin section peals should be very helpful in such studies.

Some horizons are composed almost entirely of intertwining clay coatings with varying amounts of detrital grains. In such cases the deposition of clay translocated from outside the horizon does not seem to be the processes involved. More likely it may be a combination of *in situ* weathering and reorganisation of the soil in response to some specific moisture regime. In certain cases laminations of clay very similar in appearance to coatings appear to form actually within the matrix and not as deposits on a free face. Thus, a measure of caution must be exercised when making interpretations based on the presence of clay coatings.

Sesquioxide coatings

Coating of sesquioxides often associated with organic matter are common in many soils, particularly in the middle horizons of Podzols.

Silt coatings

Certain horizons have local concentrations of detrital grains mainly of silt size. These may be randomly distributed through the soil but usually they occur on the upper surfaces of stones and peds or may form linings in fissures and cracks (Figs. 6.46 and 6.47).

Calcite coatings

Coatings of finely crystalline calcite occur on the surfaces of peds, around roots and rock fragments, and lining pores. These coatings form by the slow growth of calcite crystals from the percolating soil

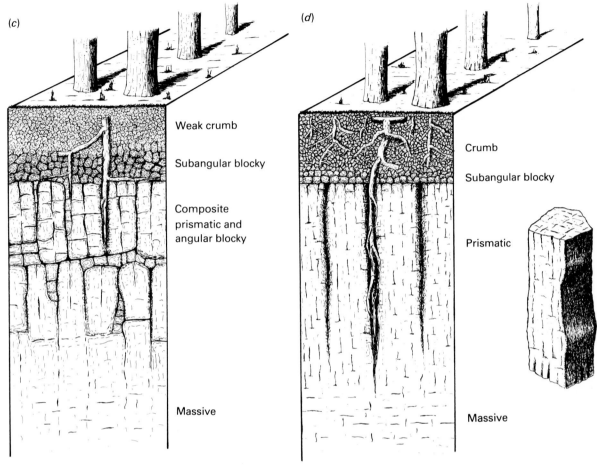

(c)

Weak crumb

Subangular blocky

Composite
prismatic and
angular blocky

Massive

(d)

Crumb

Subangular blocky

Prismatic

Massive

Fig. 4.32 (*continued*)

solution. The amount of these coatings is very variable
and in extreme cases almost the whole soil is made up
of various arrangements of calcite coatings, their
growth ultimately leading to a massive accumulation.
As this takes place the surrounding material can be
displaced to such an extent that the original material
occupies only small islands and is surrounded by
calcite.

In a number of soils the calcite coating forms a
pendant beneath stones to which it may be attached
or it may be separate.

Silica coatings

In some semiarid areas particularly in the greater part
of Australia, the soil solution contains large amounts
of silica. This may become deposited within the
matrix of some soils or may form coatings on the

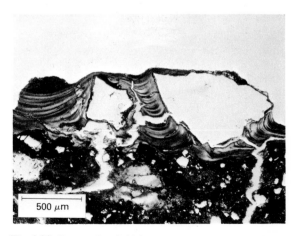

500 μm

Fig. 4.33 Clay coatings bridging pore space and surrounding
sand grains

109

surfaces of peds, fissures, and may even cement faecal material formed in a previous wetter soil-forming period. These deposits are either as chalcedony or opal.

Manganese dioxide coatings

Manganese dioxide may occur in many situations in the soil including the formation of coatings.

Organic coatings

In certain peat deposits the very highly decomposed organic matter may flow through spaces and pores in the system to form coatings. Coatings composed almost entirely of organic material also occur in the middle horizons of Podzols where they form as a result of the translocation of organic matter from above.

Whole soil coatings

The impact of raindrops and rainwater running over the surface of the soil can cause dispersion of the soil and the formation of a soil suspension. This may flow down the pore spaces and become deposited to form whole-soil coatings at various depths and in various orientations depending upon the nature of the pore space. A similar sequence can take place following irrigation or flooding when the flood waters contain suspended materials (Fig. 4.35). This can take place on a macro scale when concretions or other large size material fall down pore spaces or passages.

Ice coatings

In soils of the tundra areas most of the stones and boulders in the frozen subsoil are surrounded by a sheath of ice which is thickest on the underside.

Composite coatings

A composite coating is one in which there are two or more layers of different material deposited on the same surface and occur in some soils with polygenetic evolution. A fairly common type is composed of an inner coating of calcite overlain by a coating of crystal-line clay. In other cases the reverse may occur.

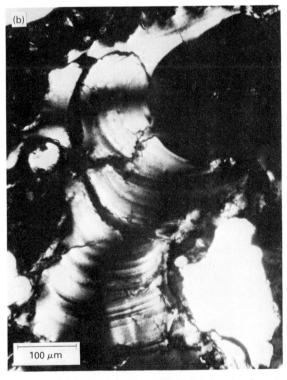

Fig. 4.34 Clay coating bridging pore space (*left*) in plane polarised light; (*right*) in crossed polarised light showing the characteristic black extinction bands

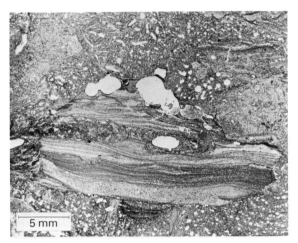

Fig. 4.35 Horizontal thin section of whole soil coatings filling drying cracks in alluvium

Compound coatings

Although many coatings appear to be relatively uniform they may be composed of two or more substances. This is particularly true of clay coatings which may be composed of two or more clay minerals together with varying amounts of iron oxides. In fact most clay coatings are a complex of more than two substances.

14. CLAY PLUGS

In a number of cases pores have been filled completely with clay and are known as clay plugs. Sometimes the form of the clay suggests that deposition has been the method of formation but in other cases other mechanisms seem to be involved.

15. SURFACE RESIDUES

Surface residues occur on the outer surfaces of peds and contain less clay and/or iron oxide than the ped interior and form by the differential removal of material by percolating water.

When water passes repeatedly over the surface of a ped, material may be deposited to form coatings or material may be removed. The type and amount of material removed will depend upon a number of factors such as the composition of the soil and the character of the percolating water. The substances most commonly removed are clay and iron oxide,

when it is the former the ped surface becomes progressively more sandy so that the peds finally develop a sandy outer surface. This is common in soils where clay is being removed from the upper horizon.

16. DENSITY

The true density of a soil is a measure of the density of the individual constituent components and varies from about 900 kg m^{-3} for the organic matter to about 2 650 kg m^{-3} for the composite of all the mineral particles. The density of the individual minerals varies from 5 200 kg m^{-3} for magnetite to about 2 600 kg m^{-3} for the feldspars. On the other hand the density of the whole undisturbed soil or *bulk density* is:

$$\frac{\text{Wt of a dry block of soil}}{\text{Volume of block when sampled}} = \frac{\text{g}}{\text{cm}^3} = x$$

This takes into account the density of the soil materials themselves and their arrangement or structure. Therefore, a loose porous soil will have a smaller bulk density than a compact soil even though the density of the individual particles in the two soils may be the same.

Increasing attention is being given to the bulk density of soils since it is a significant and distinctive characteristic of some horizons. Whereas the bulk density of most cultivated horizons is about 1.3, the extremes can vary from 0.55 for some soils developed on volcanic ash to 2.0 for some strongly compacted lower horizons. This property is also attaining some importance in fertility studies because continuous cultivation by heavy implements induces compaction which reduces percolation and root penetration. In pedological studies Oertel (1961) used bulk density in certain fundamental studies to establish whether the clay content in certain horizons was increased by translocation or weathering *in situ*.

17. MOISTURE CONTENT

There is a great need for more quantitative data about the variations in moisture regimes of soils in different environments and to relate these data to other soil characteristics. At present the annual variations in the content of moisture in soils are inferred from morphological characteristics and on limited observations made at various times of the year.

The traditional class names for differences in soil moisture include the word 'drainage' which implies rate of movement of water through the system. Probably it is better to make inferences about the degree of anaerobism, since soils may have a fairly rapid rate of water movement and still be continuously anaerobic. The traditional class names for predominantly mineral horizons are given below followed by alternative names in brackets based on inferences about anaerobism.

Excessively drained (excessively aerobic)

The horizon is usually too dry to support adequate plant growth. It has a range of colours and textures.

Well drained (aerobic)

The horizon is usually moist or dry with a sufficiently moist period to support plant growth. It has bright to very bright colours and a wide range of textures depending upon local conditions.

Imperfectly drained (weakly anaerobic)

The horizon is anaerobic for short periods and moist for long periods. Colours are less bright than the aerobic horizons and they are marbled or mottled.

Poorly drained (strongly anaerobic)

The horizon is anaerobic for long periods and otherwise moist. It is strongly mottled with large areas of grey, olive or blue which is indicative of considerable reduction.

Very poorly drained (continuously anaerobic)

The horizon is saturated with water throughout the year. It is blue, olive or grey but may have pipes of oxidised iron around root passages.

Flooded soils

Soils that are flooded annually and followed by a dry period tend to have drab or dark colours and may show some mottling.

Improved drainage

Sometimes an area may be drained but this may not show in the soil morphology which alters very slowly. Stewart and Adams (1968) have shown that there is a correlation between soil morphology and the percentage waterlogging measured three days after rain and defined as a 'field capacity day'. The percentage waterlogging is the amount of moisture present in the soil on a field capacity day expressed as a percentage of the total pore space. They showed that horizons, mottled grey and yellowish-brown were more than 70 per cent waterlogged on such a day. Brown, middle horizons were usually less than 40 per cent waterlogged while yellowish-brown horizons with faint greyish areas were about 60 per cent waterlogged. Massive surface horizons were over 70 per cent waterlogged. This is very interesting since there are few morphological features in massive upper horizons that could be ascribed to wetness apart from a little rusty mottling along root channels. These horizons usually have a large population of anaerobic bacteria correlating well with the high percentage waterlogging which in these horizons is due not only to rainfall but to higher retentivity resulting from poor structure.

Perhaps it should be pointed out that a given figure of *c.* 50 per cent waterlogged does not mean that all the pores are half-full of water but that the larger pores are almost empty while the small ones are waterlogged. Thus, in a moist environment, soils that are more than 50 per cent waterlogged three days after rain are likely to show some evidence of waterlogging, either through their morphology or in their bacterial population. The concept of percentage waterlogging cannot be overstated since soils with many small pores are likely to have higher waterlogging percentages for longer periods following rain. This will influence the microflora and determine very largely the space available for root penetration. Although these results indicate a valuable potential method for the study of soil moisture in a humid environment its usefulness has to be demonstrated for other environments.

18. pH

Soil pH is generally regarded as a very important soil property since it tends to correlate with other properties such as the degree of base saturation. In addition, many soils have characteristic vertical patterns in their pH values. Indeed Northcote (1971) uses the soil reaction trend as a major property in his system of classification.

The reaction of the soil normally ranges from pH 3 to 9 but occasionally values outside these limits are encountered. Very low values are found in soils of drained marshes and swamps that contain pyrite or elemental sulphur. At the other extreme very high values result from the presence of sodium carbonate.

Within the normal range the two principal controlling factors are organic matter and the type and amount of cations. Large amounts of organic matter induce acidity except when counterbalanced by high concentrations of basic cations. Although it is normal for peat to have low pH values of 3 to 4 they can be near to neutrality if there is a high content of bases, mainly calcium carbonate, in the water influencing their formation.

Hydrogen and aluminium ions are largely responsible for soil acidity. Aluminium is released by hydrolysis of the primary minerals or comes into solution from the exchange sites. Then each Al^{3+} ion combines with three OH^{-1} ions and precipitates leaving three free H^{1+} ions which reduce the pH value. In soils depleted of basic cations, aluminium becomes increasingly soluble because of the decreasing pH and is absorbed in preference to hydrogen on the permanent charge leaving the H^{+1} ions in solution.

pH tends to be related to rainfall. As rainfall increases the pH falls as a result of the depletion of basic cations.

19. CATION-EXCHANGE CAPACITY AND PERCENTAGE BASE SATURATION

The only exchange property which seems to be important for use as a differentiating criterion in categorising soils is the cation-exchange capacity (CEC) of the whole soil. This property is a measure of the exchange capacity or negative charges of the constituent clay minerals, allophane and humus, expressed as milligram equivalents kg^{-1} of soil. The range is from about 50 me kg^{-1} for some lower horizons up to 1 000 me kg^{-1} for upper horizons containing a high percentage of organic matter, vermiculite or montmorillonite.

The negative charges are of two types: permanent charges and the pH dependent charges. The permanent charges occur in the silicate clays and result from isomorphous replacement within the clay structure and are highest in the 2:1 type clays. The pH dependent charges are directly related to variation in the soil pH. At low pH values, the charge and thus the CEC is low but increases as the pH value rises.

Both mineral and organic materials have pH dependent charges. In mineral material the charge is thought to be due to SiOH and AlOH groups at broken edges and surfaces of clay minerals while in organic matter there are carboxyl (COOH) and phenol (−OH) groups. All of these groups contain covalent bonded hydrogen ions which can be dissociated at high pH values giving a negative charge at an exchange site.

Tightly adsorbed aluminium and iron in some 2:1 clays in acid soils block some of the negatively charged sites, thus reducing the CEC. With an increase in pH the aluminium and iron are removed and the sites become available.

The silicate clay minerals are high in permanent charge while humus and allophane have mainly pH dependent charges. Thus the CEC determined values are very largely dependent upon the pH in the technique that is used. Thus a realistic value of the CEC should be that determined at the field pH value. The difference between the CEC determined at the field pH and that determined at pH 8.2 is considered to be the pH dependent charge.

The silt fraction sometimes contributes substantially to the CEC especially when it is present in high proportions.

For many tropical soils the CEC of the whole soil is very low because of the presence of large amounts of kaolinite and/or gibbsite. In such cases the CEC of the clay fraction is used.

A further property of some importance is the percentage base saturation (BS) which is a measure of the extent to which the exchange complex is saturated with basic cations. This is used at lower category levels in many classifications to differentiate between soils that have high or low base saturation.

The general trend is for the amount of exchangeable bases to increase with decreasing rainfall and for calcium to be dominant but the nature of parent material may have a local influence. As the cation content increases there is also the tendency for sodium to become increasingly important and it may even be dominant in many soils of arid and semiarid areas. Conversely, low figures for the percentage base saturation are used as a criterion of intense leaching. However, since the amount and type of cations can be changed rather easily by cultivation they are used only occasionally as differentiating criteria.

20. SOLUBLE SALTS

Soluble salts occur in significant proportions only in the soils of arid and semiarid areas where they accumulate because the annual precipitation is insufficient to leach the soils or because the water-table is at a shallow depth and moisture is drawn to the surface by capillarity bringing with it dissolved salts which are left behind as the moisture evaporates. Flooding by sea water also causes salinity in soil but this is of minor importance except in countries such as the Netherlands which depend upon large areas reclaimed from the sea. The predominant anions are bicarbonate, carbonate, sulphate and chloride while the cations include sodium, calcium, magnesium and small amounts of potassium.

These ions occur in widely varying proportions and depending upon the particular ratio they impart a number of properties to the soil, some of which are detrimental to plant growth. Soils can be divided broadly into saline and sodic depending upon the proportions of these cations as follows:

saline: E.C. (electrical conductivity) of saturated extract >4 mmhos cm^{-1} exchangeable sodium <15 per cent, pH <8.5.

sodic: E.C. (electrical conductivity) of saturated extract <4 mmhos cm^{-1} exchangeable sodium >15 per cent, pH >8.5.

Although the above classification covers most situations there are a number of soils that are not adequately accommodated, therefore some workers (Miljkovic, 1965) prefer to have a separate scale for salinity and alkalinity as follows:

Degree of salinity

Slightly saline with E.C.	2–4 mmhos cm^{-1}
Moderately saline with E.C.	4–8 mmhos cm^{-1}
Strongly saline with E.C.	8–15 mmhos cm^{-1}
Very strongly saline with E.C.	>15 mmhos cm^{-1}

Degree of alkalinity

Slightly alkaline	<20 per cent exchangeable sodium
Moderately alkaline	20–50 per cent exchangeable sodium
Strongly alkaline	>50 per cent exchangeable sodium

In neither of these two schemes is an account taken of the proportions of the different ions present as done by some workers, particularly the variations in the amounts of anions. A further method of subdivision of the content of the soluble salts taking into account the proportions of the anions can be achieved using a ternary system in which the coordinates are per cent chloride, per cent sulphate and per cent carbonate and bicarbonate as given in Fig. 4.36.

A common feature of many soils containing soluble salts is that many microelements accumulate to toxic proportions. In the case of boron, concentrations over 1.0 ppm are toxic.

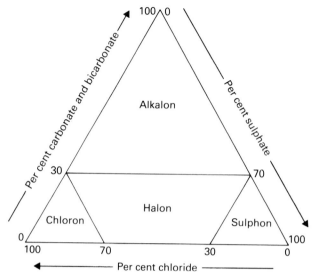

Fig. 4.36 Ternary system of anions

21. CARBONATES

Carbonates of calcium and magnesium, particularly the former, are widely distributed in soils, occurring separately or they may be associated with soluble salts. The most important properties of carbonates are: (1) They are relatively easily soluble in water containing dissolved carbon dioxide, and therefore can be quickly lost or redistributed within the soil. (2) When present in as small amounts as 1 per cent of the soil they can dominate the course of soil development, because this amount is sufficient to raise the pH value over neutrality and sustain a high level of biological activity. (3) Carbonates, particularly calcium carbonate, are the first substances to start accumulating as the climate becomes arid. (4) Both calcium and magnesium are essential plant nutrients. (5) Carbonates are regularly added to many arable soils to raise their pH values for optimum plant growth.

22. ELEMENTAL COMPOSITION

Only in very detailed studies is it necessary to determine the total amount of each element present in the soil. Where soil formation has proceeded for a long period and where there has been a considerable amount of hydrolysis and solution it is usual to perform a partial ultimate analysis to estimate the ten or so dominant elements in each horizon as well as in the underlying parent material, in order to determine the predominant chemical changes that have taken place during soil formation. These elements include silicon, aluminium, iron, sodium, potassium, calcium, magnesium manganese, titanium, zirconium, nitrogen and phosphorus. However, the determination of the total amount of certain elements is performed very frequently, particularly nitrogen and phosphorus, because of their immense importance as plant nutrients.

It is claimed that the molecular ratio of silica to alumina and iron oxide (silica:sesquioxide ratio or SiO_2/R_2O_3 where $R = Fe + Al$) is an important criterion in the study of tropical soils. Therefore, the total amount of these three substances is often determined, although the validity of this ratio has still to be demonstrated convincingly.

The silica:sesquioxide ratio is also determined for the clay fraction isolated from soils such as Cambisols and Podzols. In these soils the amount of hydrolysis is relatively small and a partial ultimate analysis does not show strong contrasts between their horizons. However, the data from the clay analyses are used as an aid in establishing distinguishing criteria, for by this means it is possible to detect differential translocation of these substances. As well as the silica:sesquioxide ratio, silica:alumina, silica:iron oxide and alumina:iron oxide ratios are used.

At present most laboratories conducting these investigations employ long and tedious methods, but it is hoped that more rapid and sophisticated techniques such as X-ray fluorescence will be used in the future so that greater amounts of data will become available, further to test the validity of some of these criteria. Whereas in the past large samples were usually required and were destroyed during the analysis, new techniques including X-ray fluorescence and electron probe microanalyses make it possible to analyse materials without destroying them and also to use very small samples such that the distribution of an element within individual minerals in soil can be determined. For such work minerals can be hand picked from bulk samples or the analysis can be made on thin sections or impregnated polished blocks.

23. AMORPHOUS AND MICRO-CRYSTALLINE OXIDES OF IRON, ALUMINIUM AND SILICON

It has been stated previously that the initial stages of hydrolysis may form amorphous materials which may then crystallise gradually to produce a number of new substances or they may be translocated within the soil. It is assumed therefore that the type, amount and

distribution of amorphous materials can be used as criteria for measuring the degree and type of soil formation. Also, they are regarded as being formed during the current phase of pedogenesis and, therefore, are valuable criteria for differentiating between relic and contemporary phenomena. When performing the determinations it is customary to treat the soil with an extracting reagent such as potassium dithionite, oxalic acid or sodium pyrophosphate and the amount of each constituent in the extract is determined. However, published data pertaining to these constituents should be treated with caution as the suitability of many of the techniques can be challenged but certain trends have been discovered which seem to be reasonably valid particularly for Podzols and related soils.

Although chemical techniques have been used almost exclusively for the study of these oxides, other methods including the study of soils in thin sections, by the electron microscope and electron probe are yielding much new and interesting data. For example it has been found that amorphous mineral materials such as ferric hydroxide are isotropic in cross polarised light, but probably the information supplied by the electron microscope is more revealing for it has been shown that silica often forms small aggregates while alumina and iron oxides form coatings on clay minerals and other surfaces.

24. SEGREGATIONS AND CONCRETIONS

Segregations and concretions are local concentrations of one or more substances in the soil. In hand specimens segregations are soft and concretions are firm to hard but this difference cannot be determined in thin section. Segregations and concretions can be classified crudely on the basis of their composition; sometimes shape and internal structure are taken into consideration but it appears that these two properties can be determined by the original structure of the soil and then are not related to the processes causing the concentration of the material.

Calcium carbonate concretions

Subspherical concretions of calcium carbonate are common features in many soils. Generally they are composed of very small crystals of calcite and appear to have grown in the pore space because many have a central space which probably represents part of the original pore space that has become enclosed as the concretion grew.

115

Iron—manganese segregations and concretions

Many soils contain concentrations composed of varying proportions of iron oxide and manganese dioxide. They vary from dark brown to black with increasing amounts of manganese dioxide and are usually subspherical but irregular shapes do occur. The outline of these varies from diffuse to sharp and the internal structure can be amorphous, concentrically ringed or cellular or any combination of these three. Most iron—manganese segregations and concretions appear to be developed in moist soils. Those with a cellular structure form in weathered rock and develop their characteristic structure through the deposition of material along old cleavages and mineral remnants (Fig. 6.8). In some soils subject to seepage there may be dendritic growths of manganese dioxide on ped and pore surfaces.

Spore-like concretions

In some soils clay particles segregate within the matrix so as to isolate subspherical volumes of soil. The segregations strongly resemble coatings but they are not deposits on a free surface. After the segregations have become well formed they split down the middle like biological sporulating process to give concretions. These concretions then are seen to have randomly arranged material as a core surrounded by a concentrically ringed shell (Fig. 6.14).

25. ORGANIC MATTER

The organic matter in soils consists of the remains and decomposition products of both plants and animals. In spite of the very large soil fauna it is usually very difficult to find their remains probably because they are vigorously decomposed by the microflora and microfauna. Occasionally one finds pieces of chitin, fragments of mollusc shells and eggs or cysts.

By far the greatest amount of organic matter in soils is derived from plants both from the above ground parts as well as from roots. Generally, leaves and needles constitute up to 80 per cent of the litter, while roots make up most of the remaining 20 per cent but may be as much as 50 per cent in grassland.

Plants are composed of the following types of tissues (Babel, 1975).

Parenchyma. This is living tissue and includes the green tissues of the leaves, the cortex of young twigs and fine roots and other material such as the centre of potato tubers; generally the cell walls are thin and contain protoplasm.

Lignified tissue. This is the dead xylem tissue and forms the mass of the woody material including the leaf veins and the bulk of the stems and roots and forms the main water conducting system.

Phlobaphene containing tissue. This includes the older cortex, bark and cork with its characteristic parallel arrangement of cells. Also, the lumens of these cells contain yellow to brown tannin oxidation products called phlobaphenes.

Epidermis. This forms the outer layer in many young parts of the plant and is similar to the parenchyma.

Collenchyma. These are living thin walled cells that occur beneath the epidermis.

Phloem. These are living thin-walled cells that form a major part of the organic conducting system in the plant.

The micromorphology and chemical composition are the two main properties that are usually used to categorise organic matter.

Micromorphology of organic matter

The principal type of organic matter seen in thin sections is the litter and its decomposition products both of which usually occur at or near the surface of the soil. Some horizons contain large numbers of roots due to the occurrence of a freely available nutrient supply or where they fan out over an impermeable horizon. In most cases the distribution of roots is similar to that of the dead and decomposing organic matter, i.e. the greatest amount occurs at the surface except for those soils in which translocation has produced a second maximum of organic matter in the middle horizon where roots may or may not occur in large numbers.

It is usually possible to trace the morphological changes during the alteration of plant material. Some plant remains are invaded by bacteria or overgrown by fungal mycelium and actinomycetes and some are partially comminuted or invaded by small arthropods whose presence is indicated by the occurrence of faecal pellets. As decomposition proceeds the plant fragments are transformed into *humus* which is a dark coloured translucent or transparent amorphous substance. Thus, the size of the individual pieces of organic matter is variable and can range from complete plant parts such as leaves, needles and fruit to very small fragments of any part of a plant or animal. Similarly, the degree of decomposition of organic

116

matter is very variable and often is not related to its size or degree of fragmentation. Generally, in an aerobic environment, organic matter becomes more decomposed as it is more fragmented, but under anaerobic conditions, as in peat, there may be very large fragments of organic materials that are in a very advanced stage of decomposition and can easily be crushed between the fingers. Depending upon the nature of the organic material and the organisms involved in its decomposition it is possible to find that the degree of decomposition varies from place to place within the same plant fragment with some types of tissue decomposing before others. Generally, the order of increasing resistance is phloem, collenchyma, parenchyma, lignified tissue, epidermis and phlobaphene containing tissue. Thus decomposition is a selective process.

Consequently, in Europe there has always been the tendency to stress the humus profile in which 80 per cent of the organic matter disappears during decomposition down to the well humified material. Some tissues such as the parenchyma disappear without leaving any residues. In others, dark spherical bodies form in the cells or as crusts on the surfaces of leaves immediately after leaf fall. This dark material or leaf browning substance is similar to the phlobaphenes produced in the living cell.

Fragments of organic matter are a common constituent of the faecal material of the soil fauna. The amount varies widely and is determined to some extent by the amount of mineral material taken in by the organism. Some plant parts pass through the alimentary system of the fauna without being markedly altered, thus faecal material will contain organic material that varies from moderately to strongly decomposed. The humus or colloidal organic matter is often translocated within the soil and later deposited to form coatings around sand grains, or on surfaces both within mineral soils and peat. In some cases microscopic fragments of cellulose can also be translocated with the clay fraction and can only be recognised by their primary fluorescence (Babel, 1975).

Other physical processes such as freezing and thawing or expansion and contraction will incorporate partially decomposed plant fragments into the mineral soil up to a depth of over a metre.

When the vegetation is accidentally or purposely burned there is usually a small residue of charred material which becomes incorporated into the topsoil. When the fragments are large they are easily recognised by their cellular structure and opaqueness in transmitted light and are black in reflected light. When the particles are small they can be confused

with magnetite, ilmenite and manganese dioxide (Fig. 6.39).

Other forms of organic matter include the skeletons of animals, phytoliths, etc. (Fig. 3.9).

Identification of organic matter

The ease with which organic matter can be identified in thin sections varies considerably. Living or freshly fallen plant material is recognised by its cellular structure and interference colours of the first order but it may not extinguish. On the other hand, humus and decomposing organic matter is isotropic except for fragments of cellulose even though it may retain some of its original cellular characteristics. Cellulose shows very marked primary fluorescence, therefore, it is very easy to identify very small fragments in the upper part of the soil as well as lower in the soil to where it may have been translocated. The amorphous and isotropic properties of humus are distinguishing characteristics but they are also properties of amorphous ferric hydroxide with which it can be confused in certain circumstances, particularly as both occur in middle horizons of certain Podzols. Perhaps the development of a staining or other technique may differentiate between these two substances in the future.

Classification of organic materials

Initially organic matter can be divided into *coarse residues* and *fine residues*. The coarse residues include all of the plant fragments that contain many cells and usually two or more different types of tissue and can usually be assigned to a particular organ. The fine residue is composed of single cells, cell wall fragments of fungal hyphae or generally cannot be identified with any degree of confidence. It is <40 μm in size and is usually in a continuous or floccular arrangement. The coarse material can further be subdivided according to the plant organ into woody tissues, herbaceous tissues and mossy tissue. A general classification might be as follows:

Living. These are plant parts that were alive at the time of sampling. Some usually show interference colours of the first order.

Fresh. This category includes very slightly decomposed plant material within the litter that was dead at the time of sampling but shows cellular structure and little or no sign of decomposition, some parts usually show interference colours of the first order and contains >75 per cent coarse residues.

117

Slightly decomposed. Plant materials showing evidence of decomposition by microflora and/or fauna, through the loss of some of the cell structure and possibly overgrown with fungal mycelium, usually show interference colours of the first order but may be masked by colouring substances and contains 60 to 75 per cent coarse residues.

Moderately decomposed. This plant material is still recognisable but shows considerable loss of structure and fragmentation. Interference colours are difficult to see but the undecomposed cellulose shows primary fluorescence and it contains 45 to 60 per cent coarse residues.

Strongly decomposed. Plant material that is just recognisable but has lost most of its structure. The undecomposed cellulose shows primary fluorescence and it contains 30 to 45 per cent coarse residues.

Amorphous. The amorphous remains of plant material, recognised by its association with other plant material and its general isotropism. The small undecomposed fragments of cellulose show primary fluorescence and it contains <30 per cent coarse residues.

Charcoal. Opaque material which will show the cellular structure of the original plant material if the particles are sufficiently large. The classification of organic matter in peat is also based largely on the degree of decomposition. Methods for measuring decomposition vary considerably. It can be measured in the field using the Von Post scale (see p. 230) or in the laboratory where the fibre content or pyrophosphate extractable material is determined. These methods have been tested by Stanek and Silc (1977) who suggest that the Von Post scale should be used.

Chemistry of soil organic matter

Despite decades of intensive and painstaking research, a detailed chemical categorisation of the humified organic matter occurring separately at the surface or mixed with mineral material has not been achieved. Therefore, in routine analyses for categorising soils only the total amounts of carbon and nitrogen are determined. The results are normally presented as percentages by weight but it might be more realistic to present the data on a volume basis so as to take account of the very wide difference in density between mineral material and organic matter. The carbon percentages usually are multiplied by the conversion factor of 1.72 to give an indication of the total amounts of organic matter present. The ratio of carbon to nitrogen (C/N ratio) is used as a measure of the degree of humification and ranges from >100 to 8. The former high figure is for the fresh and little decomposed material found in litter and peat while the latter occurs in many upper horizons containing a mixture of mineral material and well humified organic matter and shows that there is a reduction of carbon relative to nitrogen upon humification. The C/N ratio usually decreases with depth and a few middle horizons have ratios as low as 4 which is difficult to explain and may be caused by a high content of ammonium ions fixed by the clay. Another important characteristic of the C/N ratio is that it tends to increase with increasing soil acidity.

The total amount of organic matter in soils varies from <1 per cent up to >90 per cent. This latter figure is for the relatively unaltered material that occurs at the surface. Normally however, except in peats upper horizons contain <15 per cent organic matter and a very large number contain <2 per cent even when the supply of litter to the surface is high as in certain humid tropical areas.

The properties of humus are unique and very largely determine the character of many upper horizons. Firstly, humus is capable of absorbing large quantities of water thus increasing the water-holding capacity of soils and is therefore of importance in crop production. It acts as a binding substance and aids in structure formation. Also it has a cation-exchange capacity of about 3 000 me kg^{-1} and it can be dispersed or flocculated like clays depending upon the nature of the cations present, and as already stated on page 85 it influences the handling consistence.

Under natural conditions the humus content of a virgin soil is usually higher than in adjacent cultivated areas. This is caused by a higher rate of addition of organic matter by the natural vegetation accompanied by a lower rate of biological activity and lower temperatures.

The soil organic matter is derived from plant and animal tissues which can be divided into three broad chemical groups. These are the carbohydrates, proteins and lignins, there are also small amounts of other substances such as waxes, oils, pigments.

The carbohydrates include simple sugars and starches but the dominant components are polysaccharides of which cellulose is the main constituent. The polysaccharides are composed of hexose sugars, pentose sugars and uronic acids. Other carbohydrates such as glucosamine are products of microbial metabolism or derived from their cell walls. However, most of these substances are short-lived in soils because they are readily metabolised by the soil flora and fauna. Another polysaccharide-like material found in soils is chitin which is found in the exoskeleton of arthropods and in fungal cell-walls and is a condensation product of N-acetyl glucosamine.

The nitrogen-containing components in soils seem to vary little from soil to soil and are composed predominantly of only fourteen different amino acids. Thus the uniformity of composition and presence of certain non-protein amino acids suggests that soil organic-N may be predominantly of microbial orgin.

Large amounts of lignin are supplied by plant tissues but there seems to be very little in soils although similar substances have been identified.

Soil humus seems to be composed mainly of carbon, hydrogen, oxygen, nitrogen, sulphur and phosphorus with wide variations in the amount of each element from soil to soil but for agricultural soils the ratio of C:N:S:P is about 100:10:1:2.

The total nitrogen content varies from about 0.05 to 2 per cent of which about 50 per cent occurs as amino acids, a small amount as amides, 5 to 10 per cent as amino sugars and about 30 per cent unidentified.

The phosphorus occurs mainly as complex esters with molecular weights <10 000 and so far the only positively identified phosphorus compound is inositol hexaphosphate (phytin) which seems to have a relative increase down the pedo-unit.

Few compounds of sulphur have been identified, they occur mainly as sulphur-containing amino acids.

There is still a considerable amount of doubt about the true character of the black material comprising humus. Many of the earlier workers regarded humus as a ligno-protein (Waksman and Iyer, 1932) or a condensation product of tannins and proteins (Handley, 1954). It is now generally believed that part of the humus results from the reaction between polyphenoic compounds of plant or microbial origin, with amino acids, peptides and proteins also of plant or microbial origin. The result of the reaction is to produce insoluble heterogeneous polymers which are comparatively resistant to further microbiological decomposition (Flaig, 1960; Kononova, 1961). It is also suggested that humus is composed of a resistant core of complex polycyclic compounds to which are linked carbohydrates and polymers such as those mentioned above (Cheshire *et al.*, 1967). Resistance to decomposition is achieved through heterogeneity since a very wide range of enzymes is required for their decomposition.

Although the details of the compounds present in humus are still to be determined a number of functional groups have been recognised and they impart certain properties to humus. The main groups are carboxyl and pheolic hydroxyls which impart acidity and exchange reactions thus giving humus a cation-exchange capacity. Other groups include ketones and aliphatic hydroxyls.

The classical and somewhat archaic approach to the study of soil organic matter is to treat the soil with dilute sodium hydroxide (0.1 to 0.5 M) which extracts a part of the organic matter, then to acidify the filtrate which causes a part of the organic matter to precipitate. The part that stays in solution is generally known as fulvic acid and that which is precipitated as humic acid, the unextracted residue being the humin. The humic acid precipitate is further fractionated by treatment with ethanol which dissolves a part of the precipitate known as hymanomelanic acid and the remainder is the insoluble humic acid.

Although the fractions are given names they are not homogeneous for each contains substances with a wide range of molecular weights. The molecular weight of fulvic acid is generally below 10 000 and humic acid above 5 000 and going up to several million.

26. SECONDARY CRYSTALLINE MATERIAL

The secondary crystalline material in soils varies in size from submicroscopic clay particles to large crystals of material such as kaolinite or calcite that are visible in hand specimens. Only secondary crystalline substances easily visible in thin sections or in hand specimens are listed below, their properties being given in Appendix I.

Calcite	Kaolinite
Chalcedony	Lepidocrocite
Gibbsite	Manganese dioxide
Goethite	Mica
Gypsum	Montmorillonite
Haematite	Pyrite
Halite (sodium chloride)	Quartz
Ice	Siderite
Iddingsite	Vermiculite
Jarosite	

5

Nomenclature and classification

The nomenclature and classification of soils fall into two parts. On the one hand there is the nomenclature and classification of horizons and on the other there is the nomenclature and classification of whole soils. At present there is no international agreement about soil nomenclature and classification. In fact there are not less than ten different systems used in various parts of the world. The reason for this lack of agreement is because soils are very complex and there is a considerable lack of knowledge about many soils and often the aims of classification are not clear. McRae and Burnham (1976) state that five differing and, perhaps, irreconcilable aims exist as reasons for making a soil classification:

1. To communicate an impression of the nature of a soil profile or of the soil in an area in relation to others, i.e. to use as a language. In this case:
   (*a*) it must be widely accepted; a language that has wonderful elegance of grammar and syntax but is little spoken is not worth learning.
   (*b*) It must be easily taught; if it is to be used, the gist of the classification must be reasonably easy to hold in the memory and to be so it must be logical and not too complicated.
2. To simplify processing soil data. In this case:
   (*a*) more complication can be tolerated, as a specialist with a manual or even a computer can be used to assign soils to classes.
   (*b*) More precision of definition is required, or too much of the information content of the original observations is lost.
3. To reveal or study genetic relationships. In this case:

(*a*) the selection of properties used as criteria may be irrelevant to important uses of soil, e.g. plants are colour-blind.
(*b*) It must be capable of modification as knowledge grows.
4. To use as a legend in a soil map and as mapping units. In this case:
   (*a*) there may be a conflict between the criteria used, often on an *ad hoc* basis, by the field surveyor to differentiate soils and the criteria imposed by the classifier.
   (*b*) An instant decision is required to annotate the field slip.
   (*c*) Non-soil characteristics are often involved in the classification.
5. For a specific purpose. In this case:
   (*a*) the criteria used will emphasise those factors of specific interest to the classifier and may ignore other properties of the soil, e.g. a civil engineer may classify soils on their bearing capacity or permeability, but ignore the existence of different genetic horizons.
   (*b*) numerous different classifications are likely to be needed.

Therefore, it is necessary to start by examining the main properties of soils that must be considered when constructing classifications and systems of nomenclature. This is followed by soil horizon nomenclature and then an account of the following systems of soil classification:

Systems developed in the USSR (p. 124)
Systems developed in the USA (p. 126)
Soil taxonomy (USDA, 1975) (p. 126)
Kubiëna's system 1953, (p. 145)
The system of Avery 1973, (p. 147)
French system (p. 148)
Canadian system (p. 148)
South African system (p. 149)
Del Villar's system 1937, (p. 155)
Leeper's system 1956, (p. 155)
Northcote's system, (p. 156)
FitzPatrick's system (p. 159)
Some other systems (p. 158)
Numerical taxonomy (p. 169)

The FAO system (p. 170) is given in some detail since it is used in the description of the world soils in Chapter 6.

The properties of soils to be considered when constructing classifications and systems of nomenclature include:

Horizon recognition
Horizon homology
Intergrading of horizons

HORIZON RECOGNITION

It is generally accepted that soil can be classified according to the type and number of horizons present in the pedo-unit. Therefore, the recognition of horizons is of paramount importance. However, there are innumerable difficulties and it would seem fair to say that so far horizon recognition has been more of an art based on experience rather than a science based on any set of defined principles.

Many attempts have been made to define horizons and it now appears that very lengthy definitions are required in order to attain precision as shown by the USDA (1975) and FitzPatrick (1971). These definitions suggest that the recognition of horizons is not as simple as was once thought and recourse to laboratory data is often required. A further serious problem is that it is not always possible to use the intrinsic properties of the horizon; it is often necessary to compare the properties with the horizon above or below. This indicates clearly that the most fundamental aspect of soil recognition and classification is still to be resolved.

HORIZON HOMOLOGY

The usual approach to soil classification is to try to classify whole soils or full assemblages of horizons within any one soil. This approach appeared to be very sound when it was thought that soils resulted from the straightforward interaction of the present day factors, parent material, climate, organisms and topography through time, such interactions leading to soils with characteristic horizon sequences. This led to the idea that soils were homologous and showed a number of different homologous sequences that could be the basis for their classification. Whereas major divisions within the plant and animal kingdom are based on homologous relationships, it is difficult if not impossible to establish such relationships for all soils. There may be a *prima facie* case for soils since many soils with similar profiles to the Podzol described on page 1 do exist, but there are also many departures from this somewhat ideal model.

The difficulties with soil homology are represented in Fig. 5.1 which shows five horizon sequences overlying the relatively unaltered material represented by the open shading. Profiles (a) and (b) represent two soils that are homologous since both have the same three horizons and they are in the same position and vertical sequence. This situation is very common. Similarly, we find many soils similar to that shown in

Fig. 1.1. Profile (c) is partly homologous with profiles (a) and (b) since horizons 1 and 3 are present but it is not completely homologous since the vertical sequence is different because horizon 2 is absent. This situation is also very common. Profile (d) is partly homologous with profile (c) since horizons 1 and 3 are present in both and in the same vertical sequence but profile (d) has the extra horizon 4 and therefore is not homologous with profile (c). Profile (e) is partly homologous with profiles (a) and (b) but like profile (d) it has the extra horizon 4. The situation regarding soil homology is even more complex since horizons 1 and 3 can be replaced together or separately by other horizons, so that the number and type of horizon sequences become extremely large.

The situation is further complicated since many factors particularly climate change through time, and properties formed under one set of climatic conditions can be inherited into the new set of climatic conditions and often themselves determine the direction of soil development. For example, profile (b) may gradually lose its homology with profile (a) through time and may develop into yet another sequence. Thus the homology between soils may only be temporary since each may be developing in different directions. The result is that the juxtarelationships of horizons are extremely diverse. Thus, although homology exists for many soils it is by no means universal and this seems to make it impossible to create a fixed number of classes based on homologous relationships. In fact it is true to say that nearly every horizon can occur in juxtaposition with nearly every other horizon.

INTERGRADING OF HORIZONS

Figure 5.2 illustrates a relatively simple situation in which the lateral continuum in the middle horizon of a number of soils is from those containing more iron and aluminium than the horizon above to those in which the contents of iron and aluminium are similar to the horizon above. Since it is virtually impossible to communicate information in terms of continuous variation unless lengthy descriptions are used, some attempt must be made to impose boundaries in the continuum, both with regard to the areas with maximum expression of the dominant property or properties as well as with regard to the intergrading situations even though the boundaries are of necessity somewhat arbitrary. There would seem to be three methods of creating divisions in a continuum such as soil.

The first method is to set a boundary in the con-

121

(a)   (b)   (c)   (d)   (e)

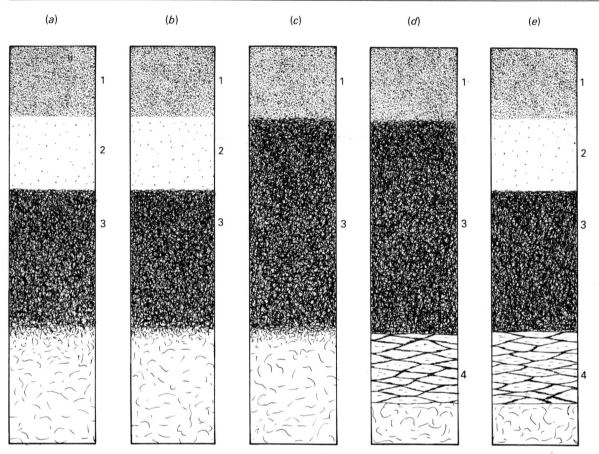

Fig. 5.1 Five profiles showing four different sequences of horizon

tinuum at point 2 in Fig. 5.2 thereby creating two segments. Each segment could then be defined in terms of the maximum expression of some specific property or properties as well as in terms of the

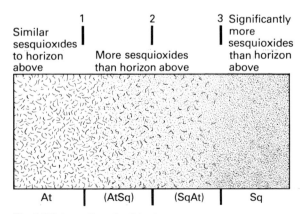

Fig. 5.2 Intergrading of soil horizons

lateral variations. This might be a practical solution if each segment formed a continuum with only one other segment, but in reality segments have continuous variations vertically as well as horizontally. Furthermore, in another situation a specific segment might form a lateral continuum into other quite different segments. The result is that segment definition would have to be written in terms of all the possible continua which would be cumbersome and could even be meaningless.

A second approach is to divide up the continuum somewhat arbitrarily and to ascribe a name and symbol to the range between each division. In this method the range between each arbitrary division is given equal status. This would tend to blur the concept of intergrades but although in reality all soils are intergrades in time, it is still necessary to give prominence to the concept that some parts of the continuum are intergrades in space while others are not.

A third and more realistic approach is to create

what would appear to be two reference segments and two intergrade segments by placing boundaries at positions 1, 2 and 3. This method produces relatively few reference segments, but there are very many intergrades since any one segment may intergrade into two or more other segments. However, it must be stated that boundaries have no fundamental meaning because any two adjacent points situated on either side of a boundary are more like each other than each is to distant points within its own segment.

The difficulties are summarised by Webster (1976) who states 'Furthermore soil populations are generally evenly distributed in character space. There seem to be few gaps and even fewer clusters. This is the nature of variation in soil, and it is against this background that the work of soil taxonomists needs to be seen and its outcome understood.'

HORIZON NOMENCLATURE

The classical approach to horizon nomenclature uses the letters, A, B and C to designate the upper, middle and lower horizons, and by the use of appropriate prefixes and suffixes adds specificity. Thus A1 is used for the upper organic mineral mixture in Podzols, A2 for the bleached horizon, and Bhfe for the horizon of accumulation of iron and humus. Although this scheme appears simple and straightforward it has not been very successful, so that each major soil survey organisation has its own set of prefixes and suffixes. However, the greatest deficiencies in this scheme are found in tropical areas where the soils are very old and have been through many contrasting phases of soil formation, and where many of the soils in the mid and lower slope positions are stratified as a result of material being washed down over the surface. The difficulties have led to workers such as Nye (1954) Watson (1964) and Webster (1965) to abandon the ABC system and to develop their own systems to suit their local needs. The problem also arises in the designation of alluvial and tundra soils. Pons and Zonneveld (1965) in discussing alluvial soils state 'It would thus seem that there is no really pure C horizon in alluvial soils.' French (1976) states 'In profile, tundra soils show considerable variation and the application of the conventional horizon designation has little meaning.' In spite of these valid and highly justified rejections of the ABC nomenclature it is still used by a considerable number of soil scientists. Therefore the system according to FAO is given on page 169.

There has always been the tendency to give names

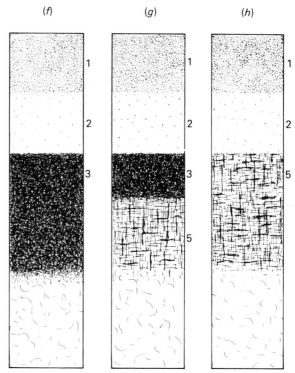

Fig. 5.3 Three horizon sequences

to horizons, such as mull for the fertile crumbly top-soil, and there are now over twenty different horizon names used in different parts of the world.

Recently, USDA (1975) has introduced the concept of the diagnostic horizon in an attempt to introduce more precision in classification.

The diagnostic horizon

In some manner not specifically stated, some horizons are termed diagnostic and are regarded as being more important than others and therefore can be used as aids to define classes more accurately.

This idea was rapidly accepted by numerous workers, and in some cases incorporated into their systems of classification, e.g. FAO (1974) and Avery (1973). We find now that great emphasis is placed on so-called diagnostic horizons for which there are lengthy descriptions, particularly by USDA (1975). Perhaps one of the most important of the diagnostic horizons is the argillic horizon which is described as an illuvial horizon containing more clay than the horizon above and possessing more than 1 per cent coatings of clay washed in from the horizon above. **123**

The definition is long and tedious and extremely difficult to understand, but that apart, the definition can be criticised on a number of important points. The description does not clearly show that an increase in clay with depth is due to clay translocation. It is well established that a marked textural change in some soils is often caused by differential weathering, differential erosion and stratification and does not result always from the washing in of fine clay. There is also the situation in which the maximum content of clay coatings is below the maximum clay percentage and seems to be caused by an *in situ* reorganisation of the clay, rather than movement from one horizon to the other. The value of 1 per cent clay coating seems very small, and often occurs in soils with no measurable clay difference between the upper and middle horizon.

Another problem with the concept of the diagnostic horizon is that often two or more diagnostic horizons appear in the same soil, thus making such a soil partly homologous with other soils that contain only one of the two diagnostic horizons. This situation is illustrated in Fig. 5.3, in which profile f is homologous with profiles (a) and (b) in Fig. 5.1. These three soils together with all of the others in Fig. 5.1 could be grouped together on the basis of horizon 3 being a diagnostic horizon. This is done for Podzols or Spodosols as they are termed by USDA. In contrast profile (h) is not homologous with profile (f) because horizon 5 replaces horizon 3. If, as is usually the case, horizon 5 is a diagnostic horizon then profile (h) is placed in a separate class. Thus profiles (f) and (h) are placed in two different classes because the middle diagnostic horizon in each is different. This seems logical until the situation is found when two or more diagnostic horizons used to create separate classes occur in the same soil, as shown in (g). This is common

in many parts of the world. Therefore, there has to be a ranking of the horizons with regard to their importance so that some diagnostic horizons are more diagnostic than others. This is another example of the absence of homology in soil science.

THE PRINCIPAL SYSTEMS DEVELOPED IN THE USSR

The work of Dokuchaev (Afanasiev, 1927) and his students focused attention on the soil profile which quickly led them to realise that many soil characteristics are profoundly affected by differences in environmental factors. It was evident also that in the USSR there is a fairly close relationship between soils and vegetation, but more particularly between soil and climate. These discoveries revolutionised soil science and inspired Dokuchaev to propose a classification of soils, the final form of which is presented in Table 5.1. Most workers in the USSR accepted these suggestions and later Sibirtsev (1914) produced a refined system by introducing the terms zonal and intrazonal soils, to replace two of the original names of the classes put forward by Dokuchaev. In both systems the members of the first two classes are defined in terms of environmental factors, while those of the third class are defined mainly on the basis of their intrinsic soil characteristics. Zonal soils have well developed pedo-units which reflect the influence of climate and vegetation. Intrazonal soils also have well developed pedo-units, but are formed as a result of the influence of some specific local factor other than climate; this may be parent material or topography. The azonal soils are poorly developed soils such as recent alluvial or stony mountain soils. Dokuchaev also emphasised that the definitions of

Table 5.1 Classification of soils by Dokuchaev

Class A. Normal, otherwise dry land vegetative or zonal soils

Zones	I. Boreal	II. Taiga	III. Forest-steppe	IV. Steppe	V. Desert-steppe	VI. Aerial or desert zone	VII. Subtropical and zone of tropical forests
Soil types	Tundra (dark brown) soils	Light grey podzolised soils	Grey and dark grey soils	Chernozem	Chestnut and brown soils	Aerial soils, yellow soils, white soils	Laterite or red soils

Class B. Transitional soils			**Class C. Abnormal soils**		
VIII. Dry land moor-soils or moor-meadow soils	IX. Carbonate containing soils (rendzina)	X. Secondary alkali-soils	X. Moor-soils	XI. Alluvial soils	XIV. Aeolian soils

soils should include a consideration of the properties of the soils themselves.

This system is well suited to the great gently undulating continental transect down through western USSR but it does not apply satisfactorily to mountainous areas, maritime conditions, or to the old land surfaces such as Africa or Australia. In the latter areas climatic changes during the Tertiary and Pleistocene Periods brought about a situation where many soils that developed under one climatic regime are present in an entirely different one today. In spite of these deficiencies this system continues to be used in the USSR, with slight modifications to include the influence of man and also some indications of the agricultural potentialities of the soils (Basinski 1959).

Among the most recent work on soil classification to emerge from the USSR is that by Kovda *et al.* (1967), who state that their scheme is an 'historical–genetic classification . . . according to their properties and characteristics which reflect their evolution in time and not according to environmental conditions as was done in some earlier classifications'. This scheme relies heavily on the interpretation of facts rather than on the facts themselves, namely the soil properties. In this respect it is identical with classifications based on environmental factors. However, it represents a substantial break with the past and many of the ideas and practices used in the USSR at present. The system of classification that is now generally used in the USSR is by Rozov and Ivanova (1968). This system has the following ten levels of categorisation: (1) class; (2) subclass; (3) ranges; (4) types; (5) subtypes; (6) genera; (7) species; (8) varieties; (9) categories; and (10) phases.

At present the type is the highest level that is used since there is as yet no full agreement about the definition of the higher levels. The subdivision of the soils below the type level is based on three groups of properties or three coordinate axes. The first group relates to the essential properties of the soils and those of its environment that can be little changed by man. These include climate, viz. seasonal soil temperature and moisture regimes, intensity of weathering, degree of leaching and organic matter turnover. The second group or axis relates to the moisture characteristics of the soil and are:

1. automorphic (aerobic)
2. weakly hydromorphic (weakly anaerobic) alluvium
3. semihydromorphic (moderately anaerobic)
4. hydromorphic (anaerobic)

The third group or coordinate relates to bio-physico-chemical soil ranges of which their are five

resulting from:

1. peculiarities of the organic matter decomposition
2. Saturation of the adsorption complex and cation composition
3. General structure of the soil profile and the presence of carbonates, gypsum and soluble salts

Soil type characteristics

The definitions of the soil types are based on soil properties including:

1. Profile morphology.
2. Mineral and chemical composition including the organic matter.
3. Physico-chemical properties.
4. Hydrothermic, gaseous and biological regimes.

In Table 5.2 the data for the 15 main soil types are given. In the first column is the list of the names of the soil types. They are usually named by combining the colour of the upper horizon with 'zem' (meaning land) to produce such names as Chernozem and Krasnozem. Other soil names such as Podzol and Solonchak that indicate the presence of a conspicuous property are also used. In the second column are data about the water regime and the depth of moistening. The third and fourth columns give data respectively for the temperature of the root-containing layer in July and the temperature gradient between the upper and lower soil horizon for the same month. The fifth column is a list of the main soil-forming processes. The sixth column gives the ABC symbols that are used in the USSR and the suffixes: humus composition f = fulvic, f–g = fulvic–humic, g = humic, exchangeable cation composition – Ca, Na, g = gley, m = cryogenic, h = always cold, og = argillification, k = calcium carbonate accumulation, s = salt accumulation. Columns 7 and 8 give the subdivisions of the subtypes and genera.

Subtype characteristics

The criteria for the definition of the subtypes include the intensity of humus and organic matter accumulation; trends and forms of translocated carbonates, gypsum, soluble salts and other substances; depth of winter cooling and of summer heating.

The nomenclature of the subtype is based on the property (podzolised, leached, etc.) or location (northern, southern) or temperature (warm, cold). Thus one gets Podzolised Chernozems which give an indication about the intergrades between soils.

Genera. This level is based on the properties of the parent material, effects of groundwater or on relic

125

features. This may cause modification in one of the horizons, a difference in the horizon sequence or the presence of an extra horizon. The nomenclature is based either on soil properties (carbonatic, solonetzic) or on some relic feature (residual hydromorphic).

Species. The genera are subdivided into species on the basis of the intensity of the main pedogenic processes. Usually three types of characteristics are used: (1) depth of a certain horizon; (2) quantity of a single or group of substances in one of the horizons; (3) content of a single or group of substances in the solum in t ha^{-1} or kg m^{-2}. In Podzols this might be the degree of leaching to give weak, medium, strong and very strong Podzols. The nomenclature of species takes into account the quantitative expression of the processes or phenomena such as deep, moderately deep, weak or strong.

Varieties. Subdivided and named on the basis of texture.

Categories. Subdivided and named on the basis of the parent material, e.g. a moderately podzolic-humus-illuvial loamy sandy soil on old alluvium.

Phases. Subdivided into four categories and named on the basis of erosion: (1) weak; (2) moderate; (3) stony; (4) very stony.

SYSTEMS DEVELOPED IN THE USA

Hilgard (1833–1906) pioneered early ideas about soil classification recognising that soils were natural phenomena and are related to their climate and vegetation. Thus, it is interesting to note that in both the USSR and USA similar ideas were being developed by soil scientists unknown to each other but they were more rapidly and further developed in the USSR. Whitney (1909) developed the first system of classification of the USA based mainly on physiography and texture and related to soil survey and mapping. Subsequent workers in various parts of the world have taken one of two paths: either they have accepted the morphogenetic approach from the USSR or they have insisted that the intrinsic properties of soils should form the basis of soil classifications. Among the first to attempt the latter approach in the USA was Coffey (1912) who produced a system purporting to be based on soil properties but his terms such as 'Dark coloured prairie soils' have genetic connotations.

Marbut can be singled out as one of the great pedologists of the first four decades of this century. Although his original ideas led him to state that soil should be classified on a morphogenetic basis, he was fully convinced at a later date that the properties of the soils themselves should be used, whereupon he presented the system given in Table 5.3.

In this system soils are divided at first into two main categories — Pedalfers and Pedocals. The Pedalfers are soils that accumulate sesquioxides while the Pedocals have an horizon of carbonate accumulation. Marbut placed great emphasis on mature, freely drained soils and so little account is taken of peat, poorly drained soils, or the initial stages of soil development. A further criticism is that difficulties are encountered when trying to accommodate soils such as the Brown Earths and Zheltozems, most of which accumulate neither iron nor carbonate.

Marbut's (1928) system was superseded quickly by that of Baldwin *et al.* (1938), who reintroduced an embellished zonal system of classification based on the USSR system but included more soil types and a few minor refinements to suit the particular conditions in the USA. This system is given in Table 5.4 but it cannot be regarded as a step forward because it does not continue the more realistic trend established by Coffey and Marbut whereby soils should be classified on their properties. Soil classifications founded on zonal concepts are very attractive to soil surveyors, geographers and ecologists and this possibly accounts for their continued use. This system has been rejected for many reasons. Firstly, some soils such as Podzols could be zonal in some areas of Europe and intrazonal in the tropics due to the influence of highly quartzose sand. Secondly, the system was based mainly on environmental factors rather than intrinsic properties of soils, and in most cases the tendency was to use virgin soils; finally, too much emphasis was placed on colour.

The most recent system of soil classification produced in the USA is entitled Soil Taxonomy and is fully discussed below.

SOIL TAXONOMY (USDA, 1975)

This system was started in 1951 and has evolved by a number of approximations, the 7th Approximation being published in 1960.

There are six levels of categorisation: (1) order; (2) suborder; (3) great group; (4) subgroup; (5) family; and (6) series.

Orders

There are 10 orders that are differentiated by the presence or absence of diagnostic horizons or features that show the dominant set of soil-forming processes that have taken place. Thus they are created in a subjective manner, i.e. there are no fixed principles involved.

Suborders

The differentiating criteria for the suborders vary from order to order and are created to suit the criteria used to establish the order in the first place.

Great groups

At the great group level the whole assemblage of horizons is considered and the most important property of the whole soil is selected.

Subgroup

There are three kinds of subgroups:

1. The central concept of the great group
2. The intergrades or transitional forms to other orders, suborders or great groups. The properties chosen include:
   (a) Horizons in addition to those definitive of the great group including an argillic horizon that underlies a spodic horizon.
   (b) Intermittent horizons.
   (c) Properties of one or more other great groups but are subordinate to the main properties.
3. Extragrades
   These have some properties that are not representative of the great group but do not indicate transitions.

Families

The grouping of soils within the families is based on the presence or absence of similar physical and chemical properties that affect their response to management and manipulation for use and may not be significant as indicators of any particular process. The properties include particle size distribution and mineralogy beneath the plough layer, temperature regime and thickness of rooting zone.

Series

This is the lowest category and the differentiating criteria are mainly the same as those used for the higher categories but the range permitted in one or more properties is less. Like the family criteria, they are related to management. The soil series are given local place names such as the Miami series.

The names of the orders are all coined words with the common ending *sol* such as *Spodosol.* A formative element is abstracted from the name and used as an ending for the names of all suborders, great groups and subgroups of one order. In the case of the Spodosols the formative element is *-od-*. Each suborder name consists of two syllables. The first indicates the property of the class and the second is the formative element of the order. Thus the *Orthods* are the common Spodosols (Gk. *Orthos* = true). The names of the great groups are produced by adding one or more prefixes to the name of the suborder. Thus the *Cryorthods* are the cold *Orthods* (Gk. *Kruos* = cold). The subgroup names consist of the name of the appropriate great group preceded by one or more adjectives as in *Typic Cryorthods* — the typical Cryorthods. The orders, suborders and great groups are given in Table 5.5 while the various formative elements are given in Tables 5.6, 5.7 and 5.8. An example of how the system operates as applied to the Spodosols is shown in Table 5.9. Also included are the equivalent horizon designations for the subgroups.

The most striking feature of this system is the overall appearance or orderliness. The subdivisions appear to follow logically one from the other and the names seem to be comprehensive and informative, but this is certainly not the case, particularly with regard to the value of the names of the subgroups. A careful inspection soon reveals that even the names at the subgroup level are either not very informative or they are not specific. For example it is shown in Table 5.9 that a Typic Fragiorthod may have a friable or hard spodic horizon (sesquon). Alfic Fragiorthods have a much wider range of possibilities; in fact the only horizon common to all the members is a fragipan which is not a diagnostic characteristic of Spodosols. The number of possibilities is even greater since the fragipan of USDA includes isons and fragons so that there may be twice as many horizon sequences if In could be substituted in every case for Fg. Thus, after descending through four levels of categorisation, very little positive information is available about the soil. This is inevitable when a presence or absence, or bifurcating system is used. An additional feature of this system is the great emphasis placed on so-called diagnostic horizons for which there are lengthy descriptions. But, as explained earlier, diagnostic horizons have doubtful value.

The authors have attempted to remove genesis as far as possible — Buol *et al.* (1973) state in reference to the definition of the classes 'genesis is not employed except as a guide to relevance and weighting of soil

127

Table 5.2 General diagnostic characteristics of main soil types of the USSR (Rozov and Ivanova, 1968)

Genetic soil types (examples)	Water regime	Thermical regime (approximate data)		Main soil type determining processes	Main genetic horizons	Subtype are subdivided according to: (processes and phenomena)	Genera are subdivided according to: (phenomena and properties)
		Main July t° (in A hor)	t° of isothermal layer				
1	2	3	4	5	6	7	8
Arctic soils	Cryogenic with a slight extra-moistening	+2° +6°	−8° −5°	Weak accumulation of weakly unsaturated humic-fulvic humus	A_f humic	Development and forms of humification and depth of active layer	Worked out
				Weak gley process in the active layer	B(g) transitional, weakly gleyed	(Desert-arctic, typical, arctic)	Not studied
				Distinct cryogen phenomena and microrelief formation of polygonal profile type depth 30–40 cm	Cm cryogenic		
Tundra soils	Cryogenic with a long period of moistening	+6° +10°	−8° 0	Superficial dry peat-formation (knobs)	A_f(g) humus gleyic	Forms of humification	Distinctness of gley process dependent on texture
				Unsaturated fulvic mobile humus formation	Bg transitional gleyic	Balance of superficial and over-cryogenic gley	
				Superficial and over-cryogenic gley		Existence of podzolisation	Existence of humic-illuvial horizon
				Cryogen phenomena and micro-relief-formation of knobbly and polygonal types with cracky thermocarst Profile depth 30–100 cm			
Cryogenic taiga soils (without pale soils)	Cryogenic percolated with a seasonal surface-water on the the perma-frost	+10° +14°	−10° 0	Accumulation of unsaturated and weakly saturated fulvic and brown fulvic and brown-humic fulvic, often mobile humus	Af humic	Existence of: Superficial (climatic) gley	Depth of summer thawing depending on rock properties
				Weak gley process and humus retention in the supercryogen layer	B(g) transitional	Podzolisation	Existence of residual carbonatic horizon
				Salt and carbonate leaching			Existence of residual signs of hydro-morphic soil formation

Table 5.2 – *(continued)*

Genetic soil types (examples)	Water regime	Thermical regime (approximate data)		Main soil type determining processes	Main genetic horizons	Subtype are subdivided according to: (processes and phenomena)	Genera are subdivided according to: (phenomena and properties)
		Main July t° (in A hor)	t° of isothermal layer				
1	2	3	4	5	6	7	8
Cryogenic taiga soils (without pale soils)				cryogen phenomena with local micro-relief formation of cracky and partly knobby types with thermo-crust. Profile depth 50–100 cm			
Podzolic soils	Permacide	+10°	−1°	Formation and partial accumulation of mobile fulvic and humic-fulvic humus	Afh humus	Existence of superficial (climatic) gley	Presence of: Iron-illuvial horizon, humus-illuvial horizon
				Podzol-forming process (podzolisation and partial lessivage) salt and carbonate leaching. Profile depth 100–200 cm	A₂ podzolic B illuvial or transitional C leached	Forms of humification (typical, sod) Depth and intensity of winter freezing and summer heating	Contact-gley horizon Residual carbonate horizon Relic 'second' humic horizon etc.
Bog-Podzolic soils (superficial moistening)	Periodical alteration of permacide and stagnant water regimes	–	–	Superficial accumulation of organic matter	A_fg organic	Development and forms of humification	Presence of: Humus-illuvial horizon
				Moblile fulvic and humic-fulvic humus formation with its partial accumulation	A_fg humic A₂g podzolic	Type of moistening (superficially and ground water gleyish and gleyed)	Iron hardpan Contact-gley horizon
				Gley podzol-forming process Gleying of the solum	Bg illuvial Cg gleyed C not gleyed, leached	Intensity of winter freezing and summer warming	Relic 'second' humus horizon etc.
Grey forest soils	Periodically permacide	+16° +20°	+1° +8°	Accumulation of weakly saturated fulvic-humic and humic-fulvic humus	Afh humus AB transitional	Development of humus accumulation (light grey, grey, dark grey soils)	Presence of: Relic 'second' humus horizon
				Podzolisation (lessivage and partly podzol-forming process)	B illuvial	Character and intensity of podzolisation	Contact-gley horizon
				Formation of illuvial-carbonate horizon at a certain depth	C leached Ck calcic	Intensity and depth of winter freezing and summer	Vertisolic horizon Residual lithogenic-

129

Table 5.2 – *(continued)*

Genetic soil types (examples)	Water regime	Thermical regime (approximate data)		Main soil type determining processes	Main genetic horizons	Subtype are subdivided according to: (processes and phenomena)	Genera are subdivided according to: (phenomena and properties)
		Main July t° (in A hor)	t° of isothermal layer				
1	2	3	4	5	6	7	8
Grey forest soils				Salt-leaching Profile depth 150–200 cm		warming	calcic horizon etc.
Brown forest soils	Permacide	+16° +22°	+4° +40°	Formation with partial accumulation brown-humic-fulvic unsaturated humus	Afh humus Bag transitional	Existence of podzolisation, lessivage and pseudogleyification	Presence of: Residual carbonate horizon
				Argilification throughout the profile (of hydro-mica-illite type)	C leached	Degree of saturation (acid, weakly unsaturated)	Ancient ferrallitic horizon etc.
				Leaching of salts and carbonates out of the solum and the crust of weathering		Intensity and depth of winter cooling and summer warming	
Cherno-zems	Imperma-cide with a profound moistening in wet years			Humic Ca-saturated humus accumulation in a deep root-containing layer	A_h^{Ca} humus $AB_{(k)}$ transitional with humus Bk carbonate-illuvial	General humus accumulation in the profile (humus reserved)	Place in the profile of the carbonate-illuvial horizon depending on the rock composition
				Carbonate migrations in the solum together with their general leaching and formation of carbonate-illuvial horizon		Existence of podzolisation and leaching Type of carbonate migration in the profile and forms of their individualisation (pseudomy-celium, farinaceous carbonates).	Presence of: Solonetzic horizon Vertisolic horizon Lithogenic-carbonate horizon Contact-gley horizon etc.
				Salt leaching or formation of deep salt illuvial horizons	Ck carbonate		
				Profile depth 150–300 cm		Depth of salt-leaching Depth and intensity of winter cooling and summer warming	
Meadow Cherno-zemic soils (ground-water moistening)	Periodical alteration of water stagnant and weakly exudative regimes			Humic Ca-saturated humus accumulation in the root-containing layer	A_k^{Ca} humus AB transitional with humus $B_{k(g)}$ carbonate illuvial gleyed Cg carbonatic gleyed, below ground water	Type of moistening (superficial, ground water) Distinctness of meadow process (meadowish or meadow-	Presence of: Podzolised horizon Solodised horizon Solonetzic horizon Solonchak horizon
				Secondary carbonate and salt migration against background of			

Table 5.2 – (*continued*)

Genetic soil types (examples)	Water regime	Thermical regime (approximate data)		Main soil type determining processes	Main genetic horizons	Subtype are subdivided according to: (processes and phenomena)	Genera are subdivided according to: (phenomena and properties)
		Main July $t°$ (in A hor)	$t°$ of isothermal layer				
1	2	3	4	5	6	7	8
Meadow Cherno-zemic soils (ground-water moistening)				total leaching Gleying of the lower part of the solum up to the ground water		Chernozemic) Intensity of winter cooling and summer warming	Vertisolic horizon etc.
Chestnut soils	Impermacide			Accumulation of humic and fulvic-humic Ca-saturated humus Carbonate migrations in the profile in certain cases of weak salt solutions Formation of salt and carbonate-illuvial horizons Profile depth 100–200 cm	A_h^{Ca} humus $B_{(k)}$ transitional carbonate Bks salt-illuvial C_s carbonate-saline	Total humus accumulation (dark chestnut, chestnut, light chestnut) Forms of migrations and individualisation of carbonates in the profile and degree of salt leaching Intensity and depth of winter cooling and summer warming	Place in the profile of the carbonate-illuvial horizon depending on the rock composition Existence of Solonetzic horizon Solodised horizon Solonchak horizon Relic humus horizon on river terraces, etc.
Meadow step chestnut Solonetz (ground-water moistening)	Impermacide with a period periodical capillary ascension			Accumulation of humic-fulvic Na, Mg and Ca-saturated humus Solonetz process Salt migrations against the background of their gradual leaching Gleying of the lower part of the profile up to the ground water	Afh^{Na-Ca} humus $B_{(Sol)}^{Na-Ca}$ B(g) transitional C_{ksg} carbonate-saline-gleyed below ground water	Forms of Solonetz process development (Solonetz-Solonchaks, typical, solodized non-natric) Distinctness of meadow process (meadowish- and meadow-steppe) Intensity of winter cooling and summer warming	Salt composi-tion (sodic, compound, chloride, sulphate) Depth of gypsum
Brown semi-desertic soils	Impermacide with a shallow spring–summer moistening	+22° +28° and more		Formation and small accumula-tion of humic-fulvic Ca-saturated humus Carbonate and weak salt-solution migra-tions in the profile	Afh^{Ca} with a low humus content $B_{(sol)}$ transitional B_k carbonate-illuvial B_{ks} carbonate-saline	Not worked out	Presence of: Solonetzic horizon Leached horizon depending upon rock composition Anciently saline horizon

131

Table 5.2 – (*continued*)

Genetic soil types (examples)	Water regime	Thermical regime (approximate data)		Main soil type determining processes	Main genetic horizons	Subtype are subdivided according to: (processes and phenomena)	Genera are subdivided according to: (phenomena and properties)
		Main July $t°$ (in A hor)	$t°$ of isothermal layer				
1	2	3	4	5	6	7	8
Brown semi-desertic soils				Profile depth 80–100 cm			Residual gypsum horizon etc.
Grey-brown desertic soils	Impermacide	+26° +32° and more		A very small accumulation of humic-fulvic Ca-saturated humus Carbonate migrations with surface enrichment Salt migrations with accumulation in the sub-humus horizon Iron oxides dehydration Profile depth about 50 cm	Afh^{Ca} with a low humus content carbonatic B_{ks} transitional carbonate-saline C_{ks} carbonate-saline with gypsum	Degree of superficial carbonate enrichment (carbonatic, poor in carbonates)	Presence of: Solonetzic horizon residual gypsum horizon
Jeltozem	Permacide	+20° +26°		Formation and accumulation of unsaturated mobile fulvic humus Argillification throughout the profile (of kaolinite and halloysite type) Salt and carbonate leaching out of the solum and crust of weathering Profile depth about 200 cm	A_f humus B transitional C leached	Saturation degree (not saturated, weakly unsaturated) Existence of podzolisation (lessivage and pseudo-gleyification Possibility of winter cooling	Presence of residual-carbonatic horizon
Cinnamonic soils	Not percolated or weakly percolated with a profound winter moistening	+22° +27°		Considerable accumulation of mainly humic Ca-saturated humus (in neutral alkaline medium) Formation of carbonate-illuvial horizons Salt leaching out of the profile Profile depth more than 200 cm	A_h^{Ca} humus B textural metamorphic C_k carbonatic	Degree of carbonate-leaching and depth of carbonate-illuvial horizon (leached, typical carbonatic)	Presence of: lithogenic-carbonate hor. Residual Solonetzic hor. Ancient saline horizon Residual meadow process indications, etc.

Table 5.2 – (*continued*)

Genetic soil types (examples)	Water regime	Thermical regime (approximate data)		Main soil type determining processes	Main genetic horizons	Subtype are subdivided according to: (processes and phenomena)	Genera are subdivided according to: (phenomena and properties)
		Main July t° (in A hor)	t° of isothermal layer				
1	2	3	4	5	6	7	8
Grey-Cinnamonic soils	Not percolated, with winter moistening	+23° +28°		Accumulation of Ca-saturated humic and humic-fulvic humus Weak in neutral medium Carbonate migrations locally to the surface and formation of carbonate-illuvial horizon Profile depth surpasses 200 cm	A_h^{Ca} humus B textural metamorphic carbonatic C_k carbonatic	Development of humus accumulation (dark, usual, light)	Presence of: Solonetzic horizon Ancient saline horizon Ancient gypsum horizon Residual signs of meadow process Pebble layers, etc.
Serozems	Not percolated, with a slight winter moistening	+24° +30° and more		Small accumulation of humic-fulvic Ca-saturated humus, crypto (in neutral alkaline medium) Carbonate migrations with ascendant movements, accumulation throughout the profile and certain leaching of humus horizon Soluble salts are present in small quantities in the lower part of the solum Profile depth about 200 cm	A_h^{Ca} with a low humus content B_k transitional crypto-textural carbonatic Ck/s carbonatic, salts are possible	Degree of superficial humus accumulation (light, typical dark) Intensity of carbonate migrations in the solum (rich or poor in carbonates)	Presence of: Ancient saline horizon Solonetzic horizon Pebble layers, etc.

properties'. Yet in their Table 14.9 the nature of the differentiating characteristics at the order level is 'Soil forming processes . . .' and at suborder level 'Genetic homogeneity'. Thus it seems that genesis is the first and second differentiating criterion and therefore must be considered to be foremost. However, the stress placed on recognising and defining a number of horizons is a major contribution and emphasises the need for greater precision and more criteria for soil categorisation.

A further severe criticism of this scheme is the absence of a simple and straightforward method for designating intergrades.

This scheme has made a considerable impact on pedologists but they are divided about its value: there are the strong adherents and there are also those who vigorously oppose the scheme (Webster, 1968). A scrutiny of this method for classifying soils does not reveal many reasons for giving it support. The proposers tacitly assumed that soils can be classified in a hierarchical manner. Adopting this well trodden and fruitless path comes as a great surprise since it

Table 5.3 Classification of soils by Marbut

	Pedalfers (VI–1)	Pedocals (VI–2)
Category VI	Soils from mechanically comminuted materials	
Category V	Soils from siallitic decomposition products	Soils from mechanically comminuted materials
	Soils from allitic decomposition products	
	Tundra	
	Podzols	Chernozems
	Grey-Brown Podzolic soils	Dark-brown soils
Category IV	Red soils	Brown soils
	Yellow soils	Grey soils
	Prairie soils	Pedocalic soils of arctic and tropical regions
	Lateritic soils	
	Laterite soils	
	Groups of mature but related soil series	Groups of mature but related soil series
	Swamp soils	Swamp soils
	Gley soils	Gley soils
	Rendzinas	Rendzinas
Category III	Alluvial soils	Alluvial soils
	Immature soils on slopes	Immature soils on slopes
	Salty soils	Salty soils
	Alkali soils	Alkali soils
	Peat soils	Peat soils
Category II	Soil series	Soil series
Category I	Soil units, or types	Soil units, or types

should be obvious that hierarchical systems have repeatedly failed to satisfy the demands of the soil continuum. The authors state that many of their decisions are pragmatic which is tantamount to saying that scientific arrangement, i.e. a hierarchical system is not valid. However, a number of very severe criticisms can be made of this system at all levels. Starting with the orders which have been created on a purely subjective basis no principles seem to be involved. One finds statements such as 'The definition of Inceptisols is unavoidably complicated'. This would seem to imply that there is something intrinsically valid about the order of Inceptisols. In fact, the authors have decided to group together a large number of very divergent soils and then found that attemping to define such wide variability is difficult. In other words the problems relative to the definition of the

Inceptisols have been created by themsleves. There are also a number of groupings that are difficult to justify. For example, the wet Histosols are subdivided into the Fibrists, Hemists and Saprists on the basis of decomposition of the constituent plant fragments, but for some reason that is not understood all Histosols that contain pyrite are grouped with the Hemists regardless of degree of decomposition.

The whole is a perfect example of creating a rigid structure without principles and then forcing the soil continuum into the various categories and at the same time having to produce the most complicated and peculiar definitions to accommodate the variability.

Alfisols

Generally these soils have an ochric epipedon, an argillic horizon and moderate to high base saturation; water is held at <15 bar tension for at least three months each year during the growing season. Other horizons that may be present include a fragipan, duripan, natric horizon, petrocalcic horizon and plinthite. Alfisols tend to form a belt between Mollisols of the grassland areas and the Spodosols and Inceptisols of the cool humid areas where they occur mainly on late Pleistocene surfaces. They also form a belt between the Aridisols and the soils of more humid and warmer areas where they tend to be red and occur on older surfaces but often a Pleistocene influence can be found.

Aqualfs
These are the grey and mottled Alfisols that are wet or artificially drained. They may have a fluctuating ground-water-table or they may have perched water within the soil. They occur on flat sites or slight depressions.
Great groups: Albaqualfs, Duraqualfs, Fragiaqualfs, Glossaqualfs, Natraqualfs, Ochraqualfs, Plinthaqualfs, Tropaqualfs, Umbraqualfs.

Boralfs
These are the dry Alfisols of cool places, where most occur naturally under coniferous forest but a few have developed under mixed or deciduous forest. Generally they have an O horizon, an albic horizon and an argillic horizon. A thin Ah may be present in some and in regions of low rainfall a ca horizon may underlie the argillic horizon. In humid areas the lower part of the albic horizon and the upper part of the argillic horizon are strongly to very strongly acid.
Great groups: Cryoboralfs, Eutroboralfs, Fragiboralfs, Glossoboralfs, Natriboralfs, Paleboralfs.

Table 5.4 Classification of soils by Baldwin, Kellogg and Thorp

Order	Suborder	Great soil groups
Zonal soils — Pedocals	Soils of the cold zone	1. Tundra soils
	1. Light coloured soils of arid regions	2. Desert soils 3. Red Desert soils 4. Sierozem 5. Brown soils 6. Reddish-brown soils
	2. Dark coloured soils of the semi-arid, subhumid, and humid grasslands	7. Chestnut soils 8. Reddish-chestnut soils 9. Chernozem soils 10. Prairie soils 11. Reddish prairie soils
Zonal soils — Pedalfers	3. Soils of the forest-grassland transition	12. Degraded Chernozem soils 13. Non-calcic brown or Shantung brown soils
	4. Light coloured podzolised soils of the timbered regions	14. Podzol soils 15. Brown Podzolic soils 16. Grey-brown Podzolic soils
	5. Lateritic soils of forested warm-temperate and tropical regions	17. Yellow Podzolic soils 18. Red Podzolic soils (and Terra Rossa) 19. Yellowish-brown Lateritic soils 20. Reddish-brown Lateritic soils 21. Laterite soils
Intrazonal soils	1. Halomorphic (saline and alkali) soils of imperfectly drained arid regions and littoral deposits	1. Solonchak or saline soils 2. Solonetz soils 3. Soloth soils
	2. Hydromorphic soils of marshes, swamps, seep areas, and flats	4. Wiesenböden (meadow soils) 5. Alpine meadow soils 6. Bog soils 7. Half bog soils 8. Planosols 9. Ground water Podzol soils 10. Ground water Laterite soils
	3. Calomorphic	11. Brown forest soils (Braunerde) 12. Rendzina soils
Azonal soils		1. Lithosols 2. Alluvial soils 3. Sands (dry)

Table 5.5 Names of orders, suborders and great groups

Order	Suborder	Great group	Order	Suborder	Great group
Alfisols	Aqualfs	Albaqualfs Duraqualfs Fragiaqualfs Glossaqualfs Natraqualfs Ochraqualfs Plinthaqualfs Tropaqualfs Umbraqualfs	Alfisols	Udalfs	Agrudalfs Ferrudalfs Fragiudalfs Fraglossudalfs Glossudalfs Hapludalfs Natrudalfs Paleudalfs Rhodudalfs Tropudalfs
	Boralfs	Cryoboralfs Eutroboralfs Fragiboralfs Glossoboralfs Natriboralfs Paleboralfs		Ustalfs	Durustalfs Haplustalfs Natrustalfs Paleustalfs Plinthustalfs

135

Table 5.5 – (*continued*)

Order	Suborder	Great group	Order	Suborder	Great group
Alfisols		Rhodustalfs	Histosols	Hemists	Luvihemists
	Xeralfs	Durixeralfs			Medihemists
		Haploxeralfs			Sulphihemists
		Natrixeralfs			Sulphohemists
		Palexeralfs			Tropohemists
		Plinthoxeralfs		Saprists	Borosaprists
		Rhodoxeralfs			Cryosaprists
Aridisols	Argids	Durargids			Medisaprists
		Haplargids			Troposaprists
		Nadurargids	Inceptisols	Andepts	Cryandepts
		Natrargids			Durandepts
		Paleargids			Dystrandepts
	Orthids	Calciorthids			Eutrandepts
		Camborthids			Hydrandepts
		Durorthids			Placandepts
		Gypsiorthids			Vitrandepts
		Paleorthids		Aquepts	Andaquepts
		Salorthids			Cryaquepts
Entisols	Aquents	Cryaquents			Fragiaquepts
		Fluvaquents			Halaquepts
		Haplaquents			Haplaquepts
		Hydraquents			Humaquepts
		Psammaquents			Placaquepts
		Sulphaquents			Plinthaquepts
		Tropaquents			Sulphaquepts
					Tropaquepts
	Arents	Arents		Ochrepts	Cryochrepts
	Fluvents	Cryofluvents			Durochrepts
		Torrifluvents			Dystrochrepts
		Tropofluvents			Eutrochrepts
		Udifluvents			Fragiochrepts
		Ustifluvents			Ustochrepts
		Xerofluvents			Xerochrepts
	Orthents	Cryorthents		Plaggepts	Plaggepts
		Torriorthents		Tropepts	Dystropepts
		Troporthents			Eutropepts
		Udorthents			Humitropepts
		Ustorthents			Sombritropepts
		Xerorthents			Ustropepts
	Psamments	Cryopsamments		Umbrepts	Cryumbrepts
		Quartzipsamments			Fragiumbrepts
		Torripsamments			Haplumbrepts
		Tropopsamments			Xerumbrepts
		Udipsamments	Mollisols	Albolls	Argialbolls
		Ustipsamments			Natralbolls
		Xeropsamments		Aquolls	Argiaquolls
Histosols	Fibrists	Borofibrists			Calciaquolls
		Cryofibrists			Cryaquolls
		Luvifibrists			Duraquolls
		Medifibrists			Haplaquolls
		Sphagnofibrists			Natraquolls
		Tropofibrists		Borolls	Argiborolls
	Folists	Borofolists			Calciborolls
		Cryofolists			Cryoborolls
		Tropofolists			Haploborolls
	Hemists	Borohemists			Natriborolls
		Cryohemists			Paleborolls
					Vermiborolls

136

Table 5.5 – *(continued)*

Order	Suborder	Great group
Mollisols	Rendolls	Rendolls
	Udolls	Argiudolls
		Hapludolls
		Paleudolls
		Vermudolls
	Ustolls	Argiustolls
		Calciustolls
		Durustolls
		Haplustolls
		Natrustolls
		Paleustolls
		Vermustolls
	Xerolls	Argixerolls
		Calcixerolls
		Durixerolls
		Haploxerolls
		Natrixerolls
		Palexerolls
Oxisols	Aquox	Gibbsiaquox
		Ochraquox
		Plinthaquox
		Umbraquox
	Humox	Acrohumox
		Gibbsihumox
		Haplohumox
		Sombrihumox
	Orthox	Acrorthox
		Eutrorthox
		Gibbsiorthox
		Haplorthox
		Sombriorthox
		Umbriorthox
	Torrox	Torrox
	Ustox	Acrustox
		Eutrustox
		Sombriustox
		Haplustox
Spodosols	Aquods	Cryaquods
		Duraquods
		Fragiaquods
		Haplaquods
		Placaquods
		Sideraquods
		Tropaquods
	Ferrods	Ferrods
	Humods	Cryohumods
		Fragihumods
		Haplohumods
		Placohumods
		Tropohumods
	Orthods	Cryorthods
		Fragiorthods
		Haplorthods
		Placorthods
		Troporthods
Ultisols	Aquults	Albaquults
		Fragiaquults
		Ochraquults
		Paleaquults
		Plinthaquults
		Tropaquults
		Umbraquults
	Humults	Haplohumults
		Palehumults
		Plinthohumults
		Sombrihumults
		Tropohumults
	Udults	Fragiudults
		Hapludults
		Paleudults
		Plinthudults
		Rhodudults
		Tropudults
	Ustults	Haplustults
		Paleustults
		Plinthustults
		Rhodustults
	Xerults	Haploxerults
		Palexerults
Vertisols	Torrerts	Torrerts
	Uderts	Chromuderts
		Pelluderts
	Usterts	Chromusterts
		Pellusterts
	Xererts	Chromoxererts
		Pelloxererts

Table 5.6 Formative elements in names of soil orders

Name of order	Formative element in name of order	Derivation of formative element	Pronunciation of formative element
Alfisol	Alf	Meaningless syllable	Pedalfer
Aridisol	Id	L. *aridus*, dry	Arid
Entisol	Ent	Meaningless syllable	Recent
Histosol	Ist	Gr. *histos*, tissue	Histology
Inceptisol	Ept	L. *inceptum*, beginning	Inception
Mollisol	Oll	L. *mollis*, soft	Mollify
Oxisol	Ox	F. *oxide*, oxide	Oxide
Spodosol	Od	Gr. *spodos*, wood ash	Odd
Ultisol	Ult	L. *ultimus*, last	Ultimate
Vertisol	Ert	L. *verto*, turn	Invert

Table 5.7 Formative elements in names of suborders

Formative element	Derivation	Mnemonicon	Connotation
Alb	L. *albus*, white	Albino	Presence of albic horizon
And	Modified from ando	Ando	Andolike
Aqu	L. *aqua*, water	Aquarium	Aquic moisture regime
Ar	L. *arare*, to plow	Arable	Mixed horizons
Arg	Modified from argillic horizon; L. *argilla*, white clay	Argillite	Presence of argillic horizon
Bor	Gr. *boreas*, northern	Boreal	Cool
Ferr	L. *ferrum*, iron	Ferruginous	Presence of iron
Fibr	L. *fibra*, fiber	Fibrous	Least decomposed stage
Fluv	L. *fluvius*, river	Fluvial	Flood plain
Fol	L. *folia*, leaf	Foliage	Mass of leaves
Hem	Gr. *hemi*, half	Hemisphere	Intermediate stage of decomposition
Hum	L. *humus*, earth	Humus	Presence of organic matter
Ochr	Gr. base of *ochros*, pale	Ocher	Presence of ochric epipedon
Orth	Gr. *orthos*, true	Orthophonic	The common ones
Plagg	Modified from Ger. *plaggen*, sod		Presence of plaggen epipedon
Psamm	Gr. *psammos*, sand	Psammite	Sand texture
Rend	Modified from Rendzina		High carbonate content
Sapr	Gr. *sapros*, rotten	Saprophyte	Most decomposed stage
Torr	L. *torridus*, hot and dry	Torrid	Torric moisture regime
Ud	L. *udus*, humid	Udometer	Udic moisture regime
Umbr	L. *umbra*, shade	Umbrella	Presence of umbric epipedon
Ust	L. *ustus*, burnt	Combustion	Ustic moisture regime
Xer	Gr. *xeros*, dry	Xerophyte	Xeric moisture regime

Udalfs

These are the brownish or reddish, more or less freely drained Alfisols that occur principally on Pleistocene surfaces but may occur on older surfaces that are underlain by calcareous material. They occur extensively in the USA and Europe and have been extensively farmed in both areas and as a consequence are much eroded.

Great groups: Agrudalfs, Ferrudalfs, Fragiudalfs, Fraglossudalfs, Glossudalfs, Hapludalfs, Natrudalfs, Paleudalfs, Rhodudalfs, Tropudalfs.

Ustalfs

These are mostly the reddish Alfisols of warm humid to semiarid regions, they have a warm rainy season and only occasionally does moisture move through to the deeper layers. They sometimes have a ca or calcic horizon beneath or in the argillic horizon. The dry season is sufficiently pronounced to allow only deciduous or xerophytic vegetation or savanna. They tend to form belts between the Aridisols and the Ultisols, Oxisols and Inceptisols of the warm humid regions. Many are on surfaces of the late Pleistocene age but many are on older surfaces where they are highly weathered possibly in a previous more humid climate since the clays are dominated by kaolinite.

Great groups: Durustalfs, Haplustalfs, Natrustalfs, Paleustalfs, Plinthustalfs, Rhodostalfs.

Xeralfs

These are mainly reddish Alfisols of regions with a Mediterranean climate. They are dry during most of the summer but in many of them water moves through the soil in the winter of most years. Generally they have an abrupt boundary between the A and B horizons and the upper horizons are hard and massive when dry. On a world scale they are not extensive but locally they are abundant. Originally they carried a mixture of grasses, herbs and woody shrubs and some formed a late Pleistocene surface but they also occur on older land surfaces that may have had a markedly different climate in the past. Most have formed on basic parent materials.

Great groups: Durixeralfs, Haploxeralfs, Natrixeralfs, Palexeralfs, Plinthoxeralfs, Rhodoxeralfs.

Aridisols

Generally these are the soils of the dry areas and during most of the year water is held at a tension >15 bar or they have a high salt content. There is no period of three months or more when water is continuously available to plants and the soil is warm. They range from cold to very hot conditions and some Aridisols occur in semiarid areas because the soils are slowly permeable and most of the water is lost by run-off.

Table 5.8 Formative elements in names of great groups

Formative element	Derivation	Mnemonicon	Connotation
Acr	Modified from Gr. *akros*, at the end	Acrolith	Extreme weathering
Agr	L. *ager*, field	Agriculture	An agric horizon
Alb	L. *albus*, white	Albino	An albic horizon
And	Modified from ando	Ando	Andolike
Arg	Modified from argillic horizon; L. *argilla*, white clay	Argillite	An argillic horizon
Bor	Gr. *boreas*, northern	Boreal	Cool
Calc	L. *calcis*, lime	Calcium	A calcic horizon
Camb	L.L. *cambiare*, to exchange	Change	A cambic horizon
Chrom	Gr. *chroma*, colour	Chroma	High chroma
Cry	Gr. *kryos*, icy cold	Crystal	Cold
Dur	L. *durus*, hard	Durable	A duripan
Dystr, dys	Modified from Gr. *dys*, ill; dystrophic, infertile	Dystrophic	Low base saturation
Eutr, eu	Modified from Gr. *eu*, good; eutrophic, fertile	Eutrophic	High base saturation
Ferr	L. *ferrum*, iron	Ferric	Presence of iron
Fluv	L. *fluvus*, river	Fluvial	Flood plain
Frag	Modified from L. *fragilis*, brittle	Fragile	Presence of fragipan
Fragloss	Compound of fra(g) and gloss		See the formative elements frag and gloss
Gibbs	Modified from gibbsite	Gibbsite	Presence of gibbsite in sheets or nodules
Gyps	L. *gypsum*, gypsum	Gypsum	Presence of a gypsic horizon
Gloss	Gr. *glossa*, tongue	Glossary	Tongued
Hal	Gr. *hals*, salt	Halophyte	Salty
Hapl	Gr. *haplous*, simple	Haploid	Minimum horizon
Hum	L. *humus*, earth	Humus	Presence of humus
Hydr	Gr. *hydor*, water	Hydrophobia	Presence of water
Luv	Gr. *louo*, to wash	Ablution	Illuvial
Med	L. *media*, middle	Medium	Of temperate climates
Nadur	Compound of na(tr) and dur		See the formative elements natr and dur
Natr	Modified from *natrium*, sodium		Presence of natric horizon
Ochr	Gr. base of *ochros*, pale	Ocher	Presence of ochric epipedon
Pale	Gr. *paleos*, old	Paleosol	Excessive development
Pell	Gr. *pellos*, dusky		Low chroma
Plac	Gr. base of *plax*, flat stone		Presence of a thin pan
Plagg	Modified from Ger. *plaggen*, sod		Presence of plaggen epipedon
Plinth	Gr. *plinthos*, brick		Presence of plinthite
Psamm	Gr. *psammos*, sand	Psammite	Sand texture
Quartz	Ger. *quarz*, quartz	Quartz	High quartz content
Rhod	Gr. base of *rhodon*, rose	Rhododendron	Dark red colour
Sal	L. base of *sal*, salt	Saline	Presence of salic horizon
Sider	Gr. *sideros*, iron	Siderite	Presence of free iron oxides
Sombr	F. *sombre*, dark	Somber	A dark horizon
Sphagn	Gr. *sphagnos*, bog	Sphagnum	Presence of Sphagnum
Sulph	L. *sulfur*, sulphur	Sulphur	Presence of sulphides or their oxidation products
Torr	L. *torridus*, hot and dry	Torrid	Torric moisture regime
Trop	Modified from Gr. *tropikos*, of the solstice	Tropical	Humid and continually warm
Ud	L. *udus*, humid	Udometer	Udic moisture regime
Umbr	L. base of *umbra*, shade	Umbrella	Presence of umbric epipedon
Ust	L. base of *ustus*, burnt	Combustion	Ustic moisture regime
Verm	L. base of *vermes*, worm	Vermiform	Wormy, or mixed by animals
Vitr	L. *vitrum*, glass	Vitreous	Presence of glass
Xer	Gr. *xeros*, dry	Xerophyte	A xeric moisture regime

Table 5.9 An illustration of the subdivisions within the US taxonomy

Order	Suborder	Great group	Subgroup	Symbols of horizons present	
				FAO	**FitzPatrick**
Spodosols	Aquods, Ferrods, Humods, Orthods	Cryorthods, Fragiorthods, Haplorthods, Placorthods	Typic Fragiorthods	Bs Cx Bs Cx	Sq Fg SqdFg
			Alfic Fragiorthods	Bs Bt Cx Bs Bt Cx Bs E Btg Cx Bs E Btg Cx	SqArFg SqdArFg SqZo(ArGs)Fg SqdZo(ArGs)Fg
			Aquic Fragiorthods	Bsg Cx ——— Bsg Cx ———	(SqGl)Fg (SqdGl)Fg
			Aquic Entic Fragiorthods	Bsg Cx ———————	(GlSq)Fg
			Cryic Fragiorthods	Bs Cx ——— Bs Cx ———	SqFg SqdFg
			Cryohumic Fragiorthods	Bh Bs Cx ——— Bh Bs Cx ———	HdSqFg HddSqFg
			Entic Fragiorthods	Bs Cx ———————	(SqFr)Fg
			Humic Fragiorthods	Bh Bs Cx——— Bh Bs Cx———	HdSqFg HddSqFg

The profile morphology of Aridisols is very varied, they usually have an ochric or anthropic epipedon which overlies an argillic, natric, calcic, petrocalcic, gypsic, petrogypsic or cambic horizon or a duripan or combinations of two or more of the above. These horizons may have formed in the present climate but often they may be inherited from a previous climatic phase which may give rise to very complex soils.

Vegetation is usually sparse and the surface is bare for long periods. If the soils contain many stones, concretions or fragments of laterite, a desert pavement may form on the surface by deflation or surface wash.

Argids
These have an argillic or natric horizon, those with an argillic horizon seem to be older than late Pleistocene when rainfall was higher while those with a natric horizon may be much younger. They may also have a calcic or petrocalcic horizon or duripan all of which vary from a few centimetres to many metres thick.
Great groups: Durargids, Haplargids, Nadurargids, Natrargids, Paleargids.

Orthids
These contain one or more horizons but do not have an argillic or natric horizon. They commonly have a salic, calcic, gypsic, petrocalcic, petrogypsic or cambic horizon or duripan or combinations of two or more.

Most Orthids have developed on surfaces of late Pleistocene or younger. A few may be older where a petrocalcic horizon has developed within an argillic horizon.
Great groups: Calciorthids, Camborthids, Durorthids, Gypsiorthids, Paleorthids, Salorthids.

Entisols

This order is distinguished by having little or no evidence of the development of middle horizons. They occur under nearly every type of climatic condition and may have an ochric, anthropic or histic epipedon, an albic horizon or sulphidic material. Most have no horizons at all due to their youth, steepness of slope, active erosion or flood plain deposition. However, some thick albic horizons are regarded as Entisols and therefore are very old.

Aquents
These are the wet Entisols and occur mainly in recent sediments in tidal marshes, deltas, margins of lakes, and flood plains. They have a characteristically mottled, greyish or bluish middle and/or lower horizon and may contain sulphidic material.
Great groups: Cryaquents, Fluvaquents, Haplaquents, Hydraquents, Psammaquents, Sulfaquents, Tropaquents.

Arents

These soils have no horizons because of deep cultivation either recently or in the past. The soil may have fragments of a former spodic or argillic horizon or duripan or other horizons. This is a unique suborder since there are no great groups.

Fluvents

These are mainly brownish or reddish soils that have formed mainly in recent flood-plain sediments. They are frequently flooded and stratification is normal; they often contain material derived from previous soils. The carbon distribution with depth is variable due to strata of different texture and forms the basis of the definition of the Fluvents.
Great groups: Cryofluvents, Torrifluvents, Trepofluvents, Udifluvents, Ustifluvents, Xerofluvents.

Orthents

These occur on recent erosional surfaces that may be geologic or induced by cultivation so that all the diagnostic horizons for the other orders are absent. Indurated material that was once plinthite may be present but if plants are absent it is regarded as rock; they occur in all climates.
Great groups: Cryorthents, Torriorthents, Troporthents, Udorthents, Ustorthents, Xerorthents.

Psamments

These soils occur in well sorted sands such as sand dunes, cover sands or material that was sorted in an earlier geological cycle but gravelly material is excluded. They occur under any climate or vegetation and on surfaces of many ages but those of old stable surfaces consist mainly of quartz sand.
Great groups: Cryopsamments, Quartzipsamments, Torripsamments, Tropofluvents, Udipsamments, Ustipsamments, Xeropsamments.

Histosols

These soils are composed predominantly of organic matter and are generally known as peats, mucks, bogs and moors. Most are usually saturated with water unless drained but a few may be unsaturated for most of the time. Histosols can form in any climate providing enough water is present. Many form in closed depressions but in very humid climates they may blanket the landscape. The determining feature is water which may come from a variety of sources. The classification of the Histosols is provisional because of the relatively small amount of work that has been carried out. The suborders are defined on their moisture regime and degree of decomposition while the great groups are partly defined on the soil temperature regime.

Fibrist

They consist predominantly of slightly decomposed plant remains but range widely in botanical composition. They tend to have the lowest bulk density and the lowest ash content except in cases where there have been additions of volcanic ash or wind blown material.
Great groups: Borofibrists, Cryofibrists, Luvifibrists, Medifibrists, Sphagnofibrists, Tropofibrists.

Folists

These are the more or less freely drained Histosols that are composed mainly of O horizons resting on rock or fragmented material and occur mainly in very humid climates from the tropics to high latitudes.
Great groups: Borofolists, Cryofolists, Tropofolists.

Hemists

These are the wet Histosols in which only about one-third of the plant material is still recognisable. The water-table is at or near to the surface at all times except after drainage. All Histosols containing a sulphuric horizon or sulphidic material are included in the Hemists irrespective of the decomposition stage of the organic material. They occur from the Equator to the tundra in closed depressions and wet flat areas.
Great groups: Borohemists, Cryohemists, Luvihemists, Medihemists, Sulphihemists, Sulphohemists, Tropohemists.

Saprists

These consist of almost completely decomposed organic materials. They are usually black and plant remains cannot easily be identified. They tend to have a fluctuating ground-water-table with advanced decomposition taking place during the season when the water-table is at its lowest. Fibric and hemic materials will decompose and change to sapric material if drained so that Fibrists and Hemists can change to Saprists.
Great groups: Borosaprists, Cryosaprists, Medisaprists, Troposaprists.

Inceptisols

These are the soils of the humid regions that have altered horizons that have lost material by leaching but also contain weatherable minerals. They do not

141

occur in arid areas and are usually not sandy throughout. They do not have shallow plinthite, a spodic, argillic, natric, oxic, gypsic, petrogypsic or salic horizon. They most commonly have an umbric or ochric epipedon, a cambic horizon, a fragipan and a duripan. The most common horizon sequences are an ochric epipedon over a cambic horizon with or without a fragipan or an umbric epipedon overlying a cambic horizon with or without an underlying duripan or fragipan. All soils with plaggen epipedons are Inceptisols. However, the full definition of Inceptisols is very complicated.

Andepts

These are more or less freely drained with high content of allophane and low bulk density and range from the tropics to high latitudes. They are formed mainly in volcanic ash but may develop on other material such as basic igneous rocks. In dry climates they have a mollic epipedon and a duripan. Those in humid climates have an umbric or ochric epipedon. Many Andepts have buried A horizons due to repeated additions of volcanic ash to the surface.
Great groups: Cryandepts, Durandepts, Dystrandepts, Eutrandepts, Hydrandepts, Placandepts, Vitrandepts.

Aquepts

These soils have poor to very poor drainage with a histic, mollic or ochric epipedon, a mottled or blue-grey cambic horizon and may have a fragipan, duripan, sulphuric horizon or permafrost.
Great groups: Andaquepts, Cryaquepts, Fragiaquepts, Halaquepts, Haplaquepts, Humaquepts, Placaquepts, Plinthaquepts, Sulphaquepts, Tropaquepts.

Ochrepts

These are mainly light-coloured, brownish, more or less freely drained and have developed mainly on late Pleistocene to Holocene surfaces with vegetation that ranges from forest to tundra. Most have an ochric epipedon and a cambic horizon and may have a calcic horizon, fragipan or duripan.
Great groups: Cryochrepts, Durochrepts, Dystrochrepts, Eutrochrepts, Fragiochrepts, Ustochrepts, Xerochrepts.

Plaggepts

These soils have a plaggen epipedon that contains little or no pyroclastic material.

Tropepts

These are brownish to reddish more or less freely

drained and occur mainly in intertropical regions. Most have an ochric epipedon and a cambic horizon, they are usually thin and occur on steep slopes of Pleistocene or Holocene age and are very extensive in hilly tropical areas with forest or anthropic savanna.
Great groups: Dystropepts, Eutropepts, Humitropepts, Sombritropepts, Ustropepts.

Umbrepts

These are acid, dark reddish or brownish, freely drained and rich in organic matter. They occur mainly in hilly to mountainous regions of the humid mid to high altitudes in late Pleistocene or Holocene deposits under coniferous forest.
Great groups: Cryumbrepts, Fragiumbrepts, Haplumbrepts, Xerumbrepts.

Mollisols

These are the very dark-coloured base rich soils of the temperate grasslands but they also occur at high latitudes, high altitudes and in the intertropical areas but generally they lie between the Aridisols and Spodosols or Alfisols. Nearly all have a mollic epipedon and many have an argillic or a natric horizon or a calcic horizon and a few have an albic horizon. A few have a duripan or a petrocalcic horizon. Most have developed in late Pleistocene or Holocene deposits but some are on older surfaces, when they normally have a reddish argillic horizon and are believed to have had a previous forest cover.

Albolls

These are the Mollisols that have an albic horizon and fluctuating ground water. There is usually an argillic horizon and rarely a natric horizon. They occur on flat or almost flat sites of late Pleistocene age and probably had a forest vegetation in the past.
Great groups: Argialbolls, Natralbolls.

Aquolls

These are the wet Mollisols that have low chromas, olive hues and high-contrast mottles below a black epipedon and commonly occur in low places.
Great groups: Argiaquolls, Calciaquolls, Cryaquolls, Duraquolls, Haplaquolls, Natraquolls.

Borolls

These are the more or less freely drained Mollisols of the cool or cold continental areas. They occur mainly in late Pleistocene or Holocene deposits.
Great groups: Argiborolls, Calciborolls, Cryoborolls, Haploborolls, Natriborolls, Paleborolls, Vermiborolls.

Rendolls

These are formed on highly calcareous material such as chalk. They have a mollic epipedon that rests on the calcareous parent material or on a cambic horizon rich in carbonates. Generally they occur under forests in humid regions.
Great groups: None.

Udolls

These are more or less freely drained and occur in the humid continental climates of the mid-latitudes. They have a mollic epipedon over an argillic or cambic horizon and have developed in late Pleistocene or Holocene deposits.
Great groups: Argiudolls, Hapludolls, Paleudolls, Vermudolls.

Ustolls

These are more or less freely drained and occur in subhumid to semiarid areas where drought may be frequent and severe. They have a mollic epipedon and ca horizon or a calcic horizon but both may be absent. They may also have a cambic, an argillic or natric horizon which if present may be overlain by an albic horizon. Most are formed on sediments of mid-Pleistocene to Holocene age.
Great groups: Argiustolls, Calciustolls, Durustolls, Haplustolls, Natrustolls, Paleustolls, Vermustolls.

Xerolls

These have a Mediterranean climate so that moisture moves down the soil during the winter. They have formed in late Pleistocene loess and generally have a relatively thick mollic epipedon, a cambic or an argillic horizon and an accumulation of carbonates in the lower part of the B horizon and neutral in most horizons.
Great groups: Argixerolls, Calcixerolls, Durixerolls, Haploxerolls, Natrixerolls, Palexerolls.

Oxisols

These are the red, yellow or grey soils of tropical and subtropical areas. They generally occur on gently sloping sites of great age. They are strongly weathered and are composed of quartz, kaolinite, oxides of iron and aluminium and small amounts of organic matter. Generally they have an oxic horizon within 2 m but they may have plinthite within 30 cm of the surface. These soils are usually very deep with very diffuse horizon boundaries, thus an arbitrary limit of 2 m is set for their depth. They occur in climates that range from arid to humid, the arid conditions have resulted from climatic change. The soils have a very low natural fertility but with modern agricultural methods they are very productive.

Aquox

These soils have continuous plinthite within 30 cm of the surface and are saturated within this depth at some time during the year and have a histic epipedon or oxic horizon.
Great groups: Ochraquox, Gibbsiaquox, Plinthaquox, Umbraquox.

Humox

These occur in relatively cool humid climates at high altitudes or high latitudes for Oxisols. Most are red, have a high content of organic matter giving a low chroma in the upper horizon and low base saturation.
Great groups: Acrohumox, Gibbsihumox, Haplohumox, Sombrihumox.

Orthox

These are developed mainly on basic rocks and are yellowish to reddish with high chromas with the oxic horizon becoming redder with depth. They have either a short dry season or none and the natural vegetation is forest.
Great groups: Acrorthox, Eutrorthox, Gibbsiorthox, Haplorthox, Sombriorthox, Umbriorthox.

Torrox

These occur in arid climates and are relics from a former wet climate. They are predominantly red with little organic matter and high base saturation, and very productive when irrigated.
Great groups: None.

Ustox

These are mainly red and dry but are moist for at least 90 days, and occur near the two tropics. The natural vegetation is savanna or deciduous forest.
Great groups: Acrustox, Eutrustox, Haplustox, Sombriustox.

Spodosols

Most of these soils have a spodic horizon below which there may be a fragipan, or another sequum that has an argillic horizon. A few Spodosols have a placic horizon that rests on a spodic horizon or on a fragipan or lies in the spodic horizon. In the undisturbed state many have an albic horizon which may be lost by cultivation if thin and a small number have a strongly cemented albic horizon. If the albic horizon is consistently >2 m the soil is excluded from the Spodosols and grouped with the Entisols.

Most Spodosols occur on late Pleistocene or Holocene surfaces and in cool humid areas but some are found in intertropical areas.

Aquods

These are the wet Spodosols that have either a shallow fluctuating ground water or occur in an extremely humid climate. They may have a histic epipedon, mottling in an albic horizon or in the upper part of the spodic horizon which may be free of iron. There may be a duripan in the albic horizon or a placic horizon that rests on a spodic horizon or on a fragipan or on an albic horizon that rests on a fragipan. *Great groups:* Cryaquods, Duraquods, Fragiaquods, Haplaquods, Placaquods, Sideraquods, Tropaquods.

Ferrods

These are freely drained soils in which the spodic horizon usually has a high chroma and there is a high percentage of iron relative to carbon. *Great groups:* None given.

Humods

These are more or less freely drained and have a large accumulation of organic carbon relative to iron in the spodic horizon. There may be an albic horizon above the spodic horizon and if there is a placic horizon it usually lies beneath the upper subhorizon. They occur mainly under coniferous forests on Pleistocene or Holocene surfaces but also in intertropical areas under rain forest. *Great groups:* Cryohumods, Fragihumods, Haplo- humods, Placohumods, Tropohumods.

Orthods

These are more or less freely drained and have a spodic horizon with about equal amounts of accumu- lated aluminium, iron and organic matter. They usually have an O horizon, an albic horizon and a spodic horizon and there may be a fragipan. They occur mainly on coarse Pleistocene or Holocene deposits under coniferous or hardwood forests. *Great groups:* Cryorthods, Fragiorthods, Haplorthods, Placorthods, Troporthods.

Ultisols

These are soils of the mid- to low latitudes that have an argillic horizon and low base saturation, the bases

occurring in the vegetation or the upper few centi- metres of the soil. These two properties are due to leaching during the wet season and their considerable age, being developed on Pleistocene or older surfaces. Kaolinite, gibbsite and interlayered aluminium are the common clays so that exchangeable aluminium is high.

Aquults

These have grey or olive colours and ground water very close to the surface during part of the year. They have an ochric or umbric epipedon and mottled argillic horizon, some have a fragipan and others have plinthite. Generally they occur on flat or gently sloping sites under forests. *Great groups:* Albaquults, Fragiaquults, Ochraquults, Paleaquults, Plinthaquults, Tropaquults, Umbraquults.

Humults

These are the more or less freely drained, humus rich Ultisols of mid- to low latitudes. In mid-latitudes they are dark in colour but in the low latitudes the content of humus is not reflected in their colour. They occur mainly on strong slopes in mountainous areas on basic rock and on surfaces that are Pleistocene or older and many are eroded with the argillic horizon at the surface. *Great groups:* Haplohumults, Palehumults, Plintho- humults, Sombrihumults, Tropohumults.

Udults

These are the more or less freely drained humus poor Ultisols of the mid- and low latitude humid climates with well distributed rainfall. Most have light- coloured upper horizons, commonly a greyish horizon that rests on a yellowish-brown to reddish argillic horizon. Some have a fragipan or plinthite or both in or below the argillic horizon. Most are formed in sediments on surfaces that are late Pleistocene or older and under forest vegetation. *Great groups:* Fragiudults, Hapludults, Paleudults, Plinthudults, Rhodudults, Tropudults.

Ustults

These are the more or less freely drained Ultisols of warm humid regions with a pronounced single or double dry season. They are generally red with a low content of organic matter. Generally they have an ochric epipedon that rests on an argillic horizon which may or may not contain plinthite. *Great groups:* Haplustults, Paleustults, Plinthustults, Rhodustults.

Xerults

These are more or less freely drained and occur in Mediterranean climates. They have a moderate or small amount of organic matter and an ochric epipedon that rests on a brownish to reddish argillic horizon. They occur on strong slopes under coniferous forest.

Great groups: Haploxerults, Palexerults.

Vertisols

These are clayey soils that have deep wide cracks at some period of the year and have high bulk density between the cracks. They generally have low organic matter but high base saturation and are dominated by montmorillonite, but other clay minerals are sometimes dominant. The surface varies from a granular surface mulch to platy, to massive and most have well developed gilgai. The middle horizons usually have a well developed wedge structure with slickensides.

Torrerts

These occur in an arid climate so that the cracks may stay open throughout the year but may become filled with soil material. Gilgai are absent or only weakly developed.

Great groups: None given.

Uderts

These occur in humid climates and have cracks that open and close irregularly according to the weather. The upper horizon is >30 cm thick and very dark grey to black and rests on grey to brownish clay.

Great groups: Chromuderts, Pelluderts.

Usterts

These occur in areas of monsoon climates with two wet and two dry seasons and in temperate regions that have low summer rainfall. The cracks open for 60 or more cumulative days but close for 60 or more consecutive days.

Great groups: Chromusterts, Pellusterts.

Xererts

These occur mainly in Mediterranean climates, the cracks open once per annum and remain open for more than two months.

Great groups: Chromoxererts, Pelloxererts.

SYSTEM OF KUBIËNA 1953

This system is a major landmark in the development of systems of soil classification. It is a fully developed scheme based on soil properties and was a major departure from the generally accepted and widely used zonal system. Kubiëna's experience in central Europe indicated that differences in the hydrological regime of soils should form the basis for their initial subdivision in a hierarchy and led to the creation of three major divisions:

1. Subaqueous or underwater soils
2. Semiterrestrial or flooding and ground water soils
3. Terrestrial or land soils

This emphasis on hydrological conditions is unique but is supported by many workers in Europe and other cool oceanic areas.

The data used in this classification include details about genesis of soils as well as about their intrinsic properties and for the first time information is included about soils in thin sections. Therefore, this system can be regarded as morphogenetic. Moreover, it goes further and contains a few features that deserve special mention. There are numerous references to intergrading situations which impart a dynamic aura, but regrettably no attempt is made to introduce a method to designate the intergrades. However, their recognition must be regarded as a positive step forward when compared with much of the preceding work. The second important feature of this work is the use of names for some horizons and the attempt to define them in terms of their macro- and micromorphology as well as by their methods of formation. The third contribution is that Kubiëna defines many more classes of soils for Europe than his predecessors. Perhaps the most serious criticism that can be made of this system is that he described only what might be called modal virgin soils. No attempt was made to define the range within the individual members of each class nor was any account taken of cultivated soils. Even with these deficiencies his work is an outstanding contribution. In Europe many workers were quick to adopt his ideas, particularly in Germany where Mückenhausen (1962) produced a modification of his scheme. The essential characteristics of Kubiëna's system are given in Table 5.11. Kubiëna also produced a grouping of soils by their profiles as shown in Table 5.12. He states 'it allows the tendency of soil development to be easily recognised which consists of a progression from the simple to the more complex. It confirms further the logic in the arrangement of the natural system which, starting with sub-

145

Table 5.10 The approximate Soil Taxonomy equivalents to the great soil groups

Table 5.10 – *(continued)*

Revised 1938 classification	Soil taxonomy
Great soil groups	Great groups mostly or partly included (mostly in bold others in roman)
Alluvial soils	**Fluvents**
Alpine meadow soils	Cryaquods, Cryaquolls, Cryumbrepts
Ando soils	**Dystrandepts**, Andepts
Bog soils	**Fibrists, Hemists and Saprists**
Brown soils	Agriustolls, Argixerolls, Haploxerolls, Haplustolls, Argids, Orthids, Durixerolls, Haploborolls
Brown forest soils	Eutrochrepts, Haploxerolls and Hapludolls
Brown Podzolic soils	**Cryandepts**, Fragiorthods, Haplorthods, Eutrochrepts, Dystrochrepts, Fragiochrepts
Brunizems (prairie soils)	Argiudolls, Hapludolls, Argixerolls, Haploxerolls, Cryoborolls, Durixerolls, Palexerolls, Argiustolls
Calcisols	**Calciborolls, Calciorthids, Calciustolls, Calcixerolls, Paleorthids,** Camborthids, Durorthids, Ustochrepts and Xerochrepts
Calcium carbonate Solonchaks	**Calciaquolls,** Calciustolls
Chernozems	**Cryoborolls,** Argiustolls, Haplustolls, Argiborolls, Haploborolls, Haploxerolls
Chestnut soils	Argixerolls, Durixerolls, Haploxerolls, Palexerolls, Argiustolls, Argixerolls, Haploxerolls, Haplustolls, Argiborolls, Haploborolls
Degraded Chernozems	Eutroboralfs, Haploborolls, Argiborolls
Desert soils	Durargids, Haplargids, Paleargids, Camborthids, Durorthids
Grey-brown Podzolic soils	**Hapludalfs,** Ochraqualfs, Fragiudalfs, Glossoboralfs, Hapludults
Grey wooded soils	**Eutroboralfs,** other Boralfs
Ground water Laterite soils	**Plinthaquults, Plinthudults, Plinthustalfs and Plinthustults**
Ground water Podzols	**Aquods,** Haplohumods
Grumusols	**Vertisols,** Haplaquepts, Haplaquolls
Half bog soils	**Aquepts,** Aquolls, Aqualfs
Humic Ferruginous Latosols	**Tropohumults,** Dystrandepts, Orthox
Humic Gley soils	**Argiaquolls, Cryaquolls, Haplaquolls, Humaquepts, Umbraquults,** Andaquepts, Calciaquolls, Fluvaquents, Haplaquepts, Ochraqualfs
Humic Latosols	**Tropohumults,** Dystrandepts, Gibbsihumox, Hydrandepts
Hydrol Humic Latosols	**Hydrandepts**
Laterite soils	Acrorthox
Latosolic brown forest soils	**Dystrandepts,** Vitrandepts
Latosols	**Humoxic Tropohumults, Tropepts,** Tropohumults, Oxisols, Hydrandepts, Rhodustults
Lithosols	Orthents, Alfisols, Aridisols, Inceptisols, Mollisols, Ultisols, other Entisols
Low Humic Gley soils	Aquults, Haplaquents, Haplaquepts, Ochraqualfs, Cryaquents, Fluvaquents, Psammaquents, other Aquepts and Aqualfs
Low Humic Latosols	Ustropepts, Haplustox
Non-calcic brown soils	**Durixeralfs, Haploxeralfs, Palexeralfs,** Xerochrepts
Planosols	**Albaqualfs, Argialbolls, Fragiaqualfs, Glossaqualfs,** Fragiaquults, Albaquults
Podzol soils	**Cryorthods, Fragiorthods, Haplorthods,** Haplohumods
Prairie soils	(see Brunizems)
Red desert soils	Camborthids, Durargids, Haplargids, Paleargids, Calciorthids
Reddish-brown Lateritic soils	Humults, Paleudalfs, Paleudults, Rhodudults
Reddish-brown soils	**Haplustalfs, Paleustalfs,** Calciorthids, Camborthids, Haplargids, Eutrandepts, Argiustolls, Calciustolls, Haplustolls
Reddish chestnut soils	**Haplustalfs, Paleustalfs,** Argiustolls, paleustolls, Haplustolls
Reddish prairie soils	Argiustolls, Paleustolls, Eutrandepts, Paleudolls
Red-yellow Podzolic soils	Fragiudults, Hapludults, Paleudults, Paleudalfs, Haplustalfs, Paleustalfs
Regosols	**Psamments,** Orthents, Vitrandepts, Haploxerolls, Hapludolls, Haplustolls, Paleudults, Psammaquents, Xerochrepts
Rendzina soils	**Rendolls,** Calciustolls, Hapludolls, Haplustolls
Sierozems	Camborthids, Durargids, Haplargids, Paleargids
Solonchak soils	**Salorthids,** Halaquepts
Solonetz soils	Alfisols, Aridisols, Mollisols
Soloth soils	Natraqualfs, natric subgroups of Cryoborolls, Duraquolls, Durixeralfs, Haploxeralfs
Sols Bruns Acides	**Dystrochrepts, Fragiochrepts, Haplumbrepts,** Udorthents, Xerumbrepts
Subarctic brown forest soils	**Cryochrepts,** Cryoborolls
Tundra soils	**Pergelic Cryaquepts,** Cryandepts, Cryochrepts, Cryumbrepts

Table 5.11 Kubiëna's system of soil classification

A. Division of the Subaqueous or Underwater Soils
   AA. Subaqueous soils not forming peat
      1. Protopedon – subaqueous soil on sediment with little development
      2. Dy – brown muddy mixture of organic and mineral material
      3. Gyttja – highly organic with much faecal material
      4. Sapropel – predominantly organic with high content of iron sulphide and strong development of hydrogen sulphide
   AB. Peat forming subaqueous soils
      5. Fen – accumulation of organic matter

B. Division of Semiterrestrial or Flooding and Ground Water Soils
   BA. Semiterrestrial raw soils
      6. Rambla – alluvium with little development
      7. Rutmark – arctic and snow basin alluvium with little development
      8. Syrogley – gley soil with poorly developed upper humus horizon
   BB. Anmoor-like soils
      9. Anmoor – gley soil with very humose upper horizon
     10. Marsh – drained marine sediments with very humose upper horizon
   BC. Semiterrestrial peat soils
     11. Carr – very wet mossy upper horizon overlying a previous soil
     12. Moss – mossy peat over a previous soil
   BD. Salt soils
     13. Solonchak
     14. Solonetz
     15. Solod
   BE. Gley soils with land humus formations
     16. Gley
   BF. Ungleyed warp soils with land humus formations
     17. Paternia – alluvium with well developed upper horizon
     18. Borovina (Rendzina-like warp soils) – calcareous alluvium with well developed upper horizon
     19. Smonitza (Chernozem-like warp soils) – Chernozem-like soil developed in alluvium
     20. Vega – alluvial soil derived from previously weathered soil material

C. Terrestrial or Land Soils
   CA. Terrestrial raw soils
     A. Climax Raw Soils
     21. Rawmark (raw soils of the cold deserts) generally with continuously frozen subsoil
     22. Yerma – dry desert raw soils
     B. Non-climax raw soils
     23. Syrosem (raw soil of the temperate zone) not to be confused with serosem
   CB. Ranker-like soils
     24. Ranker-soil with humus horizon directly on rock
   CC. Rendzina-like soils
     25. Rendzina (Eurendzina)
     26. Pararendzina – Rendzina developed on basic rock

Table 5.11 – *(continued)*

   CD. Steppe soils
     27. Serosem (grey desert steppe soil)
     28. Burosem (brown desert Steppe soil)
     29. Kastanosem (chestnut-coloured soil)
     30. Chernozem
     31. Para-Chernozem-Ranker-like
     32. Para-Serosem – similar to serosem but without carbonate horizon

   CE. Terrae Calxis
     33. Terra – terra rossa, terra fusca, terra gialla
   CF. Bolus-like silicate soils – strongly weathered (Plastosols)
     34. Braunlehm – brown loams
     35. Rotlehm – red loams
   CG. Latosols
     36. (Lateritic) Roterde – red earth
   CH. Brown earths
     37. Braunerde
   CI. Pseudogley class
     38. Pseudogley – surface water gley
   CJ. Podzol class
     39. Semipodzol – podsolized soils
     40. Podzol

aqueous soils, progresses through semiterrestrial to true terrestrial soils, in particular by the absence of the more complex profile types in the first two groups.' However, the grouping of soils only by profiles is not satisfactory as the exclusive foundation of the natural system.

SYSTEM OF AVERY 1973

This system has been produced specifically for England and Wales and therefore is an example of one produced for only a small geographical region. It is intended for use in general-purpose surveys of both cultivated and uncultivated land.

Profile classes are defined at four levels of abstraction by progressive division. They are: (1) major group; (2) group; (3) subgroup; and (4) series. The first three classes are given in Table 5.13. The classes are largely defined using properties that can be observed or easily measured in the field or inferred within certain limits from field examination.

The main innovations over previous British systems are major groups accommodating raw and Ranker-like soils and soils profoundly modified by human activity. Also the creation of groups for very sandy soils and cracking clays. Generally the higher categories are named using English words or estab-

147

Table 5.12 General grouping of soils by their profiles

1. (A)C-Soils
With soil life but without macroscopically distinguishable humus layers, and only with an upper layer colonised by organisms (with or without a plant root layer).

Subaqueous:	Underwater – raw soils (e.g. red deep sea clays, coral reefs, marine marl and marine chalk.
Semiterrestrial:	Raw gley soil.
Terrestrial:	Raw soils (e.g. nival raw soils of the Alps, arctic raw soil, desert soils, white Rendzinas).

2. AC-Soils
With distinct humus horizon but without B horizons

Subaqueous:	Underwater humus soil (e.g. dy, gyttja, sapropel, reed peat).
Semiterrestrial:	Humus-gley soils (e.g. anmoors, gleyed grey warp soils, mull gley soils).
Terrestrial:	Rendzina and Ranker-like soils (e.g. Rendzinas, Chernozems, Rankers, Para-Chernozems).

3. A(B)C-Soils
With pronounced B horizons which, however, are not real illuvial horizons built up by peptisable substances, but whose origin, in the first place is due to deep reaching weathering with sufficient aeration and oxidation.

Subaqueous:	None.
Semiterrestrial:	None.
Terrestrial:	Brown earth- and red earth-like soils (e.g. Brown Earths, Brown and Red Loams, Red Earths, Terra Rossa).

4. ABC-Soils
With B horizons which are at the same time developed illuvial horizons, i.e. having a strong enrichment of peptisable substances.

Subaqueous:	None.
Semiterrestrial:	None.
Terrestrial:	Bleached soils (e.g. Podzols, bleached brown and red loams, soloti).

5. B/ABC-Soils
Strong enrichment of illuvial substances transported to the surface in peptised state by intensive capillary rise and irreversible.

Subaqueous:	None.
Semiterrestrial:	None.
Terrestrial:	Rind and surface crust soils.

lished terms. In some cases terms have been adopted from other European systems or the USDA system. The soil series are named by giving geographical names or by appending terms to the subgroup name that gives the differentiating characteristics of the series.

A further innovation is the separation of the palaeo–argillic horizon which has properties developed before the last glaciation.

This system is strictly hierarchical and therefore it has all of the deficiencies of such systems but it does provide a better catalogue of *some* of the soils in England and Wales and probably Scotland as well.

FRENCH SYSTEM

The French system of classification produced by the Commission of Pedology and Soil Cartography (1967) is a development of that produced earlier by Aubert and Duchaufour (1956). The underlying principle used is one of soil evolution or the degree of evolution of the soil profile. They also take into account the humus type, and structure. Great importance is attached to the degree of hydromorphism which is recognised at the highest level. The higher categories are the class, subclass, group and subgroup. The classes have Roman numerals, subclasses have the class numeral followed by an Arabic digit and the groups have the subclass symbol and an Arabic digit but these are not shown separately. A synopsis of the scheme is given in Table 5.14.

CANADIAN SYSTEM

This system has six categories – order, great group, subgroup, family, series and type. The three highest categories, orders, great groups and subgroups are conceptual in nature and are based on morphological features that reflect similar pedogenic environments. The order includes soils with properties that show the broadest influence of the factors and processes in soil formation. The great group is based on the presence of diagnostic horizons which reflect their degree of development. The subgroup includes an orthic subgroup which reflects the modal great group concept and divergence and intergrading from that concept.

The three lower categories, family, series and type are the groupings that are essential in soil mapping and reflect differences in properties such as parent material and texture. The orders, great groups and subgroups are given in Table 5.15 while the orders and great groups together with their equivalents in USDA and FAO systems are given in Table 5.16.

This is a hierarchical system designed specifically for Canada and therefore has all the intrinsic deficiencies of hierarchy. The chief one being that the subgroup names tend to be long and not very informative.

SOUTH AFRICAN SYSTEM

This system was devised by MacVicar *et al.* (1977). They state that it is a two-category or binomial system having *soil forms* and *soil series.* The higher category or soil forms contain 41 members, each of which has a characteristic vertical sequence of diagnostic horizons, there being a maximum of 4. There are 5 topsoil and 15 subsoil diagnostic horizons defined principally in terms of their properties and briefly discussed in terms of genesis but they are not easy to identify in the field. The lower category or series contains 504 members differentiated on criteria such as texture and base status. The series are given geographical names and the form derives its name from one of the constituent series. For example the Griffin form derives its name from the Griffin series. Set out below are brief descriptions of the various diagnostic horizons.

Topsoil horizons

Organic O horizon: >10 per cent organic matter >30 cm thick.

Humic A horizon: >2 per cent organic matter >45 cm thick contains <0.28 me exchangeable cations per 1 per cent of clay.

Melanic A horizon: dark colour, blocky structure <15 per cent clay, >0.28 me exchangeable cations, >15 per cent clay, >30 cm thick.

Vertic A horizon: blocky structure and one or more of the following: slickensides, or cracks >25 mm wide or self-mulching surface.

Orthic A horizon: An upper horizon that does not qualify for any of the above.

Subsoil horizons

E horizon: pale-coloured subsurface eluvial or intermittently gleyed horizon.

G horizon: grey and saturated with water for the greater part of the year.

Red apedal B horizon: red with poorly developed peds dominated by 1 : 1 clays and amorphous compounds.

Yellow-brown apedal B horizon: yellow-brown with poorly developed peds and dominated by 1 : 1 clays.

Red structured B horizon: red with strong blocky structure and a substantial amount of 2 : 1 clays especially montmorillonite.

Soft plinthic B horizon: mottled red, yellow, grey, black, is vesicular or concretionary, non-indurated and can be cut with a spade and occurs as the second or third horizon of a vertical sequence, it is not diagnostic if it occurs beneath the third diagnostic horizon.

Hard plinthic B horizon: hardened soft plinthic material occurs as the third horizon in a vertical sequence, it is not diagnostic when it occurs higher in the profile.

Gleycutanic B horizon: mottled pattern with dark clay coatings, structures other than prismatic and columnar.

Prismacutanic B horizon: prismatic or columnar structure with abrupt upper boundary and marked increase in clay with clay coatings.

Pedocutanic B horizon: Well developed structure with prominent coatings on most ped surfaces.

Lithocutanic B horizon: >50 per cent weatherable material with clay coatings most often as tongues.

Neocutanic B horizon: coherent soil material but no peds, more clay than the horizon above and clay coatings, developed in recent sediments.

Ferrihumic B horizon: (Podzol B horizon)

Regic sand: dune sand.

Stratified alluvium: finely stratified material beneath a diagnostic topsoil horizon.

The Griffin form can be used as an example to illustrate how the system works. The Griffin form has an orthic A, a yellow-brown apedal B and a red apedal B. There are 12 series with variable properties as shown in Table 5.17. This form correlates in the FAO system with Acrisols, Xanthic Ferralsols, Eutric and Chromic Cambisols and in the USDA system with Typic, Humoxic and Xeric Palehumults, and Typic and Arenic Umbriorthox and Haplustox. It is interesting to note that this form like many of the others does not correlate with just one higher level of classification in either the FAO or USDA systems. In extreme cases as with the Hutton form which has an ochric A and a red apedal B the correlation is with Rhodic and Xanthic Ferralsols and Arenosols, some Cambisols, Xerosols and Yermosols in the FAO system. In the USDA system the correlation is with some Oxisols, Inceptisols, Aridisols, Entisols and Ultisols.

149

Table 5.13 Soil classification according to Avery – classes in higher categories

Major group	Group	Subgroup
1 *Terrestrial raw soils* Mineral soils with no diagnostic pedogenic horizons or disturbed fragments of such horizons, unless buried beneath a recent deposit more than 30 cm thick	1.1 *Raw Sands* Non-alluvial, sandy (mainly dune sands) 1.2 *Raw Alluvial Soils* In recent alluvium, normally coarse textured 1.3 *Raw Skeletal Soils* With bedrock or non-alluvial fragmental material at 30 cm or less 1.4 *Raw Earths* In naturally occurring, unconsolidated, non-alluvial loamy, clayey or marly material 1.5 *Man-Made Raw Soils* In artificially disturbed material, e.g. mining spoil	
2 *Hydric Raw Soils* (Raw gley soils) Gleyed mineral soils, normally in very recent marine or estuarine alluvium, with no distinct topsoil, and/or ripened no deeper than 20 cm	2.1 *Raw Sandy Gley Soils* In sandy material 2.2 *Unripened Gley Soils* In loamy or clayey alluvium, with a ripened topsoil less than 20 cm thick	
3 *Lithomorphic (A/C) Soils* With distinct, humose or organic topsoil over C horizon or bedrock at 40 cm or less, and no diagnostic B or gleyed horizon within that depth	3.1 *Rankers* With non-calcareous topsoil over bedrock (including massive limestone) or non-calcareous, non-alluvial C horizon (excluding sands)	3.11 Humic Ranker 3.12 Grey (Non-Humic) Ranker 3.13 Brown (Non-Humic) Ranker 3.14 Podzolic Ranker (with greyish E) 3.15 Stagnogleyic (fragic) Ranker
	3.2 *Sand-Rankers* With non-calcareous, non-alluvial sandy C horizon	3.21 Typical Sand-Ranker 3.22 Podzolic Sand-Ranker 3.23 Gleyic Sand-Ranker
	3.3 *Ranker-like Alluvial Soils* In non-calcareous recent alluvium (usually coarse textured)	3.31 Typical Ranker-like Alluvial soil 3.32 Gleyic Ranker-like Alluvial soil
	3.4 *Rendzinas* Over extremely calcareous non-alluvial C horizon fragmentary limestone or chalk	3.41 Humic Rendzina 3.42 Grey (Non-Humic) Rendzina 3.43 Brown (Non-Humic) Rendzina 3.44 Colluvial (Non-Humic) Rendzina 3.45 Gleyic Rendzina 3.46 Humic Gleyic Rendzina
	3.5 *Pararendzinas* With moderately calcareous non-alluvial C horizon (excluding sands)	3.51 Typical (Non-Humic) Pararendzina 3.52 Humic Pararendzina 3.53 Colluvial Pararendzina 3.54 Stagnogleyic Pararendzina 3.55 Gleyic Pararendzina
	3.6 *Sand-Pararendzinas* With calcareous sandy C horizon	3.61 Typical Sand-Pararendzina
	3.7 *Rendzina-like Alluvial Soils* In recent alluvium	3.71 Typical Rendzina-like Alluvial soil 3.72 Gleyic Rendinza-like Alluvial soil

Table 5.13 – *(continued)*

Major group	Group	Subgroup
4 *Pelosols* Slowly permeable (when wet), non-alluvial clayey soils with B or BC horizon showing vertic features and no E, non-calcareous Bg or paleoargillic horizon	**4.1** *Calcareous Pelosols* Without argillic horizon	**4.11** Typical (Stagnogleyic) Calcareous Pelosol
	4.2 *Non-Calcareous Pelosols* Without argillic horizon	**4.21** Typical (Stagnogleyic) Non-Calcareous Pelosol
	4.3 *Argillic Pelosols* With argillic horizon	**4.31** Typical (Stanogleyic) Argillic Pelosol
5 *Brown Soils* Soils, excluding Pelosols, with weathered, argillic or palaeo-argillic B and no diagnostic gleyed horizon at 40 cm or less	**5.1** *Brown Calcareous Earths* Non-alluvial, loamy or clayey, without argillic horizon	**5.11** Typical Brown Calcareous Earth **5.12** Gleyic Brown Calcareous Earth **5.13** Stagnogleyic Brown Calcareous Earth
	5.2 *Brown Calcareous Sands* Non-alluvial, sandy, without argillic horizon	**5.21** Typical Brown Calcareous Sand **5.22** Gleyic Brown Calcareous Sand
	5.3 *Brown Calcareous Alluvial Soils* In recent alluvium	**5.31** Typical Brown Calcareous Alluvial Soil **5.32** Gleyic Brown Calcareous Alluvial Soil
	5.4 *Brown Earths (sensu stricto)* Non-alluvial, non-calcareous, loamy or clayey, without argillic horizon	**5.41** Typical Brown Earth **5.42** Stagnogleyic Brown Earth **5.43** Gleyic Brown Earth **5.44** Ferritic Brown Earth **5.45** Stagnogleyic Ferritic Brown Earth
	5.5 *Brown Sands* Non-alluvial, sandy or sandy gravelly	**5.51** Typical Brown Sand **5.52** Gleyic Brown Sand **5.53** Stagnogleyic Brown Sand **5.54** Argillic Brown Sand **5.55** Gleyic Argillic Brown Sand
	5.6 *Brown Alluvial Soils* Non-calcareous, in recent alluvium	**5.61** Typical Brown Alluvial Soil **5.62** Gleyic Brown Alluvial Soil
	5.7 *Argillic Brown Earths* Loamy or clayey, with ordinary argillic B	**5.71** Typical Argillic Brown Earth **5.72** Stagnogleyic Argillic Brown Earth **5.73** Gleyic Argillic Brown Earth
	5.8 *Palaeo-Argillic Brown Earths* Loamy or clayey, with palaeo-argillic B	**5.81** Typical Palaeoargillic Brown Earth **5.82** Stagnogleyic Palaeoargillic Brown Earth
6 *Podzolic Soils* With podzolic B	**6.1** *Brown Podzolic Soils* (Podzolic Brown Earths) With Bs below an Ap or 15 cm, and no continuous albic E, thin ironpan, distinct Bhs with coated grains, or gleyed horizon at 40 cm or less	**6.11** Typical (Non-Humic) Brown Podzolic Soil **6.12** Humic Brown Podzolic Soil **6.13** Palaeo-Argillic Brown Podzolic Soil **6.14** Stagnogleyic Brown Podzolic Soil **6.15** Gleyic Brown Podzolic Soil
	6.2 *Humic Cryptopodzols* (Humic Podzolic Rankers) With very dark humose Bhs more than 10 cm thick and no peaty topsoil, thin ironpan, continuous albic E, Bs, or gleyed horizon	**6.21** Typical Humic Crypto-Podzol

Table 5.13 – *(continued)*

Major group	Group	Subgroup
6 *Podzolic Soils* – (continued)	6.3 *Podzols (sensu stricto)* With continuous albic E and/or distinct Bh or Bhs with coated grains and no peaty topsoil, bleached hardpan or gleyed horizon above, in or directly below the podzolic B or at less than 50 cm	6.31 Typical (Humo-Ferric) Podzol 6.32 Humus Podzol 6.33 Ferric Podzol 6.34 Palaeo-Argillic (Humo-Ferric) Podzol 6.35 Ferri-Humic Podzol
	6.4 *Gley-Podzols* With continuous albic E and/or distinct Bh or Bhs, gleyed horizon directly below the podzolic B or at less than 50 cm, and no continuous thin ironpan or bleached hardpan	6.41 Typical (Humus) Gley-Podzol 6.42 Humo-Ferric Gley-Podzol 6.43 Stagnogley-Podzol 6.44 Humic (peaty) Gley-Podzol
	6.5 *Stagnopodzols* With peaty topsoil and/or gleyed E or bleached hardpan over thin ironpan or Bs horizon (wet above a podzolic B)	6.51 Ironpan Stagnopodzol 6.52 Humus-Ironpan Stagnopodzol 6.53 Hardpan Stagnopodzol 6.54 Ferric Stagnopodzol
7 *Surface-Water Gley Soils* (Stagnogley *sensu lato*) Non-alluvial soils with distinct, humose or peaty topsoil, non-calcareous Eg and/or Bg or Btg horizon, and no G or relatively previous Cg horizon affected by free groundwater	7.1 *Stagnogley soils (sensu stricto ≈ Pseudogley)* With distinct topsoil	7.11 Typical (argillic) Stagnogley Soil 7.12 Pelo-Stagnogley Soil 7.13 Cambic Stagnogley Soil 7.14 Palaeo-argillic Stagnogley Soil 7.15 Sandy Stagnogley Soil
	7.2 *Stagnohumic Gley Soils* With humose or peaty topsoil	7.21 Cambic Stagnohumic Gley Soil 7.22 Argillic Stagnohumic Gley Soil 7.23 Palaeo-Argillic Stagnohumic Gley Soil 7.24 Sandy Stagnohumic Gley Soil
8 *Ground-water Gley Soils* With distinct, humose or peaty topsoil and diagnostic gleyed horizon at less than 40 cm, in recent alluvium ripened to more than 20 cm, and/or with G or relatively pervious Cg horizon affected by free ground water	8.1 *Alluvial Gley Soils* With distinct topsoil, in loamy or clayey recent alluvium	8.11 Typical (non-calcareous) Alluvial Gley Soil 8.12 Calcareous Alluvial Gley Soil 8.13 Pelo-(vertic) Alluvial Gley Soil 8.14 Pelo-Calcareous Alluvial Gley Soil 8.15 Sulphuric Alluvial Gley Soil
	8.2 *Sandy Gley Soils* Sandy, with distinct topsoil and without argillic horizon	8.21 Typical (non-calcareous) Sandy Gley Soil 8.22 Calcareous Sandy Gley Soil
	8.3 *Cambic Gley Soils* Non-alluvial, with distinct top-soil, loamy or clayey Bh horizon and relatively pervious Cg or G horizon	8.31 Typical (non-calcareous) Cambic Gley Soil 8.32 Calcaro-Cambic Gley Soil 8.33 Pelo-(vertic) Cambic Gley Soil
	8.4 *Argillic Gley Soils* With distinct topsoil and argillic (Btg) horizon over relatively pervious Cg	8.41 Typical Argillic Gley Soil 8.42 Sandy-Argillic Gley Soil
	8.5 *Humic-Alluvial Gley Soils* With humose or peaty topsoil, in loamy or clayey recent alluvium	8.51 Typical (non-calcareous) Humic-Alluvial Gley Soil 8.52 Calcareous Humic-Alluvial Gley Soil 8.53 Sulphuric Humic-Alluvial Gley Soil

Table 5.13 – *(continued)*

Major group	Group	Subgroup
8. *Ground-water Gley Soils* – (continued)	8.6 *Humic-Sandy Gley Soils* Sandy, with humose or peaty topsoil and no argillic horizon	8.61 Typical Humic-Sandy Gley Soil
	8.7 *Humic Gley Soils (sensu stricto)* Non-alluvial, loamy or clayey, with humose or peaty topsoil	8.71 Typical (non-calcareous) Humic Gley Soil 8.72 Calcareous Humic Gley Soil 8.73 Argillic Humic Gley Soil
9 *Man-made soils* With thick man-made A horizon or disturbed soil (including material recognizably derived from pedogenic horizons) more than 40 cm thick	9.1 *Man-Made Humus Soils* With thick man-made A horizon, including Plaggen soils	9.11 Sandy Man-Made Humus Soil 9.12 Earthy Man-Made Humus Soil
	9.2 *Disturbed Soils* Without thick man-made A horizon	
10 *Peat (Organic) Soils*	10.1 *Raw Peat Soils* Without earthy topsoil or ripened mineral surface layer	10.11 Raw Oligo-fibrous Peat Soil 10.12 Raw Eu-fibrous Peat Soil 10.13 Raw (unripened) Oligo-amorphous Peat Soil 10.14 Raw (unripened) Eutro-amorphous Peat Soil
	10.2 *Earthy Peat Soils* With earthy topsoil or ripened mineral surface layer	10.21 Earthy Oligo-fibrous Peat Soil 10.22 Earthy Eu-fibrous Peat Soil 10.23 Earthy Oligo-amorphous Peat Soil 10.24 Earthy Eutro-amorphous Peat Soil 10.25 Earthy Sulphuric Peat Soil

* Names in parenthesis are alternative or explanatory.

This demonstrates very clearly the great difficulties that are encountered when classifying soil in old landscapes that have one or more major climatic changes. The problem is also very evident in Australia where soils with the Hutton form are generally known as Red Earths and occur in nearly every climate. There seems to be some justification in grouping together deep, strongly weathered soils with a greyish-red topsoil that grades into a red middle horizon with slight to moderate clay increase, irrespective of other properties even if the variation is from calcareous to very acid in the middle horizon. On balance, the grouping of such divergent soils seems a little unrealistic since the current processes tend not to get sufficient prominence. This system goes further than that of USDA by recognising more middle or B horizons, thereby adding a greater degree of precision in characterising many soils. It contains fewer levels of classification than most other systems and emphasises the number and sequential arrangement of the horizons. This is a real improvement and has many similarities with that of the author (FitzPatrick, 1971). However, by using geographical names for the different forms much is lost since a geographical name does not give any indication about horizons present and their sequential arrangement. If symbols were given to the various horizons, they could be combined into a code for each form which would equate exactly with the subclass formulae of the author. Thus, it would seem that the South African system and that of the author are very similar. This system runs into great difficulties with the definition of its diagnostic horizons. The E horizon has variable morphology which it is stated is due to differences in genesis, thus horizons that are morphologically and genetically different are nevertheless regarded as the same. Soft plinthic material is a diagnostic horizon only when it occurs as the second or third diagnostic horizon, if it

153

Table 5.14 The French system of soil classification

I Sols minéraux bruts
- I.1 Sols minéraux bruts non climatique – d'érosion, alluvial, colluvial, éolien, volcanique, anthropique.
- I.2 Sols minéraux bruts climatiques des déserts froids – lithosols, cryosols
- I.3 Sols minéraux bruts des desérts chauds (ou xériques) – lithosols, xériques inorganisés, xériques organise d'ablation.

II Sols peu Évolués
- II.1 Sols peu évolués a permagel – avec ségrégation de glace, sans ségrégation de glace, bruns arctiques.
- II.2 Sols peu évolués humiféres – rankers, lithocalciques, à allophanes.
- II.3 Sols peu évolués xériques – gris subdésertiques, xerorankers.
- II.4 Sols peu évolués non climatiques – d'érosion, alluvial, colluvial, éolien, volcaniques friables, anthropique.

III Vertisols
- III.1 Vertisols à drainage externe nul ou réduit – structure arrondie, structure anguleuse.
- III.2 Vertisols à drainage externe possible – structure arrondie, structure anguleuse.

IV Andosols
- IV.1 Andosols des pays froids.
- IV.2 Andosols des pays tropicaux.

V Sols Calciomagnésiques
- V.1 Sols carbonatés – Rendzines – sol bruns calcaires – cryptorendzines.
- V.2 Sols saturés – bruns calciques, humiques carbonatés, calciques mélanisés.
- V.3 Sols gypseux – gypseux rendziniformes, bruns gypseux.

VI Sols Isohumiques
- VI.1 Sols isohumiques de pédoclimat relativement humide – Brunizems.
- VI.2 Sols isohumiques à pédoclimat trés froid – Chernozem – sols chatâins, sols bruns isohumiques.
- VI.3 Sols isohumiques à pédoclimat frais pendant les saisons pluvieuse – sols marrons, Sierozems.
- VI.4 Sols isohumiques à pédoclimat à temperature élevée en periode pluvieuse – sols bruns arides.

VII Sols Brunifiés
- VII.1 Sols brunifiés des climats tempérés humides – sols bruns, sols lessivés.
- VII.2 Sols brunifiés des climats tempérés continentaux sols gris forestiers – sols derno-podzoliques.
- VII.3 Sols brunifiés des climats boréaux – sols lessivés boréaux.
- VII.4 Sols brunifiés des pays tropicaux – sols bruns eutrophes tropicaux.

VIII Sols Podzolisés
- VIII.1 Sols podzolisés de climat tempéré
  Podzols
  Sols podzoliques

I Raw mineral soils
- I.1 Raw mineral soils non-climatic eroded, alluvial, colluvial, aeolian, volcanic, anthropic.
- I.2 Raw mineral soils of cold deserts – lithosols, cryosols.
- I.3 Raw mineral soils of hot deserts (or dry) – lithosols, dry weak development, dry with ablation.

II Soils with little development
- II.1 Soils with little development and permafrost – with and without ice segregations, arctic brown.
- II.2 Organic soils with little decomposition – Rankers, lithocalcic, Andosols.
- II.3 Dry soils with little development – grey semi-desert, xerorankers.
- II.4 Non-climatic soils with little development – eroded, alluvial, colluvial, aeolian, friable, volcanic, anthropic.

III Vertisols
- III.1 Vertisols with external drainage and no reduction – rounded structure, angular structure.
- III.2 Vertisols with possible external drainage – rounded structure, angular structure.

IV Andosols
- IV.1 Andosols of cold regions.
- IV.2 Andosols of tropical countries.

V Calcomagnesian soils
- V.1 Carbonate soils – Rendzinas, brown calcareous, cryptorendzinas.
- V.2 Saturated soils – brown calcareous, humus carbonate, clayey calcareous soils.
- V.3 Gypseous soils – gypseous rendziniform, brown gypseous.

VI Isohumic soils
- VI.1 Isohumic soils with relatively humid pedoclimate – Brunizems.
- VI.2 Isohumic soils with very cold soil climate – Chernozems, chestnut soils, brown isohumic.
- VI.3 Isohumic soils with cool, wet season – marron soils, serozems.
- VI.4 Isohumic soils with high temperatures during the wet season – brown arid soils.

VIII Brownised soils
- VII.1 Brownised soils of temperate humid climates – brown soils, sols lessivés.
- VII.2 Brownised soils of temperature continental climates – grey forest soils, derno-podzolic soils.
- VII.3 Brownised soils of boreal climates – sols lessivés boreaux.
- VII.4 Brownised soils of tropical regions – eutrophic brown tropical soils.

VIII Podzolic soils
- VIII.1 Podzolic soils of temperate climates
  Podzols
  Podzolic soils

Table 5.14 – (*continued*)

Sols ocre-podzoliques	Sols ocre-podzolique
Sols crypto-podzoliques.	Cryptopodzolic soils.
VIII.2 Sols podzolisés de climat froid	VIII.2 Podzolic soils of cold climates
Podzols boréaux	Boreal Podzols
Podzols alpins.	Alpine Podzols.
VIII.3 Sols podzolisés hydromorphes	VIII.3 Hydromorphic Podzols
Podzols à gleys	Gley Podzols
Molken-Podzols	Molken Podzols
Podzols de nappe tropicaux.	Podzols of wet tropics.

IX Sols à sesquioxydes de fer

IX.1 Sols ferrugineux tropicaux peu lessivés, lessivés, appauvris.

IX.2 Sols fersiallitiques, à réserves calcique et le plus souvent, peu lessivés, sans réserve calcique (et lessivés).

X Sols Ferrallitiques

X.1 Sols ferrallitiques faiblement désaturés en (B) typique, appauvris, remaniés, rajeunis au pénévolués.

X.2 Sols ferrallitiques moyennement désaturés en (B), typiques, humifères, appauvris, remaniés, rejeunis ou pénévolués.

X.3 Sols ferrallitiques fortement désaturés en (B), typiques, humifères, appauvries, remaniés, rajeunis au pénévolués, lessivés.

XI Sols Hydromorphes

XI.1 Sols hydromorphes organiques – tourbe fibreuse, tourbe semi-fibreuse, tourbe altérée.

XI.2 Sols hydromorphes moyennement organiques – humiques à gley, humiques à stagnogley.

XI.3 Sols hydromorphes peu humiferes (au mineraux), à gley, à pseudogleyè à stagnogley, à amphigley, à accumulation de fer en carapace ou cuirasse, à redistribution de calcaire ou du gypse.

XII Sols sodiques

XII.1 Sols sodiques à structure non dégradée – sols salins.

XII.2 Sols sodiques à structure dégradée – sols salins à alcalins, sols sodiques à horizon B (Solonetz), sols sodiques à horizon blanchi.

IX Soils with sesquioxides of iron

IX.1 Ferruginous tropical soils, little leached, leached, impoverished.

IX.2 Fersiallitic soils, reserve of calcium and very often a little leached, without calcium and leached.

X Ferrallitic soils

X.1 Ferrallitic soils with weak desaturated in (B) – typical, impoverished, altered, rejuvenated or eroded.

X.2 Ferralitic soils with moderate desaturation in (B) – typical, humified, impoverished, altered, rejuvenated or eroded.

X.3 Ferralitic soils with strong desaturation in (B) – typical, humid, impoverished, altered, rejuvenated or eroded, lessivé.

XI Hydromorphic soils

XI.1 Organic hydromorphic soils – fibrous peat, semi-fibrous peat, altered peat.

XI.2 Moderately organic hydromorphic soils, humic gley, humic stagnogley.

XI.3 Mineral hydromorphic soils with little humified organic matter – gley, pseudogley, stagnogley, amphigley, iron accumulation in hardpan or ironstone, redistribution of carbonate or gypsum.

XII Sodic soils

XII.1 Sodic soils without degraded structure – saline soils.

XII.2 Sodic soils with degraded structure, saline and alkaline, sodic in B horizon (Solonetz), sodic with bleached horizon.

occurs beneath the third diagnostic horizon then it is not diagnostic. Thus, in this system there still remains the question 'When is a diagnostic horizon diagnostic?' – only when it suits the whims of the classifiers.

SYSTEM OF DEL VILLAR 1937

This is probably the first fully developed system in the form of a key based solely upon intrinsic characteristics. Although this work seems to have

been overlooked, it is of some significance since it establishes a complete break with previous systems.

SYSTEM OF LEEPER 1956

Leeper suggested that properties should be used for classifying soils. He states that a fixed number of properties should be chosen for each horizon and ordered according to their importance and used on a presence or absence basis. He does not develop the system, so it is difficult to know exactly how it would operate, but some of his ideas seem to be

Table 5.15 Orders, great groups and subgroups of the Canadian system of soil classification

1. Chernozemic
  1.1 Brown: Orthic, Rego, Calcareous, Eluviated, Solonetzic, Solodic, Saline, Carbonated, Grumic, Gleyed, Lithic.
  1.2 Dark Brown: Orthic, Rego, Calcareous, Eluviated, Solonetzic, Solodic, Saline, Carbonated, Grumic, Gleyed, Lithic.
  1.3 Black: Orthic, Rego, Calcareous, Eluviated, Solonetzic, Solodic, Saline, Carbonated, Grumic, Gleyed, Lithic.
  1.4 Dark Grey: Orthic, Rego, Calcareous, Solonetzic, Solodic, Saline, Carbonated, Grumic, Gleyed, Lithic.

2. Solonetzic
  2.1 Solonetz: Brown, Black, Grey, Alkaline, Gleyed, Lithic
  2.2 Solod: Brown, Black, Grey, Gleyed, Lithic.

3. Luvisolic
  3.1 Grey Brown Luvisol: Orthic, Brunisolic, Bisequa, Gleyed, Lithic.
  3.2 Grey Luvisol: Orthic, Dark, Brunisolic, Bisequa, Solodic Orthic, Solodic Dark, Gleyed, Lithic.

4. Podzolic
  4.1 Humic Podzol: Orthic, Placic, Gleyed, Lithic.
  4.2 Ferro-Humic Podzol: Orthic, Mini, Sombric, Placic, Gleyed, Lithic.
  4.3 Humo-Ferric Podzol: Orthic, Mini, Sombric, Placic, Bisequa, Cryic, Gleyed, Lithic.

5. Brunisolic
  5.1 Melanic Brunisol: Orthic, Degraded, Gleyed, Lithic.
  5.2 Eutric Brunisol: Orthic, Degraded, Alpine, Cryic, Gleyed, Lithic.
  5.3 Sombric Brunisol: Orthic, Gleyed, Lithic.
  5.4 Dystric Brunisol: Orthic, Degraded, Alpine, Cryic, Gleyed, Lithic.

6. Regosolic
  6.1 Regosol: Orthic, Cumulic, Saline, Cryic, Gleyed, Lithic.

7. Gleysolic
  7.1 Humic Gleysol: Orthic, Rego, Fera, Saline, Carbonated, Cryic, Lithic.
  7.2 Gleysol: Orthic, Rego, Fera, Saline, Carbonated, Cryic, Lithic.
  7.3 Eluviated Gleysol: Humic, Low Humic, Fera, Lithic.

8. Organic
  8.1 Fibrisol: Fenno-, Silvo-, Sphagno-, Mesic, Humic, Lumino, Cumulo, Terric, Terric Mesic, Terric Humic, Cryic, Hydric, Lithic.
  8.2 Mesisol: Typic, Fibric, Humic, Lumino, Cumulo, Terric, Terric Fibric, Terric Humic, Cryic, Hydric, Lithic.
  8.3 Humisol: Typic, Fibric, Mesic, Limno, Cumulo, Terric, Terric Fibric, Terric Messic, Cryic, Hydric, Lithic.
  8.4 Folisol: Typic, Lithic.

incorporated into the more recent work of Northcote (1971) which is discussed later.

There is some merit in Leeper's scheme because it is the first attempt to give a designation to soils based purely on their intrinsic properties. Further, it requires the quantitative characterisation of each property to be used. Also, it focuses attention and concentrates on the soil itself and not upon attempts to produce a hierarchical system. However, like its predecessors it fails to make any allowance for intergrades and by ordering the properties according to their importance it is assumed that a given property always has a certain level of importance.

When due allowance is made for the above deficiencies it is questionable whether the presence or absence system constitutes a classification since it is largely devoid of versatility. After the system is established, the discovery or introduction of a new property means that the entire scheme has to be reconstructed to accommodate this new property. Therefore, this is not a classification but the formulation of a designation for each horizon, but it is worthy of a measure of praise as it is the first attempt to recognise and designate all horizons.

Leeper shows much dislike for Kubiëna's system and states that all systems are artificial since they are the creations of man. Whether classifications be natural or artificial it would seem correct to consider all of the known properties when a phenomenon such as soil is to be classified and not merely a relatively small fixed number.

SYSTEM OF NORTHCOTE

This system is entitled a 'Factual Key for the Recognition of Australian Soils' and judging from the evidence available, it accomplishes its objective in a most satisfactory manner. This is largely a bifurcating scheme using defined values of soil properties. Starting with the properties of whole 'profile form', thereafter it employs the properties of individual horizons. Tables 5.18 to 5.22 illustrate how the scheme works. Since this is a key and not a classification, soils such as Uc 2.3 and Um 2.3 that are morphologically and genetically similar will appear at widely different positions in the final column of the system but this is usually permissible in a key.

However, although this key appears to work in a highly efficient manner it can still be criticised on a number of very important points. As with all previous schemes little account is taken of intergrades. This

Table 5.16 Orders and great groups of the Canadian system of soil classification and their equivalents in the USDA and FAO systems

Canadian		USDA	FAO
Order	Great group		
1. Chernozemic		Borolls	Kastanozems
			Chernozems
	1.1 Brown	Aridic Boroll subgroups	Kastanozems (aridic)
	1.2 Dark brown	Typic Boroll subgroups	Kastanozems (typic)
	1.3 Black	Udic Boroll subgroups	Chernozems
	1.4 Dark grey	Boralfic Boroll subgroups	Greyzems
2. Solonetzic		Natric great groups	Solonetz
	2.1 Solonetz	Natric great groups	Mollic, Orthic and Gleyic Solonetz
	2.2 Solod	Glossic Natriborolls or Natriborolls	Solodic Planosols (Mollic)
3. Luvisolic	—	Alfisols, Boralfs, Udalfs	Luvisols
	3.1 Grey-Brown Luvisol	Hapludalfs or Glossudalfs	Albic Luvisols
	3.2 Grey Luvisol	Boralfs	Albic Luvisols
		a. Eutroboralfs	
		b. Cryoboralfs	
		c. Pergelic Cryoboralfs	
4. Podzolic		Spodosols	Podzols
		a. Humods	
		b. Orthods	
	4.1 Humic Podzol	Humods	Humic and Placic Podzols
		a. Cryohumods	
		b. Haplohumods	
	4.2 Ferro-Humic Podzol	2a Humic Cryorthods	Orthic Podzols
		2b Humic Haplorthods	
	4.3 Humo-Ferric Podzol	Cryorthods or Haplorthods	Orthic Podzols
5. Brunisolic		Inceptisols	Cambisols
	5.1 Melanic Brunisol	(Mollic)	Eutric Cambisols
		Eutrocrepts	
	5.2 Eutric Brunisol	2a Eutrocrepts	Eutric Cambisols
		2b (Eutric) Cryochrepts	
	5.3 Sombric Brunisol	2a Umbric Dystrochrepts	Humic Cambisols
		2b Typic Eutrochrepts	
		2c Dystric Eutrochrepts	
	5.4 Dystric Brunisol	2a Dystrochrepts	Dystric Cambisols
		2b Dystric Cryochrepts	
6. Regosolic		Entisols	Fluvisols and Regosols
	6.1 Regosol	Entisols	Fluvisols and Regosols
7. Gleysolic		Aqu-suborders	Gleysols and Planosols
	7.1 Humic Gleysol	2a Aquolls	3a Mollic Gleysols
		2b Humaquepts	3b Humic Gleysols
			3c Calcaric Gleysols (Humic)
	7.2 Gleysol	Aquents, Fluvents, Aquepts	Eutric and Dystric Gleysols
	7.3 Eluviated Gleysol	2 Albolls, Aquolls, Aqualfs	Planosols
8. Organic Order		Histosols	Histosols
	8.1 Fibrisol	Fibrists	
		2a Medifibrist	
		2b Borofibrist	
		2c Cryofibrist	
		2d Sphagofibrist	
	8.2 Mesisol	Hemist	
		2a Medihemist	
		2b Borohemist	
		2c Cryohemist	

Table 5.16 – (*continued*)

8.3 Humisol	Saprist	
	2a Medisaprist	
	2b Borosaprist	
	2c Cryosaprist	
8.4 Folisol	Folist	
	2b Borofolist	
	2c Cryofolist	

Table 5.17 Soil series in the Griffin form

Clay content of B21 horizon (%)	Dystrophic in B21 horizon		Mesotrophic in B21 horizon		Eutrophic in B21 horizon	
6–15	Burnside	10	Erfdeel	20	Runnymeade	30
15–35	Cleveland	11	Umzimkulu	21	Welgemoed	31
35–55	Griffin	12	Ixopo	22	Cradock	32
above 55	Farmhill	13	Zwagershoek	23	Slagkraal	33

Underlying material is not specified, but is usually saprolite

Table 5.18 Divisions and subdivisions of the factual key

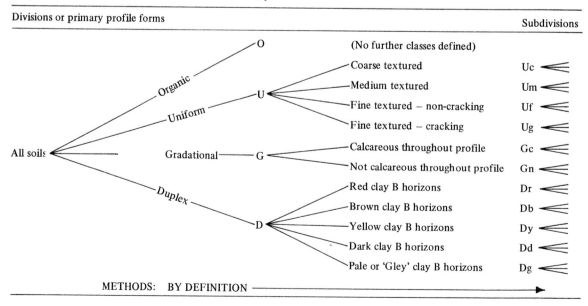

Divisions or primary profile forms — Subdivisions

O	(No further classes defined)
U — Coarse textured	Uc
U — Medium textured	Um
U — Fine textured – non-cracking	Uf
U — Fine textured – cracking	Ug
G — Calcareous throughout profile	Gc
G — Not calcareous throughout profile	Gn
D — Red clay B horizons	Dr
D — Brown clay B horizons	Db
D — Yellow clay B horizons	Dy
D — Dark clay B horizons	Dd
D — Pale or 'Gley' clay B horizons	Dg

All soils → Organic (O), Uniform (U), Gradational (G), Duplex (D)

METHODS: BY DEFINITION ⟶

158

also causes many soils that are very similar to appear at widely different points in the last column of the scheme. A further point of weakness is that the code letters and numbers are not specific at all levels, for example, Uc 1.3 has high coherence and is calcareous, whereas Uc 2.3 has a non-carbonate pan. This shows that the final code number at the class stage has more than one meaning. The same is found at the stage of 'Principal Profile Form'. Thus it is not a system that could be memorised easily. Like the Soil Taxonomy, orderliness is apparent at first, but there are the same intrinsic deficiencies. It is not very informative. For example, the pedo-unit represented by the code Uc 2.33 can either have a zolon or candon overlying a hudepon or husesquon since this scheme does not differentiate between the former two or latter two horizons. Thus, after descending through five category levels an indication is made about the presence of only two horizons. At present there are about 300 codes, and any further subdivisions aimed at increasing its precision would enlarge the scheme to such a level as to make it too cumbersome. A further examination of the code Uc 2.33 shows that it is composed of five separate symbols. This does not compare favourably with soil formulae since any formula that contains five symbols states positively the presence of five horizons which is usually enough to designate fully most soils and too many for some. However, in spite of the deficiencies mentioned above Northcote's key is superior to the Soil Taxonomy.

SYSTEM OF FITZPATRICK 1971

From the above it should be clear that the situation is very confused at present. Some workers use the ABC system together with or separate from diagnostic horizons, while others construct systems to suit local requirements, particularly in tropical areas.

A new approach is needed, and I have proposed and developed a system (FitzPatrick, 1971) that incorporates suggestions made earlier by the Russian scientists, Vilensky (1927) and Sokolovsky (1930) and by Crowther (1953).

Vilensky disagreed with the ABC system of horizon nomenclature and suggested that horizons should be designated by the letters, A, E and I qualified by a second small letter to give a more specific indication of their composition. Sokolovsky also criticised the use of the ABC designation of soil horizon and agreed with the proposals of Vilensky that a number of horizons be recognised and that the first letter of each horizon be used as the symbol for that horizon,

for example H — humus, E — eluvial horizon, I — illuvial horizon, Gl — glei horizon, G — gypsum horizon, S — horizon of accumulation of soluble salts, P — parent rock. Further, he suggested that the letters could be used in combination to indicate complex horizons and that they could be used to designate the whole soil. For example, a Podzol profile might be designated as follows:

$$HE_{14}E_{14}El_{40}I_{40}IP$$

where the letters indicate the different horizons and the subscripts number the thickness of the horizons in centimetres. These suggestions have been virtually ignored and were unknown to me at the time of the publication of my first paper on soil classification (FitzPatrick, 1967). It is very unfortunate that these ideas have largely remained dormant for over 40 years: their implementation might have allowed pedology to develop much faster.

Crowther (1953) considers that the basic concept in soil classification should be one of infinite co-ordinates in hyperspace. By selecting certain soil properties and using them as coordinates it is possible to produce diagrams such as the ternary diagrams for particle size distribution and the feldspars. In such diagrams the arbitrary divisions create spaces or segments with defined values and are regarded as the units of classification.

Soils are generally regarded as forming a continuum in space and time and although it is possible in most environments to find sequences of soils that change laterally from one to the other gradually, their spatial distributions may not be equal. Some parts of the sequence may therefore occupy a smaller area than another part. But in a different environment the soils which occupied the smaller areas may be dominant. Thus, arbitrary lines have to be drawn within the continuum. In order to achieve this, the various properties of soils are considered as coordinate axes that can be conveniently divided.

Generally, three or more coordinates in a geometric arrangement are involved as shown in Fig. 5.4 which attempts to resolve some of the difficulties in classifying tropical soils. It shows part of the range of variability among the strongly weathered soils thus four coordinates are involved.

1. Colour
2. Clay content
3. Cation-exchange capacity
4. Weatherable minerals — all the segments have >0.5 per cent Total Basic Cations — but these are not shown as a separate coordinate in the diagram.

Table 5.19 Schema for the principal profile forms of subdivision Uc.

Subdivision	Sections	Subsections	Classes	Subclasses	Principal profile forms

Underlining indicates no soils recorded

Table 5.20   Detailed schema for part of the principal profile forms of subdivision Uc

Subdivision	Section	Class	Subclass	Principal profile form

Ucl

Uc 2.1 — Carbonate pan or concretions, or other rock material including sesquioxidic layers below A2 — Soil material above the pan is not calcareous material, below is:
- Calcareous — Uc 2.11
- Not calcareous — Uc 2.12

Uc 2.2 — No pan present; B horizon with weak coherence and is:
- Whole coloured, value/chroma rating 1 — Uc 2.20
- Whole coloured, value/chroma rating 2, 4 or 5 — Uc 2.21
- Mottled, value chroma rating (background) 2 — Uc 2.22
- Whole coloured or mottled value/chroma rating (background) 3 or any grey colour — Uc 2.23

Uc 2.3 — Non carbonate pan present

Pan of uniform appearance
- Value/chroma rating 4 — Uc 2.31
- Value/chroma rating 5 — Uc 2.32
- Value/chroma rating 1 — Uc 2.33
- Value/chroma rating 2 or 3 — Uc 2.34

Pans (two) are not of uniform appearance
- Upper pan has a value/chroma rating 2 or 3, lower pan has value/chroma rating 1 — Uc 2.35
- Upper pan has a value/chroma rating of 1, lower pan has a value/chroma rating of 4 — Uc 2.36

Uc

Little, if any bleaching
Conspicuous bleached A₂ → Uc2
Sporadic bleached A₂ · Uc3
A₂ not bleached
No A₂, no peds, weak colour consistence changes → Uc5
No A₂, peds
Uc4
Uc6

Table 5.21 Schema for the principal profile forms of subdivision Dr

Subdivision	Sections	Subsections	Classes	Subclasses	Principal profile forms
Dr	Dr1	a	Dr1.1	a	Dr1.<u>11</u>, .12, .13
				b	Dr1.<u>14</u>, .15, .16
			Dr1.2	a	Dr1.21, .<u>22</u>, .23
			Dr1.3	a	Dr1.<u>31</u>, .32, .33
			Dr1.4	a	Dr1.<u>41</u>, .42, .43
			Dr1.5	a	Dr1.<u>51</u>, .<u>52</u>, .<u>53</u>
				b	Dr1.<u>54</u>, .<u>55</u>, .<u>56</u>
		b	Dr1.6	a	Dr1.<u>61</u>, .<u>62</u>, .<u>63</u>
			Dr1.7	a	Dr1.<u>71</u>, .<u>72</u>, .73
			Dr1.8	a	Dr1.<u>81</u>, .82, .83
	Dr2	a	Dr2.1		Dr2.11, .12, .13
			Dr2.2		Dr2.21, .22, .23
			Dr2.3		Dr2.31, .32, .33
			Dr2.4		Dr2.41, .42, .43
		b	Dr2.5		Dr2.51, .52, .53
			Dr2.6		Dr2.61, .62, .63
			Dr2.7		Dr2.71, .72, .73
			Dr2.8		Dr2.81, .82, .<u>83</u>
	Dr3	a	Dr3.1		Dr3.11, .12, .13
			Dr3.2		Dr3.21, .22, .23
			Dr3.3		Dr3.31, .32, .33
			Dr3.4		Dr3.41, .42, .43
		b	Dr3.5		Dr3.51, .<u>52</u>, .<u>53</u>
			Dr3.6		Dr3.61, .62, .<u>63</u>
			Dr3.7		Dr3.71, .<u>72</u>, .<u>73</u>
			Dr3.8		Dr3.81, .<u>82</u>, .83
	Dr4	a	Dr4.1		Dr4.11, .12, .13
			Dr4.2		Dr4.21, .22, .23
			Dr4.3		Dr4.<u>31</u>, .<u>32</u>, .33
			Dr4.4		Dr4.41, .42, .43
		b	Dr4.5		Dr4.<u>51</u>, .<u>52</u>, .53
			Dr4.6		Dr4.61, .<u>62</u>, .63
			Dr4.7		Dr4.<u>71</u>, .72, .73
			Dr4.8		Dr4.81, .82, .<u>83</u>
	Dr5	a	Dr5.1		Dr5.11, .12, .<u>13</u>
			Dr5.2		Dr5.21, .<u>22</u>, .23
			Dr5.3		Dr5.<u>31</u>, .32, .33
			Dr5.4		Dr5.41, .42, .43
		b	Dr5.5		Dr5.<u>51</u>, .<u>52</u>, .<u>53</u>
			Dr5.6		Dr5.<u>61</u>, .62, .<u>63</u>
			Dr5.7		Dr5.<u>71</u>, .<u>72</u>, .<u>73</u>
			Dr5.8		Dr5.81, .<u>82</u>, .<u>83</u>

Underlining indicates no soils recorded

This range of variability is large, and it seems to warrant the creation of 49 segments of both theoretical and practical significance. Objections might be raised to the above principles in that the difference between adjacent segments is usually only one of degree within one property. This seems to be inevitable but as the segments become more and more distant their differences become greater and greater.

Although it might not be difficult to accept the principles of coordinate classification the problem is to decide exactly where to draw the arbitrary lines. This is clearly demonstrated by the fact that nearly all major soil survey organisations have their own ternary diagram for particle size distribution. This does not mean that each organisation considers that its own diagram is correct and the others are wrong.

Table 5.22   Detailed schema for the principal profile forms of subdivision Dr

Subdivision	Section	Subsection	Class	Principal profile form	
	Dr 1				
		Dr 2.1	No A2 horizon present	Acid SRT Neutral SRT Alkaline SRT	Dr 2.11 Dr 2.12 Dr 2.13
		Dr 2.2	A2 horizon present but not bleached	Acid SRT Neutral SRT Alkaline SRT	Dr 2.21 Dr 2.22 Dr 2.23
		Dr 2.3 (Peds are evident in clayey B horizon)	A2 horizon present and sporodically bleached	Acid SRT Neutral SRT Alkaline SRT	Dr 2.31 Dr 2.32 Dr 2.33
		Dr 2.4	A2 horizon present and conspicuously bleached	Acid SRT Neutral SRT Alkaline SRT	Dr 2.41 Dr 2.42 Dr 2.43
	Dr 2 (A horizons are hard setting, B horizons are whole coloured)	Dr 2.5	No A2 horizon present	Acid SRT Neutral SRT Alkaline SRT	Dr 2.51 Dr 2.52 Dr 2.53
		Dr 2.6	A2 horizon present but not bleached	Acid SRT Neutral SRT Alkaline SRT	Dr 2.61 Dr 2.62 Dr 2.63
		Dr 2.7 (Few if any peds evident in the clayey B horizon)	A2 horizon present and sporodically bleached	Acid SRT Neutral SRT Alkaline SRT	Dr 2.71 Dr 2.72 Dr 2.73
		Dr 2.8	A2 horizon present and conspicuously bleached	Acid SRT Neutral SRT Alkaline SRT	Dr 2.81 Dr 2.82 Dr 2.83*

Surface crust when dry

Dr — (A horizons are hard setting, B horizons are mottled) Dr 3

(A horizons not hard setting, B horizons whole coloured) Dr 4

(A horizons not hard setting, B horizons are mottled) Dr 5

* Dr 2.83 is not known to occur

It means that the diagrams have been constructed to accommodate local requirements and are not designed to be universally acceptable. This is not ideal, but it does demonstrate clearly that soil classifiers are usually influenced very strongly by local conditions. It is hoped that eventually agreement will be reached with regard to putting in the arbitrary divisions within coordinate diagrams.

Indeed it is generally agreed that the whole of soil classification is concerned with drawing arbitrary divisions but it is easier to see the problem when set in a coordinate framework. By using a coordinate system, spaces are created, each with a given range of variability for the properties that have been chosen. Previously these spaces have been called horizons, but it now seems inappropriate to use the term horizon, both for a somewhat finite phenomenon that can be observed in the field as well as for a conceptual spatial entity delimited by coordinates. Whereas a specific horizon in the field will have a fairly narrow range of variability, the conceptual space of which it forms a part will have a much wider range. In fact, it is possible to find two horizons that do not have the same range in the chosen properties but are within the same conceptual spatial entity. Therefore, it seems that two terms are required: one for the conceptual space and the other for the observed phenomenon in the field. The term horizon could be retained and used in the field while the term 'segment' might be used for the volume in conceptual space.

Thus, the basic theoretical unit becomes the segment and it may be defined as follows: 'A segment .

163

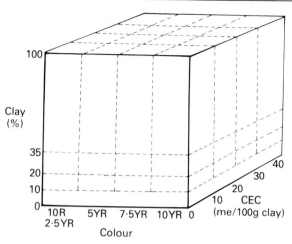

Fig. 5.4 Some soil segments in the humid tropics

is a part of the conceptual model of soils having defined ranges of properties created on coordinate principles and embracing a number of individual horizons'. In this system it is suggested that the starting point is an attempt to create the various segments using coordinates. At present two basic types of segments are recognised.

1. *Reference segment:* one that has a single unique dominating property or a unique combination of dominating properties and formed principally by a single set of processes.
2. *Intergrade segment:* one that contains properties that grade between two reference segments.

From the above it is possible to say that a segment is a class and a horizon is a member of the class and to recognise four types of horizons:

Reference horizons
Intergrade horizons
Compound horizons
Composite horizons

A reference horizon: one that has properties that fall within a reference segment.

An intergrade horizon: one that contains properties that fall within an intergrade segment.

A compound horizon: one that contains a combination of properties of two or more reference segments. These properties are usually contrasting and develop as a result of either contrasting seasonal processes or they have one set of properties superimposed upon an earlier set. This may be as a result of climatic change or progressive soil evolution.

A composite horizon: one that contains discrete volumes of two or more segments.

Designation of reference segments and horizons

Each and every reference segment that is created is given a name and symbol. The name is based on some conspicuous or unique property of the segment; this may be its colour or some substance present in significant proportions. The symbols are formed by using the first letter of the segment name plus one other which is usually part of the name.

All of the names of segments are coined words ending with 'on', the same as for the word horizon. When creating these new words every attempt was made to modify words that already form part of pedological terminology so that the new names appear to be somewhat familiar. For example the name *mullon* is the well known term *mull* with an *-on* ending, and designated by the symbol *Mu; fermenton* and *humifon* are derived from fermentation and humification respectively. Each horizon within a segment is given the same symbol. All of the reference segments recognised up to the present are given in Appendix II.

Designation of intergrade segments and horizons

The intergrades are not given names but designated by the use of the reference segment symbols. Referring again to Fig. 5.2 the intergrades are designated by combining the appropriate symbols in round brackets according to the proportion of each set of properties, the dominant set being placed first. The intergrade (AtSq) displays properties of an alton At and sesquon Sq but is more like an alton. Each intergrade horizon within an intergrade segment is given the same symbol.

Designation of compound horizons

Compound horizons are designated in a similar manner to intergrades but the symbols are put in curley brackets thus {InFg}.

Designation of composite horizons

For a composite horizon the symbols for the discrete volumes of each horizon are placed in round brackets and separated by a stroke. One of the best examples is found in Chernozems which have crotovinas and are designated (Ch/-2Cz) where Ch is the symbol for a chernon and -2Cz is the symbol for calcareous loess.

Designation of a selected number of segments and their intergrades

An example of the designation of a number of reference segments and their intergrades is given in Fig. 5.5 which is the 'exploded' and fully annotated version of Fig. 5.4. In this diagram five reference segments are recognised krasnon — Ks, zhelton — Zh, rosson — Ro, flavon — Fv and arenon — Ae. The diagram should be read as follows: at the top Ks intergrades to Zh, Ro and Fv so that adjacent to Ks there are (KsZh) (KsRs) and (KsFv). All of the top segments intergrade to the arenon, so that Ks intergrades to Ae through (KsAe) and (AeKs): similarly (KsFv) intergrades to Ae through (KsFvAe) and (AeKsFv) with Ae being subordinate in the first place and dominant in the second. The relationship between segments is also shown in Fig. 5.6 where krasnons are shown as having a range of variability in four properties and as each property changes a new segment is created.

These systems of notation are attempts to recognise and preserve the reality of the soil continuum and to accommodate soils with polygenetic evolution. It must be emphasised that segment names and symbols are merely a means of communicating information in a concise form and are not a substitute for a full and comprehensive characterisation which involves a detailed description of the soil and supporting analytical data. In many cases a specific horizon may be very thick and it may be necessary for it to be subdivided for analytical purposes in which case the sub-horizons are designated with the aid of Arabic numbers — a sesquon which has three subdivisions would be symbolised as follows: 1Sq, 2Sq, 3Sq.

Creating boundaries and attempting to introduce the concept of reference segments is not as easy as it might appear at first, because each segment has a wide variety of properties. Therefore, the first step is to delimit the various reference segments based on their known intrinsic properties, although in a few cases a comparison with another segment may be necessary. The lack of intrinsic differentiating criteria for some segments is probably due to a lack of knowledge rather than to lack of criteria. It is hoped that further research will supply these criteria. At present the limits are set on a purely subjective basis and, wherever possible, segments have been created using coordinate principles. With present computing methods involving the use of ordination techniques it may be possible to set objective standards and create more exact limits in the future.

In the field and in the general description of soils the names and symbols of the segments are used to designate the horizons which form part of the segment.

Soil designation, nomenclature and classification

Soils are composed of horizons and as a whole form continua in space and time, which implies that any divisions or boundaries created within soils are similarly as arbitrary as the boundaries that separate segments. The divisions do not delimit discrete entities. THIS SITUATION DEFIES CLASSIFICATION. The best that is attainable for soil, is to give each soil a designation and to have one or two higher levels of grouping using *ad hoc* methods. This procedure recognises that it is impossible to create a fundamental system of soil classification and at the same time places greater emphasis on the soil itself. The designation of the soil is achieved by using a formula which is produced by writing in order the horizon symbols as they occur in vertical sequence starting with the uppermost horizon and including the relatively unaltered underlying material which is shown by the last set of symbols. The thickness of each horizon in centimetres is also indicated by means of a subscript number. Two examples are given below:

1. $Lt_2Fm_3Hf_2Mo_7Zo_5Sq_{40}$–As– Podzol
2. $Lt_1Mu_{20}At_{35}$–Bl– Altosol (Cambisol)

When the soil is of the type that the relatively unaltered underlying material does not show in the profile, the parent material can be indicated by placing a point after the lowest horizon followed by the parent material symbol, e.g.

$$Hy_{20}Gl_{25}Cu_{20+} \cdot Icl$$

The point after the cerulon, Cu, indicates that it continues below the depth of the exposed profile and may continue for several metres. In addition the symbol Icl indicates that the unaltered underlying material is an intermediate clay loam.

The texture patterns in soils are known to be very varied. In a number of cases they are inherited from the stratification of the parent material while in others they are produced by pedogenic processes.

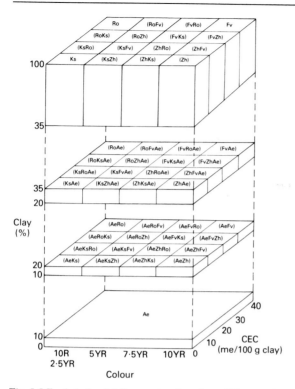

Fig. 5.5 Exploded and fully annotated version of Figure 5.4 showing all the reference segments and their intergrades

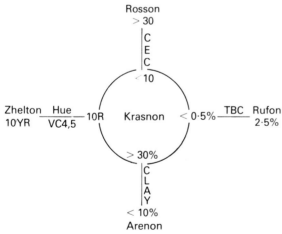

Fig. 5.6 The intergrading of krasnons to four other segments

Although the origin may be obvious in some cases, there are numerous cases where it is not possible to be certain about the reason for the texture change between horizons. In order to be able to show the various patterns in the soil formula without invoking

any process, it is suggested the $<$ and $>$ symbols should be placed between horizon symbols and used singly or in pairs as follows:

   $<$ the horizon above has slightly less clay or coarser texture than the horizon below

   $\ll$ the horizon above has much less clay or much coarser texture than the horizon below

   $>$ the horizon above has slightly more clay or finer texture than the horizon below

   $\gg$ the horizon above has much more clay or much finer texture than the horizon below

The use of these symbols is shown in the formula given below, and illustrates the situation in which the sesquon Sq has a finer texture than the ison In.

$$Lt_2 Fm_3 Hf_2 Mo_{10} Zo_{10} Sq_{25} > In_{20+} \; As$$

Each set of symbols in each formula characterises a group of soils, all of which have the same horizon sequence but there may be variations in the thickness of the horizons, particle size distribution and many other features. However, all the members of the group are sufficiently alike to be placed in the same category. Each distinctive member of each group is given a name, usually a local place name which is also given to the series (mapping unit) in which this group member is dominant. Perhaps it should be stated that mapping units are not necessarily classes of pedo-units. Most mapping units are defined as having a certain range of variability within each property. Some times the range may fall within the defined limits of the constituent segments but as often happens the range overlaps the arbitrary divisions separating the segments. Therefore, the mapping unit may span two pedo-unit classes in the same way that mapping units regularly span two arbitrarily created particle size classes.

Horizon key

Probably the ultimate in soil taxonomy and soil recognition is the production of a key that could be used in the field for the identification of soils. Two major attempts in this direction have been made, one by Northcote (1971) and the other by Kubiëna (1953). Kubiëna produced four keys in all, three for soils and one for humus forms. Northcote's key is used extensively in Australia, but Kubiëna's keys have been little used. The problem with most keys is that they cannot cope with the wide range of horizon assemblages. It would seem that for a key to succeed it must be constructed to operate at the horizon level as attempted by Kubiëna for humus forms, thus allowing each horizon to be identified separately. Furthermore, it is desirable to construct the key using coordinate principles in a similar manner to those used to draw the arbitrary lines and creating segments in the continuum.

A key is like a flora and produced to serve local needs. It would be ideal to have one for all soils, but this is probably beyond the capabilities of a single person, since a considerable amount of detailed local knowledge is required before such a key can be constructed.

A key has been prepared for the identification of Scottish soil horizons, and below selected portions are reproduced to show how it is used to identify horizons. The terminology is that of FitzPatrick (1971) and FAO (1974).

The properties of the horizons used in the key are principally those that are recognisable in the field but in some cases recourse is made to laboratory data. The properties include texture, structure, consistence, water content and colour. Apart from colour the use of the properties follows very closely generally accepted principles. For colour, the V/C (Value/Chroma) rating introduced by Northcote (1971) is also used (see page 92).

Consider the examination of a Podzol in which it is required to identify the brown horizon beneath the bleached horizon. At first it is necessary to decide whether the horizon is in the upper, middle or lower position. By examining the possible positions under 1 below it is a middle horizon. The next step is to go to 25 and examine the possibilities there, it would be 'Rock structure absent . . . fragments', then it is necessary to go to 27. Since the horizon is 'friable' the next step is to 28 and because it is 'uniformly coloured' the next move is to 29 and then to 31 because it is '10YR 4/4'. Since it has a medium colour with VC4 the next move is to 33 where because it has more sesquioxides than the horizon below the move is to 37. As the increase in sesquioxides is significant it is a SESQUON, Sq or Bs horizon.

1. Position of horizon
   UPPER HORIZONS   2
   1. Organic horizons
   2. Organic–mineral mixtures at or near to the surface
   3. Weakly to strongly bleached horizons – moderate to strong colour contrast with horizon below

4. Uppermost mineral horizons – periodically or permanently saturated with water

MIDDLE HORIZONS
1. Beneath organic–mineral mixtures which are at or near to the surface
2. Accumulated sesquioxides, clay and humus by translocation
3. Accumulated sesquioxides and clay by weathering *in situ*

25. MIDDLE OR LOWER HORIZON
    Dominant angular rock fragments, fine earth not forming continuous phase   LITHON, Lh or compounds
    Rock structure partially preserved   26
    Rock structure absent, fine earth forms continuous phase with variable frequency of rock fragments   27

26. Rock structure partially preserved
    Physically weathered angular fragments with little fine earth   LITHON, Lh
    Disaggregated – minerals singly or in small clusters   Type of weathered rock such as AKlw
    Predominantly weathered to secondary minerals   SAPRON, Sa C

27. Rock structure absent fine earth forms continuous phase with variable frequency of rock fragments
    Loose or friable   28
    Firm   49
    Hard   56
    Plastic when wet – fine texture   69

28. Loose or friable
    Uniform colour or

167

slight colour pattern
— not mottled   29
Clear or obvious
colour pattern present   38

29. Uniform colour or slight
colour pattern — not
mottled
Hues 10YR or redder   31
Hues of Y, GY, G, N,
BG or B   30

30. Hues Y, GY, G, N, BG
or B
Above a placon — Bs   CANDON, Co Eg
Above other horizons   CERULON, Cu Cr

31. Hues 10YR or redder
VC 1 — very dark
colours   32
VC 2, 4, 5 — medium
colours   33

32. VC 1 — very dark
colours
Contains more humus
than upper mineral
horizons, may have a
small increase in
sesquioxides   HUDEPON, Hd Bh
Contains more humus
and amorphous
sesquioxides than
upper mineral
horizon   HUSESQUON, Hs Bhs
Contains black
manganese dioxide   ORON, Or Bw

33. VC 2, 4, 5 — medium
colours
Similar or less silicate
clay minerals and
sesquioxides than
upper horizon   34
More silicate clay
minerals than horizon
above   35
More sesquioxides
than horizon below   37

34. Similar or less silicate
clay minerals and
sesquioxides than upper
horizon
Massive, blocky,
prismatic, fine
texture, colour
usually from parent
material   PELON, Pe Bw

Blocky, loamy   ALTON, At Bw

35. More silicate clay
minerals than horizon
above
Clay content decreas-
ing significantly with
depth, clay coatings
on ped surfaces   ARGILLON, Ar Bt
Clay content remain-
ing uniform or
increasing   36

36. Clay content remaining
uniform or increasing
Massive, prismatic
fine texture, some
slickensides, colour
usually from parent
material   PELON, Pe Bw
Blocky, loamy   ALTON, At Bw

37. More sesquioxides than
horizon above
Significantly more
sesquioxides than
horizon above or
below   SESQUON, Sq Bs
Slightly more sesqui-
oxides than horizon
above or below   (AtSq) or (SqAt) Bs
Beneath husesquon
Bhs hudepon — Bh,
or sesquon — Bs   (AtSq) or (SqAt) Bs
                       or SESQUON, Sq Bs

It should be apparent that this key follows the conventional presentation but attempts are being made to produce a set of coordinate diagrams which can be arranged sequentially so that they will form a key as well as showing the spacial relationships between horizons.

SOME OTHER EARLIER SYSTEMS OF SOIL CLASSIFICATION

A number of other classifications based on a variety of features have been tried. Set out below is a list in order of date showing the basis on which each system is established and also the authors concerned.

Mode of weathering — von Richtofen (1886)
Climate and weathering — Ramann (1911)
Climate — Hilgard (1906)
Maturity of profile and intensity of leaching —
    Glinka (1914)

Climate – Lang (1915)
Climate – Meyer (1926)
Factors of formation – Vilensky (1927)
Climate and parent material – Neustruev (1926)
Character of the absorbing complex – Gedroiz (1929)
Climate and parent material – Stebutt (1930)
Chemistry of the soil – De'Sigmond (1933)
Environment and process – Zakharov (1931)

NUMERICAL METHODS OF SOIL CLASSIFICATION

More recently numerical taxonomy has been applied to soil classification by a number of workers including Bidwell and Hole (1964a and b). Their results are interesting but further refinements of their methods seem to be necessary. These authors were able to show that Reddish prairie soils are very similar to Chernozems thereby implying that colour is not a safe criterion for differentiating soils at a high level. This is now generally accepted by many workers hence the use of the term Altosol to replace the term Brown Earth. A further set of results shows that some Planosols appeared similar to Brunizems. Bidwell and Hole recognise this as a serious defect and state that this 'suggests the need for reviewing the characteristics and re-evaluating them in order that the Planosols might segregate more distinctly in keeping with our present concept of these soils'. In other words their desire is to express in numerical terms what is already established by subjective methods; further they state that some weighting might be necessary. This reintroduces the subjective element which does not seem to constitute an improvement upon our present system.

Sneath and Sokal (1973) agree that there is no general agreement on the optimal classification except in cladistics where the optimal classification is one best representing the branching pattern of organisms through evolutionary history. For soils there is clearly no such criterion.

Arkley (1976) has reviewed most of the techniques used at present in numerical taxonomy and showed no one is universally acceptable. Numerical methods would seem to have their greatest potential at lower levels of classification, particularly in helping to order intergrades.

FAO SYSTEM

This system is described as a monocategorical system of soil classification and differs from taxonomy in that it is a list of soil units and not grouped into higher categories at different levels of generalisation. The units are designed primarily for the legends of the Soil Map of the World and do not correspond exactly to categories in other systems but they are generally comparable at the 'great group' level.

The desire was to create map units that are sufficiently broad to have general validity and contain sufficient elements to reflect as precisely as possible the soil pattern of a large region. The legend consists of about 5 000 different units as individual units or as associations of units. These units are subdivided on the basis of textural class, slope class and phase with climatic variants. Generally, the map units consist of an association of soils so that the information given on the map relates to the dominant soil.

Slight modifications of the traditional ABC horizon nomenclature are used and also the concept of the diagnostic horizon so that grouping is based on measurable as well as observable properties. The only separations based on climate are the Yermosols and Xerosols, otherwise soil temperature and moisture are only used where they correlate with other properties such as in Gleysols and the separation of areas of permafrost.

There are also a number of characteristics which are used to define soil units but are not horizons. They are diagnostic properties of horizons or of soils. The different elements of the legend, the system of horizon designation, the definition of the diagnostic horizons and diagnostic properties are given below.

Elements of the legend

Soils
There are 106 soil units described in chap 6. As far as possible traditional names have been used but many new names have been coined.

The soil units adopted were selected on the basis of present knowledge of the formation, characteristics and distribution of the soils covering the earth's surface, their importance as resources for production and their significance as factors of the environment.

Textural class
Three textural classes are recognised:
Coarse textured: sands, loamy sands and sandy loams with less than 18 per cent clay and more than 65 per cent sand.

169

Medium textured: sandy loams, loams, sandy clay loams, silt loams, silt, silty clay loams and clay loams with less than 35 per cent clay and less than 65 per cent sand; the sand fraction may be as high as 82 per cent if a minimum of 18 per per cent clay is present.

Fine textured: clays, silty clays, sandy clays, clay loams and silty clay loams with more than 35 per cent clay.

Slope class

Three slope classes are distinguished:

Level to gently undulating: dominant slopes ranging between 0 and 8 per cent.

Rolling and hilly: dominant slope ranging between 8 and 30 per cent.

Steeply dissected to mountainous: dominant slopes are over 30 per cent.

Phases

This subdivision is based on characteristics which are significant to use or management. They are:

Stony phase: mechanised agricultural equipment impracticable.

Lithic phase: hard rock within 50 cm of surface.

Petric phase: a layer with >40 per cent oxidic concretions >25 cm thick, the upper part occurring within 100 cm of the surface.

Petrocalcic phase: petrocalcic horizon within 100 cm of surface.

Petrogypsic phase: petrogypsic horizon within 100 cm of the surface.

Petroferric phase: petroferric horizon within 100 cm of the surface.

Phraetic phase: ground-water-table between 3 and 5 m from the surface.

Fragipan phase: fragipan within 100 cm of the surface.

Duripan phase: duripan within 100 cm of the surface.

Saline phase: electrical conductivity of >4 mmho cm^{-1} within 100 cm of the surface.

Sodic phase: more than 6 per cent exchangeable sodium saturation within 100 cm of the surface.

Cerado phase: areas with tall grass and low contorted trees.

Climatic variants

Not given on the soil maps but given in the explanatory text as a necessary requirement for interpretation and evaluation for development purposes.

Soil horizon designation

A soil is usually characterised by describing and defining the properties of its horizons. Abbreviated horizon designations, which have a genetic connotation, are used for showing the relationships among horizons within a profile and for comparing horizons among different soils.

Horizon designations are therefore an element in the definition of soil units and in the description of representative profiles.

The symbols used to designate soil horizons are as follows:

Capital letters. H, O, A, E, B, C and R indicate master horizons, or dominant kinds of departure from the assumed parent material. Strictly, C and R should not be labelled as 'soil horizons' but as 'layers', since their characteristics are not produced by soil-forming factors. They are listed here with the master horizons as important elements of a soil profile. A combination of capital letters is used for transitional horizons.

Lower case letters. These are used as suffixes to qualify the master horizons in terms of the kind of departure from the assumed parent material. The lower case letters immediately follow the capital letter. Two lower case letters may be used to indicate two features which occur concurrently.

Arabic figures. These are used as suffixes to indicate vertical subdivision of a soil horizon. For A and B horizons the suffix figure is always preceded by a lower case letter suffix.

Arabic figures. These are used as prefixes to mark lithological discontinuities.

Master horizons

H: An organic horizon formed or forming from accumulations of organic material deposited on the surface, that is saturated with water for prolonged periods (unless artificially drained). It contains 30 per cent or more organic matter if the mineral fraction contains more than 60 per cent of clay, 20 per cent or more organic matter if the mineral fraction contains no clay, or intermediate proportions of organic matter for intermediate contents of clay.

O: An organic horizon formed or forming from accumulations of organic material deposited on the surface, that is not saturated with water for more than a few days a year and contains 35 per cent or more organic matter.

O horizons are the organic horizons that develop on top of some mineral soils — for example, the 'raw humus' mat which covers certain acid soils. The organic material in O horizons is generally poorly decomposed and occurs under naturally well-drained conditions. This designation does not include horizons formed by a decomposing root mat below the surface of the mineral soil which is characteristic of A horizons. O horizons may be buried below the surface.

A: A mineral horizon formed or forming at or adjacent to the surface that either:

(*a*) shows an accumulation of humified organic matter intimately associated with the mineral fraction, or

(*b*) has a morphology acquired by soil formation but lacks the properties of E and B horizons.

The organic matter in A horizons is well decomposed and is either distributed as fine particles or is present as coatings on the mineral particles. As a result A horizons are normally darker than the adjacent underlying horizons. In warm, arid climates where there is only slight or virtually no accumulation of organic matter, surface horizons may be less dark than adjacent underlying horizons. If the surface horizon has a morphology distinct from that of the assumed parent material and lacks features characteristic of E and B horizons, it is designated as an A horizon on account of its surface location.

E: A mineral horizon showing a concentration of sand and silt fractions high in resistant minerals, resulting from a loss of silicate clay, iron or aluminium or some combination of them.

E horizons are eluvial horizons which generally underlie an H, O or A horizon from which they are normally differentiated by a lower content of organic matter and a lighter colour. From an underlying B horizon an E horizon is commonly differentiated by colours of higher value or lower chroma, or by coarser texture, or both.

B: A mineral horizon in which rock structure is obliterated or is but faintly evident, characterised by one or more of the following features:

(*a*) an illuvial concentration of silicate clay, iron, aluminium, or humus, alone or in combinations;

(*b*) a residual concentration of sesquioxides relative to source materials;

(*c*) an alteration of material from its original condition to the extent that silicate clays are formed, oxides are liberated, or both, or granular, blocky, or prismatic structure is formed.

B horizons may differ greatly. It is generally necessary to establish the relationship between overlying and underlying horizons and to estimate how a B horizon has been formed before it can be identified. Consequently, B horizons generally need to be qualified by a suffix to have sufficient connotation in a profile description. A 'humus B' horizon is designated as Bh, an 'iron B' as Bs, a 'textural B' as Bt, a 'colour B' as Bw. It should be stressed here that the horizon designations are qualitative descriptions only. They are not defined in the quantitative terms required for diagnostic purposes. B horizons may show accumulations of carbonates, of gypsum or of other more soluble salts. Such accumulations, however, do not by themselves distinguish a B horizon.

C: A mineral horizon (or layer) of unconsolidated material from which the solum is presumed to have formed and which does not show properties found in any other master horizon.

Traditionally, C has been used to designate 'parent material'. It is seldom possible to ascertain that the material underlying the A, E and B horizons and from which they are assumed to have developed is unchanged. The designation C is therefore used for the unconsolidated material underlying the solum that does not meet the requirements of the A, E or B designations. This material may, however, have been altered by chemical weathering below the soil and may even by highly weathered ('preweathered').

Accumulations of carbonate, gypsum or other more soluble salts may be included in C horizons if the material is otherwise little affected by the processes which contributed to the formation of these interbedded layers. When a C horizon consists mainly of sedimentary rocks such as shales, marls, siltstones or sandstones, which are sufficiently dense and coherent to permit little penetration of plant roots but can still be dug with a spade, the C horizon is qualified by the suffix m for compaction.

R: A layer of continuous indurated rock. The rock of R layers is sufficiently coherent when moist to make hand digging with a spade impracticable. The rock may contain cracks but these are too few and too small for significant root development. Gravelly and stony material which allows root development is considered as C horizon.

Transitional horizons

Soil horizons in which the properties of two master horizons merge are indicated by the combination of two capital letters (for instance AE, EB, BE, BC, CB, AB, BA, AC and CA). The first letter marks the master horizon to which the transitional horizon is most similar.

171

Mixed horizons that consist of intermingled parts, each of which is identifiable with different master horizons, are designated by two capital letters separated by a diagonal stroke (for instance E/B, B/C). The first letter marks the master horizon that dominates. It should be noted that transitional horizons are not marked by suffix figures.

Letter suffixes

A small letter may be added to the capital letter to qualify the master horizon designation. Suffix letters can be combined to indicate properties which occcur concurrently in the same master horizon (for example, Ahz, Btg, Cck). Normally no more than two suffixes should be used in combination. In transitional horizons no use is made of suffixes which qualify only one of the capital letters. A suffix may be used, however, when it applies to the transitional horizon as a whole (for example, BCk, ABg).

The suffix letters used to qualify the master horizons are as follows:

- b. Buried or bisequal soil horizon (for example, Btb).
- c. Accumulation in concretionary form; this suffix is commonly used in combination with another which indicates the nature of the concretionary material (for example, Bck, Ccs).
- g. Mottling reflecting variations in oxidation and reduction (for example, Bg, Btg, Cg).
- h. Accumulation of organic matter in mineral horizons (for example, Ah, Bh); for the A horizon, the h suffix is applied only where there has been no disturbance or mixing from ploughing, pasturing or other activities of man (h and p suffixes are thus mutually exclusive).
- k. Accumulation of calcium carbonate.
- m. Strongly cemented, consolidated, indurated; this suffix is commonly used in combination with another indicating the cementing material (for example, Cmk marking a petrocalcic horizon within a C horizon, Bms marking an iron pan within a B horizon).
- n. Accumulation of sodium (for example, Btn).
- p. Disturbed by ploughing or other tillage practices (for example, Ap).
- q. Accumulation of silica (Cmq, marking a silcrete layer in a C horizon).
- r. Strong reduction as a result of ground water influence (for example, Cr).
- s. Accumulation of sesquioxides (for example, Bs).
- t. Illuvial accumulation of clay (for example, Bt).
- u. Unspecified; this suffix is used in connection with A and B horizons which are not qualified by another suffix but have to be subdivided vertically by figure suffixes (for example, Au1, Au2, Bu1, Bu2). The addition of u to the capital letter is provided to avoid confusion with the former notations A1, A2, A3, B1, B2, B3 in which the figures had a genetic connotation. If no subdivision using figure suffixes is needed, the symbols A and B can be used without u.
- w. Alteration *in situ* as reflected by clay content, colour, structure (for example, Bw).
- x. Occurrence of a fragipan (for example, Btx, Cx).
- y. Accumulation of gypsum (for example, Cy).
- z. Accumulation of salts more soluble than gypsum (for example, Az or Ahz).

When needed, i, e and a suffixes can be used to qualify H horizons composed of fibric, hemic or sapric organic material respectively.

Letter suffixes can be used to describe diagnostic horizons and features in a profile (for example argillic B horizon: Bt; natric B horizon: Btn; cambic B horizon: Bw; spodic B horizon: Bhs, Bh or Bs; oxic B horizon: Bws; calcic horizon: k; petrocalcic horizon: mk; gypsic horizon: y; petrogypsic horizon: my; petroferric horizon: ms; plinthite: sq; fragipan: x; strongly reduced gleyic horizon: r; mottled layers: g. But it should be emphasised that the use of a certain horizon designation in a profile description does not necessarily point to the presence of a diagnostic horizon or feature since the letter symbols merely reflect a qualitative estimate.

Figure suffixes

Horizons designated by a single combination of letter symbols can be vertically subdivided by numbering each subdivision consecutively, starting at the top of the horizon (for example, Bt1 − Bt2 − Bt3 − Bt4). The suffix number always follows all of the letter symbols. The number sequence applies to one symbol only so that the sequence is resumed in case of change of the symbol (for example, Bt1 − Bt2 − Btx1 − Btx2). A sequence is not interrupted, however, by a lithological discontinuity (for example, Bt1 − Bt2 − 2Bt3).

Numbered subdivisions can also be applied to transitional horizons (for example, AB1 − AB2), in which case it is understood that the suffix applies to the entire horizon and not only to the last capital letter.

Numbers are not used as suffixes of undifferentiated A or B symbols, to avoid conflict with the old notation system. If an otherwise unspecified A or B horizon should be subdivided, a suffix u is added.

Figure prefixes

When it is necessary to distinguish lithological discontinuities, Arabic (replacing former Roman) numerals are prefixed to the horizon designations concerned (for instance, when the C horizon is different from the material in which the soil is presumed to have formed the following soil sequence could be given: A, B, 2C. Strongly contrasting layers within the C material could be shown as an A, B, C, 2C, 3C . . . sequence).

Diagnostic horizons

Soil horizons that have a set of quantitatively defined properties which are used for identifying soil units are called 'diagnostic horizons'. Since the characteristics of soil horizons are produced by soil-forming processes, the use of diagnostic horizons for separating soil units ensures that the classification system is based on general principles of soil genesis. Objectivity is secured, however, in that the processes themselves are not used as criteria but only their effects, expressed quantitatively in terms of morphological properties that have identification value.

The definitions and nomenclature of the diagnostic horizons used here are drawn from those adopted in Soil Taxonomy (USDA, 1975). The definitions of these horizons have been summarised and sometimes simplified in accordance with the requirements of the FAO/UNESCO legend. Reference to be made to Soil Taxonomy for additional information on the concepts underlying the definitions of the diagnostic horizons and for detailed descriptions of their characteristics. Where there was compatibility between horizon designations and diagnostic horizons the ABC terminology has been combined with the diagnostic qualification. A few horizons of Soil Taxonomy are not used in the FAO system but they are included here for the sake of completeness. They are the plaggen epipedon, the anthropic epipedon, the sombric horizon, the agric horizon, the duripan, petrocalcic horizon and petrogypsic horizon.

Histic H horizon

The histic H horizon is an H horizon which is more than 20 cm but less than 40 cm thick. It can be more than 40 cm but less than 60 cm thick if it consists of 75 per cent or more, by volume, of sphagnum fibres or has a bulk density when moist of less than 0.1.

A surface layer of organic material less than 25 cm thick also qualifies as a histic H horizon if, after having been mixed to a depth of 25 cm, it has 28 per cent or more organic matter and the mineral fraction contains more than 60 per cent clay, or 14 per cent or more organic matter and the mineral fraction contains no clay, or intermediate proportions of organic matter for intermediate contents of clay. The same criteria apply to a plough layer which is 25 cm or more thick.

A histic H horizon is eutric when it has a pH (H_2O, 1 : 5) of 5.5 or more throughout; it is dystric when the pH (H_2O, 1 : 5) is less than 5.5 in at least a· part of the horizon.

Mollic A horizon

The mollic A horizon is an A horizon which, after the surface 18 cm are mixed, as in ploughing, has the following properties:

1. The soil structure is sufficiently strong, so that the horizon is not both massive and hard or very hard when dry. Very coarse prisms larger than 30 cm in diameter are included in the meaning of massive if there is no secondary structure within the prisms.
2. Both broken and crushed samples have colours with a chroma of less than 3.5 when moist, a value darker than 3.5 when moist, and 5.5 when dry; the colour value is at least one unit darker than that of the C (both moist and dry). If a C horizon is not present, comparison should be made with the horizon immediately underlying the A horizon. If there is more than 40 per cent finely divided lime, the limits of colour value dry are waived; the colour value, moist, should then be 5 or less.
3. The base saturation is 50 per cent or more (by the NH_4OAc method).
4. The organic matter content is at least 1 per cent throughout the thickness of mixed soil, as specified below. The organic matter content is at least 4 per cent if the colour requirements are waived because of finely divided lime. The upper limit of organic carbon content of the mollic A horizon is the lower limit of the histic H horizon.
5. The thickness of 10 cm or more if resting directly on hard rock, a petrocalcic horizon, a petrogypsic horizon or a duripan; the thickness of the A must be at least 18 cm and more than one-third of the thickness of the solum where the solum is less than 75 cm thick, and must be more then 25 cm where the solum is more than 75 cm thick.
6. The content of P_2O_5 soluble in 1 per cent citric acid is less than 250 ppm, unless the amount of P_2O_5 soluble in citric acid increases below the A horizon or when it contains phosphate nodules,

as may be the case in highly phosphatic parent materials. This restriction is made to eliminate plough layers of very old arable soils or kitchen middens. Such a horizon is an anthropic A horizon.

Umbric A horizon

The requirements of the umbric A horizon are comparable to those of the mollic A horizon in colour, organic matter and phosphorus content, consistency, structure and thickness. The umbric A horizon, however, has a base saturation of less than 50 per cent (by the NH_4OAc method). The restriction against a massive and hard or very hard horizon when dry is applied only to those A horizons that become dry. If the horizon is always moist, there is no restriction on its consistency or structure.

Horizons which have acquired the above requirements through the slow addition of materials under cultivation are excluded from the umbric A horizon. Such horizons are plaggen A horizons.

Ochric A horizon

An ochric A horizon is one that is too light in colour, has too high a chroma, too little organic matter, or is too thin to be mollic or umbric, or is both hard and massive when dry.

In separating Yermosols from Xerosols a distinction is made between very weak and weak ochric A horizons:

1. A very weak ochric A horizon has a very low content of organic matter with a weighted average percentage of less than 1 per cent in the surface 40 cm if the weighted average sand/clay ratio for this depth is 1 or less; or less than 0.5 per cent organic matter if the weighted sand/clay ratio is 13 or more; for intermediate sand/clay ratios the organic matter content is intermediate. When hard rock, a petrocalcic horizon, a petrogypsic horizon or a duripan occurs between 18 and 40 cm, the contents of organic matter mentioned above are respectively less than 1.2 and 0.6 per cent in the surface 18 cm of the soil.

2. A weak ochric A horizon has a content of organic matter which is intermediate between that of the very weak ochric A horizon and that required for the mollic A horizon.

Argillic B horizon

An argillic B horizon is one that contains illuvial layer-lattice clays. This horizon forms below an eluvial horizon, but it may be at the surface if the soil has been partially truncated. The argillic B horizon has the following properties:

1. If an eluvial horizon remains, the argillic B horizon contains more total and more fine clay than the eluvial horizon, exclusive of differences which may result from a lithological discontinuity. The increase in clay occurs within a vertical distance of 30 cm or less:
   (a) if any part of the eluvial horizon has less than 15 per cent total clay in the fine earth (less than 2 mm) fraction, the argillic B horizon must contain at least 3 per cent more clay (for example, 13 per cent versus 10 per cent);
   (b) if the eluvial horizon has more than 15 per cent and less than 40 per cent total clay in the fine earth fraction, the ratio of the clay in the argillic B horizon to that in the E horizon must be 1.2 or more;
   (c) if the eluvial horizon has more than 40 per cent total clay in the fine earth fraction, the argillic B horizon must contain at least 8 per cent more clay (for example, 50 per cent versus 42 per cent).

2. An argillic B horizon should be at least one-tenth the thickness of the sum of all overlying horizons, or more than 15 cm thick if the eluvial and illuvial horizons are thicker than 150 cm. If the B horizon is sand or loamy sand, it should be at least 15 cm thick, if it is clayey, it should be at least 7.5 cm thick. If the B horizon is entirely composed of lamellae, the lamellae should have a thickness of 1 cm or more and should have a combined thickness of at least 15 cm.

3. In soils with massive or single grained structure the argillic B horizon has oriented clay bridging the sand grains and also in some pores.

4. If peds are present, an argillic B horizon either:
   (a) shows clay coatings on some or both the vertical and horizontal ped surfaces and in the pores, or shows oriented clays in 1 per cent or more of the cross-section;
   (b) if the B has a broken or irregular upper boundary and meets requirements of thickness and textural differentiation as defined under 1 and 2 above, clay coatings should be present at least in the lower part of the horizon;
   (c) if the B horizon is clayey with kaolinitic clay and the surface horizon has more than 40 per cent clay, there are some clayskins on peds and in pores in the lower part of that horizon having blocky or prismatic structure; or
   (d) if the B horizon is clayey with 2 to 1 lattice clays, clay coatings may be lacking, provided there

are evidences of pressure caused by swelling; or if the ratio of fine to total clay in the B horizon is greater by at least one-third than the ratio in the overlying or the underlying horizon, or if it has more than 8 per cent more fine clay; the evidences of pressure may be occasional slickensides or wavy horizon boundaries in the illuvial horizon, accompanied by uncoated sand or silt grains in the overlying horizon.

5. If a soil shows a lithologic discontinuity between the eluvial horizon and the argillic B horizon, or if only a plough layer overlies the argillic B horizon, the horizon need show clay coatings in only some part, either in some fine pores or, if peds exist, on some vertical and horizontal ped surfaces. Thin sections should show that some part of the horizon has about 1 per cent or more of oriented clay bodies, or the ratio of fine clay to total clay should be greater by at least one-third than in the overlying or the underlying horizon.

6. The argillic B horizon lacks the set of properties which characterise the natric B horizon.

Natric B horizon

The natric B horizon has the properties 1 to 5 of the argillic B horizon as described above. In addition, it has:

1. A columnar or prismatic structure in some part of the B horizon, or a blocky structure with tongues of an eluvial horizon in which there are uncoated silt or sand grains extending more than 2.5 cm into the horizon.

2. A saturation with exchangeable sodium of more than 15 per cent within the upper 40 cm of the horizon; or more exchangeable magnesium plus sodium than calcium plus exchange acidity (at pH 8.2) within the upper 40 cm of the horizon if the saturation with exchangeable sodium is more than 15 per cent in some subhorizon within 200 cm of the surface.

Cambic B horizon

A cambic B horizon is an altered horizon lacking properties that meet the requirements of an argillic, natric or spodic B horizon; lacking the dark colours, organic matter content and structure of the histic H, or the mollic and umbric A horizons; showing no cementation, induration or brittle consistence when moist, having the following properties:

1. Texture that is very fine sand, loamy very fine sand, or finer.

2. Soil structure or absence of rock structure in at least half the volume of the horizon.

3. Significant amounts of weatherable minerals reflected by a cation-exchange capacity (by NH_4OAc) of more than 16 me per 100 g clay, or by a content of more than 3 per cent weatherable minerals other than muscovite, or by more than 6 per cent muscovite.

4. Evidence of alteration in one of the following forms:

   (*a*) higher clay content than the underlying horizon;

   (*b*) stronger chroma or redder hue than the underlying horizon;

   (*c*) evidence of removal of carbonates (when carbonates are present in the parent material or in the dust that falls on the soil) reflected particularly by a lower carbonate content than the underlying horizon of calcium carbonate accumulation; if all coarse fragments in the underlying horizon are completely coated with lime, some in the cambic horizon are partly free of coatings; if the coarse fragments are coated only on the underside, those in the cambic horizon should be free of coatings;

   (*d*) evidence of reduction processes or of reduction and segregation of iron reflected by dominant moist colours on ped faces, or in the matrix if peds are absent, is as follows:

   (i) chromas of 2 or less if there is mottling,

   (ii) if there is no mottling and the value is less than 4, the chroma is less than 1; if the value is 4 or more the chroma is 1 or less,

   (iii) the hue is no bluer than 10Y if the hue changes on exposure to air.

5. Enough thickness that its base is at least 25 cm below the soil surface.

Spodic B horizon

A spodic B horizon meets one or more of the following requirements below a depth of 12.5 cm, or, when present, below an Ap horizon:

1. A subhorizon more than 2.5 cm thick that is continuously cemented by a combination of organic matter with iron or aluminium or with both.

2. A sandy or coarse-loamy texture with distinct dark pellets of coarse silt size or with sand grains covered with cracked coatings.

3. One or more subhorizons in which:

   (*a*) if there is 0.1 per cent or more extractable iron, the ratio of iron plus aluminium (elemental)

175

extractable by pyrophosphate at pH 10 to percentage of clay is 0.2 or more, or if there is less than 0.1 per cent extractable iron, the ratio of aluminium plus carbon to clay is 0.2 or more; and

(*b*) the sum of pyrophosphate-extractable iron plus aluminium is half or more of the sum of dithionite-citrate extractable iron plus aluminium; and

(*c*) the thickness is such that the index of accumulation of amorphous material (CEC at pH 8.2 minus one-half the clay percentage multiplied by the thickness in centimetres) in the horizons that meet the preceding requirements is 65 or more.

Oxic B horizon

The oxic B horizon is an horizon that is not argillic or natric and that:

1. Is at least 30 cm thick.
2. Has a fine-earth fraction that retains 10 me or less ammonium ions per 100 g clay from an unbuffered M NH_4Cl solution or has less than 10 me of bases extractable with M NH_4OAc plus aluminium extractable with M KCl per 100 g clay.
3. Has an apparent cation-exchange capacity of the fine earth fraction of 16 me or less per 100 g clay by NH_4OAc unless there is an appreciable content of aluminium-interlayered chlorite.
4. Does not have more than traces of primary aluminosilicates such as feldspars, micas, glasses, and ferromagnesian minerals.
5. Has texture of sandy loam or finer in the fine earth fraction and has more than 15 per cent clay.
6. Has mostly gradual or diffuse boundaries between its subhorizons.
7. Has less than 5 per cent by volume showing rock structure.

Calcic horizon

The calcic horizon is a horizon of accumulation of calcium carbonate. The accumulation may be in the C horizon, but it may also occur in a B or in an A horizon.

The calcic horizon consists of secondary carbonate enrichment over a thickness of 15 cm or more, has a calcium carbonate equivalent content of 15 per cent or more and at least 5 per cent greater than that of the C horizon. The latter requirement is expressed by volume if the secondary carbonates in the calcic horizon occur as pendants on pebbles, or as concretions or soft powdery forms; if such a calcic horizon rests on very calcareous materials (40 per cent or more calcium carbonate equivalent), the percentage of carbonates need not decrease with depth.

Gypsic horizon

The gypsic horizon is a horizon of secondary calcium sulphate enrichment that is more than 15 cm thick, has at least 5 per cent more gypsum than the underlying C horizon, and in which the product of the thickness in centimetres and the per cent of gypsum is 150 or more. If the gypsum content is expressed in me per 100 g of soil, the percentage of gypsum can be calculated from the product of the me of gypsum per 100 g of soil and the me weight of gypsum, which is 0.086. Gypsum may accumulate uniformly throughout the matrix or as nests of crystals; in gravelly material gypsum may accumulate as pendants below the coarse fragments.

Sulphuric horizon

The sulphuric horizon forms as a result of artificial drainage and oxidation of mineral or organic materials which are rich in sulphides. It is characterised by a pH less than 3.5 (H_2O, 1 : 1) and jarosite mottles with a hue of 2.5Y or more and a chroma of 6 or more.

Albic E horizon

The albic E horizon is one from which clay and free iron oxides have been removed, or in which the oxides have been segregated to the extent that the colour of the horizon is determined by the colour of the primary sand and silt particles rather than by coatings on these particles.

An albic E horizon has a colour value moist of 4 or more, or a value dry of 5 or more, or both. If the value dry is 7 or more, or the value moist is 6 or more, the chroma is 3 or less. If the parent materials have a hue of 5YR or redder, a chroma moist of 3 is permitted in the albic E horizon where the chroma is due to the colour of uncoated silt or sand grains.

An albic E horizon may overlie a spodic B, an argillic or natric B, a fragipan, or an impervious layer that produces a perched water-table.

Plaggen epipedon

An upper horizon that is >50 cm thick and formed by the continuous addition of soil material to the surface by man.

Anthropic epipedon

An upper horizon similar to the mollic A horizon but contains >250 ppm of citric acid soluble P_2O_5 and is largely the result of man's activities.

Sombric horizon

A middle dark-coloured horizon containing illuvial humus but does not have associated aluminium as in spodic B horizons or sodium as in natric B horizons. It has a low cation-exchange capacity and base saturation and seems to occur in the cool moist upland areas in the tropics and subtropics.

Agric horizon

A dark horizon that occurs immediately beneath the plough layer. It is characterised by having >15 per cent thick coating of humus and clay.

Duripan

A middle or lower horizon that is at least 50 per cent cemented by silica and does not slake in water or HCl.

Petrocalcic horizon

A hard and indurated calcic horizon >50 per cent of which will dissolve in acid but does not slake in water. Some silica may be present.

Petrogypsic horizon

A hard gypsic horizon that usually contains >60 per cent gypsum.

Diagnostic properties

A number of soil characteristics which are used to separate soil units cannot be considered as horizons. They are diagnostic features of horizons or of soils which when used for classification purposes need to be quantitatively defined.

Abrupt textural change

An abrupt textural change is a considerable increase in clay content within a very short distance in the zone of contact between an A or E horizon and the underlying horizon. When the A or E horizon has less than 20 per cent clay, the clay content of the underlying horizon is at least double that of the A or E horizon within a vertical distance of 8 cm or less. When the A or E horizon has 20 per cent clay or more the increase in clay content should be at least 20 per cent (for example, from 30 to 50 per cent clay) within a vertical distance of 8 cm or less, and the clay content in some part of the underlying horizon (B horizon or impervious layer) should be at least double that of the A or E horizon above.

Albic material

Albic materials are exclusive of E horizons, and have a colour value moist of 4 or more, or a value dry of 5 or more, or both. If the value dry is 7 or more, or the value moist is 6 or more, the chroma is 3 or less. If the parent materials have a hue of 5YR or redder, a chroma moist of 3 is permitted if the chroma is due to the colour of uncoated silt or sand grains.

Aridic moisture regime

The concept of aridic moisture regime is used to characterise Yermosols and Xerosols and to separate them from soils, outside arid areas, which have a comparable morphology. In most years these soils have no available water in any part of the moisture control section more than half the time (cumulative) that the soil temperature of 50 cm is about 5 °C (the moisture control section lies approximately between 10 and 30 cm for medium to fine textures, between 20 and 60 cm for medium to coarse textures, and between 30 and 90 cm for coarse textures). There is no period as long as 90 consecutive days when there is moisture in some or all parts of the moisture control section while the soil temperature at 50 cm is continuously above 8 °C. In most years the moisture control section is never moist in all parts for as long as 60 consecutive days during 3 months following the winter solstice if mean summer and mean winter temperatures differ by 5 °C or more and mean annual temperature is less than 22 °C.

Exchange complex dominated by amorphous material

An exchange complex that is dominated by amorphous material shows the following characteristics:
1. The cation-exchange capacity of the clay at pH 8.2 is more than 150 me per 100 g measured clay, and commonly is more than 500 me per 100 g clay. The high value is, in part, the result of poor dispersion.
2. If there is enough clay to give a 15 bar water content of the soil of 20 per cent or more, the pH of a suspension of 1 g soil in 50 ml M NaF is more than 9.4 after 2 minutes.
3. The ratio 15 bar water content to measured clay is more than 1.0.
4. The amount of organic carbon exceeds 0.6 per cent.

177

5. Differential thermal analysis shows a low temperature endotherm.
6. The bulk density of the fine earth fraction is less than 0.85 g cm^{-3} at 1/3 bar tension.

Ferralic properties

The term ferralic properties is used in connection with Cambisols and Arenosols which have a cation-exchange capacity (from NH$_4$Cl) of less than 24 me per 100 g clay in, respectively, at least some sub-horizon of the cambic B horizon or immediately underlying the A horizon.

Ferric properties

The term ferric properties is used in connection with Luvisols and Acrisols showing one or more of the following: many coarse mottles with hues redder than 7.5YR or chromas more than 5, or both; discrete nodules, up to 2 cm in diameter, the exteriors of the nodules being enriched and weakly cemented or indurated with iron and having redder hues or stronger chromas than the interiors; a cation-exchange capacity (from NH$_4$Cl) of less than 24 me per 100 g clay in at least some subhorizon of the argillic B horizon.

Gilgai microrelief

Gilgai is the microrelief typical of clayey soils that have a high coefficient of expansion with distinct seasonal changes in moisture content. This microrelief consists of either a succession of enclosed microbasins and microknolls in nearly level areas, or of micro-valleys and microridges that run up and down the slope. The height of the microridges commonly ranges from a few cm to 1 m. Rarely does the height approach 2 m.

High organic matter content in the B horizon

For Ferralsols and Nitosols of low base saturation the terminology 'high organic matter content in the B horizon' refers to an organic matter content (weighted average of the fine earth fraction of the soil) of 1.35 per cent or more to a depth of 100 cm (exclusive of an O horizon if present); for Acrisols a high organic matter content in the B horizon means one or both of 1.5 per cent or more organic matter in the upper part of the B horizon, or an organic matter content (weighted average of the fine earth fraction of the soil) of 1.35 per cent or more to a depth of 100 cm (exclusive of an O horizon if present).

High salinity

The term 'high salinity' applies to soils which have an electric conductivity of the saturation extract of more than 15 mmhos cm^{-1} at 25 $^\circ$C at some time of the year, within 125 cm of the surface when the weighted average textural class of the surface is coarse, within 90 cm for medium textures, within 75 cm for fine textures, or of 4 mmhos within 25 cm of the surface if the pH (H$_2$O, 1 : 1) exceeds 8.5.

Hydromorphic properties

A distinction is made between soils which are strongly influenced by ground water, the Gleysols, and the soils of which only the lower horizons are influenced by ground water or which have a seasonally perched water-table within the profile, the 'gleyic' groups. The Gleysols have a reducing moisture regime virtually free of dissolved oxygen due to saturation by ground water or its capillary fringe. Since hydromorphic processes are dominant, the occurrence of argillic, natric, spodic and oxic B horizons is excluded from Gleysols by definition.

The morphological characteristics which reflect waterlogging differ widely in relation to other soil properties. For the sake of brevity, the expression 'hydromorphic properties' is used in the definition of Gleysols and gleyic groups. This term refers to one or more of the following properties:

1. Saturation by ground water, that is when water stands in a deep unlined bore hole at such a depth that the capillary fringe reaches the soil surface; the water in the bore hole is stagnant and remains coloured when dye is added to it.
2. Occurrence of a histic H horizon.
3. Dominant hues that are neutral N, or bluer than 10Y.
4. Saturation with water at some period of the year, or artificially drained, with evidence of reduction processes or of reduction and segregation of iron reflected by:
4.1 in soils having a spodic B horizon, one or more of the following:
   (*a*) mottling in an albic E horizon or in the top of the spodic B horizon;
   (*b*) a duripan in the albic E horizon;
   (*c*) if free iron and manganese are lacking, or if moist colour values are less than 4 in the upper part of the spodic B horizon, either:
   (i) no coatings of iron oxides on the individual grains of silt and sand in the materials in or immediately below the spodic horizon where-ever the moist values are 4 or more and, unless

an Ap horizon rests directly on the spodic horizon, there is a transition between the albic E and spodic B horizons at least 1 cm in thickness; or

(ii) fine or medium mottles of iron or manganese in the materials immediately below the spodic B horizon;

(d) a thin iron pan that rests on a fragipan or on a spodic B horizon, or occurs in an albic E horizon underlain by a spodic B horizon.

4.2 in soils having a mollic A horizon
If the lower part of the mollic A horizon has chromas of 1 or less, either:

(a) distinct or prominent mottles in the lower mollic A horizon; or

(b) colours immediately below the mollic A horizon or within 75 cm of the surface if a calcic horizon intervenes, with one of the following:

(i) if hues are 10YR or redder and there are mottles, chromas of less than 1.5 on ped surfaces or in the matrix; if there are no mottles chromas of less than 1 (if hues are redder than 10YR because of parent materials that remain red after citrate-dithionite extraction, the requirement for low chromas is waived)

(ii) if the hue is nearest to 2.5Y and there are distinct or prominent mottles, chromas of 2 or less on ped surfaces or in the matrix; if there are not mottles, chromas of 1 or less

(iii) if the nearest hue is 5Y or yellower and there are distinct or prominent mottles, chromas of 3 or less on ped surfaces or in the matrix; and if there are no mottles, chromas or 2 or less

(iv) hues bluer than 10Y

(v) any colour if the colour results from uncoated mineral grains

(vi) colours neutral N.

If the lower part of the mollic A horizon has chromas of more than 1 but not exceeding 2, either:

(a) distinct or prominent mottles in the lower mollic A horizon; or

(b) base colours immediately below the mollic A horizon that have one or more of:

(i) values of 4 and chromas of 2 accompanied by some mottles with values of 4 or more and chromas of less than 2

(ii) values of 4 and chromas of less than 2

(iii) values of 5 or more and chromas of 2 or

less accompanied by mottles with high chroma.

4.3 in soils having an argillic B horizon immediately below the plough layer or an A horizon that has moist colour values of less than 3.5 when rubbed, one or more of the following:

(a) moist chromas of 2 or less;

(b) mottles due to segregation of iron;

(c) iron–manganese concretions larger than 2 mm, and combined with one or more of the following:

(i) dominant moist chromas of 2 or less in coatings on the surface of peds accompanied by mottles within the peds, or dominant moist chromas of 2 or less in the matrix of the argillic B horizon accompanied by mottles of higher chromas (if hues of redder than 10YR because of parent materials that remain red after citrate-dithionite extraction, the requirement for low chromas is waived)

(ii) moist chromas of 1 or less on surface of peds or in the matrix of the argillic B horizon

(iii) dominant hues of 2.5Y or 5Y in the matrix of the argillic B horizon accompanied by distinct or prominent mottles.

4.4 in soils having an oxic B horizon:

(a) plinthite that forms a continuous phase within 30 cm;

(b) if free of mottles, dominant chromas of 2 or less immediately below an A horizon that has a moist colour value of less than 3.5; or if mottled with distinct or prominent mottles within 50 cm of the surface, dominant chromas of 3 or less.

4.5 in other soils:

(a) in horizons with textures finer than loamy fine sand:

(i) if there is mottling, chromas of 2 or less

(ii) if there is no mottling and values are less than 4, chromas of less than 1; if values are 4 or more, chromas of 1 or less;

(b) in horizons with textures of loamy fine sand or coarser:

(i) if hues are as red as or redder than 10YR and there is mottling, chromas of 2 or less; if there is no mottling and values are less than 4, chromas of less than 1; or if values are 4 or more, chromas of 1 or less

(ii) if hues are between 10YR and 10Y and there is distinct or prominent mottling, chromas of 3 or less; if there is no mottling, chromas of 1 or less.

Interfingering

Interfingering consists of penetrations of an albic E horizon into an underlying argillic or natric B horizon along ped faces, primarily vertical faces. The penetrations are not wide enough to constitute tonguing, but form continuous sandy residues (ped coatings of clean silt or sand, more than 1 mm thick on the vertical ped faces). A total thickness of more than 2 mm is required if each ped has a coating of more than 1 mm. Because quartz is such a common constituent of soils, the sandy residues are usually white when dry, and light grey when moist, but their colour is determined by the colour of the sand or silt fraction. The sandy residues constitute more than 15 per cent of the volume of any subhorizon in which interfingering is recognised. They are also thick enough to be obvious, by their colour, even when moist. Thinner sandy residues that must be dry to be seen as a whitish powdering on a ped are not included in the meaning of interfingering.

Permafrost

Permafrost is a layer in which the temperature is perennially at or below 0 °C.

Plinthite

Plinthite is an iron-rich, humus-poor mixture of clay with quartz and other diluents, which commonly occurs as red mottles, usually in platy, polygonal or reticulate patterns, and which changes irreversibly to an ironstone hardpan or to irregular aggregates on exposure to repeated wetting and drying. In a moist soil, plinthite is usually firm but it can be cut with a spade. When irreversibly hardened the material is no longer considered plinthite but is called ironstone.

Slickensides

Slickensides are polished and grooved surfaces that are produced by one mass sliding past another. Some of them occur at the base of a slip surface where a mass of soil moves downward on a relatively steep slope. Slickensides are common in swelling clays in which there are marked changes in moisture content.

Smeary consistence

The term 'smeary consistence' is used in connection with Andosols characterised by thixotropic soil material, that is material that changes under pressure or by rubbing from a plastic solid into a liquified stage and back to the solid condition. In the liquified stage the material skids or 'smears' between the fingers.

Soft powdery lime

Soft powdery lime refers to translocated authigenic lime, soft enough to be cut readily with a finger nail, precipitated in place from the soil solution rather than inherited from a soil parent material. It should be present in a significant accumulation.

To be identifiable, soft powdery lime must have some relation to the soil structure or fabric. It may disrupt the fabric to form spheroidal aggregates, or white eyes, that are soft and powdery when dry, or the lime may be present as soft coatings in pores or on structural faces. If present as coatings, it covers a significant part of the surface; commonly, it coats the whole surface to a thickness of 1 to 5 mm or more. Only part of a surface may be coated if little lime is present in the soil. The coatings should be thick enough to be visible when moist and should cover a continuous area large enough to be more than filaments. Pseudomycelia which come and go with changing moisture conditions are not considered a soft powdery lime in the present definition.

Sulphidic materials

Sulphidic materials are waterlogged mineral or organic soil materials containing 0.75 per cent or more sulphur (dry weight), mostly in the form of sulphides, and having less than three times as much carbonate ($CaCO_3$ equivalent) as sulphur. Sulphidic materials accumulate in a soil that is permanently saturated, generally with brackish water. If the soil is drained the sulphides oxidise to form sulphuric acid; and the pH, which normally is near neutrality before drainage, drops below 3.5. Sulphidic material differs from the sulphuric horizon in that it does not show jarosite mottles with a hue of 2.5Y or more or a chroma of 6 or more.

Takyric features

Soils with takyric features have a heavy texture, crack into polygonal elements when dry and form a platy or massive surface crust.

Thin iron pan

A thin iron pan is a black to dark reddish layer cemented by iron, by iron and manganese, or by an iron—organic matter complex, the thickness of which ranges generally from 2 to 10 mm. In spots it may be as thin as 1 mm or as thick as 20 to 40 mm, but this is rare. It may, but not necessarily, be associated with stratification in parent materials. It is in the solum, roughly parallel to the soil surface, and is commonly within the upper 50 cm of the mineral soil. It has a pronounced wavy or even convolute form. It normally occurs as a single pan, not as multiple sheets underlying one another, but in places it may be bifurcated. It is a barrier to water and roots. It is used here as a diagnostic property of Placic Podzols.

Tonguing

As used in the definition of Podzoluvisols, the term tonguing is connotative of the penetration of an albic E horizon into an argillic B horizon along ped surfaces, if peds are present. Penetrations to be considered tongues must have greater depth than width, have horizontal dimensions of 5 mm or more in fine textured argillic horizons (clay, silty clay and sandy clay), 10 mm or more in moderately fine textured argillic horizons, and 15 mm or more in medium or coarser textured argillic horizons (silt loams, loams, very fine sandy loams, or coarser), and must occupy more than 15 per cent of the mass of the upper part of the argillic horizon.

With Chernozems, the term tonguing refers to penetrations of the A horizon into an underlying cambic B horizon or into a C horizon. The penetrations must have greater depth than width, and must occupy more than 15 per cent of the mass of the upper part of the horizon in which they occur.

Vertic properties

The term 'vertic properties' is used in connection with Cambisols and Luvisols which at some period in most years show cracks that are 1 cm or more wide within 50 cm of the upper boundary of the B horizon and extend to the surface or at least to the upper part of the B horizon.

Weatherable minerals

Minerals included in the meaning of weatherable minerals are those that are unstable in a humid climate relative to other minerals, such as quartz and 1 : 1 lattice clays, and that, when weathering occurs, liberate plant nutrients and iron or aluminium. They include:

1. Clay minerals: all 2 : 1 lattice clays except aluminium-interlayered chlorite. Sepiolite, talc and glauconite are also included in the meaning of this group of weatherable clay minerals, although they are not always of clay size.
2. Silt- and sand-size minerals (0.02 to 0.2 mm in diameter): feldspars, feldspathoids, ferromagnesian minerals, glasses, micas, and zeolites.

DISCUSSION AND CONCLUSIONS

It is clear that attempts to produce hierarchical systems of soil classification based on morphogenetic principles have failed, and the newer systems of Northcote and USDA are little more than incomplete and cumbersome keys, while numerical taxonomy has as yet not contributed any fundamentally new method, but merely tries to establish by mathematical methods what has already been achieved by trial and error.

The most pertinent criticism that can be made of nearly all systems of soil classification is the paucity of information conveyed by the particular system employed, whether it involves the use of names or a code. This is a serious deficiency in these newer schemes that employ long names and codes. There is also inadequate provision for intergrades in spite of the large amount that has been written about them by many authors. In the final analysis an intergrading situation is not regarded as sufficiently important to be classified separately, consequently it is pushed into one pigeon hole or the other. This procedure amounts to depriving soils of one of their basic and fundamental properties — their continuum — Jones (1959) like many other workers has already stated this.

A new approach is necessary. Soils must be examined as separate and distinct phenomena in order to establish those features common to all soils, then to determine how these distinct and specific features can be used to construct a system or organisation for soils. The first feature possessed by all soils is their profile but as Jones (1959) rightly pointed out profiles have only two dimensions which make them unsuitable for any system of organisation. The second feature which is common to all soils are horizons which exist in three tangible dimensions, length, breadth and depth. The third feature is that horizons intergrade, laterally, vertically and in time with other horizons. Sometimes the lateral variations extend over a short distance; in other cases horizons intergrade gradually over long distances. These lateral intergrades are possibly the most important since there are many situations in which all the horizons do not intergrade laterally at the same rate. Further, detailed examination has shown that because of the continuum of horizons, soils have no properties other than horizons which can be used for their organisation. Any attempt to organise soils above the level of their constituent horizons can only be *ad hoc* and lacking in fundamental meaning, since it leads to the sacrifice of intergrading situations, by doing this the spatial continuum is ignored, and equally, the concept of soil evolution is abandoned.

Because of this a fundamental system of classification is not possible. This inescapable reality and basic truth seems to have escaped the notice of pedologists who for the last 70 to 80 years have been striving to produce a hierarchical system of classification. These attempts must spring from the false assumption and incredible misconception that soils can be treated like discrete entities which have finite boundaries.

When the various branches of science are examined with classification in mind, one aspect is prominent and conspicuous. It is only in the biological sciences that one finds a hierarchical system, and this is firmly based on Darwin's concept of evolution. There is no similar thread or theme linking soils together.

Repeated attempts have been made in previous chapters to demonstrate that soils form continua in space and time and that such a situation defies classification. However, it is necessary to divide up the continuum into arbitrary units for ease of communicating information. The best method of achieving this has been demonstrated by trying to establish reference segments and to use the segment symbols to produce formulae for soils. This method focuses attention on the pedo-unit or the soil itself, and achieves almost unlimited versatility while at the same time, the amount of nomenclature is reduced to a minimum. Intergrades are easily accommodated by using combined symbols, such as (SqAt). This is not a classification but a designation of each soil stating in a precise and concise manner its full characteristics, including the nature of the relatively unaltered underlying material. It should be pointed out that even the most complicated names of USDA do not give a full characterisation of the pedo-unit. Furthermore, it is well known that any discussions about soils, whether they be in writing or oral exchanges, quickly centre on the number and type of horizons. Therefore, it seems logical to produce a system which, in the first instant, gives in a concise form, the number and type of horizons, and also some indication of their thickness. One of the principal aims of pedologists should be to produce a designation of soils and a simple and general but *ad hoc* system of classification. At the present state of knowledge a formula appears to be the best approach.

The work of Northcote has clearly demonstrated the need for a key to aid the identification of soils. Ideally this should be produced for all soils but the data available at present are insufficient to accomplish this aim. As a first step in this direction the abstract given on page 167 is included.

The formula system used like any other which sets out to designate the soil, is suitable for punched cards and for the production of a dictionary of formulae. Information contained on punch cards as formulae is easily retrieved, it also allows relationships to be demonstrated with great ease as well as simplifying any *ad hoc* grouping or the construction of simple hierarchies.

In conclusion, it should be pointed out that systems like those of Kubiëna, Northcote and USDA are all descending or bifurcating systems operating on a presence or absence basis. The formulae system is an ascending system in which the soil formulae can be regarded as the starting point of a classification, which can be built up in a variety of ways to suit particular situations and allows almost unlimited versatility. But a number of different hierarchies can be constructed from a set of formulae, each hierarchy being produced to suit some specific need.

Soil classes of the world

The intention in this chapter is to give information about a restricted number of soils, concentrating on those which illustrate important pedological principles. Some classes are included only to give a balanced treatment and therefore are treated in a rather superficial manner.

In an attempt to make this book useful to as wide a range of workers as possible the terminology has been kept to an absolute minimum.

The principal nomenclature is that of FAO but every attempt has been made to accommodate other systems of nomenclature and classification. Within the text the first time that a soil or horizon name is mentioned using FAO terms it is followed by the USDA and FitzPatrick terms in brackets.

The soils are listed alphabetically for simplicity and ease of finding them. Table 7.1 shows the occurrence of soils in various climates on a frequency basis, therefore it is possible to get an indication of the type and frequency of soils in a particular area. The data for the classes start with a statement about the origin of the name of the class. This is followed by the general characteristics of the classes given by FAO and the characteristics of the subdivisions. A table of the approximate equivalents in other classes is given and then a fairly detailed account of one or more of the subdivisions follows. These accounts include the field morphology, micromorphology, analytical data, genesis, hydrology, variations in parent material, climate, vegetation and topography and also their distribution and utilisation.

ACRISOLS

Derivation of name: from the Latin word *acris* = very acid; connotative of low base status.

General characteristics

Soils having an argillic B horizon with a base saturation of less than 50 per cent (by NH_4OAc) at least in the lower part of the B horizon within 125 cm of the surface; lacking a mollic A horizon; lacking an albic E horizon overlying a slowly permeable horizon, the distribution pattern of the clay and the tonguing which are diagnostic for Planosols, Nitosols and Podzoluvisols respectively; lacking an aridic moisture regime. There are five subdivisions:

Orthic Acrisols. These have an ochric A horizon, an argillic B with a cation exchange capacity of >24 me/100 g clay and without the properties that distinguish the other subdivisions.

Ferric Acrisols. These have an ochric A horizon, and argillic B horizon with a cation-exchange capacity of <24 me/100 g clay and without the properties that distinguish the other subdivisions.

Humic Acrisols. These have an umbric A horizon or a high content of organic matter in the B horizon or both and without the properties that distinguish the other subdivisions.

Plinthic Acrisols. This includes all of the Acrisols that have plinthite within 125 cm of the surface. Therefore, they may have strong similarities with the other subdivisions.

Gleyic Acrisols. These show hydromorphic properties within 50 cm of the surface and without plinthite within 125 cm of the surface.

The subdivisions and their equivalents in some other systems of classification are given in Table 6.1.

Orthic Acrisols

General characteristics

This class includes a large number of somewhat diverse soils that have a clear to marked clay increase with depth but there is usually a low frequency of clay coatings in the middle horizon. It is difficult therefore to ascribe the clay increase to vertical clay translocation. In many cases there is evidence for clay translocation but this is usually rather small. In addition, the clay content in many cases remains

183

Table 6.1 Subdivisions of Acrisols and their equivalents in some other classifications

	Australia		Brazil	FitzPatrick	France
	Handbook	Northcote			
Acrisols			Red-yellow Podzolic soils	Luvosols	Soils Lessivés
Orthic Acrisols	Yellow Podzolic soils, Yellow earths	Gn2.7 Gn3.8 Gn2.3 Gn2.4			
Ferric Acrisols					
Humic Acrisols	Brown earths Xanthozems	Gn3.5 Gn3.7	Rubrozems		
Plinthic Acrisols					
Gleyic Acrisols	Humic Gleys	Gn3.0 Gn3.9 Dg4.1 Gn2.8 Gn2.9			

uniform or diminishes only slightly after the initial increase. Although the range is large they can be differentiated one from the other on the basis of their horizon sequences. In addition, many of these soils seem to be polygenetic or to have developed in sediments derived from strongly weathered soils.

Morphology (see Plate Ia)

There may be a loose leafy litter at the surface but this is usually very thin so that the bare soil may be exposed. The upper mineral horizon is usually dark greyish-brown, granular, sandy loamy Ah horizon (ochric epipedon; modon – Mo) which varies in thickness from <10 cm to >30 cm. In thin sections the matrix is isotropic and there are many partly decomposed plant fragments. This horizon grades sharply into a greyish-brown or brown sandy clay loam E horizon (albic horizon; zolon – Zo). This changes gradually into a red, clay loam argillic B horizon (argillic horizon; RoAr)) with granular structure and thin clay coatings on many ped surfaces. The matrix contains abundant small domains and some anisotropic aureoles. With depth this horizon grades into weathered rock.

Analytical data

The smallest amount of clay occurs in the Ah horizon followed by a slight increase in the E horizon; and a further increase in the argillic B horizon, it then decreases gradually with depth into the weathered rock.

In the Ah horizon there may be up to 10 per cent organic matter with a C/N ratio of 15, therefore it is not in a very advanced stage of humification. The cation-exchange capacity has two maxima, one of about 35 me per cent occurs in the Ah horizon due to the presence of organic matter while the higher of about 45 me per cent occurs in the argillic B horizon due to the greater amount of clay. The amount of exchangeable cations is low throughout the soil increasing from about 15 per cent base saturation in the Ah horizon to a maximum of about 35–40 per cent in the argillic B horizon with calcium as the dominant cation. The pH values are about 5.5 and are relatively uniform throughout the soil, but there is usually a slight increase with depth. The silica : sesquioxide ratios for the total soil decrease with depth, indicating that there has been a considerable loss of iron and aluminium from the primary minerals in the Ah and E horizons.

South Africa	USA		USSR
	Old	New	
	Red-yellow Podzolic soils	Ultisols	Yeltozems
		Hapludults	
		Haplustults	
		Haploxerults	
Magwa		Palexerults	
Constantia		Paleustults	
Kranskop		Humults	
Nomanei			
Constantia			
Shepstone			
Westleigh		Plinthaquults	
Avalon		Plinthudults	
Bainsvlei		Plinthustults	
Pinedene		Aquults	

Genesis

These soils form most commonly from rocks, acid or extremely acid sediments of Pleistocene age or they evolve from previous soils. The principal process in their formation is the reduction of the clay content in the upper horizons but there is no general agreement about the precise processes involved, however, it seems likely that the hypothesis of Simonson (1949) is correct. He suggested that clay is destroyed and removed from the system but some authors are of the opinion that clay translocation is the dominant process (McCaleb, 1959). It is doubtful whether this latter hypothesis can be sustained since the thin section morphology shows that in many cases there is little evidence for clay accumulation in the middle horizons.

When these soils develop in a previous soil they usually show an interesting sequence of evolutionary stages that often can be traced back to the Tertiary Period. A high proportion lie outside the limits of Pleistocene glaciation but some occur within the areas influenced by periglacial conditions therefore the Tertiary soil was in part eroded by solifluction. On flat and gently sloping sites the effect was small so that a considerable part of the Tertiary soil remained. On sloping sites most of the old soils were removed

and deposited in the valleys below, but where the movement was slow the material deposited in the valleys differed little from the original soil. Also, the amount of material remaining in place gets progressively thicker away from the influence of periglacial activity. This is very well demonstrated by a transect through the eastern USA from New Jersey to west Virginia. Acrisols also occur in areas not strongly influenced by glacial or periglacial processes but some of these areas were also subjected to climate change. This is seen in Australia where conditions for these soils are cooler at present than when original soil formation took place. This new set of conditions seems to induce greater acidity and more vigorous hydrolysis causing decomposition of the clay minerals and oxides in the Ah and E horizons and their removal from the system. This accounts for the sharp change in texture between the upper and middle horizons and the uniform texture through much of the remaining part of the soil except when there is a change to the lower horizon.

Principal variations in the properties of the class

Parent material

Acrisols usually show some measure of polygenesis, therefore the relatively unaltered underlying material may have little direct influence upon their formation. They are often found on fine or medium textured drift deposits derived from a previous soil which was transported only a short distance as mentioned above or they may be developing directly in a previous soil; this gives rise to the wide variety of middle horizons.

Climate

The climate ranges from Wet Equatorial to Tropical Wet-Dry and in all cases precipitation is heavy and fairly uniformly distributed through the year, so that water is constantly moving through these soils. There is a tendency for the texture to become coarser with increasing temperature.

Vegetation

The natural vegetation varies with the climate; under humid continental conditions there is usually a poor deciduous forest. In contrast, coniferous forests dominated by pine, form one of the principal communities in the humid subtropics of south-eastern USA and in the tropics there are usually rain forest communities. When the natural vegetation is removed, regeneration is often difficult because of their low fertility; thus a plagioclimax often results.

Topography
These soils develop on stable sites that range from flat.
to steeply sloping, providing that water can percolate
freely, but their most common occurrence is on flat
or undulating landscape.

Age
The development of Acrisols has extended over most
of the Holocene Period but in some areas these soils
seem to be very much older, for example, in part of
southern USA they can be traced from the present
surface to beneath late Pleistocene loess.

Distribution (see Fig. 6.1)
The principal areas where these soils occur are in
south-eastern USA, some of the Mediterranean
countries, south-eastern Australia and extensively
throughout humid tropical areas.

Utilisation
The thick, acid and strongly leached upper horizons
create many problems for utilisation. After the
removal of the natural vegetation the soils must be
limed heavily, followed by the addition of large
amounts of fertilisers particularly nitrogen and
phosphorus. Even so their productivity is low and
small areas are unable to provide a livelihood, therefore they are often devoted to plantation agriculture.
   Deficiencies in a number of microelements are
common and their application can bring about
dramatic results. The crops grown include fruit,
maize, tobacco and sweet potatoes.

ANDOSOLS

Derivation of name: from the Japanese *An* = dark
and *Do* = soil; connotative of soils formed from
materials rich in volcanic glass and commonly having
a dark surface horizon.

General characteristics
Soils having a mollic or an umbric A horizon possibly
overlying a cambic B horizon, or an ochric A horizon
and a cambic B horizon; having no other diagnostic
horizons (unless buried by 50 cm or more new
material); having to a depth of 35 cm or more one or
both of:
   (*a*) a bulk density (at 1/3 bar water retention) of
        the fine earth (less than 2 mm) fraction of the
        soil less than 0.85 g cm^{-3} and an exchange
        complex dominated by amorphous material;
   (*b*) 60 per cent or more vitric volcanic ash, cinders,

or other vitric pyroclastic material in the silt,
sand and gravel fractions; lacking hydromorphic
properties within 50 cm of the surface; lacking
the characteristics which are diagnostic for
Vertisols; lacking high salinity. There are four
subdivisions of Andosols:

Ochric Andosols. These have an ochric A horizon
and a cambic B horizon, a smeary consistence and/or
having a texture which is silt loam or finer.

Mollic Andosols. These have a mollic A horizon, a
smeary consistence and/or a texture which is silt loam
or finer.

Humic Andosols. These have an umbric A horizon
a smeary consistence and/or a texture which is silt
loam or finer.

Vitric Andosols. These lack a smeary consistence
and/or having a texture which is coarser than silt
loam.

   The subdivisions and their equivalents in some
other systems of classification are given in Table 6.2.

Humic Andosols

Morphology
Under natural conditions there is a loose leafy litter
at the surface. This rests on the very humose, dark
brown to black, crumb or granular upper Ah horizon
(umbric epipedon; kuron — Ku) which may be up to
30 cm thick and is isotropic in thin section. This
horizon grades into the middle, brown or yellowish-
brown cambic B horizon (cambic horizon; andon —
An) which is 20 to 30 cm thick and has an angular or
subangular blocky structure, also with an isotropic
matrix. With depth the middle horizon grades into
the relatively unaltered volcanic ash. Both horizons
are fluffy and when rubbed in the wet state have a
smeary as distinct from sticky consistence. In most
cases the middle horizon is thixotropic, i.e. it becomes
plastic when rubbed, yielding moisture but rehardens
when the rubbing is stopped. In most situations there
may be several buried soils due to successive showers
of volcanic ash.

Analytical data
The content of clay is low or very low and usually
does not exceed 20 to 25 per cent. The greatest
amount occurs in the upper horizon and decreases

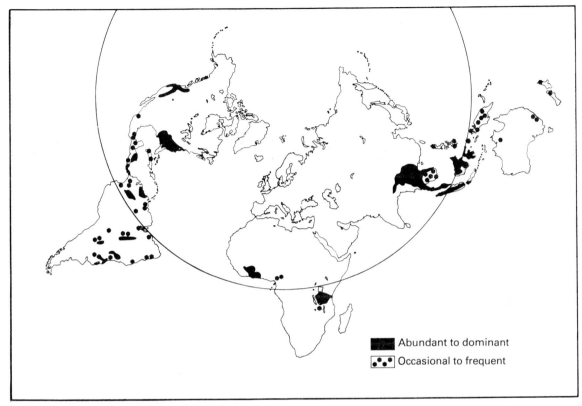

Fig. 6.1 Distribution of Acrisols

with depth to less than 5 per cent in the relatively unweathered parent material. Two important and characteristic features of all the horizons are their fluffiness and high porosity which may exceed 70 per cent in both the upper and middle horizons. Both of these properties are attributed to the presence of allophane which is the main product of hydrolysis. These soils vary from moderately to strongly acid with pH values as low as 4.5 in the surface, however, there is a steady increase with depth up to pH 6.0 or over in the relatively unaltered ash.

The content of organic matter is usually high and commonly there are values of over 20 per cent in the upper horizon. Although the material seems to be in a fairly advanced state of decomposition and appears to form a stable complex with allophane, the C/N ratio may be up to 15 which is considerably higher than most other highly humified horizons.

Due to the high content of humus the cation-exchange capacity is high in the upper horizon and may be over 35 me per cent, below, in the middle horizon, it decreases sharply to about 10 to 15 me per cent. However, these ion-exchange data have to

be interpreted with caution since allophane does not behave like other clay minerals, the determined values varying markedly with technique. The base saturation is normally lowest in the upper horizon, increasing with depth, but there may be high values at the surface due to cultivation. In some Andosols manganese dioxide is a product of weathering and may be present in sufficient amounts to give a fairly vigorous reaction with hydrogen peroxide.

Genesis

The formation of Andosols is a very rapid process resulting from the high surface area of the volcanic ash parent material which behaves in a unique manner under humid conditions. The principal process is hydrolysis which weathers the volcanic ash initially to yellow, brown or orange palagonite which is presumed to be an amorphous alumino-silicate containing calcium, magnesium and potassium; however, this changes quickly into allophane. Amorphous and microcrystalline iron and aluminium oxides are also formed following hydrolysis and are fairly uniformly distributed through the soil. The other main process

187

Table 6.2 Subdivisions of Andosols and their equivalents in some other classifications

	Canada	FitzPatrick	France	Japan	New Zealand
Andosols	Acid brown wooded soils, Acid brown forest soils	Andosols	Andosols	Kuroboku	Yellow-brown loams, Yellow-brown pumice soils
Ochric Andosols		[TnAn]			
Mollic Andosols		[KuAn]			
Humic Andosols		[KuAn]			
Vitric Andosols		[Ku(An-Is1)]			

is the partial humification of the organic matter and the formation of the stable complex with allophane.

Hydrology
These soils develop under aerobic conditions and have a downward flow of water but because of the high content of allophane they have a high water retaining capacity.

Principal variations in the properties of the class
Parent material
This is mainly volcanic material in the form of vitric volcanic ash, cinders or other vitric pyroclastic material. It may have accumulated from the atmosphere following a volcanic eruption, or it may be reworked material such as alluvium. There are some differences in the chemical and mineralogical composition of these materials, but the main variations are in the texture, however, it is usually of medium texture.

Climate
Because the development of these soils is determined mainly by the nature of the parent material, they are found under humid conditions ranging from the arctic to the tropics, but development is fastest under humid tropical conditions. Volcanish ash does not give rise to Andosols in dry or very dry climates.

Vegetation
The nature of the vegetation is as variable as the climate with a tendency for natural plant communities to be more luxuriant than those on adjacent sites where the soils have developed from other parent materials. This is due to faster weathering of the fine grain volcanic material giving a more plentiful supply of plant nutrients in these soils.

Topography
The sites on which these soils develop vary from flat to steeply sloping, but their best development is on the more stable, flat or gently sloping situations.

Age
Most of the soils of this class, occurring within areas that were glaciated or subjected to periglacial conditions, started to develop since the beginning of the Holocene Period. In tropical areas the starting dates vary from middle Holocene to late Pleistocene. Where the ash is older the original Andosols have evolved into other soils typical of the environment.

Distribution
Since these soils are associated primarily with volcanic ash their distribution is approximately the same as recent volcanic ash. They occur principally in Japan, New Zealand, north-western USA, East Indies, West Indies, Hawaii, northern and southern extremities of the Andes and East Africa.

Utilisation
Andosols are reputed to be very infertile when cultivated, however, they respond well to amelioration and can be highly productive. For example, in New Zealand, grass–clover mixtures will provide adequate grazing for livestock if sufficient lime and fertilisers are applied. Phosphorus is particularly important because the allophane in these soils has a very high capacity for adsorbing and fixing this element. It is

USA		USSR
Old	New	
Andosols	Andepts	Volcanic soils
	Dystrandepts	
	Eutrandepts	
	Dystrandepts	
	Hydrandepts	
	Vitandepts	

advantageous in warm climates to plant certain crops such as sweet potatoes (*Ipomaea batatas*) which show little response to the addition of phosphorus. Rice will grow well because the soils have a high water holding capacity.

ARENOSOLS

Derivation of name: from the Latin word *arena* = sand; connotative of weakly developed coarse-textured soils.

General characteristics
Soils from coarse-textured unconsolidated materials, exclusive of recent alluvial deposits, consisting of albic material occurring over a depth of at least 50 cm from the surface or showing characteristics of argillic, cambic or oxic B horizons which, however, do not qualify as diagnostic horizons because of textural requirements; having no diagnostic horizons (unless buried by 50 cm of more new material) other than an ochric A horizon; lacking hydromorphic properties within 50 cm of the surface; lacking high salinity.
There are four subdivisions of Arenosols:

Cambic Arenosols. These show colouring or alteration characteristic of a cambic B horizon immediately below the upper horizon.

Luvic Arenosols. These have lamellae of clay accumulation within 125 cm of the surface; not consisting of albic material in the upper 50 cm of the soil.

Ferralic Arenosols. These show ferralic properties, i.e. a cation-exchange capacity of less than 24 me per 100 g clay.

Albic Arenosols. These have albic material to a depth of at least 50 cm from the surface.

The subdivisions and their equivalents in some other systems of classification are given in Table 6.3.
Arenosols vary in colour from yellow to red as determined by the origin of the material and the internal drainage status. They grade gradually into Ferralsols (Oxisols; Krasnozems and Zheltozems) as the clay content increases. There are also sequences to Podzols (Spodosols) or Solonchaks (Salorthids; Halosols) depending upon climate.

Cambic Arenosols

Morphology
The colours of these soils range from reddish-yellow to yellowish-brown to red as determined in part of the origin by the material and moisture status. The upper mineral Ah horizon (ochric epipedon; tannon Tn) is usually not more than 10 to 15 cm thick and stained slightly by the presence of organic matter. Typically it is a loose, coarse sand with single grain structure which is clearly seen in thin sections (Fig. 4.27). Faecal pellets and worm casts may also be present, but they are usually very rare because very sandy soils are not ideal habitats for the meso-fauna. The middle horizon is a uniformly-coloured coarse sandy cambic B horizon (cambic horizon, arenon — Ae) which may exceed 2 m in thickness. In spite of its coarse texture it is often massive and hard. In thin section the individual sand grains usually have a thin isotropic coating which increases in thickness as the clay content increases then there may be a few small random domains. In many Arenosols some of the quartz grains may have red stainings in cracks (wusten-quartz) indicating their origin from a previous phase of soil formation.

Analytical data
The outstanding physical property is the high content of coarse sand, which imparts a high permeability and causes the formation of the single grain or massive structure.
In spite of the high content of quartz these soils are often only weakly acid with pH values between 6.0 and 7.0. This is coupled with a high base saturation but the cation-exchange capacity is very low because of the low clay content with kaolinite being

189

the principal clay mineral. The thin upper-horizon usually has <2 per cent organic matter which may have a C/N ratio of <12 indicating an advanced level of humification.

Genesis

These soils illustrate a small part of the complexity that is found in tropical and subtropical areas. Since they contain a small amount of clay there is little horizon differentiation except for the formation of the upper horizon. Arenosols appear to form in a number of distinct phases, the first often being a Ferralsol (Oxisol; Krasnozem or Zheltozem) or a similar strongly weathered soil composed mainly of clay and the resistant residue. This is then subjected to insidious erosion so that the finest material is preferentially removed leaving a residue of coarse particles. This may have taken place at a specific period during the Pleistocene, or alternatively the accumulation of the sand may represent the normal relationship between soil formation and landscape development. In some cases the evidence for a period of erosion is clear when these soils are formed in alluvial terraces, sand dunes or in stratified colluvial deposits. Therefore, they may occupy different parts of the landscape and have various relationships with other soils as discussed on page 300.

Termite activity has been evoked as playing a significant part in their formation. When termitaria are evacuated they collapse, then differential erosion takes place during which the fine material is washed into the valley leaving the coarse fraction on the slopes. Arenosols may also form from coarse textured sediments such as alluvium or dune sands particularly on the edges of deserts. Recent evidence indicates that progressive leaching and removal of the clay may

lead to the formation of Arenosols. Thus it seems that there are a number of different processes that can lead to their formation.

Hydrology

These soils are continuously aerobic and have very rapid through drainage. Their water-holding capacity is very low and plants usually suffer moisture stress except in very humid areas where the soils seldom occur.

Principal variations in the properties of the class

Parent material

Unconsolidated deposits of aeolian, colluvial or alluvial origin are common parent materials. In many cases they seem to have developed from a previous soil or by the leaching of clay from a soil formed from a material such as granite that contains a high content of quartz.

Climate

The greatest extent of these soils seem to be in the Tropical Wet-Dry and Tropical Desert and Steppe climates. This is probably because it is under these conditions that suitable parent material forms over large areas of the landscape.

Vegetation

The variation is from deciduous tropical forest to tropical and subtropical savanna.

Topography

This varies from flat to moderately sloping. On flat sites they are usually developed in Pleistocene or Holocene sediments, whereas on sloping sites they are

Table 6.3 Subdivisions of Arenosols and their equivalents in some other classifications

	Australia		Brazil	FitzPatrick	France
	Handbook	Northcote			
Arenosols			Red and yellow sands	Arenosols	Sols mineraux bruts
Cambic Arenosols	Siliceous sands Podzols Lateritic Podzolic soils	Uc 4.2		[TnAe]	
Luvic Arenosols				[TnAeArAeAr]	
Ferralic Arenosols	Earthy sands	Uc 5.2		[TnAe]	
Albic Arenosols	Lateritic Podzolic soils Podzols Lithosols	Uc 2.1 Uc 2.2 Um 2.1		[ZoAe]	

common in cols at the heads of some small valleys where sandy material has accumulated as a result of differential erosion.

Age
When these soils have formed through progressive weathering of the rock and differential erosion or leaching, they are very old and date back to the Tertiary Period. On the other hand many of those on sediments may have started their development during the middle to late Pleistocene Period.

Distribution
These soils are found in tropical and subtropical areas. They occur principally throughout east and south-east Africa, southern Sahara, central and western Australia and in various parts of equatorial South America.

Utilisation
In spite of their high content of sand these soils give good crop yields. Their reserve of nutrients is low as is their capacity to adsorb and hold them, therefore they have to be supplied constantly with fertilisers supplemented wherever possible by the addition of organic matter. When the pedo-units are thick, they have a large storage capacity for moisture which is readily available to plants but there may be a water shortage if there is a prolonged dry season.

CAMBISOLS

Derivation of name: from the Latin word *cambiare* change; connotative of changes in colour, structure and consistence resulting from weathering *in situ*.

South Africa	USA	
	Old	New
	Sands	**Psamments**
Hutton Clovelly		Psamments
		Alfic Psamments
		Oxic Quartzipsamments
		Spodic Udipsamments

General characteristics
Soils having a cambic B horizon and (unless buried by more than 50 cm or more new material) no diagnostic horizons other than an ochric or an umbric A horizon, a calcic or a gypsic horizon; the cambic B horizon may be lacking when an umbric A horizon is present which is thicker than 25 cm; lacking high salinity; lacking the characteristics diagnostic for Vertisols or Andosols; lacking an aridic moisture regime; lacking hydromorphic properties within 50 cm of the surface.

There are nine subdivisions of Cambisols:

Eutric Cambisols. These have an ochric A horizon and a base saturation of 50 per cent or more at between 20 and 50 cm from the surface but which are not calcareous at this depth.

Dystric Cambisols. These have an ochric A horizon and a base saturation of less than 50 per cent between 20 and 50 cm from the surface.

Humic Cambisols. These have an umbric A horizon which is thicker than 25 cm when a cambic B horizon is absent.

Gleyic Cambisols. These have an ochric or umbric A horizon and show hydromorphic properties between 50 and 100 cm of the surface.

Gelic Cambisols. These have permafrost within 200 cm of the surface.

Calcic Cambisols. These have an ochric A horizon, a cambic B horizon and one or more of the following: a calcic horizon, a gypsic horizon or soft powdery lime within 125 cm of the surface; calcareous between 20 and 50 cm.

Chromic Cambisols. These have an ochric A horizon and a base saturation of 50 per cent or more at between 20 to 50 cm from the surface but are not calcareous at that depth, have a strong brown to red cambic B horizon.

Vertic Cambisols. These have an ochric A horizon and a cambic B horizon showing vertic properties.

Ferralic Cambisols. These have an ochric A horizon and a cambic B horizon showing ferralic properties, i.e. having a cation-exchange capacity of less than 24 me per 100 g clay.

Many of the natural chemical properties can be altered by cultivation. This applies particularly to the pH and distribution of base cations, both of which are often higher in the surface due to liming and the application of fertilisers.

The subdivisions and their equivalents in some other systems of classification are given in Table 6.4.

Dystric Cambisols

General characteristics
Cambisols usually have a fairly uniform brown pedo-units with a total thickness of about 1.5 m. There is an upper moderately humose granular A horizon that grades into a blocky middle cambic B horizon. Dystric Cambisols occur mainly in temperate oceanic and continental conditions beneath deciduous vegetation and develop in a wide range of parent materials but have their best development on basic or intermediate drifts. They occur also in other humid areas where they tend to represent an early stage in soil development.

Morphology (see Plate Ib)
Under natural conditions there is usually a loose leafy litter at the surface resting on a humose greyish-brown granular Ah horizon (umbric epipedon; mullon — Mu) which varies from about 5 to 20 cm in thickness. In thin section the granular peds have an isotropic clay—humus matrix and are complete or fragments of earth-worm faecal material and sometimes they contain small fragments of partially decomposed plant material. This upper horizon grades gradually with depth into a brown middle Bw horizon (cambic horizon; alton — At) with angular or subangular blocky structure and which may be more than 30 cm thick. The matrix is usually isotropic or there may be rare small random domains (Fig. 6.2).

The middle horizon grades gradually into the underlying material which is very variable as determined by the evolutionary history of the soil, but often it is basic or intermediate drift. A conspicuous feature of many of these soils is the presence of numerous primary unweathered detrital grains throughout the soil as seen in thin sections.

Analytical data (see Fig. 6.3)
These soils are usually of medium texture with the upper horizon having the maximum content of clay which decreases in amount gradually with depth or it may remain fairly uniform throughout the pedo-unit. Where there is a marked textural contrast between horizons it is usually due to stratification of the original parent material. The pH value of the upper horizon varies from 5.0 to 6.5 and increases with depth to about neutrality in the underlying material. The organic matter content of the upper horizon varies from about 3 to 15 per cent with a C/N ratio of 8 to 12 and is indicative of a high degree of humification. The cation-exchange capacity is usually about 15 to 30 me per cent in the surface, decreasing with depth as the organic matter and clay contents decrease. Calcium is usually the principal exchangeable cation but magnesium may predominate when there is a high content of this element in the minerals in the soil. The percentage base saturation is very variable as determined by many factors particularly climate; it usually increases at the drier end of the climate range for this soil. A high content of ferromagnesian minerals can also lead to a higher base saturation. When the base saturation of the middle horizon is >50 per cent it is regarded as being high.

The amount and distribution of pyrophosphate extractable iron oxide are considered to be important criteria for differentiating between Cambisols and Podzols. In Cambisols the content of extractable Fe^{3+} should be uniformly distributed throughout the upper and middle horizon. In addition the content in the middle horizon is usually less than 10 per cent of the total clay fraction. In some cases this value is exceeded due to a high content of iron in the parent material.

Although many workers no longer use the $SiO_2 : Fe_2O_3 + Al_2O_3$ ratio of the clay fraction for categorising these soils and Podzols, it is probably a better criterion than the pyrophosphate extractable iron because it takes into account the total composition of the clay fraction. In these soils the ratio follows a similar trend to that of the extractable iron by remaining uniform or being slightly smaller in the middle horizon than in the upper horizon.

Genesis
At the surface the litter is rapidly decomposed both by microorganisms and the mesofauna which also incorporate some organic matter into the mineral soil and are largely responsible for the formation of a crumb or granular structure which is their faecal material. In the middle horizon the formation of the subangular blocky structure seems to be due largely to normal wetting and drying.

Hydrolysis is fairly active throughout the soil with

most of the iron and aluminium released being precipitated fairly close to the point of release with only a small amount being removed in the drainage or redistributed by leaching. This accounts for the fairly uniform distribution of free iron oxide and the uniform $SiO_2 : Fe_2O_3 + Al_2O_3$ ratios throughout the pedo-unit.

The clay maximum at the surface indicates that this is the position of maximum hydrolysis which is aided by the acid decomposition products of the organic matter. Leaching of the basic cations is one of the principal processes; whereas most of the sodium, potassium and magnesium ions released by weathering are lost completely from the soil in the drainage water, calcium can be precipitated as calcium carbonate in the lower part of the soil at the dry end of the soil's climatic range.

The isotropic nature of the clay in thin sections seems to be due on the one hand to the presence of organic matter and faunal mixing in the upper horizon and on the other to microcrystalline oxides in the middle horizon which either mask or inhibit the formation of domains.

Hydrology

Cambisols form under aerobic conditions in which there is usually free and rapid movement of water through at least the upper and middle parts of the soil. Although they develop where precipitation exceeds evapotranspiration they often show moisture stress during the dry period of the year.

Principal variations in the properties of the class

Parent material

Unconsolidated deposits of silty and loamy textures are the usual parent materials. These include loess, glacial drift, alluvium and solifluction deposits. In most cases the material is of intermediate composition but it is often basic, ultrabasic or calcareous. Of interest is the development of Cambisols in acid parent materials providing there is a high content of silt which gives a large surface area from which cations are released. It is essential that the release of basic cations by weathering keeps pace with their removal by leaching otherwise the soil will become progressively more acid and a Podzol or Luvisol would form.

Climate

These soils develop most easily either in a marine west coast or a humid continental climate. Variations in the character of these two climatic types, particularly the former, lead to variations in the nature of these soils especially in the degree of leaching which is greater in a marine climate with high rainfall. These soils occur also in most other humid environments where they seem to represent an evolutionary stage of short duration.

Vegetation

Deciduous forest is the most common natural plant community found on Cambisols. The precise species vary from place to place with communities dominated

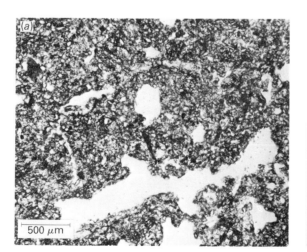
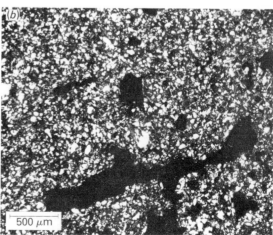

Fig. 6.2 Thin section of the B horizon of a Cambisol (a) in plain transmitted light, (b) in crossed polarised light showing the isotropic matrix and anisotropic sand grains which are mainly quartz

Table 6.4 Subdivisions of Cambisols and their equivalents in some other classifications

	Australia		Belgium	Canada	FitzPatrick
	Handbook	Northcote			
Cambisols			Sols bruns		Altosols
Eutric Cambisols		Um4.2 Um4.3		Orthic Brown Forest soils	[TnAth] [MuAth]
Dystric Cambisols				Acid Brown Forest soils Acid Brown Wooded soils	[MuAt] [Tn(AtSq)] [TnAt]
Humic Cambisols	Alpine humus soils	Um7.1			[MuAt] [Mu-Isl]
Gleyic Cambisols					[TnGl] [HyGl] [MuGl] [TnGlh] [MoGl] [MuGlh]
Gelic Cambisols				Cryic Brunisols	[MoAtCy] [MoCy] [GlCy]
Calcic Cambisols					[TnAthCk] [TnAthCkGy] [MuAthCk] [MuAthGy] [TnAthGy] [MuAthCkGy]
Chromic Cambisols	Alpine humus soils	Um7.1			[TnRoh] [TnFvh]
Vertic Cambisols	Chocolate soils				[Tn(AtVe)]
Ferralic Cambisols		Um4.4 Uf4.4			[TnRo] [MuRo] [TnFv] [MuFv]

by oak (*Quercus* spp.), beech (*Fagus sylvatica*), hickory (*Carya* spp.) and hazel (*Corylus avellana*) being among the most common.

Topography
Cambisols develop on sites that vary from flat to strongly sloping, their best development being on stable flat or gently sloping sites. In areas of moderate to high rainfall they usually occur on the lower parts of moderate to steep slopes where their presence is due to the lateral movement of moisture carrying dissolved cations which maintain a fairly high base status.

Age
Generally these soils occur within areas that were glaciated or subjected to periglacial conditions during the Pleistocene and therefore started their development about 10 000 years B.P.

Although in a number of cases their formation seems to have taken place predominantly in the Holocene Period the parent material may be weathered rock or old soil material formed in the Tertiary Period.

Distribution
These soils occur principally throughout central and western Europe, east and central North America, and eastern New Zealand. Often they do not form large continuous areas but small areas as determined by the occurrence of suitable parent material. They occur in many parts of the tropics in hilly areas where the slopes are too steep for deep soils to develop or as an early stage in an evolutionary sequence.

Utilisation
Cambisols are highly prized because of their fairly high inherent fertility. When the natural deciduous

France	Germany Braunerden	Kubiëna	South Africa	USA Old	USA New	USSR
		Braunerden		Brown Earths	Inceptisols	
Sols bruns eutrophes tropicaux	Typische Braunerde		Glencoe Oakleaf Clovelly Griffin Glenrosa		Eutrochrepts Ustochrepts Xerochrepts Eutropepts	
Sols bruns acides	Saure Braunerde				Dystrochrepts Dystropepts	
			Kranskop Magwa Inanda Nomanci		Haplumbrepts Humitropepts	
					Aquic Dystrochrepts Aquic Eutrochrepts	
					Pergelic Cryochrepts	
Sols bruns calcaires Sols bruns calciques	Kalk-braunerde		Oakleaf Clovelly Glenrosa		Eutrochrepts Ustochrepts	Cinnamonic Soils
Sols fersiallitiques non lessivés			Griffin Hutton		Xerochrepts	
Sols bruns eutrophes tropicaux			Arcadia		Vertic Tropepts	
					Oxic Tropepts	

vegetation is removed and the slopes are not too steep they can be adapted to a variety of systems of land use, more usually it is to mixed farming but large areas are used for dairying, orchards and other types of land use. When agriculture is practised annual applications of artificial fertilisers and periodic liming are necessary. Sometimes these soils are replanted with deciduous forests.

CHERNOZEMS

Derivation of name: from the Russian words *chern* = black and *zemlja* = earth; connotative of soils rich in organic matter having a black colour.

General characteristics

Soils having a mollic A horizon with a moist chroma of 2 or less to a depth of at least 15 cm; having one or more of the following: a calcic or gypsic horizon or concentrations of soft powdery lime within 125 cm of the surface; lacking a natric B horizon; lacking the characteristics which are diagnostic for Rendzinas, Vertisols, Planosols or Andosols; lacking high salinity; lacking hydromorphic properties within 50 cm of the surface when no argillic B horizon is present; lacking bleached coatings on structural ped surfaces.

There is a fairly narrow range of variability within this class of soils. This is in part due to the relatively short period during which they have been forming and also to their restricted environmental conditions.

195

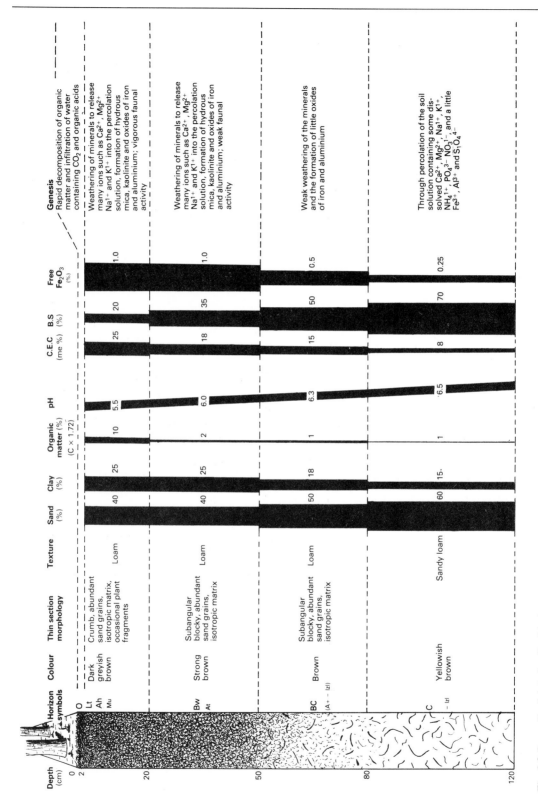

Fig. 6.3 Generalised data for Cambisols

Within the class one of the principal variations is in the thickness and humus content of the upper horizon. The thickness varies from about 50 cm to over 2 m while the humus content varies between 3 and 15 per cent. Some workers have used the variability in the amounts of organic matter and thickness of the upper horizon as criteria to subdivide the class. The clay mineralogy also shows some variation, whereas in the USSR mica is predominant, montmorillonite is most common in the USA. There are four subdivisions of Chernozems:

Haplic Chernozems. These soils have a mollic A horizon but without the distinguishing properties for the other subdivisions.

Calcic Chernozems. These have a mollic A horizon and a calcic or gypsic horizon but without the properties that distinguish the other subdivisions.

Luvic Chernozems. These have a mollic A horizon and an argillic B horizon, a calcic or gypsic horizon may be present.

Glossic Chernozems. These have a mollic A horizon which tongues into the cambic B horizon or C horizon and without the properties that distinguish the other subdivisions.

Chernozems grade gradually in at least seven principal directions into:
Cambisols, Luvisols, Luvic Kastanozems, Kastanozems, Solonetz, Solonchaks and Gleysols. The spatial change to the first four is due to a gradual change in climate, on the other hand the change to the latter three is due mainly to topography.

The subdivisions and their equivalents in some other systems of classification are given in Table 6.5.

Calcic Chernozems

Morphology (see Plate Ic)
At the surface there may be a thin, loose, leafy litter resting on a root mat up to 5 cm thick. Below is the dark-coloured upper Ah horizon (mollic epipedon; chernon – Ch) which may exceed 2 m in thickness but is normally 50 to 100 cm thick. This horizon has a marked granular or vermicular structure which is composed very largely of earthworm casts that are clearly seen in thin section (Fig. 4.30). In the lower part of the upper horizon calcium carbonate is deposited in the form of a white thread-like pseudo-mycelium (Fig. 6.4).

With depth the content of organic matter decreases and the amount of carbonate deposition increasing forming a calcic horizon (calcic horizon; calconlk). In many cases the carbonate accumulates in the form of concretions and occasionally forms a massive horizon. The lowest part of the profile is comprised of relatively unaltered material which is most often a calcareous loess.

Crotovinas are common phenomena in most Chernozems, and where they are well developed there is a thick mixed layer of upper horizon and relatively unaltered material from beneath. The organisms responsible for mixing the soil vary from place to place: in Europe the mole rat (*Spalax typhluon*) and gopher (*Sitellus suslyka*) are mainly responsible. Throughout the soil there are earthworm passages which even occur at depths of over 2 m in the relatively unaltered material and are clearly seen particularly when their passages are filled with dark material brought down from above.

Fig. 6.4 Pseudomycelium of tightly packed acicular crystals of calcite

Analytical data (Fig. 6.5)
The distribution pattern of clay down the pedo-unit is either uniform or shows a small maximum at about 50 to 200 cm from the surface, probably representing slight eluviation of fine material or weathering *in situ*. In view of the relatively low rainfall it is difficult to explain the slight accumulation of clay under contemporary conditions but there could have been some migration during the later part of the Pleistocene and early Holocene Periods, when the climate was cooler and probably more moist.

197

Perhaps the most important chemical property is the steady increase with depth in the amount of carbonate. Usually when the parent material contains only a small amount of carbonate most of the upper horizon has been decalcified and acid with pH values as low as 6.0, therefore it is not possible to estimate the rate of carbonate removal. In a Chernozem at Novi Sad in Yugoslavia the parent material contains over 50 per cent carbonate but the surface only has 15 per cent carbonate, therefore 35 per cent carbonate has been lost from the uppermost part of the soil during 10 000 years which gives a rate of removal of 0.035 mg g^{-1} p.a. This is a very crude estimate for there is some uncertainty about the time factor and further it is assumed that the climate has remained unchanged during the period. However, it does give some indication of the rate of removal of carbonate under Humic Continental warm summer conditions.

The organic matter content in the top of the upper horizon varies from about 3 to 15 per cent with a C/N ratio of 8 to 12. With depth the organic matter content decreases steadily to <1 per cent in the horizon of carbonate accumulation. Afanaseva (1927) has demonstrated that a Chernozem may have about 300 g m^{-2} of plant material above ground and about 1 000 g m^{-2} below, i.e. there is three times more organic material in the soil than in the above ground foliage, therefore a very high proportion of the organic matter in the soil is derived from decomposing roots.

The pH value varies from 5.5 to 8.0 in the upper part of the soil as determined by the amount of decalcification and leaching that has taken place, increasing with depth to values up to 8.0 and over in the relatively unaltered material.

The cation-exchange capacity tends to be rather low because the clay fraction is normally composed mainly of mica but there is usually a significant amount of montmorillonite.

Genesis

The principal processes taking part in the formation of Chernozems are the rapid incorporation of organic matter into the soil accompanied by its humification and the leaching of soluble salts and carbonates. In Europe the incorporation and decomposition of the organic matter is accomplished mainly by earthworms, the activity of which has been so vigorous and sustained for such a long period that all of the material in the upper 50 cm seems to have passed through the alimentary system at some time or another. However, the great thickness of the upper horizon is due to the small vertebrates that constantly churn the soil in pursuit of earthworms for food. Leaching completely removes the easily soluble salts from the soil but in most cases carbonate is deposited as pseudomycelia or concretions in the middle of the soil.

Hydrology

Chernozems develop under aerobic conditions in which there is free movement of water through the soil. The thick dark upper horizon with its well developed vermicular structure is unique and has excellent moisture relationships for most of the growing season. It is capable of storing large quantities

Table 6.5 Subdivisions of Chernozems and their equivalents in some other classifications

	Canada	FitzPatrick	France	Germany	Kubiëna	South Africa
Chernozems		Chernozems	Chernozems	Chernozems	Chernozems	
Haplic Chernozems	Rego Black Orthic Black	[Ch-lCzl]	Chernozem modal			Bonheim Tambankulu Inhoek Mayo
Calcic Chernozems	Calcareous Black soils	[ChCk] [ChGy]				
Luvic Chernozems	Eluviated Black soils	[ChAr] [ChArCk] [ChArGy]	Chernozem à B textural			
Glossic Chernozems		[(Ch(Ch-lCzl)]				

of moisture within the peds, at the same time excess water drains away freely and allows good aeration. Since these soils develop where evapotranspiration exceeds precipitation they show moisture stress during the middle and latter part of the growing season, hence the development of a natural grass plant community.

Principal variation in the properties of the class

Parent material

Chernozems are developed almost exclusively in loess but they also occur on other sediments which are calcareous or otherwise contain easily weatherable minerals that release a high proportion of calcium. Where quartzose sands or coarse textured materials occur within areas of Chernozems, Podzols or Luvisols may develop.

Climate

These soils are confined largely to continental conditions where the range is from Humid Continental to Middle Latitude Steppe. The important features of the climate are the cold winters, hot summers and an excess evapotranspiration over precipitation.

Vegetation

Tall grass is almost the only plant community found on these soils but deciduous woodland dominated by species of oak is common in certain transition situations.

USA		USSR
Old	New	
Chernozems	Mollisols	Chernozems
	Haploborolls	Typic
	Vermiborolls	Chernozems
	Calciborolls	
	Argiborolls	Podzolised
		Chernozems
		Tonguing
		Chernozems of
		Siberia

Topography

These soils are found on sites that vary from flat to moderately undulating. They are absent or only poorly developed on steeper slopes.

Age

Most Chernozems started their development during the last 10 000 years. Some may have been forming throughout the whole of that period while others may only be about half that age.

Distribution (see Fig. 6.6)

The total area occupied by these soils is relatively small. There is a belt in the south central part of western USSR, and one area in the central and north-central USA. Elsewhere in the middle latitudes they occupy very small areas.

Utilisation

These soils have a high nutrient status, excellent structure and high water-holding capacity, all of which impart a natural high fertility that makes them eminently suitable for agriculture. Indeed it was once thought that they were inexhaustible, however, over-cropping can lead to deficiencies particularly in nitrogen and phosphorus which have to be added as fertilisers.

Since these soils are found in areas of relatively low rainfall summer drought is sometimes a hazard so that moisture conservation is extremely important. This can be achieved by retaining snow in furrows or through the use of shelter belts.

Wheat, barley and maize are the principal crops grown and where there is a plentiful supply of water vegetables can be grown under irrigation. In recent years there has been a tendency to rear livestock and to practice dairying, but the lack of water is again a limiting factor.

FERRALSOLS

Derivation of name: from the Latin word *ferrum* = iron and aluminium; connotative of a high content of sesquioxides.

General characteristics

Soils having an oxic B horizon.

The variations in the properties of this class are wide and as a consequence offer a number of problems with regard to their designation and classification. Probably the most important variable is the thickness of the oxic B horizon (oxic horizon; krasnon — Ks) which may range from < 1 m to > 10 m.

200

Soil classes of the world

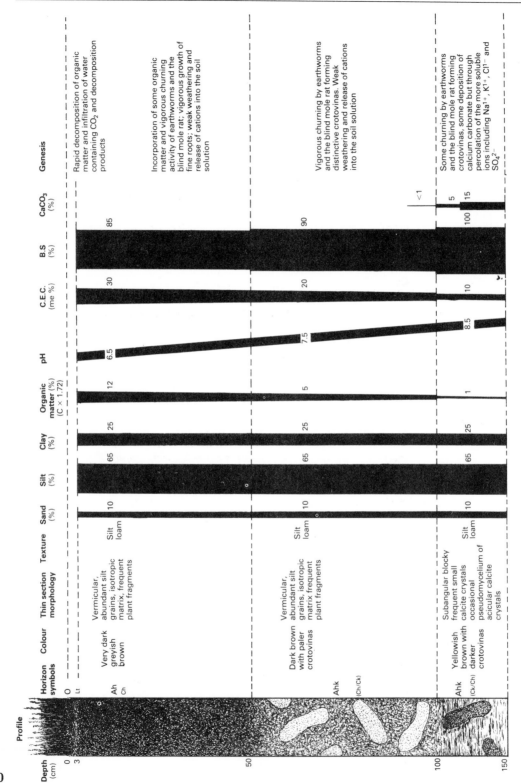

Fig. 6.5 Generalised data for Chernozems

This range can be due to time, the thicker ones being older but the thin ones are often due to erosion during the pluvial periods. Ferralsols also vary in colour with hues ranging from 10YR to 10R as determined by the nature of the parent material, the redder colours developing on material rich in ferro-magnesian minerals or on limestone. The presence of a hard horizon generally known as laterite is common in many of these soils; this is discussed below.

There are six subdivisions of Ferralsols:

Orthic Ferralsols. These have an ochric A horizon overlying an oxic B horizon that is neither red to dusky red nor yellow to pale yellow and has a cation-exchange capacity >1.5 me per 100 g clay.

Xanthic Ferralsols. These have an ochric A horizon overlying a pale yellow to yellow oxic B horizon with a cation-exchange capacity of >1.5 me per 100 g clay.

Rhodic Ferralsols. These have an ochric A horizon overlying a red to dusky red oxic B horizon and having a cation-exchange capacity of >1.5 me per 100 g clay.

Humic Ferralsols. These have an umbric A horizon or high organic matter in the oxic B horizon or both.

Acric Ferralsols. These have an ochric A horizon and an oxic B horizon with a cation-exchange capacity of 1.5 me or less per 100 g clay.

Plinthic Ferralsols. These have plinthite within 125 cm of the surface.

The subdivisions and their equivalents in some other systems of classification are given in Table 6.6.

Rhodic Ferralsols

Morphology (see Plates Id, IIa)
Under natural conditions the surface of the ground may be bare or there may be a very sparse litter. In some cases there are termitaria particularly when the plant communities are open forest or grassland. The uppermost mineral horizon seldom exceeds 20 cm in thickness and is a red or greyish-red ochric A horizon slightly stained with organic matter (ochric epipedon; tannon — Tn). This changes with depth into the red

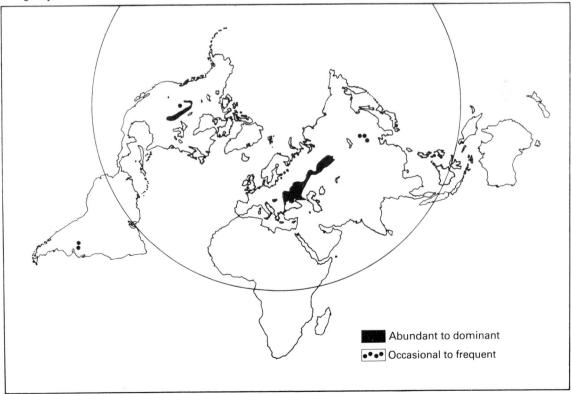

Abundant to dominant

Occasional to frequent

Fig. 6.6 Distribution of Chernozems

201

or brownish-red middle oxic B horizon which may be over 3 m thick (oxic horizon; krasnon – Ks). With depth there is a change into a number of horizons as discussed below, often there is a distinctive horizon with red and cream mottling and that hardens on exposure to the atmosphere. This is generally known as mottled clay or plinthite (plinthite; flambon – Fb) and usually grades into chemically weathered rock with core stones which in turn grades into the solid rock.

When considering this sequence in detail it is better to start with the rock and work upwards through the progressively more weathered material and finally into the red middle and upper horizons. A development sequence on granitic gneiss will be described because this seems to provide good illustrations of the important features. Within the first metre, disaggregation of the rock takes place along the surfaces of the joint blocks but the greater part of the rock is hard. In the next stage which occupies 1 to 2 m or more, disaggregation penetrates deeper into the joint block to form core stones accompanied by marked exfoliation of biotite (see Fig. 1.2). In hand specimens many of the feldspars appear fresh but in thin section they are in a fairly advanced stage of weathering being replaced by secondary products, mainly kaolinite, gibbsite and a little hydrous mica. Generally, the degree of hydrolysis increases upwards with some of the secondary products occurring as coatings around grains and in pores, and the whole mass is stained yellow by hydrated goethite. This stage grades into the mottled clay with its characteristic cream and red mottled pattern which is usually reticulate or laminar. In the field some of the rock structure is evident but this is more obvious in thin section particularly when the rock was a coarse grained, acid, metamorphic rock such as a granitic gneiss with a high content of quartz and a marked crystal orientation (Fig. 6.7). In thin section the mottled clay exhibits some of the most fascinating and complex phenomena to be found in soils since it represents an advanced but not complete stage of hydrolysis. The less resistant minerals have either been completely decomposed or are in an advanced stage of decomposition. In some cases the feldspars are decomposed along their cleavages and being replaced first by a pale yellow clay and later by a dark brown material which remains to give a cellular structure when the remainder of the feldspar has weathered away. This in turn becomes encapsulated to form a concretion (Fig. 6.8). The formation of macrocrystalline kaolinite (Fig. 3.10) seems to be at its maximum in this horizon, occurring as pseudomorphs after

feldspars but usually it has grown in spaces left after the decomposition of more than one mineral. Perhaps the most important and dominant feature is the thick and continuous clay coatings that cover nearly every surface and partly fill all the pore space (Fig. 6.9). On sedimentary rocks with a lower content of hydrolysable silicates the frequency of clay coatings is lower.

The mottled clay grades gradually upwards into the red middle oxic B horizon through a zone that appears in the field to be of uniform red or reddish-brown colour but in thin section it is seen to be composed of some uniform areas but a considerable proportion is almost identical to the mottled clay (Fig. 6.10).

The oxic B horizon has a sandy clay to clay texture and massive to incomplete subangular blocky structure. In thin section it is usually composed of rare to occasional sand grains set in a red or reddish-brown matrix. In sections of normal thickness the matrix appears dense and isotropic but in very thin sections (5 to 10 μm) it is seen to be dominantly anisotropic and composed of abundant small domains arranged randomly or in thin zones. Sometimes there are thin, rare clay coatings occurring on the surfaces of peds or as concentric deposits in pores.

Finally, at the surface there is the reddish-brown ochric A horizon with granular or subangular blocky structure but it often has a complex structure of subangular blocky and granular, the latter component being due to earthworms or termites. In thin section it is isotropic or there may be rare to frequent small random domains.

Analytical data (see Fig. 6.11)

These soils have a clayey texture and in a number of cases they have well developed aggregates which may inhibit dispersion for particle size analysis, therefore it is not possible to determine accurately their true clay content. When the aggregates are well developed, the term pseudosand is applied to this type of structure. It is commonly assumed that iron oxides are the cementing agents but DesPande *et al.* (1964), have shown that aluminium oxide is responsible. Where dispersion is achieved it is the fine clay (<0.5 μm) that is the dominant fraction <2 μm. Generally, this shows a gradual increase from the surface downwards to a maximum in the middle horizon followed by a considerable decrease to very low percentages in the chemically weathered rock.

In most cases kaolinite accounts for more than 80 per cent of the clay fraction, the remainder being goethite and gibbsite but the latter may be absent.

Hydrous mica sometimes occurs in small amounts.

A characteristic feature of these soils is the very low silt : clay ratio, due to the profound weathering that has taken place. This ratio varies from about 0.15 for soils developed on igneous rocks to 0.2 for soils on sedimentary rocks; the latter higher ratio being due to the higher initial content of resistant minerals in sedimentary rocks.

The acidity is remarkably uniform with pH values about 5.5 in the surface, decreasing to about 4.5 in the middle horizon then increasing to 5.0 in the mottled clay.

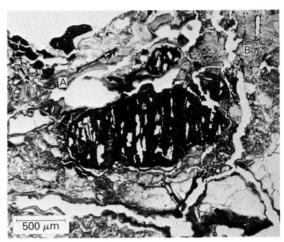

Fig. 6.8 A concretion in the mottled clay. The concretion has a characteristic cellular structure which apparently is determined by the original cleavages of the feldspar crystal. The spaces between the cell walls are empty. Also present are: (A) clay coatings and (B) macrocrystalline kaolinite

Fig. 6.7 Partial retention of the rock structure in the mottled clay as shown by the rectangular grains of quartz

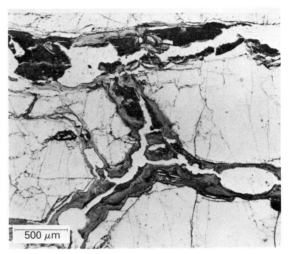

Fig. 6.9 Clay coatings between grains of quartz in the mottled clay. The grains of quartz partly retain the original rock structure

Because kaolinite is the principal clay mineral th cation-exchange capacity is very low with values of <15 me per cent per 100 g of clay, which give tota soil values of about 12 me. The amount of exchang able cations and the base cation saturation are also low but the latter can be high in some situations. Calcium is the principal exchangeable cation with smaller amounts of magnesium, potassium and sodium.

The surface horizon seldom contains more than about 5 per cent organic matter, usually there is only 2 to 3 per cent decreasing to <1 per cent with depth. The C/N ratio is 8 to 12 which indicates an advanced stage of humification.

Partial ultimate analysis reveals that from the mottled clay upwards the main mineral constituents are silicon, aluminium and iron with lesser amounts of titanium and practically no basic cations, thus the reserve of basic cations is extremely low, see Table 3.1.

Genesis

These soils are formed by the progressive hydrolysis and complete transformation of the rock into clay minerals, oxides, concentrations of the resistant residue and the loss in the drainage of much material, particularly basic cations and silica. The particular course of development seems to follow a number of different pathways as determined by the nature of the parent material and drainage characteristics. On

Table 6.6 Subdivisions of Ferralsols and their equivalents in some other classifications

	Australia		Brazil	FitzPatrick	France
	Handbook	Northcote			
Ferralsols			**Latosols**	**Krasnozems**	**Sols ferrallitiques**
Orthic Ferralsols			Red-Yellow Latosols	[Tn(KsZh)]	Sols ferrallitiques moyennement à fortement désatur
Xanthic Ferralsols	Xanthozems Yellow earths Yellow Podzolic soils	Uf5.2, Gn3.5, Gn3.7 Gn2.2, Gn2.6 Gn2.7	Pale Yellow Latosols	[TnZh]	Sols ferrallitiques jaunes fortement désaturés
Rhodic Ferralsols	Krasnozems Red earths	Uf5.2, Gn3.1, Gn4.3, Gn4.1 Gn2.1	Latosols Roxo	[TnKs]	Sols ferrallitiques faiblement à moyennement désaturés
Humic Ferralsols	Krasnozems	Gn3.1	Humic Latosols	[MuKs]	Sols ferrallitiques fortement désaturés humiques
Acric Ferralsols				[TnKs] [TnZh]	
Plinthic Ferralsols	Xanthozems Red earths	Uf5.2 Gn2.1		[TnKsFb] [TnFb] [TnKsVs] [TnVs]	

Fig. 6.10 Decomposition of clay coatings in the mottled clay to form the oxic B horizon

100 μm

very basic parent material the stages may be similar to those given on page 68 but in the example given above the stages of decomposition are more gradual and decomposition products are dominated by goethite and kaolinite. The latter may form pseudomorphs but more commonly appears to be re-

distributed within the weathering zone to form coatings which become progressively large and more numerous as decomposition proceeds, attaining their maximum development in the mottled clay. A satisfactory explanation has not been offered for the formation of the mottling. It would appear to be due to the segregation of compounds of iron to form the red areas but partial ultimate analysis shows no difference in the amount of iron in the two areas, therefore it seems to be due to the state of the iron, it being oxidised to the ferric state in the red areas. The change in colour from yellowish-brown in the initial stages of weathering to red or reddish-brown in the mottled clay suggests that the latter contains a less hydrated form of goethite. The development of the oxic B horizon seems to result from the homogenisation of the mottled clay by a mechanism that is not understood. Termites and other organisms may be responsible but expansion and contraction may be the controlling mechanism since the coatings in the mottle horizon seem to merge and disappear into the matrix to form the oxic B horizon.

Hydrology

The upper parts of these soils have free drainage but may become saturated with water during the rainy

Kubiëna	South Africa	USA	
		Old	New
Latosols		Latosols	Oxisols
			Orthox
			Torrox
			Ustox
Braunlehm	Hutton		Orthox
Rotlehm	Hutton		Orthox
			Torrox
			Ustox
	Kranskop		Humox
	Magwa		
	Inanda		
			Acrox
	Avalon		Plinthaquox
	Bainsvlei		

season but this state does not seem to exist for a very long period since there is no evidence of reduction. In the lower part of the soil the mottled clay may derive its colour pattern from prolonged periods of wetness.

Principal variations in the properties of the class

Parent material
All types of consolidated rocks are the predominant parent materials, but these soils also form on un-consolidated deposits such as Tertiary sediments and old volcanic ash.

Climate
Characteristically these are one of the main soils of the Wet Equatorial, Trade Wind Littoral and Tropical Wet-Dry climates. The optimum conditions for their development appear to be in areas with a mean annual temperature of $>25\ ^{\circ}$C and mean annual precipitation of over 1 000 to 1 200 mm. Where precipitation is much higher or temperatures slightly lower they tend to be confined to the more basic parent materials such as basalt. Pleistocene and Holocene climatic changes have caused some of these soils to occur outside their normal climatic range; particularly good examples of this are seen in parts of East Africa.

Vegetation
Rain forest and semideciduous tropical forests are the principal plant communities. When these soils occur outside their normal climatic range they may carry communities such as thorn woodland, deciduous tropical forests and savanna communities.

Topography
Ferralsols are soils of the lowlands, they are not found above about 1 200 to 1 500 m where they give way to soils of cooler, wetter conditions. They occur on sites that vary from flat to steeply sloping but they tend to be more common on moderate slopes.

Age
The majority of these soils are found in materials that are at least of mid-Tertiary age and therefore they are very old. Some occur on late Tertiary or Pleistocene materials but these usually have a low content of weatherable minerals.

Distribution (See Fig. 6.12)
These soils are of widespread occurrence in tropical and subtropical areas. They occupy large areas in the north of Australia and throughout much of India. In Africa they occupy large areas on both sides of the Equator and have a similar distribution in South America particularly in Brazil. They are found also in a number of subtropical countries.

Utilisation
The low nutrient status and low content of organic matter give these soils a very low fertility but they often support high forest. The relationship between Ferralsols and their natural forest vegetation is a good example of the delicate balance of nature in which nutrients are constantly recycled to maintain the natural forest community. When the forest is removed and agriculture is practised, fertility is quickly exhausted and crop failure is the normal result. This accounts for the practice of shifting cultivation so highly developed in parts of Africa. This system of land use involves felling the forest, cultivating the ground for a few years during which the nutrients are exhausted and then moving on to a fresh site leaving the area to develop a new forest community and rebuild the fertility for another period of cultivation. In recent years, modern practices including liming and the applications of fertilisers have made possible a more stable system of agriculture but as yet a completely satisfactory system of land use has not been developed for many of these soils. In many

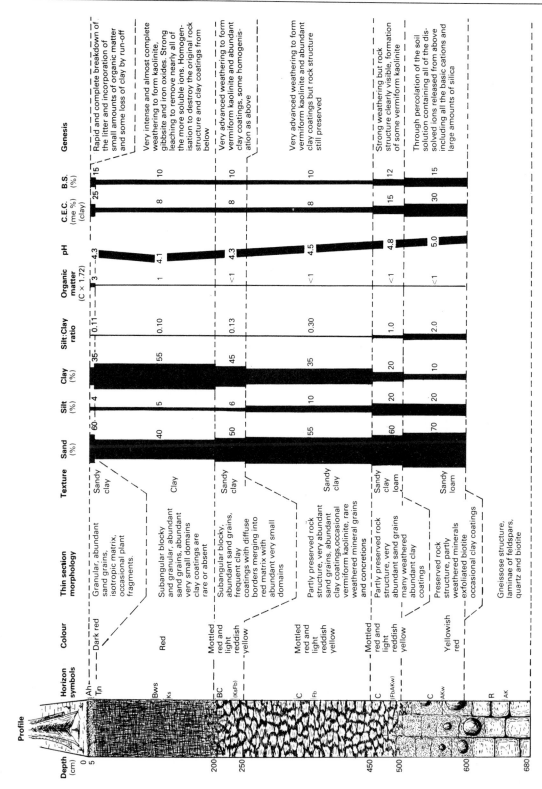

Fig. 6.11 Generalised data for Ferralsols

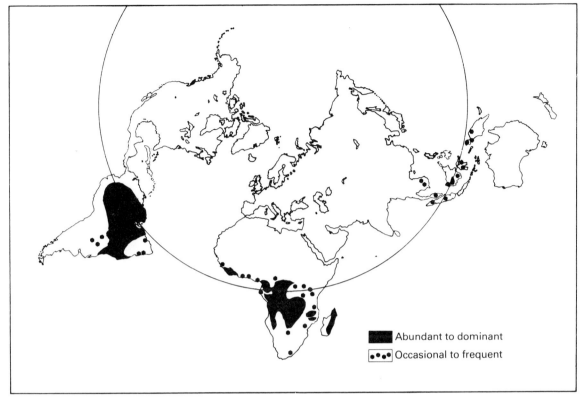

Fig. 6.12 Distribution of Ferralsols

cases the greatest success has been achieved with tree crops such as, cocoa, oil palm, tea, coffee and rubber. Very hardy plantation crops such as sugar cane have also been extremely successful. However, these soils seldom sustain a high standard of living except when their utilisation is on an extensive scale because the profit per hectare from carbohydrate crops is small.

Laterite

The term laterite appears to have been introduced by Buchanan in 1807, although some workers credit Babbington (1821) as being the first to use the term scientifically. In either case the term was used in India to refer to the vesicular mottled, red and cream, clay which was dug out of the ground, shaped into bricks and allowed to dry in the sun. The early workers did not describe the material in great detail, nor did they carry out careful chemical analyses, so that the term has been used rather indiscriminately for material which was hard or would harden on exposure. In addition, the term was often stretched

and applied to soils so that red, strongly weathered, tropical soils became known as lateritic soils and the term applied whether they contain true laterite or not. Subsequently many workers have tried to tidy up the definition by more careful descriptions and chemical analysis. However, it seems that to most people the term laterite refers to a soil horizon which is hard or will harden on exposure and is composed mainly of the oxides of iron and/or aluminium with varying amounts of kaolinite and quartz and sometimes oxides of manganese. This broad definition includes quite a range of material, since horizons that are hard or harden, can be vesicular, pesolitic, concretionary, massive or mixtures of these types. Although vesicular laterite is very common the greatest variability seems to be in the number and type of concretions in any one formation. Some workers divide the concretions into two main categories – pesolitic and nodular. Here there is also confusion, since some define pesolites as having concentric layers, while to others they are not layered. Further, in some cases small rounded fragments of detrital vesicular laterite are termed pesolites. In spite of the above the following

207

types of concretions are easily recognised:

(*a*) Black 1 to 10 mm with little or no internal structure and composed of iron oxides. May be magnetic or non-magnetic and may include quartz grains. These two varieties may occur together.

(*b*) As above with brown or reddish-brown, concentrically ringed skin. Seems to represent a two-stage process, the concretion forming first followed by the concentric deposit.

(*c*) Black or dark brown 1 to 10 mm with cellular internal structure. Composed mainly of goethite and seems to be formed as a replacement of feldspars, the cell walls representing the infilled space left by the weathering of one of the feldspar twins (Fig. 6.8).

(*d*) Black 1 to 10 mm with an amorphous porous centre (Fig. 6.18).

(*e*) Yellowish-brown 5 to 20 mm with central red or brown gibbsitic core having preserved rock structure and surrounded by very distinctive concentric layering of gibbsite (Figs. 6.13 and 6.14).

In some soils concretions of types *a*, *b* or *c* may occur in low concentrations in the soil but may become concentrated due to differential erosion and then cemented to form laterite.

There appears to be a large number of profile and landscape positions in which laterite will form. In the majority of cases laterite seems to be a specific soil horizon formed above or within the mottled clay. Such a type of laterite occurs predominantly on flat or gently undulating landscapes and is attributed by some to a fluctuating water-table.

Since laterite hardens on exposure, we find that in many landscapes it now forms a surface capping following upon uplift and aeration of the topsoil, so that the landscape has a number of mesas and escarpments below which are long slopes (Fig. 6.15). With time the laterite weathers, breaks off into fragments of various sizes and forms colluvium on the slope and at the bottom of the slope, and in a number of circumstances this colluvium becomes cemented to give a secondary type of laterite. In fact, it is the belief of a number of workers that a high proportion of laterites are of this recemented variety.

Thus the morphology, mineralogy and genesis of laterites are extremely variable but generally the following types can be recognised.

1. Mottled red and cream material with weak vesicular structure: composed of kaolinite, iron oxides, gibbsite with partially preserved rock structure; will harden on exposure. This is derived

Fig. 6.13 Gibbsite concretions in a profile in Western Australia

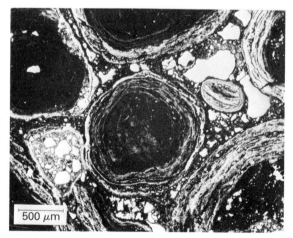

Fig. 6.14 Gibbsite concretions similar to those in Fig. 6.13 having a central core with partially preserved rock structure and outer concentric rings

from the characteristic mottled clay of the tropics (Figs. 6.16 and 6.17).

2. Vesicular, composed of a hard continuous phase

Fig. 6.15 A mesa in the desert of Western Australia, caused by the presence of a laterite capping

of dark brown, iron impregnated material surrounding cream kaolinitic or gibbsitic clay sometimes with crudely hexagonal cross-section.

3. Similar to the above but containing small black concretions like type *a*.

4. Very dense massive material dominated by subspherical black iron oxide concretions embedded in a reddish-brown matrix. The iron oxide concretions may or may not be magnetic. In some cases the matrix may contain much gibbsite and shows little organisation, but in others there is concentric ringing of material around the concretions suggesting that the concretions first formed and were later surrounded and cemented by the red material.

5. Reddish-brown massive with abundant black concretions and thin veins of well crystalline kaolinite between the concretions (Fig. 6.18).

6. Yellowish-brown to reddish-brown, hard, scoriaceous and composed dominantly of gibbsite.

7. Yellowish-brown, massive, dominated by gibbsitic concretionary material like type *e* (Fig. 6.19).

8. Black or very dark brown massive containing much quartz sand, gravel and small rock fragments. This occurs in lower slope positions and is cemented by iron oxide and manganese dioxide from seepage waters (Fig. 6.20).

9. Black or very dark brown, fused nodular concretions, very high bulk density, composed mainly of iron oxide and manganese dioxide. This forms in wet lower slope and depression situations, particularly in subtropical areas. This seems to equate with bog iron ore of the higher latitudes, and is sometimes called ortstein.

10. Yellowish-brown to red cemented black concretions, very high bulk density. This is a high porosity mass of fused concretions that have been concentrated by differential erosion and then cemented (Fig. 6.21).

11. Massive, dark brown to black, containing abundant black subspherical units and rounded rock fragments. This is a cemented colluvial deposit in which the black units are rounded fragments of vesicular laterite that have been broken off an escarpment and moved down slope. The rock fragments are usually strongly weathered with the feldspars being largely removed or altered to kaolinite or gibbsite.

209

Amphiboles and pyroxenes are usually altered to iron oxides.

Sometimes two different layers of laterite occur one above the other in the same profile. Since most laterites are very old, they are often influenced by later pedogenic process so that some have accumulations of calcite, gypsum and may even be silicified.

3 cm

Fig. 6.16 A block of vesicular laterite, the outer weathered surface shows the characteristic pitted surface

3 cm

Fig. 6.17 A block of vesicular laterite in cross-section showing the distinctive vesicular structure

FLUVISOLS

Derivation of name: from the Latin word *fluvius* = river; connotative of flood plains and alluvial deposits.

General characteristic
Soils developed from recent alluvial deposits having no diagnostic horizons other than (unless buried by 50 cm or more new material) an ochric or an umbric A horizon, a histic H horizon, or a sulphuric horizon. As used in this definition, recent alluvial deposits are

fluviatile, marine, lacustrine, or colluvial sediments characterised by one or more of the following properties:
  (a) having an organic matter content that decreases irregularly with depth or that remains above 0.35 per cent to a depth of 125 cm (thin strata

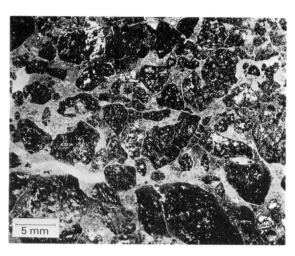

5 mm

Fig. 6.18 Thin section of hardened pesolitic laterite composed of concretions set in a matrix containing thin accumulations of finely crystalline kaolinite

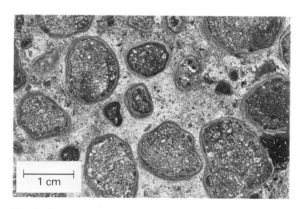

1 cm

Fig. 6.19 Laterite containing mainly gibbsitic concretions

of sand may have less organic matter if the finer sediment below meets the requirements);
  (b) receiving fresh material at regular intervals and/or showing fine stratification;
  (c) having sulphidic material within 125 cm of the surface.

There are four subdivisions of Fluvisols:

1 cm

Fig. 6.20 Massive laterite containing quartz sand and rock fragments cemented by iron and manganese oxides

5 cm

Fig. 6.21 A concentration of concretions formed by differential erosion and then cemented together

Eutric Fluvisols. These have a base saturation of >50 per cent between 20 to 50 cm from the surface but not calcareous at this depth.

Calcaric Fluvisols. These are calcareous between 20 to 50 cm from the surface.

Dystric Fluvisols. These have a base saturation of <50 per cent between 20 to 50 cm from the surface.

Thionic Fluvisols. These have a sulphuric horizon or sulphidic material or both at less than 125 cm from the surface.

The subdivisions and their equivalents in some other systems of classification are given in Table 6.7.

Utilisation

Many Fluvisols require a considerable amount of amelioration before they can be utilised. Generally, flooding has to be prevented so that they can dry out and become suitable for growing crops. These processes are generally known as ripening.

Soil ripening

The changes that take place during ripening can be divided into:

    Chemical ripening
    Biological ripening
    Physical ripening

Chemical ripening

This includes primarily the oxidation of organic matter, iron and manganese. The oxidation of the organic matter usually causes a reduction in volume and settling particularly in the cooler areas such as the Netherlands. Some of the iron is also oxidised to give yellow, brown and red mottles while some of the manganese is oxidised to give black manganese dioxide but this reaction does not take place when jarosite is formed.

Biological ripening

As soon as the water is removed the areas are rapidly colonised by plants such as reeds. *Phragmites communis* may be planted as is done in the Netherlands. The plants remove large quantities of water thus drying the soil to a much greater extent than drainage because of the low hydraulic conductivity in these fine textured compact muds.

The growth and expansion of the plant roots and the burrowing of earthworms cause homogenisation of the topsoil as well as creating passages. If pyrite is present the formation of jarosite and sulphuric acid prevent the penetration of roots and the cessation of ripening even though a good drainage system may be present.

Physical ripening

With the loss of water and organic matter there may be a 50 per cent loss of volume resulting in settling, shrinking and cracking and the formation of a prismatic structure. During the rainy season the soil will swell slightly but never to the same extent as before. Also, the impact of the rain drops at the surface cause some dispersion of the soil and the formation of a slurry which runs down the surfaces of the prisms to form whole soil coatings (Fig. 4.35). With time the surface develops a good structure and the shrink–swell process is reduced to a minimum except where the clay content is very high.

Thus the soils go through a progression of changes from the raw, unripe deposit to the fully ripened soil. This can be divided into five stages based on the

211

n-value of the soil. The *n*-value is the weight of water in grams absorbed by 1 g of clay in the soil. This determination has to be made in the laboratory but a field system based on consistence has been developed to aid mapping as given in Table 6.8.

Although ripening is generally a man-induced process, it can take place naturally by such mechanisms as a change in sea level.

When the soils are fully ripe they can be utilised for a number of purposes but the tendency is to use them for labour intensive crops, such as the growth of bulbs in the Netherlands.

Thionic Fluvisols

General characteristics

These soils occur predominantly in the naturally or artificially drained alluvium of mangrove estuaries and deltas of tropical rivers. At the surface there is an organic–mineral mixture which overlies a middle horizon with characteristic yellow mottles composed predominantly of jarosite. At the base of the soil is a dark grey or bluish-grey completely anaerobic horizon containing pyrite.

Morphology

There is an upper grey to very dark grey organic mineral mixture with massive or coarse angular blocky structure and with occasional to frequent brown or reddish-brown mottles. This is followed by a greyish-brown horizon with massive or coarse angular blocky structure. There are common yellowish-brown and prominent red mottles and some brown root pipes. Many of the ped faces have thin coatings of the upper horizon. The third horizon is again brown with yellowish-brown mottles and in addition much yellow (2.5Y 6/8) jarosite filling passages up to 1 cm in diameter. There is also a well developed prismatic structure with topsoil coatings particularly in the upper part. The fourth horizon is greyish-brown, massive with fewer yellowish-brown mottles and no jarosite, this is followed by a massive dark grey permanently anaerobic horizon containing pyrite.

Analytical data

The chemical and physical analyses show that these soils have a number of unique properties related in most cases to the initial presence of pyrite throughout the system. There are no significant trends in the particle size distribution, there is usually a high content of silt and clay that remains uniform with depth. Where there are variations they are attributed to stratification of the original alluvium. The pH values of the fresh soils show an interesting and unique pattern. At the surface the values are about 4.5 decreasing to about 3.5 in the third horizon containing jarosite, this is followed by an increase to pH 5 or over in the lowest completely anaerobic horizon. If the soil is aerated prior to determining the pH values then the values decrease from about 4.5 in the surface to <3.0 and often <2.5 in the lowest horizon. These low values are due to the oxidation of pyrite and the formation of sulphuric acid as will be discussed below. The data for iron show that both total and dithionite extractable Fe_2O_3 are at a maximum in the third horizon where there is jarosite and are due to the translocation of iron from the upper parts of the soil. Pyrite occurs only in the lowest horizon.

Few micromorphological studies have been conducted on these soils, however, Eswaran (1967) has shown that the lowest horizon is massive with few pores. The matrix contains abundant, random, large domains and occasional small crystals of pyrite often associated with decaying roots (Fig. 6.22). The horizon with jarosite has no pyrite and the matrix has fewer domains.

Genesis

These soils are developed mainly in sediments which are usually bluish-grey or grey marine muds, but they can be inland alluvium.

The following are the stages in their development:

1. Reduction of sulphates to hydrogen sulphide by sulphate reducing bacteria under anaerobic conditions. The sulphates are usually supplied by sea water but the reduction of elemental sulphur or organic sulphur can also be accomplished by microorganisms. In all cases an adequate supply of organic matter is necessary as a food supply for the bacteria.
2. Reaction of hydrogen sulphide with iron compounds present in the soil to form ferrous sulphide – FeS and pyrite – FeS_2. If the soil remains anaerobic the reaction stops here and there is a gradual build up of pyrite under the naturally wet conditions.
3. With drainage for cultivation, oxidation of the iron sulphide by oxygen of the atmosphere takes place resulting in the formation of ferric sulphate and sulphuric acid.

4. Hydrolysis of the ferric sulphate to form the straw coloured basic ferric sulphate (jarosite) and more sulphuric acid. Although these two final stages can be accomplished by simple chemical processes, microorganisms may also play a part.

Sulphuric acid has a number of very adverse effects on soils and plants; in addition to increasing the acidity, the large number of dissociated hydrogen ions replace the base cations in the exchange complex. It

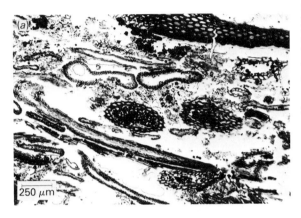

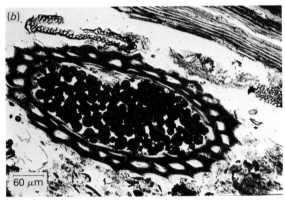

Fig. 6.22 A cluster of pyrite grains associated with a decomposing root

    (a) Lower part of a Histosol profile containing pyrite
    (b) Enlarged portion of a) showing cluster of pyrite crystals in a plant root

attacks the primary minerals causing essential plant nutrients such as calcium and potassium to be released by hydrolysis, brought into solution and lost in drainage water. The clay minerals are also attacked with the result that large amounts of aluminium sulphate are formed followed by the precipitation of aluminium hydroxide. Since the acidity only develops

upon oxidation, the pH in the soil may vary from less than 3 in the oxidised surface up to 7 in the lowest horizon.

Although the sequence of events outlined above usually takes place after drainage, Moormann (1963) has reported the effect of the natural occurrence of drainage on old elevated river terraces.

Principal variations in the properties of the class

Since Thionic Fluvisols are associated mainly with coastal conditions, variations in the factors of soil formation have little influence upon their formation. However, there are some differences in the nature of the pedo-unit. The upper horizon may vary widely in thickness and content of organic matter and so may the horizon with jarosite which increases in thickness as drainage improves and oxygen penetrates deeper into the soil.

Distribution

These soils occur to a greater or lesser extent in most tropical areas adjacent to the sea. Extensive areas occur in South-East Asia particularly in Vietnam, Thailand, Indonesia, Sumatra and Borneo; they occur in East Pakistan and in East as well as West Africa. In the New World they have been reported from Surinam and doubtless they occur elsewhere. They also occur in temperate areas as in Holland where they were first described and intensively studied.

Utilisation

Thionic Fluvisols are utilised only when there is great need for land, otherwise they are left in their natural state. The poor structure, low base saturation and wide C/N ratio are obvious limiting factors to plant growth but it is the high acidity following drainage that is the main factor preventing land use, for it is necessary to apply extremely large amounts of lime to raise the pH and this may not be economic. The extreme acidity causes iron, aluminium and possibly magnesium to be present in toxic proportions also the high content of iron causes phosphate fixation which reduces their fertility still further. However, some success has been achieved, for example, by keeping the soil saturated throughout the year, oxidation is prevented and the acidity is kept to a minimum, then some varieties of rice can be grown. When the soils are drained and become very acid some   213

Table 6.7 Subdivisions of Fluvisols and their equivalents in some other classifications

	Australia		Canada	FitzPatrick	France	Germany	Kubiëna
	Handbook	Northcote					
Fluvisols	Alluvial soils	Om soils	Regosols	Fluvisols	Sols minéraux bruts d'apport alluvial ou colluvial, Sols peu évolués non climatiques d'apport alluvial ou colluvial	Auenböden	Warp soils
Eutric Fluvisols				[Mu-Acl]			Paternia
Calcaric Fluvisols				[Muc-2Csc]			Borovina
Dystric Fluvisols				[Mo-Acl]			Paternia
Thionic Fluvisols				Thiosols [LnTo]		Brackmarsch	

Table 6.8 Stages of physical ripening

n-Value	Stage	Consistence for clay soils
0.7	ripe	firm, does not stick to the hands or only slightly and cannot be squeezed through the fingers
0.7–1.0	nearly ripe	fairly firm, tends to stick to the hands and cannot be easily squeezed through the fingers
1.0–1.4	half ripe	fairly soft, sticks to the hands and can be easily squeezed through the fingers
1.4–2.0	practically unripe	soft, sticks fast to the hands and can be easily squeezed through the fingers
>2.0	unripe	liquid mud cannot be kneaded

acid tolerant species such as pineapples, can sometimes be grown. Another difficulty encountered when drainage is carried out is the rapid silting or clogging of the drains by $Al(OH)_3$.

The most expedient method of amelioration seems to be a gradual improvement of drainage, coupled with the addition of small amounts of lime. In this way there is only a small annual production of sulphuric acid which is neutralised by the lime and the resulting salts, particularly aluminium sulphate, are lost in the drainage and not precipitated as aluminium hydroxide.

GLEYSOLS

Derivation of name: from the Russian local name *gley* = mucky soil mass; connotative of an excess of water.

General characteristics

Soils formed from unconsolidated materials exclusive of recent alluvial deposits, showing hydromorphic properties within 50 cm of the surface; having no diagnostic horizons other than (unless buried by 50 cm or more new material) an A horizon, a histic H horizon, a cambic B horizon, a calcic or a gypsic horizon; lacking the characteristics which are diagnostic for Vertisols; lacking high salinity; lacking bleached coatings on structural ped surfaces when a mollic A horizon is present which has a chroma of 2 or less to a depth of at least 15 cm.

There are seven subdivisions of Gleysols:

Eutric Gleysols. These have a base saturation of >50 per cent between 20 and 50 cm from the surface but are not calcareous within this depth; having no diagnostic horizon other than an ochric A horizon and a cambic B horizon.

Calcaric Gleysols. These have a calcic or gypsic horizon within 125 cm of the surface and/or are calcareous at least between 20 and 50 cm from the surface; having no diagnostic horizons other than an ochric A horizon and a cambic B horizon.

South Africa	USA		USSR
	Old	New	
	Alluvial soils	Fluvents	Alluvial soils
Dundee			
Dundee			
Dundee			
		Sulphaquepts	Mangrove soil
		Sulphic	
		Haplaquepts	

Dystric Gleysols. These have a base saturation of <50 per cent at least between 20 to 50 cm from the surface; have no diagnostic horizon other than an ochric A horizon and a cambic H horizon.

Mollic Gleysols. These have a mollic A horizon or a eutric histic H horizon.

Humic Gleysols. These have an umbric A horizon or a dystric histic H horizon.

Plinthic Gleysols. These have plinthite within 125 cm from the surface.

Gelic Gleysols. These have permafrost within 200 cm of the surface.
The subdivisions and their equivalents in other systems of classification are given in Table 6.9.

Humic Gleysols

The principal horizon in this class of soils is the mottled grey or olive middle, cambic B horizon which forms as a result of prolonged periods of anaerobism. Therefore these soils occur in moist situations particularly in cool climates where they are associated with plant communities containing wet habitat species.

Morphology (see Plate IIb)
At the surface under natural conditions there is usually a spongy or matted litter about 1 cm thick often resting on well decomposed organic matter

10 to 20 cm thick. This is very plastic, contains abundant fine roots and in thin section it is dense, massive and isotropic. There is a sharp change into the Ah which is a dark grey, mixture of organic and mineral material (umbric epipedon; modon — Mo) which is commonly a loam about 15 cm thick. In thin section it is seen to be composed of small, amorphous, black areas of organic matter and sand grains some of which have dark coatings. This horizon changes fairly sharply into the mottled grey or olive Bg horizon (cambic horizon; gleyson-Gl). The mottling may form an irregular pattern or it may be associated with pores and ped surfaces. When the pores are old root passages they often develop iron pipes which are clearly seen by the concentration of brown or yellowish-brown iron oxide. The matrix is usually anisotropic with large random domains or random zones of domains and anisotropic aureoles around sand grains and pores. Clay coatings vary in amount and are often occasional to frequent on ped surfaces. Weathered sand grains are sometimes quite common in this horizon. With depth the Bg horizon grades into a grey, olive or blue completely anaerobic Cg horizon (cambic horizon; cerulon — Cu).

Analytical data
The content of organic matter in the O horizon commonly exceeds 50 per cent with a C/N ratio of more than 20. Beneath there is a marked decrease in the Ah to about 10 per cent with a C/N ratio of 15 to 20. This is followed by a further decrease to <1 per cent in the Bg horizon. The pH value in the surface is about 4.5 increasing to about neutrality with depth. The low values in the surface are due in part to the acid litter produced by the vegetation and to its decomposition under cool conditions. The base saturation is commonly high and in many of these soils the ratio of exchangeable calcium : magnesium is <4, and lower than the associated dry soils. This is more common when these soils develop in depressions receiving drainage, for magnesium is more readily released by hydrolysis and transported in solution. Usually the clay fraction is fairly uniformly distributed throughout the pedo-unit or there may be a tendency for a little more to be present in the upper mineral horizon.

Genesis
The mottled horizon is regarded as forming in a zone in the soil that is saturated with water for a part of the year but partially or completely aerated during the **215**

Table 6.9 Subdivisions of Gleysols and their equivalents in some other classifications

	Australia		Canada	FitzPatrick	France	Germany
	Handbook	Northcote				
Gleysols			**Gleysols**	**Subgleysols**		
Eutric Gleysols			Gleysols	[MuGlh]	Sols à gley peu profond peu humifères	Gley
Calcaric Gleysols	Ground water Rendzina		Carbonated Gleysols	[MucGlc]	Sols hydromorphes à redistribution du calcaire	Borowina
Dystric Gleysols			Gleysols	[MoGl]	Sols à gley peu profond peu humiferes	Gley
Mollic Gleysols			Humic Gleysols	[MuGl]	Sols humiques à gley	Tschernozem-artige, Auenböden
Humic Gleysols	Humic Gleysols	Uf6.4 Uf6.6	Humic Gleysols	[HyGl]		
Plinthic Gleysols				[GlFb]		
Gelic Gleysols			Cryic Gleysols	[GlCy]		

summer or drier period of the year. As a result, oxidation of iron takes place locally along those surfaces that freely receive a supply of oxygen. These are usually the larger fissures or cracks formed upon drying as well as down old root channels. In contrast, the lowest grey, olive or blue horizon is formed in a zone that is permanently saturated, hence the absence of mottling.

In addition to the wet conditions it appears that organic matter is necessary for iron to occur in the ferrous state, for there are many subsurface geological strata that remain continuously saturated with water without developing grey or olive colours. When horizons at or near the surface are saturated and the products of organic matter decomposition are dissolved in the water, then the characteristic colour patterns form. It does not appear that any specific organic substance is responsible, for Bloomfield (1964) has demonstrated that a variety of organic substances including leaf leachates can accomplish the mobilisation of iron and its transformation to the ferrous state. The most dramatic reduction of iron is seen in parent materials originating from certain Triassic or Devonian sediments in which the original bright red colour of the material has been transformed into olive and grey colours.

In the presence of such large amounts of moisture, hydrolysis and solution proceed at a faster rate than in aerobic soils or parent material of a similar age. This has caused more rapid rock disintegration, for it is common to find in these soils, stones that are easily broken down to their constituent minerals with a light blow from a hammer but similar stones are quite hard in the adjacent aerobic soils. Probably this is better illustrated by the distribution of total phosphorus which is lower in these soils than in their aerobic neighbours.

Principal variations in the properties of the class
Parent material
The greatest extent of Humic Gleysols is found on glacial deposits, consequently, they usually have a medium or coarse texture and contain large amounts of feldspars and other primary minerals. Elsewhere they are developed on a wide range of sediment mainly of Pleistocene or Holocene age. Since the

Kubiena	South Africa	USA		USSR
		Old	New	
Gley soils		Gley soils		Meadow soils
Mull Gley	Katspruit		Haplaquents, Psammaquents, Tropaquents, Andaquepts, Fragiaquepts, Haplaquepts, Tropaquepts	Meadow soils
Borowina				
Moder Gley	Katspruit Fernwood		Haplaquents, Psammaquents, Tropaquents, Andaquepts, Fragiaquepts, Haplaquepts, Tropaquepts	Meadow soils
Mull Gley			Haplaquolls	Meadow soils
	Champagne Willowbrook		Humaquepts	Meadow soils
	Fernwood Longlands		Plinthaquepts	Lateritic Gleysoils
		Tundra soils	Pergelic Cryaquepts	Tundra Gleysoils

drainage status of the soil is the most important factor in their formation, their mineralogy varies from basic to acidic and in some instances there are high proportions of carbonates. However, the mineralogy does influence some of the properties such as the type and amount of cations which will increase as the material becomes more basic or contains more carbonates.

Climate
The atmospheric climate has little affect upon the formation of these soils because their development is controlled largely by topography, however, they tend to be more common in areas of high rainfall or where evapotranspiration is much less than precipitation. These latter conditions are most common where the climate is Marine, Tundra or Humid Continental cool summer, but they are also widespread in certain situations in the wet and dry tropics, rainy tropics and monsoon tropics.

Vegetation
The plant communities on these soils usually differ strongly from those of the adjacent drier sites. In a

cool temperate climate, species of *Sphagnum*, *Eriophorum* and *Juncus* predominate.

Topography
These soils usually form where the water-table comes near to the surface, this is on flat sites, depressions or at the lower ends of slopes. However, they can form over the surface of gentle undulations in some areas with high rainfall.

Age
The greatest areas of Humic Gleysols have formed during the Holocene Period. Even when they are found associated with strongly weathered soils in old landscapes it is usual to find that their parent materials are late Pleistocene or early Holocene superficial deposits. This is shown also by their degree of weathering which is usually small as compared with some of the surrounding soils which can be Ferralsols in tropical regions.

Pedo-unit
Some of the most important variations are found in the upper horizons. Under natural conditions an

217

umbric A horizon or histic A horizon is common but where cultivation has been practised for some time or where the parent material is basic or contains carbonate it is usual to find a mollic A horizon.

The pattern of mottling shows important variations which seems to result from differences in the texture of the material. In sands and loams there are usually small irregular areas which as stated above are associated with large pores especially old root passages. In fine textured material with a well developed structure the surfaces of the peds are usually grey due to maximum reduction of iron. This results because moisture moves freely down the pore spaces. Part of the iron that is reduced diffuses into the inner part of the ped where it is oxidised during the succeeding dry season. Thus the pattern from the outside to the inside of the ped is grey, yellow and reticulate mottling on the inside often with black staining of manganese dioxide. Humic Gleysols intergrade into a number of other soils which may be freely draining such as Podzols or Cambisols or they may grade into Histosols. Commonly they grade into Solonchaks as the climate becomes more arid.

When the parent material is very basic or contains carbonates they are usually saturated with basic cations and the pH values are slightly above neutrality.

Distribution
These soils occur throughout a large part of the earth's surface. Their greatest extent is in the cool humid parts of the world particularly in Canada and northern Eurasia.

Utilisation
The major obstacle to the utilisation of Humic Gleysols is the large amount of water which has to be removed by installing an adequate system of drainage. This may be in the form of open drains but tile drains are usually more effective and it is often advantageous to place the drains at some depth in order to intercept the ground water thus preventing its rise to the surface. Where the content of organic matter at the surface is high and pH values are low, these properties can be altered by the addition of liming material and ploughing. The latter improves aeration while the former raises the pH values, providing a better habitat for the soil micro- and mesoorganisms so that the decomposition of the organic matter will proceed. Where continuous agriculture is practised it is necessary to add fertilisers particularly phosphorus, potassium and nitrogen. A further hazard is that ethylene is produced under anaerobic conditions and is very toxic to plants.

When adequate drainage is accomplished these soils can be adapted to the systems of agriculture of the adjacent sites. This may be a form of arable crop rotation, dairying or horticulture. In low lying situations drainage is extremely difficult so that cultivation is almost impossible except in the driest years. When an attempt is made to cultivate soils that are very wet, the implements may become stuck or large clods may be produced instead of a good seed bed, similar difficulties may be encountered at harvest time. Often wet soils are kept under permanent grass but the sward may be badly damaged by the hooves of the grazing animals, during the wettest period of the year.

Gelic Gleysols

General characteristics
The distinctive feature of this class is the permanent frozen lower horizon or permafrost which is impermeable to moisture, often causing periodic waterlogging of the overlying horizons.

Morphology (see Plate IIc)
At the surface there is a thin litter which rests on a partially humified O horizon that may be up to 15 cm thick. The plant fragments are relatively fresh but are being decomposed by fungi whose mycelia are clearly seen in thin section. The organic matter is underlain by a massive, loamy, grey cambic B horizon (cambic horizon; gleyson – Gl) about 25 cm in thickness, with the characteristic ochreous or brown mottling. This grades fairly quickly into a horizon containing discrete areas of organic matter and mottled mineral material. In thin section the features of the mottled horizon are typical for that horizon, it is massive with rare, circular pores, abundant small random domains and anisotropic aureoles around many sand grains. The ochreous mottled areas are more dense but they have a similar fabric. In the mixed horizon the fabric of the mottled part is similar to the horizon above while the organic matter is composed of slightly decomposed plant fragments.

There is a sharp change to the permafrost with its distinctive thin horizontal veins of clear ice and lenses of frozen soil (Fig. 6.23). In many cases there are inclusions of frozen organic matter indicating that this has formed in material similar to the horizon above. When stones are present in the permafrost they have a complete sheath of clear ice.

Fig. 6.23 The polished surface of a horizontal core of permafrost showing the characteristic lenticular structure. The dark areas are veins of clear transparent ice and the light areas are lenses of frozen soil

Analytical data

These soils usually have a loamy texture with uniform particle size distribution throughout the pedo-unit, however, the very low temperature and the impermeable lower horizon are the two most important physical properties. The former severely restricts the type of vegetation to a few grasses, sedges, lichens and mosses, while the latter inhibits drainage causing wetness.

The surface horizon is usually moderately acid but the pH values increase with depth. The cation-exchange capacity is low when the content of organic matter is low; similarly the base saturation is low but this is very variable as determined by the nature of the mineral material. The organic matter is usually at a maximum in the surface, but there is often a second maximum just above or just within the permafrost. The C/N ratio is usually wide due to the low degree of humification. When stones are present in the upper or middle horizons they have a marked tendency to be oriented vertically.

Genesis

These soils have formed either as a result of a progressive change from warm to cold climatic conditions or they have developed following the deposition of un-consolidated sediments in a cold environment. At the onset of a climatic change from warm to cool conditions the heat lost from the soil during the winter is not balanced by the incoming heat received during the summer period. Gradually the stage is reached when the soil freezes during the winter period and thaws during the summer. This process may take place in a variety of soils and gradually leads to changes in them. With continued change to cooler conditions the depth of freezing is greater than the depth of thawing because the heat lost during the winter is greater than the incoming summer heat. The result is that a lower layer of frozen soil — permafrost extends from one winter to the next. If conditions become even colder the permafrost becomes thicker and thicker and may be >300 m as encountered in certain parts of Siberia. Thus the cyclic conditions are attained in which the upper part of the soil freezes every winter and thaws every summer and is generally known as the active layer. When this process is continuous for long periods, the attendant churning and wetness eliminates the previous horizons that were present at the initial stage. On a sloping site solifluction takes place and any previous horizonation is also destroyed. Ultimately, the Gelic Gleysol is formed in the previously underlying material.

When unconsolidated materials are deposited in a cold environment they usually rest on material which itself has permafrost so that a single winter period is sufficient for it to form in the new deposit but if the new deposit is thick it may require more than one winter. When a thick deposit is laid down on a previous fully developed soil, it may be preserved. This happened periodically during the Pleistocene Period so that buried soils of various ages and types are found in Pleistocene deposits.

After the formation of the permafrost, annual freezing and thawing causes a considerable amount of churning of the soil and various surface patterns develop and are discussed below. The presence of large lumps of partially decomposed organic matter just above and within the permafrost suggests that churning in some of these soils was more active in the past when the freeze—thaw horizons were thicker.

The wet or moist conditions above the permafrost have caused the reduction of iron to the ferrous state and the development of a mottled pattern.

Hydrology

Although the lateral movement of water through the active layer may be fairly rapid its vertical movement is severely restricted by the permafrost. On flat sites and gentle to moderate slopes this leads to partial or complete anaerobism so that most tundra soils show some evidence of reduction. It is only on moderately steep to steep slopes that there is little evidence of reduction.

Principal variations in the properties of the class

Parent material
Unconsolidated deposits including glacial drift, alluvium, marine deposits and solifluction deposits, are the usual parent materials. The mineralogy and particle size distribution vary considerably from place to place, but they do not seem to play a major role in the formation of these soils. However, the mineralogy does exercise a little influence on the base status, particularly if carbonates are present.

Climate
The range of climate under which these soils are found is fairly narrow, always having a mean annual temperature below −1 °C. They are found predominantly under the Taiga conditions of central Siberia, Canada and Alaska but they also occur under cold moist oceanic conditions on the west coast of Alaska.

Vegetation
Grasses, sedges, lichens and mosses are the principal members of most plant communities, but in many places dwarf *Salix* spp. are frequent. Where these soils grade into those of warmer areas the tree species increase in frequency and the plant communities are dominated by various conifers such as *Larix dahurica* in Siberia.

Topography
Flat or gently sloping sites are the normal situations for these soils but they are found also on moderate slopes particularly those with a cold northerly aspect. Although they are associated usually with low elevations, they can occur on flat or gently sloping sites up to 1 000 m in the Tundra but above this height vegetation is sparse.

Age
Most of these soils started to develop towards the end of the Pleistocene as shown by the ^{14}C date of 11 000 years B.P. for the organic matter within the soil at Point Barrow in Alaska. However, some areas like central and northern Alaska, that have similar soils, were not glaciated during the Pleistocene so it is possible that their development may have started about 125 000 years B.P., i.e. at the beginning of the last major cold period or glaciation.

Pedo-unit
The variations in the soils of this class occur mainly in the middle and upper horizons. Usually there is some evidence of wetness but on sloping situations this may only be weakly developed immediately above the permafrost so that the horizons above are aerobic. In such cases they are usually brown and have a well developed lenticular structure.

In the absence of a covering of superficial deposits frost action produces a thick covering of angular rock fragments from the rock. This is particularly common in the higher and lower latitudes. In some cases this may be the only horizon over the permafrost.

In very wet situations there may be a thin cover of peat overlying a completely reduced horizon which in turn overlies the permafrost. Frequently the active layer is underlain by massive, somewhat pure ground ice many metres thick. In the drier parts of the tundra where vegetation is sparse or absent there is usually a well developed stony layer at the surface produced by a combination of stones being heaved upwards and deflation by wind. In addition, there may be well developed surface pattern of stone circles.

In nearly every case where there is a cover of vegetation some organic matter is incorporated into the mineral soil below. This can vary from a few leaves scattered throughout the pedo-unit to a fairly large amount as described above.

Distribution (see Fig. 6.24)
About one-fifth of the world's land area contains permafrost with half of Canada and the USSR being affected. It is also extensive in Alaska and Antarctica.

Utilisation
The low temperatures of these soils, their wetness and the presence of the permafrost prevent their utilisation for agriculture or forestry. In the Northern Hemisphere they produce herbage for the natural herds of reindeer and caribou, but in some places the herds are crudely managed.

The various phenomena found in the tundra include:
Wedges
Palsas
Pingos
Thermokarst
Thufur
Mud polygons
Stone polygons

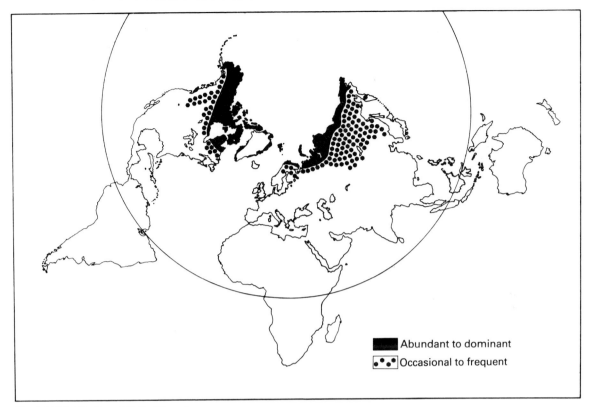

Fig. 6.24 Distribution of Gelic Gleysols

Wedges

Ice wedges are the dominant and most conspicuous type of wedge but there are also sand wedges and soil wedges.

Ice wedges

During the winter of polar areas when temperatures fall to −40 to −60 °C and the top surface of the permafrost cools to −15 to −20 °C, the whole soil contracts sufficiently to cause the ground including the permafrost to crack into a hexagonal or rectangular pattern about 10 to 15 m in diameter. Hoar frost then accumulates in these cracks so that during the following summer when the ground expands a thin vertical vein of ice is trapped in the permafrost. This process is repeated many times, possibly not annually but there may be a biennual or triennual cycle, so that the thin veins of ice become progressively thicker ultimately forming a wedge (Fig. 6.25). Black (1952, 1960 and 1973) estimates that the rate of growth is about 0.5 to 1 mm/year. Ice wedges are usually 1 to 1.5 m wide and 3 to 4 m deep but they can be up to 10 m wide and 50 m deep in parts of Siberia. The tendency is for the largest wedges to occur in the lower parts of the tundra and for the size and frequency to decrease in the high Arctic.

The thin vein of ice in the active layer melts during the summer so that the mineral soil reunites. Thus the soil profile is one of mineral soil overlying ice. As the wedges enlarge the surface pattern becomes progressively better developed giving the characteristic polygonal or rectangular pattern (Fig. 6.26).

Ice wedges are ubiquitous on the flat and gently sloping sites throughout the Arctic, being largest in the old landscapes such as northern Alaska and Siberia. In some cases when the cracks develop, organic matter such as leaves, lemming droppings, etc. fall into the cracks and become trapped and can be used as a means of dating. In most situations it seems that ice wedges form after the deposition of material, be it loess, alluvium or a glacial deposit. Such wedges are known as epigenetic wedges. In some cases the wedges develop as the sediment is being deposited. This leads to the formation of tiers of wedges which may have a honeycomb pattern. Such wedges are syngenetic.

221

Fig. 6.25 An ice wedge 4 m deep on Gary Island, northern Canada

Sand wedges

In the very dry tundra environments, particularly in Antarctica, the wedges become filled with mineral material to form sand wedges (Péwé, 1959; Ugolini *et al.*, 1973).

Soil wedges

Within the active layer cracks other than those mentioned above develop and become filled with soil but the details of their formation are still obscure.

Palsas

These are mounds of various shapes in a peat landscape. They are 1 to 7 m high and 10 to 50 m in diameter and are composed mainly of mineral material with an outermost layer of peat up to 2 m in thickness. They seem to form by the simultaneous accumulation of peat and the formation of thick segregations of ice in the underlying mineral material

and in the peat itself. In some cases palsas coalesce to form a peat plateau.

Pingos

A pingo is a small conical hill with an ice core. A high proportion are about 10 to 20 m high but can vary from <5 m to over 60 m high and up to 300 m in diameter. Pingos form by the slow growth of ice which is continuously fed from a source of water. It is believed that there are two types of pingos – open system and closed system. Open system pingos occur as isolated features and are common in the southern fringes of the tundra where the permafrost is thin or discontinuous. Under such conditions water moves upwards through openings in the permafrost and freezes causing an up doming of the surface.

Closed system pingos occur almost exclusively in areas of continuous permafrost and in contrast to open system pingos they often occur in clusters. One such conspicuous occurrence is on the island of Tuktoyaktuk in northern Canada. They form in alluvium or old lake beds which were previously unfrozen but freeze upon drainage. Then large ice segregations form and the surface is forced upwards.

There is still some uncertainty about the rate of growth of pingos but evidence from northern Canada suggests that initially they grow at the rate of 1.5 m p.a. and then decrease but continue to grow for up to 1 000 years.

The upward growth of the pingo causes cracks to form in the overburden near to the top and a small crater develops which becomes filled with water during the summer (Fig. 6.27).

Thermokarst

This is irregular hummocky terrain formed by an increase in the thickness of the active layer leading to the melting of ground ice which includes permafrost and buried glacier ice but more usually permafrost. This can form naturally by the change in an environmental factor such as an increase in temperature or the death of the vegetation following a fire or flooding. Often in the forested tundra it is man induced due to deforestation. In central Yakutia in Siberia over 40 per cent of the land surface is affected.

A distinctive type of thermokarst is the alas which is a large depression formed by the deep melting of the permafrost. Often alases become filled with water to give a lake which may in turn become filled with sediment in which a new generation of periglacial features develop including ice wedges. In the fossil state the recognition of thermokarst is very difficult.

Fig. 6.26 Oblique aerial view of tundra polygons in northern Alaska. Each polygon is 15 to 20 m in diameter.

Thufur, mud polygons and stone polygons

These are small but very distinctive surface phenomena that develop in response to the freezing and
thawing of the surface (Figs. 6.28, 6.29 and 6.30).

GREYZEMS

Derivation of name: from the English word grey and
the Russian word *zemlja* = earth or land; connotative
of white silica powder which is present in layers rich
in organic matter.

General characteristics
Soils having a mollic A horizon with a moist chroma
of 2 or less to a depth of at least 15 cm and showing
bleached coatings on structural ped surfaces; lacking
a natric and oxic B horizon; lacking the characteristics
which are diagnostic for Rendzinas, Vertisols,
Planosols or Andosols; lacking high salinity.

The bleached coatings on ped surfaces in the mollic
A horizon indicate that many of these soils are intergrades between Chernozems and Luvisols.

These soils develop in warm continental areas
beneath grassland vegetation and are generally
confined to gentle slopes.

There are two subdivisions of the Greyzems:

Orthic Greyzems. These have a mollic A horizon
with moist chroma of 2 or less to a depth of at least
15 cm and show bleached coatings on structural ped
surfaces.

223

Fig. 6.27 A pingo about 20 m high in Svalbard

Gleyic Greyzems. These have a mollic A horizon with moist chroma of 2 or less to a depth of at least 15 cm and show bleached coatings on structural ped surfaces. Hydromorphic properties occur within 50 cm of the surface.

The subdivisions and their equivalents in some other systems of classification are given in Table 6.10.

HISTOSOLS

Derivation of name: from the Greek word *histos* = tissue; connotative of soils rich in fresh or partly decomposed organic matter.

General characteristics
Histosols or peat soils as they are generally known have an H horizon of 40 cm or more (60 cm or more

224   **Fig. 6.28** Thufur in Alaska

Morphology

At the surface there is not a clear distinction between the live organic matter and the dead material, there seems to be a gradation, particularly when there is a large amount of sphagnum in the plant community. The upper horizon which may be up to 40 cm or more is composed of easily identifiable fibrous plant remains (fibric material; fibron — Fi) showing little sign of decomposition. This grades gradually with depth into progressively more decomposed pseudofibrous material (hemic material; pseudofibron — Pd) which appears fibrous but is easily broken down when rubbed between the fingers. This horizon can also be 40 cm or more in thickness and grades with depth into material that is at a very advanced stage of decomposition and appears as a black massive mass which is quite amorphous and gelatinous when rubbed between the fingers (sapric material; amorphon — Ap). This may not be the final stage of transformation for under certain circumstances the organic material is slowly and steadily replaced by siderite. In a hand specimen it is difficult to distinguish between organic material in an advanced state of decomposition and its replacement by siderite but in thin sections siderite is seen as subrounded or large masses of needle shaped crystals.

Horizons in Histosols

Histosols contain a number of specific horizons as determined by the original character of the material, degree of decomposition and the conditions of formation. Set out in Table 6.11 are the names, symbols and brief definitions of the main horizons in peat according to USDA and FitzPatrick.

Genesis

Histosols will form continuously on those surfaces that remain wet and to which organic matter is constantly added. Since such surfaces occur in most areas with a humid climate it is difficult to make generalisations because of the widely differing species

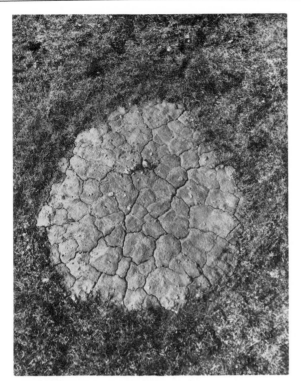

Fig. 6.29 A mud polygon 120 cm wide in northern Alaska

if the organic material consists mainly of sphagnum or moss or has a bulk density of less than 0.1) either extending down from the surface or taken cumulatively within the upper 80 cm of the soil; the thickness of the H horizon may be less when it rests on rocks or on fragmental material of which the interstices are filled with organic matter. Some workers question the inclusion of Histosols as a soil, preferring to regard such organic accumulations as parent material, even so it is possible to recognise a number of different horizons or layers which have specific properties and methods of formation.

Table 6.10 Subdivisions of Greyzems and their equivalents in some other classifications

	Canada	FitzPatrick	USA		USSR
			Old	New	
Greyzems	Dark grey	Greyzems	Chernozems	Mollisols	
Orthic Greyzems		[(ChZo)Ar]		Argiborolls	Grey Forest soils
Gleyic Greyzems		[(ChZo)Gl]		Aquolls	Meadow Grey Forest soils

Fig. 6.30 Stone polygons 2 to 3 m wide on Baffin Island (photo by F. M. Synge)

Table 6.11   Horizons in Histosols

USDA		FitzPatrick		Description
Fibric material	Oi	Fibron	Fi	fibrous and partially decomposed
Hemic material	Oe	Pseudofibron	Pd	appears fibrous but decomposes when rubbed − pseudofibrous
Sapric material	Oa	Amorphon	Ap	very advanced stage of decomposition
Marl	Lca	Limon	Lm	deposit of the skeletons of small animals
Coprogenous earth	Lco			
Diatomaceous earth	Ldi			diatomaceous earth
Frozen organic material	f	(FiCy) (PdCy) (ApCy)		various types of frozen organic material
Ferrihumic material	cn	Oron	Or	black accumulation of iron and manganese
		Sideron	Sd	replacement of amorphon by siderite
		(FiTo) (ApTo)		various types of peat with pyrite
		(FiJa) (PaJa) (PdJa)		various types of peat with jarosite

being added to the surface and to the variations in the chemical composition of the water causing the wetness. Generally, there are two principal causes of anaerobism. It may be due to the accumulation of water in a natural depression or it may result from high precipitation and high humidity. The former conditions lead to the formation of basin Histosols and the latter to blanket Histosols. Sometimes an impermeable layer such as a thin iron pan (placic horizon; placon − Pk) or permafrost may cause water to accumulate at the surface causing Histosols to form, such formations are included with the blanket Histosols.

Basin Histosols

This form of Histosol commonly occurs on a very flat, low-lying landscape, on valley floors, between moraines and in lagoons often situated in deltas. These latter two may be closed depressions or they may be fed and drained by very slow flowing streams as are the former two. Most basin Histosols started to accumulate at some period during the last 10 000 years and usually have a distinctive layered structure produced by the successive plant communities that have colonised the area. There may have been a straight accumulation under uniform climatic conditions as given above, or the various layers may be composed of contrasting species caused by climatic changes. An example of this in Europe is the characteristic *grenze* layer containing tree species, and resulting from the warmer and drier subboreal period which extended from 5 000 to 2 500 years B.P. Subsequently, conditions became cooler which increased the wetness of these Histosols resulting in their recolonisation by wet habitat species. As the thickness of basin Histosols increases it steadily rises above the level of the ground water but development continues to take place because of very slow drainage and the presence of plants such as sphagnum that have high water retaining capacity. Eventually the surface rises to form a domed outline and it is then dry enough to support certain tree species, such as birch forests in the USSR. This is the hochmoor or highmoor stage as compared with the low moor stage when the surface is flat. Examples of basin Histosols include the Everglades of Florida, the peat bogs of northern Europe and the peat deposits of the many coastal situations in the humid tropics such as in Guyana and western Borneo. Figure 6.31 illustrates the developmental stages in the formation of a Histosol deposit.

Basin Histosols sometimes exceed 10 m in thickness and vary widely in their mineral content.

The lower part of these deposits usually contains a fairly high content of mineral material and above are the various layers of organic material, some more decomposed than others.

The content of cations and pH values of Histosols vary widely as determined by the composition of the water causing the anaerobism. When the water flows from areas of acid crystalline rocks the Histosol is acid but if it flows from an area containing limestone or chalk then the pH of the peat may be around neutrality and calcareous skeletons of aquatic organisms may accumulate to form lake marl which may sometimes be interstratified in the peat. The fens of eastern England are among the better known examples of a base rich peat.

Blanket Histosols

This form of Histosol occurs in areas where precipitation is high but more particularly where the humidity is high producing a constantly moist or wet surface, thus reducing the rate of decomposition of plant remains by aerobic soil flora and fauna. Such conditions are common throughout a considerable part of a wide belt that runs through northern Europe and central Canada where the muskeg of Canada is probably the largest single area in the world. Similar conditions occur at the higher elevations in the humid tropics.

Whereas the mineral content of basin Histosols is very variable, depending upon the nature of the water entering the deposit, blanket Histosols are invariably acid because the water is derived directly from precipitation. There is also a greater tendency for this form of Histosol to be more decomposed because it may dry to a limited extent during a very dry year. In addition, there is a constant movement through the Histosol of moisture which will contain some dissolved oxygen thereby allowing a limited amount of decomposition to take place.

These formations are usually not more than 3 m thick and rest upon rock but more usually upon an old soil especially a Placic Podzol (Placaquept; Placosol). In large parts of north-western Europe the major phase of blanket Histosol formation started about 7 500 years B.P. so that they may rest on soils that developed during the early postglacial period.

Since climatic conditions are the principal reasons for the formation of blanket Histosols they are found in a number of topographic situations including slopes up to 25°. In such situations Histosols can be eroded easily and they tend to flow *en masse* down the slope particularly in the cooler areas where they may form solifluction lobes. As Histosols increase in

227

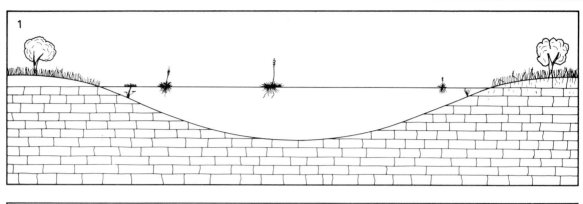

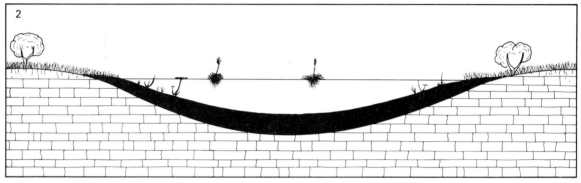

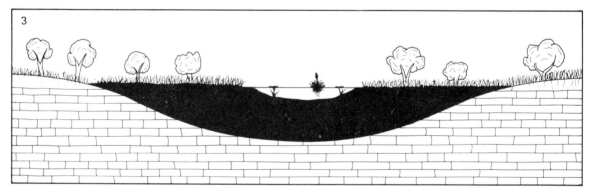

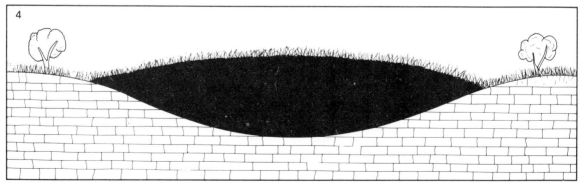

thickness any material that is deposited on their surface becomes trapped. Excellent examples of this, in the form of volcanic ash, occur in Iceland where there are two or more ash strata within the Histol deposits. Such stratification is often very useful in determining the history of the area particularly if the ages of the strata can be determined.

Botanical composition of peat

The study of the composition of the plant-remains in Histosols reveals the character of the communities that once grew in the area. The principal studies are conducted on the amount and distribution of pollen in the individual layers. This provides a fuller set of data because the pollen grains have an outer coating of cutin which protects them from decomposition, except under neutral or alkaline conditions, so that they remain preserved in the deposit very often long after other plant tissue has been altered beyond recognition. The pollen present in the Histosols is derived from the plants at a distance of over 1 km as well as those in the immediate neighbourhood thus the information which is obtained refers to the area in general.

In contrast to mineral soils it is possible to determine the age of Histosol deposits fairly accurately by using the technique of radio-carbon dating. Thus by combining the ^{14}C data and pollen data for a peat deposit it is possible to reconstruct the vegetative history of the area and indirectly the soil history.

Pedological variability in Histosols

Eutric Histosols.   These have a pH value of 5.5 or more between 20 and 50 cm from the surface.

Dystric Histosols.   These have a pH value of less than 5.5 between 20 and 50 cm from the surface.

Gelic Histosols.   These have permafrost within 200 cm of the surface.

These subdivisions and their equivalents in some other classifications are given in Table 6.12.

Fig. 6.31 Stages in the formation of Histosols
1. A small pond or lake which could be situated between two moraines
2. At the bottom of the pond there is a thin accumulation of organic matter from the plants growing in the pond and on the surrounding soil
3. Considerable increase in the thickness of the organic matter and spread of vegetation on to its surface
4. Continued thickening of the peat to develop the characteristic domed form of the final stage

Another approach to the variability in Histosols is to determine the degree of decomposition. There are many methods for making this determination but the scale introduced by Von Post (1922) has the widest acceptance (see Table 6.13).

Distribution (see Fig. 6.32)

The total area of Histosols in the world is about 150 000 000 hectares of which the greatest occurrences are in the cooler humid parts of the world. Of the total area, countries having more than 5 000 000 hectares each are the British Isles, Canada, Finland, Germany, USA and the USSR.

Utilisation

Peat has been the traditional fuel in many parts of the world but it is really too valuable for horticultural purposes to be burned. However, its use as a fuel is likely to continue in certain areas, indeed mechanised peat harvesting followed by its use to generate steam for large turbines is practised in parts of the USSR, Germany and Ireland.

The use of Histosols for agriculture offers a number of problems but when these are overcome consistently high crop yields can be obtained. If a forest is present the trees and tree stumps must be removed followed by drainage which is probably the most critical operation. Overdraining can produce excess drying resulting in the formation of hard granular structural units which do not rehydrate when wet. This is particularly characteristic of those horizons containing much silt or appreciable amounts of iron. The result is rapid percolation of water and drought for the plants. In most cases the application of lime and fertilisers is essential and it is usually necessary to add microelements, mainly copper, cobalt, magnesium and boron. After these operations arable cultivation or dairying can proceed normally, the only limitations being those common to all agricultural practices, such as climate. Peat is used extensively for crops such as onions, carrots, celery and cauliflower, but it is essential to have the correct varieties.

A common and sometimes serious problem following prolonged cultivation is subsidence which may lower the area below the possibility of drainage by gravity.

The growth of trees on Histosols is probably a more tenuous operation particularly in the case of blanket peat which is often in highly exposed situations making wind-throw an additional potentially serious hazard. However, many areas have been drained, ploughed, fertilised and planted to trees in

Table 6.12 Subdivisions of Histosols and their equivalents in some other classifications

	Australia		Canada	FitzPatrick	France	Germany	Kubiëna
	Handbook	Northcote					
Histosols	Peat		Organic soils	Peat	Sols hydromorphes organiques	Moorböden	Peat
Eutric Histosols	Neutral to alkaline peat	0, neutral 0, alkaline		[Aph]			
Dystric Histosols	Acid peats	0, acid		[FiAp]			
Gelic Histosols				[Fi(FiCy)]			

Table 6.13 Von Post scale of peat decomposition

H1	Entirely undecomposed plant remains. Squeezing in palm produces clear water.
H2	Practically undecomposed plant remains. Squeezing in palm produces almost clear, yellowish-brown water.
H3	Little decomposed peat. Squeezing in palm produces dark coloured water, but the peat which is a very fibrous mass does not protrude between the fingers.
H4	Poorly decomposed peat. Squeezing in palm produces dark coloured soil-water suspension. The plant remains are a little granulated.
H5	Somewhat decomposed peat. The structure of the plant remains is distinct to the naked eye, yet somewhat eroded. A little peat protrudes between the fingers on squeezing in palm together with water in which large amounts of soil particles are suspended.
H6	Fairly well decomposed peat; the structure of the plant remains is indistinct. On squeezing in palm not more than one-third of the sample passes between the fingers. That part which remains in palm is granular and loose, and the structure of the plant remains is more distinct than in the wet and unsqueezed sample.
H7	Well decomposed peat; the structure of the plant remains is still partially discernible. On squeezing in palm about half of the sample passes between the fingers.
H8	Very well decomposed peat; the structure of the plant remains is very indistinct. On squeezing in palm about two-thirds of the sample passes between the fingers.
H9	Almost completely decomposed peat; the structure of the plant remains may only occasionally be recognised. On squeezing in palm most of the sample passes between the fingers as a homogeneous soil-water mixture.
H10	Completely decomposed peat with no visible plant remains. On squeezing in palm the entire sample passes between the fingers as a homogeneous mass.

recent years with encouraging results. Peat is widely used as a potting substance in horticultural operations and it finds a few applications in the chemical industry.

KASTANOZEMS

Derivation of name: from the Latin word *castaneo* = chestnut and the Russian word *zemlja* = earth or land; connotative of soils rich in organic matter having a brown or chestnut colour.

General characteristics

Soils having a mollic A horizon with a moist chroma of more than 2 to a depth of at least 15 cm; having one or more of the following: a calcic or gypsic horizon or concentrations of soft powdery lime within 125 cm of the surface; lacking a natric B horizon; lacking the characteristics which are diagnostic for Rendzinas, Vertisols, Planosols or Andosols; lacking high salinity; lacking hydromorphic properties within 50 cm of the surface when no argillic B horizon is present.

There are three subdivisions of Kastanozems:

Haplic Kastanozems.   These have a mollic A horizon with a moist chroma or more than 2 to a depth of at least 15 cm.

Calcic Kastanozems.   These have a mollic A horizon with a moist chroma or more than 2 to a depth of at least 15 cm and a calcic or gypsic horizon.

Luvic Kastanozems.   These have a mollic A horizon with a moist chroma of more than 2 to a depth of at least 15 cm and an argillic B horizon, a calcic or gypsic horizon may be present.

The subdivisions and their equivalents in other systems of classification are given in Table 6.14.

South Africa	USA		USSR
	Old	New	
	Bog soils	Histosols	Bog soils
Champagne			
Champagne			

Morphology

Under natural conditions at the surface there may be a loose leafy litter resting on a dark brown, granular mollic A horizon which is about 50 cm in thickness (mollic epipedon; kastanon — Kt). The upper part of this horizon contains abundant fine roots which decrease sharply in frequency with depth and are almost absent in the lower part. Thin sections show that it is composed largely of granular faecal material and that there are frequent fragments of organic matter. With depth the structure usually changes from granular to fine prismatic before grading into a firm, massive or prismatic accumulation of calcium carbonate with pseudomycelia and concretions. In thin section the secondary formation of carbonate is very evident in the form of microcrystalline calcite on pores and tightly packed acicular crystals in the pseudomycelia (cf. Fig. 6.4). Between 1 to 2 m there is usually marked accumulation of gypsum beneath which is the relatively unaltered material. Crotovinas occur in Kastanozems but their frequency is much less than in Chernozems.

Analytical data

A high proportion of Kastanozems are of uniform texture throughout the pedo-unit and are dominated by particles <50 μm indicating the loessic nature of the parent material. There are a few of these soils in which the clay content is at a maximum in the lower part of the upper horizon.

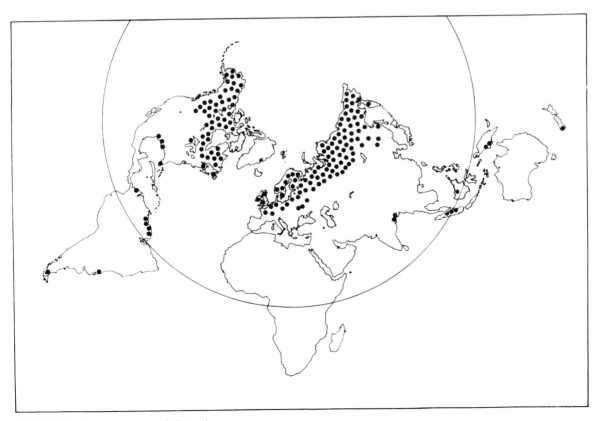

Fig. 6.32 Principal occurrences of Histosols

Table 6.14 Subdivisions of Kastanozems and their equivalents in some other classifications

	Australia		Canada	FitzPatrick
	Handbook	Northcote		
Kastanozems			Brown and dark brown soils	Kastanozems
Haplic Kastanozems	Chernozems Rendzinas Prairie soils Prairie soils	Um6.1 Um6.2 Um6.3 Um6.4	Rego Brown and Orthic Brown soils. Rego Dark Brown and Orthic Dark Brown soils	[Kt-lCzl]
Calcic Kastanozems	Chernozems	Um6.1	Calcareous Brown soils. Calcareous Dark Brown soils	[KtCk] [KtCkGy] [KtGy] [Kt(CkGy)]
Luvic Kastanozems			Eluviated Brown soils Eluviated Dark Brown soils	[KtAr] [KtArCk] [KtArCkGy] [KtAr(CkGy)]

The content of organic matter varies from 3 to 6 per cent in the uppermost part of the soil with a C/N ratio of 8 to 12 thus indicating a high degree of humification. The pH values are usually above neutrality increasing from about 7.0 in the surface to over 8.0 in the carbonate accumulation. The cation-exchange capacity in the upper horizon varies from about 20 to 30 me per cent and is usually saturated with basic cations of which calcium is dominant followed in amount by magnesium, potassium and sodium. Carbonates are usually absent or present in only small amounts in the upper 60 cm of the soil but they can be present in large amounts if the content in the parent material was high while there is 10 to 20 per cent more carbonate in the horizon of accumulation than in the underlying material. Soluble salts are present in very small amounts and never exceed 0.1 per cent.

Genesis

One of the principal processes taking place in these soils is the decomposition and incorporation of organic matter into the mineral soil. This process is fairly complete because the supply of organic matter is small while the activity of the soil fauna is high due to the basic nature of the soil and high summer temperatures. The second major process is leaching of the more soluble constituents, ions such as chloride, sulphate and sodium are either removed completely into underground waters or are deposited at great depths. Calcium is translocated as the bicarbonate which is deposited as calcium carbonate in the lower part of the soil. When calcium and sulphate ions are translocated they often combine to form calcium sulphate below the layer of calcium carbonate accumulation.

Table 6.15 Subdivisions of Lithosols and their equivalents in some other classifications

	Australia		FitzPatrick	French	German	Kubiëna	South African
	Handbook	Northcote					
Lithosols	Lithosols	Uc1.4, Um1.4, Um4.1, Um5.4, Um6.2, Uf1.4	Lithosols	Lithosols	Rankers	Rankers	
	Grey-brown and Red calcareous soils						Mispah
		Um1.3					

France	Kubiëna	South Africa	USA		USSR
			Old	New	
Sols châtains	Kastanozems		Chestnut soils	Mollisols	Chestnut soils
Sols châtains modaux		Inhoek Mayo		Haplustolls Aridic Haploborolls	
Sols châtains encroûtés		Milkwood		Calciustolls Aridic Calciborolls	
		Bonheim		Argiustolls Aridic Argiborolls	

The reason for the increase in clay in the middle of some Kastanozems is not clear and may result from at least four different mechanisms. It may be due to the translocation of clay since it appears that clay coatings are present in some cases. Weathering *in situ* in the middle part of the soil might be responsible or destruction of clay at the surface could be a third possibility. The fourth possibility is that the soil may be polygenic, the previous phase being a Luvisol. Until more evidence is available the increase in clay will be regarded as due in part to weathering *in situ*. In certain places Kastanozems have developed in previous soils of early Pleistocene or Tertiary age, the so-called Reddish Chestnut soils of the south central USA probably represent polygenetic development extending back to the Samgamonian interglacial or to an earlier period, since the underlying material is sometimes strongly weathered.

USA		USSR
Old	New	
Lithosols	Lithic subgroups	Shallow mountain soils

Hydrology
Kastanozems develop under aerobic conditions in which there is free movement of water through the soil. Since these soils develop where evapotranspiration exceeds precipitation they show moisture stress during the middle and latter part of the growing season, hence the development of a natural grass plant community.

Principal variations in the properties of the class
Parent material
Unconsolidated deposits including glacial drift, alluvium, and solifluction deposits, are the usual parent materials. The mineralogy and particle size distribution vary considerably from place to place but they do not seem to play a major role in the formation of these soils. However, the composition of the parent material does exercise a little influence on the base status particularly if carbonates are present.

Climate
These soils are found predominantly in the Middle Latitude Steppe conditions.

Vegetation
Grassland of medium height is the main plant community. In Europe the principal plants are *Stipa*, *Festuca* and *Poa*, spp. In the USA mixed short and

233

tall grass form the main members of the plant communities.

Topography

These soils are found on sites that vary from flat to moderately undulating and are absent or only poorly developed on steeper slopes. They are usually absent from depressions which tend to be saline.

Age

Most of these soils started their development during the last 10 000 years, some have been forming throughout the whole of that period while others may be only about half of that age.

Pedo-unit

There is a fairly wide range of variability within this class as determined by differences in parent material, age and topographic situation. Thus it is possible to find Kastanozems that vary from high in clay to those containing a fairly high content of boulders. Similarly, the carbonate content of the parent material largely determines the amount that will be deposited in the horizon of carbonate accumulation.

Usually these soils are found on elevated situations but they form a continuum to Solonetz or Solonchaks in depression, they also form a continuum to Chernozems.

Distribution

Kastanozems occupy fairly large areas in the semiarid regions of the world particularly in the mid-western part of the USA and Canada, and southern USSR.

Utilisation

The low rainfall imposes severe restrictions on the utilisation of these soils. Generally, they are used for growing grain on an extensive scale but drought is frequent at the drier end of the climatic range where dry farming is fairly common. When irrigation is possible, these soils have proved to be extremely fertile for a variety of crops such as maize and vegetables. Wind erosion is a constant hazard, so that preventative methods should always be in operation.

LITHOSOLS

Derivation of name: from the Greek word *lithos* = stone; connotative of soils with hard rock at very shallow depth.

General characteristics

Soils which are limited in depth by continuous coherent hard rock within 10 cm of the surface. They occur principally in mountainous areas but can occur in many other places such as on flat rock surfaces scraped bare by ice or on inselberge.

The equivalent names in some other classifications are given in Table 6.15.

LUVISOLS

Derivation of name: from the Latin word *luo* = to wash; connotative of illuvial accumulations of clay.

General characteristics

Soils having an argillic B horizon which has a base saturation of 50 per cent or more (by NH_4OAc) at least in the lower part of the B horizon within 125 cm of the surface; lacking a mollic A horizon; lacking an albic E horizon overlying a slowly permeable horizon, the distribution pattern of the clay and the tonguing which are diagnostic for Planosols, Nitosols and Podzoluvisols respectively; lacking an aridic moisture regime.

There are eight subdivisions of Luvisols:

Orthic Luvisols. These have an argillic B horizon which is not strong brown or red, without an albic E horizon.

Chromic Luvisols. These have a strong brown to red argillic B horizon without an albic horizon.

Calcic Luvisols. These have an argillic B horizon and a calcic horizon within 125 cm of the surface.

Vertic Luvisols. These have an argillic B horizon and show vertic properties but without an albic E horizon.

Ferric Luvisols. These have an argillic B horizon and show ferric properties but without an albic E horizon.

Albic Luvisols. These have an albic E horizon over an argillic B horizon.

Gleyic Luvisols. These have an argillic B horizon and hydromorphic properties within 50 cm of the surface.

Plinthic Luvisols. These have an argillic B horizon and plinthite within 125 cm of the surface.

The subdivisions and their equivalents in some other systems of classification are given in Table 6.16.

Albic Luvisol

Morphology (see Plate IId)

At the surface there may be a loose leafy litter resting directly on the mineral soil or their may be a thin mat of humified plant remains. This passes sharply into a dark grey or dark greyish-brown, granular ochric A horizon (ochric epipedon; modon — Mo) about 10 cm thick and which is a mixture of mineral grains and fragments of humified organic matter. In this horizon and the one above, there may be fruiting bodies of fungi and some mycelia (Fig. 6.33). There is a sharp change into a leached grey or brownish-grey albic E horizon which appears loose with single grain structure in the field (albic horizons; zolon). In thin section the structure is alveolar and contrasts sharply with the apparent single grain structure seen in the field.

The bleached horizon which may exceed 20 cm in thickness usually changes fairly sharply into the underlying brown or yellowish-brown argillic B horizon (argillic horizon; argillon — Ar) with a marked increase in the content of clay and angular blocky or subcuboidal structure. The surfaces of the peds and many pores are normally shiny with waxy clay coatings which may occur also within the peds where there are varying amounts of domains (Fig. 4.10), the majority of which are randomly oriented but some are oriented parallel to the surfaces of peds and sand grains.

With depth the content of clay and clay coatings decreases and there is a gradation into the relatively unaltered material.

Analytical data (see Fig. 6.34)

These soils usually develop in material of medium texture with a high proportion forming on loess, therefore they are composed mainly of material in the size range 2 to 60 μm. The content of clay (<2 μm) is at a minimum in the two uppermost mineral horizons increasing to a maximum in the

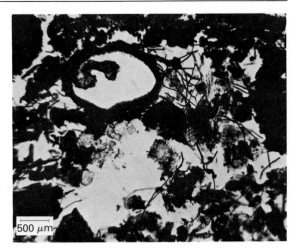

Fig. 6.33 Fungal mycelia decomposing plant fragments in the upper organic horizons

middle horizon where there are prominent and often continuous clay coatings. The texture usually becomes finer by at least one texture class but the clay can increase by as much as 20 per cent or more.

The pH values have an interesting pattern; they tend to vary from about 5.5 to 6.5 in the upper horizon decreasing to about 4.5 to 5.0 where there is the clay maximum. There is then a steady increase into the relatively unaltered material where the value may be over 7.5 if it is calcareous. Apart from the surface, the maximum amount of organic matter occurs in the uppermost mineral horizon where it varies from 5 to 10 per cent and has a moderately high C/N ratio of 12 to 18 reflecting the partially decomposed state of the incorporated organic matter. Throughout the rest of the soil the content of organic matter and C/N ratio are very low due to the advanced state of decomposition of the organic matter. Sometimes C/N ratios of less than 8 coincide with the clay maximum and are probably due to ammonia fixed by the clay fraction. The cation-exchange capacity normally has two maxima: one is associated with the high content of organic matter in the surface and the second is in the middle horizon where the clay content is also at its maximum. In a similar way the individual exchangeable cations dominated by calcium often have two maxima, which correspond to the CEC maxima.

The total amount of extractable Fe_2O_3 is usually small throughout the soil but it increases from about 1 per cent in the leached horizon to about 2 per cent at the clay maximum, thereafter it decreases slightly with depth. This suggests that the extractable iron is

Table 6.16   Subdivisions of Luvisols and their equivalents in some other classifications

	Australia		Canada
	Handbook	Northcote	
Luvisols			**Luvisolic**
Orthic Luvisols	Yellow Podzolic soils, Soloths, Lateritic Podzolic soils, No suitable group.	Dy2.2, Dy2.3, Dy3.2, Dy3.3, Dy3.6 Dy2.3, Dy3.3, Dd1.1, Dd1.3, Dd2.3 Dy2.6 Dy3.1, Dy2.1, Db2.3	Grey-Brown Podzolic soils
Chromic Luvisols	Non-calcic brown soils, Red Podzolic soils, Soloths, Red earths, Calcareous red earths, Euchrozems, Chocolate soils, No suitable group.	Dr2.1, Dr2.2, Dr3.2, Db1.2, Db1.1 Dr2.2, Dr2.3, Dr3.2, Dr3.3, Dr4.2, Db1.1, Db1.2, Dy3.3 Dr2.2, Dr2.3, Dr3.3, Db1.3, Db2.3 Gn2.1 Gn2.1 Dr4.1 Dr4.1, Db3.1 Db1.1	
Calcic Luvisols	Red-brown earths, Solodised Solonetz, Euchrozems, Solodic soils, Chocolate soils, No suitable group.	Dr2.1, Dr2.2, Dr2.3, Dr3.2 Dr3.3, Db1.3, Db2.3, Dy2.3 Dr4.1 Db1.3, Dy2.3 Dr4.1, Db3.1, Dr4.3, Dy2.1, Dy2.2, Dy3.1, Dy3.2, Dy3.6	
Vertic Luvisols			
Ferric Luvisols	Lateritic Podzolic soils	Dy2.6, Dy3.6	
Albic Luvisols	Red Podzolic soils	Dr4.4, Dr4.3	Grey-Brown Luvisols, Grey Luvisols
Plinthic Luvisols			
Gleyic Luvisols			

not completely associated with the clay but appears to be lost from the upper part of the soil.

Genesis

These soils appear to be formed in large part by progressive migration of material downwards. Initially, any soluble salts and carbonates are removed by the moderate amount of precipitation. This is followed by the gradual translocation of the clay from the upper horizon or horizons to form the middle horizon. Here the clay is deposited as clay coatings on the surfaces of peds and pores. When translocation has been proceeding for a long period of time the clay coatings may occupy more than 5 per cent of the soil as seen in thin section. The very finest clay <0.5 μm is removed first and then follows the coarser particles up to 2 μm. Any free oxides of iron, aluminium or silica associated with the clay are translocated simultaneously. The precise mechanism of clay deposition is not understood but some workers including Brewer and Haldane (1956) have shown that when a suspension of clay is passed through a column of soil the clay is deposited, and in thin sections exhibits similar characteristics to those seen in soils.

A comparison of the data on the clay coatings with

FitzPatrick	France	Germany	Kubiëna	South Africa	USA		USSR
					Old	New	
Argillosols					**Grey brown Podzolic soils**	**Alfisols**	
[MoAr]	Sols lessivés modaux	Parabraunerde				Hapludalfs Haploxeralfs	Podzolised Brown Forest Soils
[Mo(ArRo)]	Sols fersiallitiques lessivés.	Terra Rossa Terra Fusca	Leached Terra Rossa Leached Terra Fusca	Swartland Valsriver Shortlands		Rhodoxeralfs Haploxeralfs	Cinnamonic soils
[MoArCk]						Haplustalfs	
[Mo(ArVe)]						Vertic Haploxeralfs	
[(Tn(ArFv)]	Sols ferrugineux tropicaux lessivés			Shortlands			
[ZoAr]	Sols podzoliques			Shepstone Vilafontes		Eutroboralfs	
[MoArFb]				Westleigh Avalon Bainsvlei		Plinthustalfs Plinthoxeralfs	
[ZoArGl]	Sols lessivés hydromorphes sols à gley lessivés	Gley braunerde		Cartref Pinedene		Aqualfs	Podzolic Gley soils

the particle size analysis data suggests that the amount of coatings is somewhat less than to be expected from the amount of clay removed from the overlying horizons. Therefore, it seems probable that the textural contrast may be enhanced by some clay destruction in the upper horizons and its complete removal from the system.

The precise time at which translocation has taken place is still open to question. It may be taking place at present but it may have formed at an earlier phase in the development of the soil.

One of the important climatic requirements for these soils is a distinct to marked dry season; hence their occurrence mainly in mildly to strongly continental climates. Under such conditions material is translocated in the wet season but during the dry period the fine particles on the surfaces of peds and pores become partly dehydrated and firmly attached. With annual repetition of this cycle layers of material gradually accumulate to form the coatings.

Hydrology

Luvisols form under aerobic conditions in which there is usually free movement of water through at least the upper and middle parts of the soil. It appears that a distinct dry season is required for the soils to develop.

237

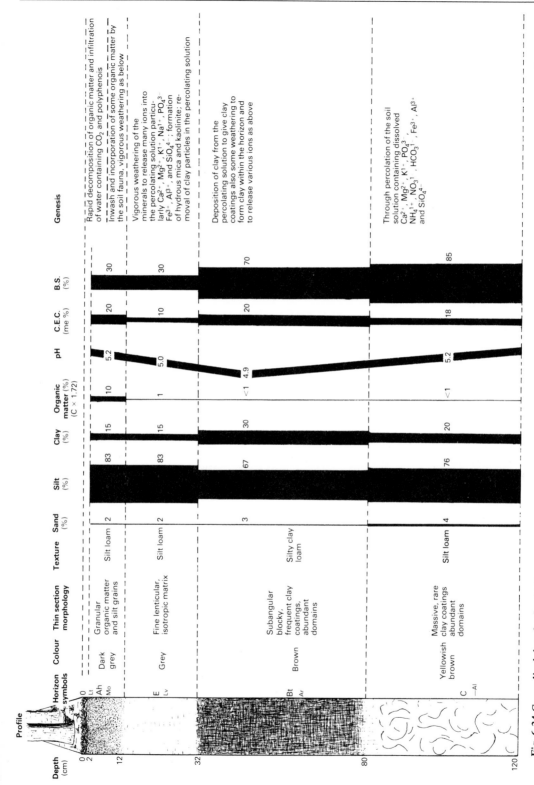

Fig. 6.34 Generalised data for Luvisols

Principal variations in the properties of the class

Parent material

Unconsolidated deposits of medium to fine texture including loess, glacial drift, alluvium and solifluction deposits are the usual parent materials, but they can occur on coarser materials. These deposits are often calcareous but they can be acid or intermediate.

Climate

The maximum development of this class invariably takes place under humid conditions with a marked dry season. The precise temperature regime and total precipitation are variable for they are found under the following climate types:

Tropical Wet-Dry
Humid Subtropical
Mediterranean
Humid Continental

In addition, they occur in arid areas where they are regarded as having formed under more humid conditions and are now fossilised by a change to drier climatic conditions.

Vegetation

Deciduous forest is the most common plant community but the species composition varies from place to place as determined by environmental factors, particularly climate. Oak and beech are usual under cool conditions while mixed forests of deciduous and coniferous species are associated with these soils in certain parts of the USSR. Luvisols can occur under grassland particularly when intergrading to other soils such as Phaeozems.

Topography

Although members of this class sometimes occur on moderate or steep slopes they are found most often on flat or gently sloping sites.

Age

The differentiation of most argillic B horizons seems to have taken place during the Holocene Period but in certain soils of Europe they appear to be a feature of the late Pleistocene when they would have formed under relatively cold conditions.

Distribution

Luvisols are of widespread distribution, being found more especially in west-central USSR, east-central and north-central USA, the central belt of Europe, southern Australia and elsewhere.

Utilisation

The potential of these soils for agriculture varies from moderate to good. Soils with a thick albic E horizon fall into the former category whereas those with an umbric A horizon are included among the world's most productive soils. Because they occur under moist conditions they are frequently used for mixed farming, dairying or horticulture but wheat, maize and oats can also be grown. Fertility is maintained by the normal procedure of liming and fertiliser application, only in exceptional circumstances is it necessary to adopt any special techniques. Erosion is a common feature and rigorous control methods must be maintained at all times.

NITOSOLS

Derivation of name: from the Latin word *nitidus* = shiny; connotative of shiny ped surfaces.

General characteristics

Soils having an argillic B horizon with a clay distribution where the percentage of clay does not decrease from its maximum amount by as much as 20 per cent within 150 cm of the surface; lacking a mollic A horizon; lacking an albic E horizon; lacking the tonguing which is diagnostic for the Podzoluvisols; lacking ferric and vertic properties; lacking plinthite within 125 cm of the surface; lacking an aridic moisture regime.

This class has been created fairly recently and has not yet been fully categorised. They are strongly weathered kaolinitic soils, the principal feature being the steady increase in clay with depth to a maximum in the middle horizon; thereafter remaining uniform for some depth. Then there is a steady decrease as the material becomes progressively less weathered. The middle horizon with the clay maximum is generally regarded as an argillic B horizon but often the absence or low frequency of clay coatings and the gradually grading upper boundary make such a designation somewhat tenuous. In many cases the intrinsic properties of the individual horizons are similar to Ferralsols or Ferric Cambisols but with the clay increase in the middle horizon. Since the content of clay coatings is often small and the clay content remains uniform for some thickness it does not seem that clay translocation is the primary mechanism in the formation of these soils. It would seem that

239

progressive weathering and hydrolysis of the clay in the upper horizon is the main process responsible for the reduction in the clay content.

Soils of this type are very common in tropical and subtropical areas and were first considered to need a separate designation by Northcote who places them in his 'G' Primary Profile Form.

There are three subdivisions of Nitosols:

Eutric Nitosols. These have a base saturation of 50 per cent or more throughout the argillic B horizon.

Dystric Nitosols. These have a base saturation of less than 50 per cent in at least a part of the argillic B horizon.

Humic Nitosols. These have a base saturation of less than 50 per cent in at least part of the argillic B horizon, have an umbric A horizon or high organic matter in the B horizon or both.

The subdivisions and their equivalents in some other systems of classification are given in Table 6.17.

PHAEOZEMS

Derivation of name: from the Greek word *phaios* = dusky and the Russian word *zemlja* = earth or land.

General characteristics
Soils having a mollic A horizon; lacking a calcic horizon, a gypsic horizon and concentrations of soft powdery lime within 125 cm of the surface; lacking a natric and an oxic B horizon; lacking the characteristics which are diagnostic for Rendzinas, Vertisols, Planosols or Andosols; lacking high salinity; lacking hydromorphic properties within 50 cm of the surface when no argillic B horizon is present; lacking bleached coatings on structural ped surfaces when the mollic A horizon has a moist chroma of 2 or less to a depth of at least 15 cm.
There are four subdivisions of Phaeozems:

Haplic Phaeozems. These have a mollic A horizon.

Calcaric Phaeozems. These have a mollic A horizon and are calcareous between 20 and 25 cm of the surface.

Luvic Phaeozems. These have a mollic A horizon and an argillic B horizon.

Gleyic Phaeozems. These have a mollic A horizon and an argillic B horizon and show hydromorphic properties within 50 cm of the surface.

The subdivisions and their equivalents in some other systems of classification are given in Table 6.18.

Luvic Phaeozems

Morphology (see Plate IIIa)
At the surface there may be a thin, loose, leafy litter resting on the mineral soil or there may be an intervening thin root-mat. The upper mineral horizon is a very dark grey mollic A horizon with granular or vermicular structure and may be up to 50 cm thick (mollic epipedon; chernon – Ch). In thin section it is seen to contain many faecal pellets and to have a clay–humus matrix that is isotropic and weakly translucent to opaque (cf. Fig. 4.30). This grades into a dark brown, angular or subangular blocky argillic B horizon with clay coatings on the surfaces of the peds (argillic horizon; argillon – Ar). In thin section the isotropic clay–humus mixture of the upper horizon gradually changes to anisotropic in the middle horizon due to the presence of common, medium random domains. The surfaces of the peds and pores have varying amounts of clay coatings which are usually at a maximum in the middle or lower part of this horizon (cf. Fig. 4.33). With depth the clay content and frequency of clay coatings decrease and the middle horizon gradually grades into the underlying relatively unaltered material.

Analytical data (see Fig. 6.35)
A high proportion of these soils are developed in loess, therefore they are dominated by particles $<60 \mu m$ in diameter. The maximum amount of clay occurs in the middle horizon where values of between 30 and 40 per cent are common and are 10 to 20 per cent higher than the horizon above.

The pH values have a fluctuating pattern which may be due in part to cultivation, to differences in the composition of the parent material or to pedogenic processes. The pH values at the surface can be >7 but decrease to about 5 to 7 in the middle horizon and coincide with the clay maximum. Then follows a steady increase through the lowest part of the middle horizon and into the relatively unaltered material.

The organic matter steadily decreases from about 5 per cent in the upper horizon to 1 to 2 per cent in the lower part of the middle horizon which may contain 3 per cent organic matter in its upper part. This is high for a middle horizon but it is a characteristic feature for many members of this class and is largely responsible for the dark colour.

The narrow C/N ratio of 10 to 12 in the upper horizon indicates that the organic matter is well humified while the thin section morphology shows that there is an intimate blend of the organic and mineral material. The cation-exchange capacity is variable; in the USA it is about 25 me per cent and 30 me per cent in the upper and middle horizons respectively due to montmorillonite being the principal clay mineral in that country. In the USSR the values are lower because mica is the predominant clay mineral. The base saturation is normally high, exceeding 80 per cent in most situations, with calcium being the predominant exchangeable ion. In some soils the lowest percentage base saturation occurs in the middle of the soil and tends to coincide with the minimum pH value.

Genesis

Luvic Phaeozems seem to have developed in two or more distinct stages and it appears that they may be out of phase with their present environment. Evidence from pollen analysis (West, 1961) for the central USA indicates that successive waves of vegetation have colonised areas of these soils following the climatic trends of the Pleistocene. There was spruce, followed by hardwood and now prairies. Therefore, it seems likely that the first stage was the development of a Luvisol with the formation of a middle horizon with a clay maximum and clay coatings. Superimposed upon this has been the formation of the highly organic upper horizon induced by the prairie vegetation and churning organisms. Further, it is suggested that the present vegetation should be deciduous woodland, the prairie communities being a relic from the climatic optimum but the deciduous forests now find it difficult to compete with the vigorous growth of the grasses (Walscher *et al.*, 1960). However, in the USSR these soils carry deciduous forests so it would appear that the highly organic layers are due to the present conditions rather than to a grassy vegetation. In any case the middle horizon is regarded generally as partially if not totally a relic horizon.

Hydrology

Phaeozems develop under aerobic conditions in which there is free movement of water through the soil. Since they develop where evapotranspiration exceeds precipitation they show moisture stress during the middle or latter part of the growing season, hence the development of a natural grass plant community.

The granular structure in the upper horizon and subangular blocky structure in the middle horizon impart a high porosity to these soils. This allows easy penetration of roots and moisture, while at the same time the peds have high moisture retaining powers. Consequently, these soils have excellent moisture relationships allowing the excess moisture to percolate freely, and at the same time retaining a large amount in the porous peds.

Principal variations in the properties of the class

Parent material

Unconsolidated deposits including glacial drift, loess and alluvium are the usual parent materials. The texture is usually a silt, silty clay loam or clay loam with variable mineralogy, normally ranging from acid to strongly calcareous.

Climate

Phaeozems are confined largely to continental conditions where the variation in temperature is fairly wide. Perhaps one of the widest ranges is found in North America where these soils extend from Alberta in Canada to Missouri in the USA. Within this range there is a steady decrease in mean annual temperature from 18 °C in Mississippi to 2 °C in Alberta, accompanied by a decrease in precipitation from 1 200 mm to 400 mm. This is a good example of the same type of soil extending through contrasting atmospheric climate types. In all cases the amount of moisture passing through the soil is similar because evapotranspiration decreases with decreasing temperatures so that the increased precipitation does not produce a significant increase in percolation.

Vegetation

In many cases the natural communities are dominated by tall grasses such as big bluestem (*Andropogon furcatus*) in North America; in Europe there are often deciduous forest communities dominated by species such as oak (*Quercus* spp.).

Topography

These soils are confined almost exclusively to flat or gently undulating situations and are almost completely absent from moderate or steep slopes.

241

Table 6.17   Subdivisions of Nitosols and their equivalents in some other classifications

	Australia		Brazil	FitzPatrick
	Handbook	Northcote		
Nitosols				**Luvosols**
Eutric Nitosols	Euchrozems Brown Earths	Gn3.1 Gn3.5	Terra Roxa estruturada, medium to high base status	[Tn(ArRo)], [Tn(ArFv)], [Mu(ArRo)], [Mu(ArFv)], [Tn Ro], [Tn < Fv].
Dystric Nitosols	Krasnozems Red Podzolic soils	Gn3.1, Gn4.1, Gn4.3 Gn3.1	Terra Roxa estruturada, low base status	[Tn(ArKs)], [Tn(ArZh)], [Tn < Ks], [Tn < Zh]
Humic Nitosols	Brown Earths	Gn3.5		[Mu(ArZh)], [Mu(ArKs)] [Mu < Fv], [Mu < Ro]

Age

Most of these soils are developed in deposits of late Pleistocene age, therefore they are relatively young.

Pedo-unit

These soils form a continuum with many others particularly Chernozems and Luvisols. As they grade into Chernozems the thickness of the upper horizon increases while the clay differential and amount of clay coatings in the middle horizon diminish. The intergrades to Luvisols show a progressive decrease in the thickness of the upper horizon and an increase in the clay differential and amount of clay coatings in the middle horizon.

The lower horizon is often a hard and brittle Cx horizon (fragipan; fragon — Fg) or there may be a second sequence of horizons, composed of a bleached horizon overlying a second horizon of clay accumulation.

In a number of cases these soils are developed in a relatively thin deposit of loess overlying till so that the horizon of clay accumulation may rest directly on the till. Alternatively, where the till is impermeable the horizon above may become periodically saturated with water and become mottled.

The type of clay minerals in this class varies fairly widely, whereas in the USA the principal component is montmorillonite, in the USSR mica generally predominates. This suggests that the type of clay mineral

Table 6.18   Subdivisions of Phaeozems and their equivalents in some other classifications

	Australia		Canada	FitzPatrick	France	Germany
	Handbook	Northcote				
Phaeozems				**Brunizems**	**Brunizems**	**Tschernozem**
Haplic Phaeozems	No suitable groups Prairie soils Terra Rossa soils	Um6.1, Uf6.2 Um6.3 Uf5.3	Rego Dark Grey soils	[Ch-lCzl]	Brunizem modal	Tschernozem
Calcaric Phaeozems				[Ch-lCzl]		
Luvic Phaeozems	Brown earths Prairie soils Chernozems Terra Rossa soils Chocolate soils Humic Gleys	Gn3.2 Gn3.4 Gn3.4 Gn4.1, Gn4.3 Gn4.4 Dy5.1	Orthic Dark Grey soils	[ChAr]	Brunizem à B textural	Parabraunerde- Tschernozem
Gleyic Phaeozems	Humic Gleys	Dy5.1	Gleyed Dark Grey soils	[Ch(GlPn)]		Gley- Tschernozem

USA		USSR
Old	New	
Latosols		
	Tropudalfs Paleudalfs, Rhodustalfs	
	Tropudults, Rhodudults, Rhodustults, Palexerults	Krasnozems
	Tropohumults, Palehumults	

is determined by the original composition of the parent material rather than by pedogenic processes.

Utilisation

These soils have a high natural fertility and give good crop yields, but often these can be increased by the application of phosphorus, and when intensive cultivation is practised it is necessary to apply other fertilisers and lime. Traditionally, these soils have been used for growing grain crops including maize, wheat and oats, other crops such as soya beans are now extensively grown. The utilisation of the crops varies from place to place, sometimes they are used exclusively for human consumption whereas in others they are fed to animals on the farms where they are grown. Both wind and water erosion are extremely

South Africa	USA		USSR
	Old	New	
	Brunizems		
Inhoek Mayo Milkwood		Hapludolls	Degraded Chernozems
		Vermudolls	
Bonheim		Argiudolls	Podzolised Chernozems
Willowbank		Argiaquolls	

serious hazards and rigorous control methods have to be maintained at all times.

PLANOSOLS

Derivation of name: from the Latin word *planus* = flat, level; connotative of soils generally developed in level or depressed topography with poor drainage.

General characteristics

Soils having an albic E horizon overlying a slowly permeable horizon within 125 cm of the surface (for example, an argillic or natric B horizon showing an abrupt textural change, a heavy clay, a fragipan), exclusive of a spodic B horizon; showing hydromorphic properties at least in a part of the E horizon. There are six subdivisions of Planosols:

Eutric Planosols. These have an ochric A horizon and a base saturation of 50 per cent or more throughout the slowly permeable horizon but having <6 per cent exchangeable sodium throughout.

Dystric Planosols. These have an ochric A horizon and a base saturation of less than 50 per cent in at least part of the slowly permeable horizon within 125 cm of the surface but having no more than 6 per cent exchangeable sodium throughout.

Mollic Planosols. These have a mollic A horizon or an eutric histic H horizon; and have no more than 6 per cent sodium in the exchange complex of the middle horizon.

Humic Planosols. These have an umbric A horizon or a dystric histic H horizon and have no more than 6 per cent sodium in the exchange complex of the middle horizon.

Solodic Planosols. These have no more than 6 per cent sodium in the exchange complex of the middle horizon.

Gelic Planosols. These have permafrost within 200 cm of the surface.

The subdivisions and their equivalents in some other systems of classification are given in Table 6.19. **243**

Eutric Planosols

General characteristics (see Plate IIIb)
The most conspicuous feature of this class is the marked and often abrupt increase in the clay content on passing from the upper to the middle horizons. The former may be a mollic, umbric, or ochric A horizon (mollic, umbric or ochric epipedon; mullon — Mu, modon — Mo, tannon — Tn) but there is often an ochric A horizon followed by an albic E horizon (albic horizon; zolon — Zo or luvon — Lv) and then the sharp increase in clay to the mottled brown, grey or olive argillic B horizon (argillic horizon; planon — Pn). This horizon has a coarse angular blocky, prismatic or massive structure and in some cases clay coatings are present but the change in texture appears to be due in part to strong weathering *in situ* under wet conditions for there are usually more weathered minerals in these soils than in the adjacent freely drained soils. The texture differential is probably enhanced by clay destruction and removal from the albic E horizon and other upper horizons containing organic matter or receiving its acid decomposition products.

The initial wetness in the argillic B horizon can result from a number of reasons: the soils may be in a depression or on a poorly drained flat or gently sloping site where water will accumulate; alternatively there may be an impermeable horizon or initially a high content of clay in the parent material causing slow permeability.

Planasols appear to evolve from Gleysols (Albaqualfs; Supragleysols) and have their maximum distribution under Humid Continental conditions. They occur sometimes in the Humid Subtropical and Marine West Coast climate where the soils are warm and moist during the summer.

Solodic Planosols (see Plate IIIc)

Solodic Planosols can be regarded as strongly leached Solonetz. Their morphology shows a thin litter at the surface resting on an ochric A horizon (ochric horizon; modon — Mo) 5 to 10 cm thick containing up to 20 per cent organic matter and having a C/N ratio of about 15. Below is the clay-depleted albic E horizon (albic horizon; minon — Mi) which may be 20 to 50 cm thick but in extreme cases it may be over a metre in thickness and has thickened through the removal of clay from the upper part of the underlying natric B horizon (natric horizon; solon — Sl) with its columnar or prismatic structure. Frequently, the rounded outline of the columns of the natric B horizon are still visible within the albic E horizon which grades sharply into the finer textured natric B horizon. With depth there is a change to a gypsic or salic horizon (gypson — Gy or halon — Hl) or gradation to the underlying material. The thin section characteristics are similar to those given on other pages for ochric A, albic E and natric B horizons.

Clay depletion is probably due to a combination of clay translocation and clay destruction in the albic E horizon with the latter probably being the dominant process. Since Solodic Planosols are formed by continued leaching of Solonetz they become progressively more acid causing the albic E horizon to have pH values as low as 5.0 and less than 30 per cent base saturation. Similarly, the natric B horizon which originally had a pH value of 8.5 or over may now have a lower pH value and may have lost some of the original high content of exchangeable sodium and magnesium.

Many Solodic Planosols occur in shallow depressions which receive run-off water, consequently they are moister for longer periods of the year than the adjacent soils, the result is greater leaching and often a denser plant community. For example in the USSR it is common to find communities dominated by birch or aspen whereas grass communities occur on the adjacent slightly elevated and drier sites.

The acid, strongly leached albic E horizon imparts an extremely low fertility to these soils, therefore it is necessary to add large amounts of liming materials and fertilisers, and where possible organic material followed by deep ploughing. Since Solodic Planosols occur mainly in arid and semiarid areas, improved utilisation can be achieved by irrigation which will also remove salts and help to reduce the pH values in the natric B horizon thereby producing more root-room for plants.

PODZOLS

Derivation of name: from the Russian words *pod* = under and *zola* = ash; connotative of soils with strongly bleached horizon.

General characteristics
Soils having a spodic B horizon.
There are six subdivisions of Podzols:

Orthic Podzols. These have a spodic B horizon which has a ratio of free iron to carbon of less than 6 but contains sufficient free iron to turn redder on ignition. May have an albic E horizon.

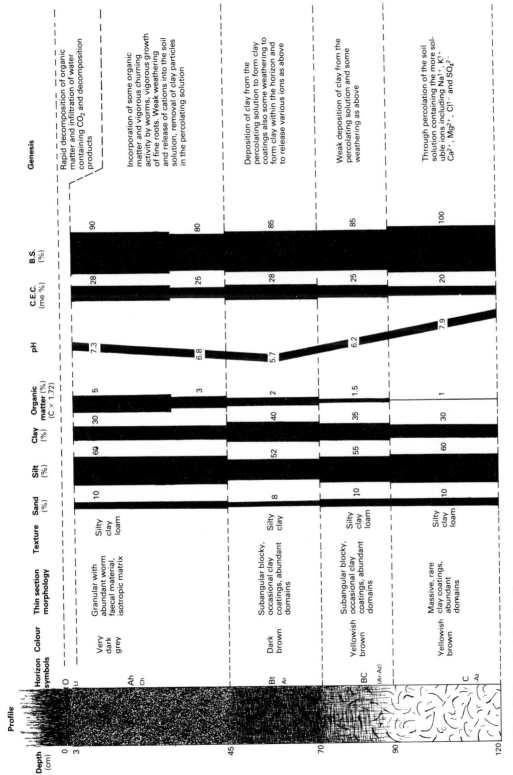

Fig. 6.35 Generalised data for Phaeozems

Table 6.19 Subdivisions of Planosols and their equivalents in some other classifications

	Australia		Canada
	Handbook	Northcote	
Planosols			
Eutric Planosols	Soloths	Dr2.4, Dr3.4, Db1.4, Db2.4	Gleysolic
	Red Podzolic soils	Dr2.4, Dr3.4	
Dystric Planosols	Soloth	Dr2.4, Dy3.4, Dy5.4, Dy5.8	Eluviated
	Red Podzolic soils	Dr2.4	Gleysol
	Yellow Podzolic soils	Dy2.4, Dy3.4, Dy5.4, Dy5.8	
Mollic Planosols			
Humic Planosols	Soloths	Dg2.4, Dg2.8	
	Gleyed Podzolic soils	Dg2.4, Dg2.8	
Solodic Planosols	Soloths,	Dr2.4, Dr3.4, Db1.4, Db2.4, Dy3.4, Dy5.4, Dy5.8, Dd1.4	Solods
	Red Podzolic soils	Dr2.4, Dr3.4	
	Solodised Solonetz	Dr2.4, Dr3.4, Db1.4, Db2.4, Dy3.4, Dy5.4, Dy5.8, Dd1.4, Dg2.4, Dg2.8	
	Yellow Podzolic soils	Dy2.4, Dy3.4, Dy5.4, Dy5.8	
	Gleyed Podzolic soil	Dg2.4, Dg2.8	
	Solodic soils	Dr2.4, Dr3.4, Db1.4, Db2.4, Dy2.5, Dy3.4, Dy3.8, Dy5.4, Dy5.8, Dd1.4, Dg2.4, Dg2.8	
Gelic Planosols			

Table 6.20 Subdivisions of Podzols and their equivalents in some other classifications

	Australia		Canada	FitzPatrick	France
	Handbook	Northcote			
Podzols	**Podzols**		**Podzolic**	**Podzols**	**Podzols**
Orthic Podzols	Podzols	Uc2.2 Uc2.3	Ferro-Humic Podzols	[ZoSq], [ZoSqIn] [ZoHs], [ZoHsIn]	Podzols humo-ferrugineux
Leptic Podzols	Brown Podzolic soils			[MoSq], [MoHs]	
Ferric Podzols				[ZoFr]	Podzols ferrugineux
Humic Podzols	Humus Podzols	Uc2.2 Uc2.3	Humic Podzols	[ZoHd]	Podzols humiques
Placic Podzols			Humic Podzols	[PkSq], [PkHs]	
Gleyic Podzols	Podzols	Uc2.2 Uc2.3		[Zo(SqGl)]	Podzols à gley

FitzPatrick	German	Kubiëna	South Africa	USA Old	USA New	USSR
Planosols	Pseudogley Stagnogley			Planosols		Podbels
[MiPnh], [TnPnh] [ZoArh], [ZoArhCk]					Albaqualfs, Paleargids Palexeralfs, Paleustalfs	
[MiPn]			Estcourt Kroonstad		Albaquults	
[MuMiPn]					Argialbolls, Mollic Albaqualfs	
[HyMiPn]						
[MiSl] [MiSlCk]			Estcourt		Natraqualfs and natric subgroups of Cryobololls, Daraqualfs Durixeralfs, Haploxeralfs	Solods
[MiPnCy]						

Germany	Kubiëna	South Africa	USA Old	USA New	USSR
Podzols	Iron Podzols		Podzols	Spodosols	Podzols
Eisenhumuspodzol	Iron Podzoic			Orthods	Humic-Ferrous Illuvial Podzols
	Semi Podzols		Brown Podzolic soils		
Eisenpodzol	Iron Podzols	Constantia		Ferrods	Iron-Illuvial Podzols
Humuspodzol	Humus Podzols	Houwhoek Lamotte		Humods	Humus-Illuvial Podzols
				Placorthods Placohumods	
Gley – Podzols	Iron Gley Podzols Humus Gley Podzols			Aquods	Podzolic Swampy Humic-Illuvial soils

Leptic Podzols. These have a spodic B horizon which has a ratio of free iron to carbon of less than 6 but contains sufficient iron to turn redder on ignition. Without an albic E horizon.

Ferric Podzols. These have a spodic B horizon which has a ratio of free iron to carbon of 6 or more. May have an albic E horizon.

Humic Podzols. These have a spodic B horizon that contains dispersed organic matter and without sufficient free iron to turn redder on ignition.

Placic Podzols. These have a thin iron pan in or over the spodic B horizon.

Gleyic Podzols. These have hydromorphic properties within 50 cm of the surface.

The subdivisions and their equivalents in some other systems of classification are given in Table 6.20.

Orthic Podzols

General characteristics

The fully developed Orthic Podzol has an upper, pale grey, strongly leached horizon which underlies surface organic horizons and overlies a brown to very dark brown horizon in which iron, aluminium and/or humus have accumulated. These soils are associated mainly with coniferous vegetation and a cool humid climate, but they do occur under other sets of conditions.

Morphology (see Plate IIId)

One of the more common types which will be described contains a very dark middle horizon in which iron, humus and aluminium have been deposited. Generally, there is a surface litter 1 to 5 cm in depth or more, this is loose and spongy and grades with depth into partly humified organic matter – the O1 horizon (O1; fermenton – Fm) which is usually of similar thickness. Here the plant fragments are still recognisable but they have been attacked by many organisms. In thin section decomposition is seen to be well advanced and many of the plant fragments have overgrowths of fungal mycelia and the softer inner parts of the stems and needles are partly replaced by the faecal pellets of arthropods (Figs. 6.36 and 6.37). In addition, some of the live roots show well developed mycorrhizae. With depth the organic matter is in a more advanced state of decomposition and forms a black amorphous O2 horizon with little evidence of plant fragments

(O2; humifon – Hf). In thin section it is isotropic and forms a dense mass of poorly formed granules which are interpreted as faecal pellets. There are usually a few fragments of charcoal and many sclerotia of the fungi (Fig. 6.38).

There is then a sharp change from the strongly decomposed organic matter into a very dark grey mixture of organic matter and mineral material (ochric epipedon; modon – Mo). This organic matter is similar to the horizon above while the mineral material is mainly quartz with many of the grains having black or dark brown isotropic coatings. This horizon grades fairly quickly into the bleached pale grey E horizon (albic horizon; zolon – Zo) with its characteristic alveolar or single grain structure (Fig. 6.39). In regions where the soil is frozen in winter this pale horizon has a marked lenticular structure.

The next horizon is a very dark brown to black accumulation of humus, iron and aluminium – the Bs horizon (spodic horizon; husesquon – Hs). This varies from loose through firm subangular blocky to very hard and massive. In thin section the clay is isotropic and arranged in granules or clusters of granules and around most of the sand grains there is an isotropic coating of the same brown material (Fig. 6.40). With depth the content of organic matter decreases and there is a colour change to strong brown, this is also a type of Bs horizon (spodic horizon; sesquon – Sq). There is then a gradation to relatively unaltered material which is usually alluvium or glacial drift having varying degrees of stratification and varying proportions of boulders. In some cases there is a sharp change to a hard compact lower horizon, the Cx horizon (fragipan; ison – In).

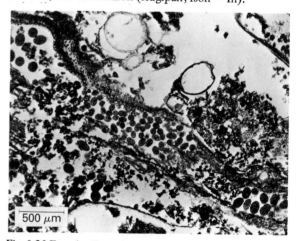

Fig. 6.36 Faecal pellets of an arthropod replacing the central part of a decomposing twig

500 μm

Analytical data (see Fig. 6.41)

The content of clay is usually low or very low and seldom exceeds 10 per cent in the upper bleached horizon. It often shows an increase in the middle horizon due to leaching and deposition of aluminium and iron hydroxides rather than to the translocation of discrete clay particles.

The organic matter distribution has two maxima; the greater of >70 per cent occurs at the surface and the lesser in the middle horizon where it has accumulated through leaching from above. The C/N ratios have a similar trend, with the maximum value of 25 to 30 in the surface horizons; in the bleached horizon the value normally decreases to 10 to 15 and then increases to 15 to 25 in the middle horizon. The cation-exchange capacity also has two maxima coincident with the distribution of the organic matter, which is almost entirely responsible because of the small amount of clay. The content of exchangeable cations is low throughout; the greatest amount occurs in the decomposing organic matter where humification is vigorously releasing basic cations. The base cation saturation is also very low in all horizons, the maximum being in the lowest horizon. The upper layers are very acid with pH values ranging from 3.5 to 4.5. This is followed by a steady increase with depth up to a maximum of about 5.5 in the relatively unaltered underlying material. These low values are due to the acid parent material, the acid litter and to the removal of bases by leaching.

The distribution of silica, alumina and iron oxide in the $<2 \mu m$ fraction is an important distinguishing property and has characteristic trends. The maximum amount of free Fe_2O_3 occurs in the middle Bs horizon and the smallest amount is in the upper E horizons. The molecular ratio of $SiO_2/Fe_2O_3 + Al_2O_3$ in the clay fraction is often more revealing and shows a marked increase in the amounts of iron and aluminium in the middle horizon. The Al_2O_3/Fe_2O_3 ratio shows that there is really more aluminium than iron but since ferric hydroxide is a strong colouring substance it is more conspicuous.

Genesis

The organic matter at the surface is undergoing progressive decomposition by the soil organisms particularly fungi and small arthropods, the evidence being clearly seen in thin sections. Some of the acid decomposition products are dissolved in the percolating rain water charged with CO_2 from the atmosphere and organic compounds such as polyphenols released by decomposition and washed off the foliage of the vegetation. Thus, the solution entering the mineral soil is acid and causes profound weathering of many of the primary silicates, release of various cations and the formation of mobile complex ions of iron and aluminium to give the bleaching of the upper part of the soil. Most of the basic cations are washed through the system but some are taken up by plant roots. Some of the silica is lost from the system but a little is deposited as white powdery material in the bleached horizon. Some of the iron and aluminium released is also lost by leaching but a considerable proportion is deposited accompanied by humus in the middle horizon. The details of this process are not clear but it appears that polyphenols chelate inorganic substances, chiefly iron and aluminium, and together they travel down and are deposited in the middle horizon. Alternatively, the oxides are precipitated as the soil solution percolates into a zone with a higher pH value. A further possibility is that the translocation and deposition is related to the concentration of organic matter and metal cations in the soil solution. As it percolates it complexes the iron and aluminium until a critical concentration of humus and ions is reached when precipitation takes place. The increase in concentration can result both from the accumulation of ions in solution and by loss of water through evapotranspiration; it is probable that both take place but the latter could be more important.

The whole soil system including the middle horizons is constantly weathering and losing material by the continual movement of water straight through the system but the amount lost is relatively small except in the upper horizons. As well as the release of ions into the soil solution, some clays are formed. These are mainly hydrous mica and kaolinite in the upper horizons and hydrous mica and vermiculite in the middle horizons.

A satisfactory explanation has not been given for the formation of the granular structure in the middle horizon seen in thin section. De Coninck and Laruelle (1964) have suggested that the individual granules are the faecal pellets of arthropods. This seems unlikely since it would require a high level of biological activity for which there might not be an adequate supply of food. Alternatively, the precipitation and flocculation of the oxides and humus in a porous medium may be responsible.

Hydrology

Podzols form under aerobic conditions in which there is usually free and rapid movement of water through at least the upper and middle parts of the soil. In a few cases where the bleached horizon or the middle

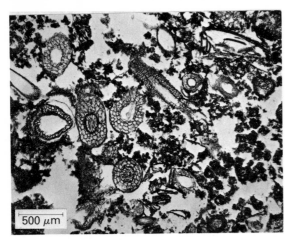

Fig. 6.37 Granular and amorphous partially decomposed organic matter with numerous live roots

Fig. 6.39 The characteristic alveolar structure in the albic E horizon (albic horizon; zolon). There are also many fragments of charcoal

Fig. 6.38 Fungal hyphae decomposing the organic matter

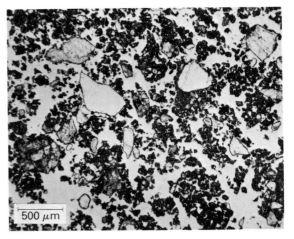

Fig. 6.40 The distinctive small granular structure in the spodic B horizon (spodic horizon; sesquon)

horizon is hard, water movement may be restricted.

Although Podzols develop where precipitation exceeds evapotranspiration, they often show moisture stress and the development of plant communities that are adapted to such conditions.

Principal variations in the properties of the class

Parent material

This is nearly always a medium to coarse textured unconsolidated deposit such as alluvium, glacial drift, dune sand or a solifluction deposit, often containing a high proportion of stones and boulders. The mineralogy is somewhat variable but usually there is a high content of quartz which may exceed 95 per cent in some deposits. However, in cool humid, maritime conditions where leaching is intense, the parent material may be of intermediate or even basic composition. In these areas a bleached horizon seldom develops; there is only a thick dark mixture of organic and mineral material.

Climate

The range of climatic conditions under which Podzols form is quite wide. They are most widespread under a Tundra or Marine climate with rainfall variation from 450 to 1 250 mm per annum. They occur also under Humid Continental cool summer climate and Humid

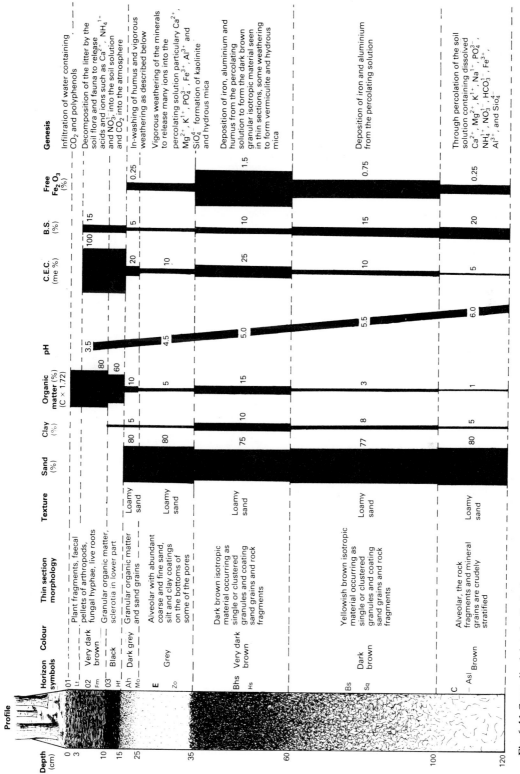

Fig. 6.41 Generalised data for Podzols

Tropical climate, but in both of these environments, particularly the latter, the parent material is highly siliceous producing unique conditions which favour Podzol formation. Thus, as the climate becomes warmer, Podzol formation requires a more quartzose parent material.

Vegetation

These soils are associated with plants that produce an acid litter, these include many species of *Pinus*, *Picea*, and the **Ericaceae** particularly *Calluna vulgaris*. In tropical areas Podzols occur under heathland.

Topography

These soils develop in any topographic situation where aerobic conditions prevail and water can percolate freely through the upper part of the soil. Consequently, their occurrence ranges from flat to steeply sloping sites. They tend to be absent from depressions that become waterlogged.

Pedo-unit

Probably because this class of soils has been studied extensively and intensively it appears to contain more variation than many other classes.

Within the upper horizons there is great variation in the intensity of bleaching as well as in the thickness of the bleached horizon. In some cases there is no distinctly bleached horizon, there is just a dark organic mineral mixture, but it does contain numerous bleached sand grains. Generally, however, there is a bleached horizon of variable depth in the higher latitudes of the taiga and the coniferous belt, but it seldom exceeds 15 cm in thickness and is more often less than 10 cm thick. In the warmer middle latitudes, on highly quartzose parent material, the bleached horizon is commonly up to 1 m in thickness and is often hard at its base presumably due to silica cementation.

The foremost variation is in the nature of the middle horizon which has variable amounts of iron, aluminium and humus. At the drier end of the climatic range for these soils the middle horizon has a high chroma with the accumulation of mainly iron and aluminium. As the soil becomes wetter the middle horizon becomes darker due to an increase in the amount of humus deposited. Where the parent material is predominantly quartz the deposition may be exclusively organic matter. As the humus content increases there is a change in the micromorphology of the middle horizon which no longer has a granular structure but complete black coatings around quartz grains (Fig. 6.42).

In high latitudes that were affected by periglacial processes there is often a very well developed indurated layer or Cx horizon (fragipan; ison – In) in the lower part of the soil.

An interesting type of Podzol is one that has developed entirely in the bleached horizon of a Luvisol so that the profile shows a bleached horizon which overlies a spodic B horizon that in turn overlies an argillic B horizon.

Distribution (see Fig. 6.43)

Podzols occur over a large part of northern Asia, Europe, North America and New Zealand but they also occur elsewhere with a suitable relationship between climate and parent material as in parts of Borneo, Brazil, Malaysia and Ghana.

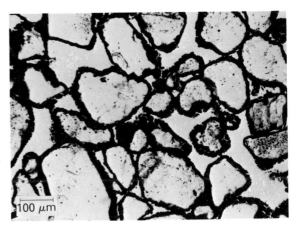

Fig. 6.42 Organic matter forming complete coatings around quartz grains in a spodic B horizon (spodic horizon: hudepon)

Utilisation

The low nutrient status, sandy texture and low pH values usually make the utilisation of these soils for agriculture difficult or impossible; often they are used for forestry or rough grazing. Where agriculture is practised, the natural or semi-natural vegetation has to be removed and the soil ploughed, followed by the addition of an adequate amount of liming material to raise the pH. In addition, fertilisers containing nitrogen, phosphorus and potassium and possibly other elements have to be added. For the first few years the area may be devoted to a grass—legume mixture which allows the large amount of surface organic matter to be decomposed and the grass roots to develop a good structure. Subsequently, these areas can be adapted to the system of land use on the adjacent better soils. In the cool temperate areas some form of mixed farming may be practised, but in the tropics amelioration is often very difficult and special hardy crops such as cashew nuts and coconuts are grown. Wind erosion may be a hazard in some areas.

Placic Podzols

Derivation of name: From the Greek word *plax* = plate; connotative of the presence of a thin iron pan.

General characteristics

A thin hard continuous iron pan is the principal feature of this class. It is relatively impermeable causing moisture to accumulate at the surface and in some instances a Histosol may form as a consequence. These soils develop mainly in cool oceanic areas with high atmospheric humidity and usually carry a heathy vegetation.

Morphology (see Plate IVa)

Commonly there is a thin litter at the surface over-lying a black, plastic organic horizon up to about 30 cm thick (O2; hydromoron — Hy). In thin section it contains a few fragments of decomposing organic matter but the greater part is composed of closely packed granular aggregates of organic matter. These are interpreted as faecal pellets but they could be due to the flocculation of organic matter. This horizon usually changes sharply into the greyish-brown or loamy Eg horizon (albic horizon; candon — Co), which varies from 10 to 50 cm in thickness and may have fine ochreous mottles. In thin section there are frequent, weathered sand grains, very frequently randomly oriented domains and rare reddish-brown, isotropic clay coatings.

Next comes the thin iron pan (placic horizon; placon — Pk) which is dark brown at the top becoming yellowish-brown below, and in some cases there is a thin root mat on its upper surface. This horizon usually follows an irregular path through soil and of special interest is its continuity through stone and boulders (Fig. 6.44). Thin sections show that it is composed of thin, alternating bands of reddish-brown and dark reddish-brown material which is dense and isotropic. Some soils have multiple pans there being as many as three or four within about 10 cm or they may be 10 to 20 cm apart.

The horizons occurring beneath the thin iron pan are very varied and appear to belong to a preceding phase of pedogenesis. There is often a spodic B horizon (spodic horizon; sesquon — Sq) or there may be a Cx horizon (fragipan; ison — In). The latter is very common in the British Isles and eastern Canada where these soils are extensive.

Analytical data (see Fig. 6.45)

The most important physical property of Placic Podzols is the impermeability of the pan which inhibits the penetration of moisture and severely restricts root development and their potential for cultivation.

These soils usually have medium to coarse texture which is fairly uniform through the pedo-unit and in addition they have occasional to abundant stones. They are very acid particularly in the surface organic horizons where the pH values may be as low as 3.5 but they increase with depth to about 5.0. The exchangeable cations, dominated by calcium and the percentage base saturation, are very low, further indicating low potential fertility. The organic matter is high in the surface with a wide C/N ratio indicating a low degree of humification.

Partial ultimate analysis of the total soil as well as the clay fraction shows that the thin iron pan has the greatest amount of iron whereas aluminium has accumulated beneath in the fragipan. The thin iron pan may contain 10 per cent more iron and 15 per cent less aluminium than the horizon below.

Genesis

These soils display interesting evolutionary sequences in which the formation of the thin iron pan is probably the most significant development. Under the cool oceanic conditions where the soils are most widespread the evolutionary sequence is often from a Gelic Gleysol of the Pleistocene period to an Orthic Podzol of the early part of the Holocene and then to Placic Podzol of the middle to late Holocene Period, but in some cases the evolutionary sequence can be traced back to the Tertiary Period as discussed on page 304 *et seq.* In all cases the thin iron pan has formed in a fully developed soil.

The precise reason for the formation of the thin iron pan is still obscure. A recent hypothesis by Koppi (1976) suggests that when the soil solution percolates through the soil, more and more iron is chelated by the organic matter in solution until a critical concentration is attained then precipitation of the iron and organic matter takes place. Further-more, it is suggested that this is likely to be at or about the same place each year. However, this hypo-thesis does not explain satisfactorily the presence of

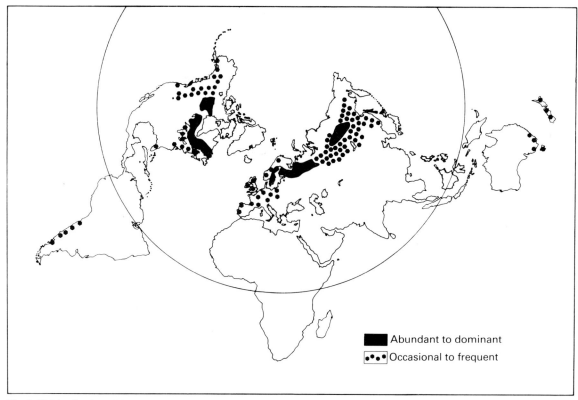

Fig. 6.43 Distribution of Podzols

Abundant to dominant

Occasional to frequent

25 mm

Fig. 6.44 A thin iron pan within a granite boulder

the pan in stones. Although Paton (1976) could not find any iron fixing organisms in an incipient pan it is possible that they might be involved. The formation of the thin iron pan has prevented vertical drainage thus causing anaerobism and the reduction of the

compounds of iron in the upper mineral horizons to give olive and grey colours as well as the formation of mottles. Where the formation of the thin iron pan is a recent phenomenon, traces of the previous horizons can still be seen, but where anaerobism has extended over a long period all evidence of previous horizons appears to be destroyed; even Bs (spodic B horizons; sesquons – Sq) have had their iron reduced to the ferrous state. The occurrence of weathered sand grains, formation of clay coatings, and many domains seems to be due to wet conditions. Since the formation of the thin iron pan, very little alteration has taken place in the underlying horizon.

Initially the fragipan (fragipan; ison) was permafrost; with the characteristic horizontal veins of ice and lenses of frozen soil which were compacted by the growth of the ice; also present were sheaths of ice around stones and boulders. With the amelioration of the climate at the end of the Pleistocene Period the ice melted leaving behind compacted mineral material. As the ice disappeared from around the stones they would gradually settle thus creating a large pore space above them. Subsequently, silt and clay filled this space and colloidal material, particularly aluminium hydroxide washed in from above, has partly

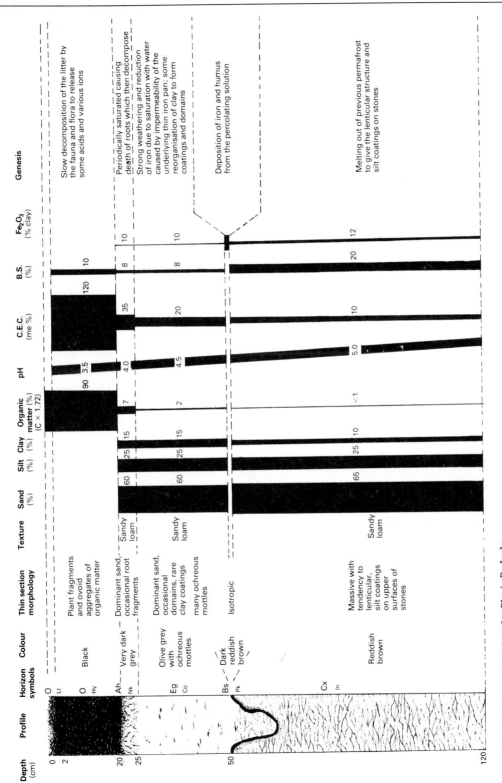

Fig. 6.45 Generalised data for Placic Podzols

cemented and preserved the lenticular structure
(Figs. 6.46 and 6.47). The result is material with a
high bulk density and firm to hard consistence. In
some cases fine material has accumulated in the
horizontal pore space previously occupied by the ice
as in this particular soil.

Hydrology

These soils have impeded drainage because of the
presence of the thin iron pan above which in most
cases there is severe anaerobism and reducing
conditions. Beneath the pan is usually aerobic and
shows bright colours of an oxidising environment but
it may be indurated with low porosity.

Principal variations in the properties of the class

Parent material
Various coarse textured sediments such as glacial
drift, wind blown sands, alluvium and highly
siliceous Tertiary deposits are the normal parent
materials. However, these soils occasionally form in
intermediate and even basic materials. In the tropics
they are a recent evolutionary stage in a number of
soils.

Climate
These soils are found mainly under maritime conditions and have their greatest known extent in the
areas of a marine West Coast climate. However, they
also occur under oceanic tropical conditions at high
elevations as in certain parts of South-East Asia.

Vegetation
These soils usually have a heathy vegetation dominated
by species of the **Ericaceae** or sedges such as *Tricho-
phorum caespitosum*. Under tropical conditions the
principal plant community is evergreen upper montane
forest containing coniferous species, members of the
Ericaceae and mosses, particularly *Sphagnum*.

Topography
The slope is usually gentle or moderate but the soils
occur also on slopes up to 20°.

Age
It would appear that the thin iron pan can form in
about 100 to 200 years and that many formed about
7 500 years B.P. Others developed at various times
since then and some are developing at present.

Pedo-unit
These soils vary mainly in the thickness and colour
of the Eg (albic horizon; candon — Co). When it is
very well developed due to a large amount of surface
moisture it is olive grey and there is usually a Histosol
at the surface.

In a number of cases the thin iron pan occurs above
a spodic B horizon; or it may rest on the relatively
unaltered material. It is characterised by its irregular-
ity in depth from the surface and sometimes occur in
the most unexpected places such as within the plough
layer of old cultivated fields that have been abandoned
for several centuries.

Distribution

These soils occur in the cooler oceanic areas of
north-western Europe including Scandinavia,
northern Germany, northern France, eastern Canada
and the British Isles. They are also found in Alaska,
Malaysia and the Solomon Islands.

Utilisation

The low nutrient status, sandy texture and thick
surface organic matter, make these soils particularly
unadaptable to utilisation. Until recently little use
was made of them but with the introduction of
powerful tractors it has been possible to rupture the
thin iron pan by deep ploughing. This improves
drainage and aeration thus rendering some of these
soils suitable for forestry but even when trees are
planted the application of fertilisers is essential,
particularly phosphorus. The role of phosphorus in
these soils is not fully understood, it seems to act
directly as a plant nutrient and also as a nutrient to
the soil microorganisms causing them to proliferate
and break down the organic matter thereby releasing
nitrogen and other plant nutrients.

PODZOLUVISOLS

Derivation of name: from Podzol and Luvisol.

General characteristics

Soils having an argillic B horizon showing an irregular
or broken upper boundary, resulting from deep
tonguing of the E into the B horizon, or from the
formation of discrete nodules (ranging from 2 to
5 cm up to 30 cm in diameter) the exteriors of which

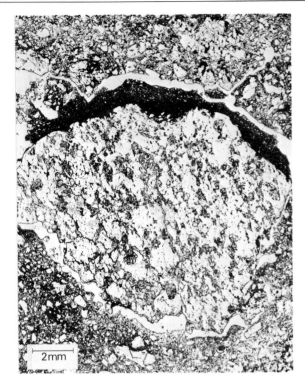

Fig. 6.46 Silt coatings on the upper surface of a rock fragment in the fragipan (fragipan; ison)

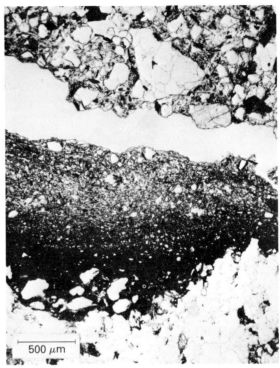

Fig. 6.47 Enlarged portion of Fig. 6.46

are enriched and weakly cemented or indurated with iron and having redder hues and stronger chromas than the interiors; lacking a mollic A horizon. There are three subdivisions of Podzoluvisols:

Eutric Podzoluvisols.   These have an argillic B horizon with tongues and a base saturation of 50 per cent or more throughout the argillic B horizon within 125 cm of the surface.

Dystric Podzoluvisols.   These have an argillic B horizon with tongues and a base saturation of 50 per cent or less throughout the argillic B horizon within 125 cm of the surface.

Gleyic Podzoluvisols.   These have an argillic B horizon and show hydromorphic properties within 50 cm of the surface.

The subdivisions and their equivalents in some other systems of classification are given in Table 6.21.

Gleyic Podzoluvisols

Morphology (see Plate IVb)

Under natural conditions the surface may have a thin litter followed by a thin humified material which rests on a dark grey ochric A horizon which grades into the pale coloured, mottled albic E horizon (albic horizon; minon — Mi) with massive or coarse sub-angular blocky structure. In thin sections clay coatings are rare around pores while the matrix is predominantly anisotropic with abundant random domains. Sesquioxide concretions are rare to occasional and randomly distributed. This horizon passes fairly sharply into the brown argillic B horizon (argillic horizon; glosson — Gs) with coarse prismatic structure, characteristic light coloured tongues and frequent manganiferous concretions. These tongues have a characteristic pattern and can be regarded as vertical continuations of the albic E horizon. In plan they form a subhexagonal pattern and have a thin ochreous border on either side. In thin section the matrix in the brown areas is predominantly aniso-tropic because of abundant, small, random domains

257

and occasional zones of domains. The manganiferous concretions are subspherical, opaque and have diffuse borders. The grey central part of the tongues have occasional clay coatings while the matrix in the yellowish-brown border is strongly anisotropic due to abundant, medium, random domains. This horizon often grades gradually into the relatively unaltered underlying material which is usually a sediment of Pleistocene age.

Analytical data

The clay content increases slightly to moderately in the argillic B horizon. The content of organic matter in the ochric A horizon is usually less than 10 per cent with a C/N ratio of 14 to 16. The organic matter decreases sharply to about 2 to 3 per cent in the albic E horizon and there is also a fall in the C/N ratio indicating a higher degree of humification. The pH values increase steadily from about 5.0 to 5.5 at the surface to 6.0 in the lowest horizon. The percentage base saturation is usually <20 per cent in the surface increasing to over 50 per cent in the lowest horizon. The minerals in the clay fraction are for the most part derived from the parent material and therefore are very variable. The usual tendency is for mica, kaolinite and vermiculite to predominate.

Genesis

Podzoluvisols are formed by the accumulation of moisture within the pedo-unit during winter, spring and the early part of summer. This causes iron to be reduced to the ferrous state aided by dissolved organic substances and to be reorganised within the soil. Some of the products of hydrolysis may be lost in the drainage water but much of the iron and manganese that is mobilised during the wet period seems to be redistributed to form the small concretions. Some redistribution of clay also seems to take place as indicated by the presence of coatings, particularly on the faces of the large prismatic units in the argillic B horizon.

The formation of the subhexagonal pattern of grey tongues may have resulted from shrinking and cracking upon drying or by the development of ice lenses during a cold phase of the Pleistocene. The latter explanation appears to be applicable to parts of Europe which have many well authenticated periglacial phenomena. When the ice disappeared the spaces that were left formed natural channels for water movement. Gradually the iron has been mobilised from their surfaces, some has been removed in solution but some iron has migrated into the soil forming the thin reddish-brown zone on either side of the light coloured streaks. Presumably migration has taken place in the ferrous state during the wet period of the year followed by oxidation during the dry summer period.

Hydrology

Since these soils are developed in medium or fine textured material with poor structure they become very wet during the rainy season, particularly the upper horizons, so that water tends to accumulate or move laterally over the argillic B horizon. The total amount of water that moves through the soil is very small and largely confined to the vertical tongues. However, the soils are sufficiently leached to remove carbonates and even to reduce the base saturation to <50 per cent.

Table 6.21 Subdivisions of Podzoluvisols and their equivalents in some other classifications

	FitzPatrick	France	Germany	Kubiëna	USA Old	USA New	USSR
Podzoluvisols	Supragleysols	Sols lessivés glossiques		Pseudogley	Grey-brown Podzolic soils	Mollisols	
Eutric Podzoluvisols	[Zo < Gsh]		Fahlerde, Braunerde-Pseudogley			Glossudalfs, Glossoboralfs	Derno-Podzolic soils
Dystric Podzoluvisols	[Zo < Gs]					Glossudalfs, Glossoboralfs	Grey Forest soils
Gleyic Podzoluvisols	[Mi < Gs]					Glossaqualfs, Aquic Glossudalfs, Aquic Glossoboralfs.	Meadow Grey Forest soils

Principal variations in the properties of the class

Parent material

This is usually material of medium or fine texture. In many cases it is loess or alluvium but these soils will also develop on loamy glacial drift. In a number of situations an old soil appears to form the parent material.

Climate

These soils are found mainly in cool moist continental areas where the upper part of the soil becomes wet during the spring and early summer but dries out during the summer.

Vegetation

The original plant communities on these soils were dominated by deciduous forests in which oak (*Quercus* spp.) was probably one of the principal species. Many sites have been cleared for agriculture which often failed. Now many Podzoluvisols have permanent pasture or a scrub community; in Europe these latter are dominated by species such as heather (*Calluna vulgaris*) and bracken (*Pteridium aquilinum*).

Topography

Podzoluvisols are confined to flat or gently sloping situations where moisture can accumulate in the upper part of the soil as a result of slow run-off and low permeability.

Age

The development of the albic E horizon has taken place mainly during the Holocene Period but some Podzoluvisols may have been developing continuously from the late Pleistocene Period. In many cases they show polygenetic evolution from a Tertiary soil or sediments derived from Tertiary soils.

Distribution

These soils occur mainly in a cool, humid, continental area; they are common in the central and eastern part of Europe, central USA, various parts of Australia and New Zealand.

Utilisation

The annual saturation of the middle part of the soil severely restricts the utilisation of Podzoluvisols but with drainage this problem can be overcome. It may be difficult to achieve adequate drainage because the zone of maximum moisture accumulation is near to the surface. When tile drains are used they are liable to be disrupted by cultivation, frost or plant roots.

The low pH values at the surface and the massive structure of the horizons are also limiting factors requiring liming and thorough cultivation. When these ameliorative methods have been carried out and fertilisers applied these soils give good yields of a variety of crops.

RANKERS

Derivation of name: from the Austrian *Rank* = steep slope; connotative of shallow soils from siliceous material.

General characteristics

Soils, exclusive of those formed from recent alluvial deposits, having an umbric A horizon which is not more than 25 cm thick, having no other diagnostic horizons (unless buried by 50 cm or more new material); lacking hydromorphic properties within 50 cm of the surface; lacking the characteristics which are diagnostic for Andosols.

These soils are regarded as the second stage in soil evolution and are extremely variable since nearly every type of upper horizon can occur directly on unaltered material. Therefore, it is impossible to give modal characteristics for this class, similarly they have few genetic processes in common with each other. However, there would appear to be a few Rankers that are more widespread than others and have formed in areas where recently fresh rock surfaces have been exposed or where deposits such as glacial drift or loess have been deposited. In large parts of North America and Eurasia the rocks are scraped bare by glaciers which left varying thicknesses of till. Where this is thin or absent Rankers have often developed. One of the principal types has the usual organic and leached horizons of the Podzol resting on the underlying rock. A second type of Ranker has a distinctive very dark humose upper Ah horizon (umbric epipedon; mullon) resting on the underlying sediments.

The subdivisions and their equivalents in some other systems of classification are given in Table 6.22.

Principal variations in the properties of the class

Parent material, climate, vegetation, topography and age

Rankers develop in association with a wide range of parent materials, climate, vegetation and topography, therefore they have very diverse properties. It is the degree of consolidation of the parent material that has the most marked influence on their properties and development. Where there is a thin deposit of drift over hard rock, upper horizons develop quickly in the drift, but further differentiation involving the rock is very slow and the soil may remain in this state for thousands of years. On the other hand in a thick deposit of drift or in weathered material, differentiation of middle or lower horizons rapidly follows those at the surface and the Ranker stage may last only for a few centuries; this is true particularly for material like volcanic ash which weathers rapidly.

Pedo-unit

Most upper horizons that develop quickly form the upper horizons of Rankers. The precise type is usually related to the adjoining deep soils but modifications are sometimes introduced particularly in those developed on rock which often affects the drainage in such a way that the soils are very wet during the wet period of the year and very dry when precipitation is small. Rankers with upper horizons similar to Podzols are usually found associated with Podzols while others have upper horizons similar to the associated Cambisols (Ochrepts; Altosols). Rankers on deep unconsolidated sediments progressively develop into the soils typical of the environment. For example, in a humid temperate area Rankers on deep acid sands will gradually and progressively develop into a Podzol (see p. 303).

A common type of Ranker in many tropical and subtropical areas occurs on deeply weathered rock, the original topsoil having been removed by erosion, often man induced. This type of Ranker would have been more widespread at the end of a pluvial period during which large areas were severely eroded.

Distribution

Rankers occur in all parts of the world; the largest extent is in mountainous areas where rock is normally near to the surface.

Utilisation

Where these soils are developed in deep, loose, unconsolidated material they can be cultivated, particularly if of medium texture. Shallow Rankers in mountainous areas can be afforested but the forests seldom have economic value, such areas are best left to develop a natural vegetation and wild life and to be used for recreative pursuits.

REGOSOLS

Derivation of name: from the Greek word *rhegos* = blanket; connotative of mantle of loose material overlying the hard core of the earth.

General characteristics

Soils from unconsolidated materials, exclusive of recent alluvial deposits, having no diagnostic horizons (unless buried by 50 cm or more new material) other than an ochric A horizon; lacking hydromorphic properties within 50 cm of the surface; lacking the characteristics which are diagnostic for Vertisols and Andosols; lacking high salinity; when coarse textured. lacking lamellae of clay accumulation, features of cambic or oxic B horizons or albic material which are characteristic of Arenosols.

They have a wide range of textures and occur in all climatic areas and thus form the initial stage of formation of a large number of soils principally Podzols, Luvisols, Cambisols, Chernozems, Kastanozems, Xerosols and Yermosols. The ochric A horizon forms quickly but is often a transitional phase to a mollic or umbric A horizon.
There are four subdivisions of Regosols:

Eutric Regosols.   These have an ochric A horizon and a base saturation of 50 per cent or more between 20 to 50 cm from the surface.

Calcaric Regosols.   These soils have an ochric A horizon and are calcareous between 20 to 50 cm from the surface.

Dystric Regosols.   These have an ochric A horizon and a base saturation of less than 50 per cent between 20 to 50 cm from the surface.

Table 6.22   Rankers and their equivalents in some other classifications

	FitzPatrick	France	Germany	Kubiëna	USA	
					Old	New
Rankers	Rankers	Rankers	Rankers	Rankers	Regosols	Lithic Haplumbrepts

Gelic Regosols. These have an ochric A horizon and permafrost within 200 cm of the surface.

The subdivisions and their equivalents in other systems of classification are given in Table 6.23.

RENDZINAS

Derivation of name: from the Polish word *rzedzic* = noise; connotative of noise made by plough over shallow stony soil.

General characteristics

Soils having a mollic A horizon which contains or immediately overlies calcareous material with a calcium carbonate equivalent of more than 40 per cent; lacking hydromorphic properties within 50 cm of the surface; lacking the characteristics which are diagnostic for Vertisols; lacking high salinity. Since these soils form due to the presence of large amounts of limestone in the parent material, they are distributed throughout a wide range of climate and carry many different plant communities.

The equivalent names in some other systems of classification are given in Table 6.24.

Morphology (see Plate IVc)

The characteristics of the surface horizons vary some-what with climate and vegetation. A common type has a loose leafy litter resting on the dark brown or black calcareous organic mineral mixture – a mollic A horizon (mollic epipedon; mullon – Mu) which is usually speckled with white fragments of the parent rock. This horizon has a well developed crumb, granular or vermicular structure with abundant earth-worm casts and in thin section each aggregate is seen to be outlined by a thin layer of calcite crystals. Fragments or complete mollusc shells are rare to occasional.

Below there is usually a sharp change to the under-lying rock or there may be a narrow transitional horizon.

Analytical data

Perhaps the most important physical properties of these soils are their shallowness, medium to fine texture and well developed granular to fine sub-angular blocky structure. Together these features allow rapid percolation of moisture which can cause drying out and in some years there may be a period of drought.

The calcareous mollic A horizon usually contains up to 80 per cent calcium carbonate giving pH values over 8.0. The organic matter content varies from 5 to 15 per cent and is usually in an advanced state of humification as indicated by the C/N ratios of 8 to 12. The fine texture and high organic matter content have resulted in cation-exchange capacity values of up to 50 me per cent. The main exchangeable cation is calcium, or magnesium on dolomites and there is complete base cation saturation.

Genesis

The principal process taking place in Rendzinas is the solution and removal of the carbonate in the drainage water, leaving behind a relatively small residue. This is mixed with the humifying organic matter and small fragments of undissolved material by the very active mesofauna to give a granular structure but expansion and contraction probably account for the blocky structure when it is present. The characteristic dark colour is due to a calcium–humus complex and is very similar to that found in Chernozems. When the residue accumulates to a very great thickness and the carbonate is below the depth of root or faunal penetration, decalcification pro-ceeds. This is followed by the formation of soils similar to those developed on fine grained igneous rocks or sediments viz. Cambisols (Ochrepts; Altosols) or Luvisols (Alfisols; Argillosols).

Hydrology

Rendzinas develop under aerobic conditions since there is usually free and rapid movement of water through at least the upper part of the soil. Although they usually occur where precipitation exceeds evapotranspiration they usually show moisture stress during the dry period of the year because they are so shallow and have little capacity for storing moisture.

Principal variations in the properties of the class

Parent material

This is an example of a class of soils whose character-istics are determined almost entirely by their parent materials which are always composed of material containing a high proportion of calcium and/or magnesium carbonate. They are usually consolidated rocks such as limestone and chalk but they can be unconsolidated sediments or drift deposits.

Climate

Since the parent material is the controlling factor in the formation of these soils they are found under

261

Table 6.23 Subdivisions of Regosols and their equivalents in some other classifications

	Australia		Canada	FitzPatrick	France
	Handbook	Northcote			
Regosols					
Eutric Regosols	Lithosols Terra Rossa soils Grey-brown and red calcareous soils	Uc6.1 Uc6.1 Um1.3	Orthic Regosols	[Tn-Bcl] [Mu-Bcl]	Sols minéraux bruts d'apport éolien ou volcanique Sols peu évolués régosolique d'erosion Sols peu évolués d'apport éolien au volcaniques friables
Calcaric Regosols	Calcareous sands Siliceous sands	Uc1.1 Uc1.3 Uc5.1		[Tn-2Czl] [Muc-2Czl]	
Dystric Regosols	Siliceous sands Lithosols	Uc1.2, Uc5.1 Uc4.1, Um4.1 Um1.4		[Tn-Als] [Tn-As]	
Gelic Regosols			Cryic Regosols	[(MoGn)Cy]	

nearly every type of climate outside the very cold and the very arid areas.

Vegetation
The natural vegetation varies with the climate but usually it contrasts strongly with adjacent plant communities on soils developed from different materials. Generally Rendzinas have a richer flora except where they are very shallow, then drought may become a limiting factor to plant growth. Earthworms, enchytraeid worms, arthropods and bacteria are the chief members of the soil organisms but again limitations may be imposed if there is a shortage of water.

Topography
The areas where Rendzinas occur are usually characterised by karst phenomena and vary from flat to steeply sloping (Fig. 2.34). The old landscapes in many tropical and Mediterranean areas with marked polygenetic development often have Rendzinas on the recently eroded slopes while older soils such as Chromic Luvisols (Rhodoxeralfs; Rossosols) occupy the flatter sites.

Age
When these soils occur in areas that were subjected to considerable erosion during the Pleistocene Period they are about 10 000 years old but in some tropical areas they appear to be much older.

Pedo-unit
The members of this class have a relatively small degree of variation which is confined largely to the thickness and regularity of the upper horizon. When they develop under fairly dry conditions they are lighter in colour and lower in organic matter; in addition, some carbonate is deposited in the upper part of the underlying material to form a calcic horizon (calcic horizon; calcon – Ck). In contrast the surface may have an accumulation of organic matter, particularly under cool humid conditions. In moist situations, montmorillonite develops and gradually these soils intergrade to Vertisols. Where the parent material is low in carbonates they tend to be browner in colour and intergrade to Cambisols (Ochrepts; Altosols).

Distribution
Rendzinas occur in small areas throughout much of the world where environmental conditions are suitable. In some places, such as in the countries bordering the Mediterranean they occupy large areas.

Utilisation
Since Rendzinas develop under a wide variety of conditions their utilisation is governed by local practices, therefore it is possible to find them producing a wide variety of crops including sugar cane,

Germany	Kubiëna	South Africa	USA		USSR
			Old	New	
Rohböden	Syrosems		Regosols	Orthents Psamments	Regosols
	Loess Syrosem	Fernwood			
		Fernwood			
				Pergelic Cryothents Pergelic Cryopsamments	

wheat and cocoa, while some of the largest areas are devoted to vineyards.

Severe limitations are imposed by their shallowness and high permeability. The former often prevents the use of large implements whereas the latter can cause drought even in a humid environment. The high content of calcium can induce microelement deficiencies by replacing them on the exchange sites. In spite of these deficiencies these soils have a high natural fertility and are highly prized by farmers but when they are shallow or occur on steep slopes they are devoted to forestry although the quality of the trees is sometimes poor.

SOLONCHAKS

Derivation of name: from the Russian word *sol* = salt; connotative of soil having a high content of salts.

General characteristics
Soils, exclusive of those formed from recent alluvial deposits, having a high salinity and having no diagnostic horizons other than (unless buried by 50 cm or more new material) an A horizon, a histic H horizon, a cambic B horizon, a calcic or a gypsic horizon.

There are four subdivisions of Solonchaks:

Orthic Solonchaks. These have an ochric A horizon and lack hydromorphic properties within 50 cm of the surface.

Mollic Solonchaks. These have a mollic A horizon and lack hydromorphic properties within 50 cm of the surface.

Takyric Solonchaks. These have takyric features and lack hydromorphic properties within 50 cm of the surface.

Gleyic Solonchaks. These have hydromorphic properties within 50 cm of the surface.

The subdivisions and their equivalents in some other systems of classification are given in Table 6.25.

Gleyic Solonchaks

Morphology
Generally these soils show only weak contrasts between horizons. The whole soil is normally grey or greyish-brown often with mottling, the greatest being in the middle of the soil. Usually the upper horizon is slightly darker due to staining by organic matter and in many cases it has a thin crust of salt at the surface (Fig. 6.48) or there may be takyrs which are

Table 6.24 Rendzinas and their equivalents in some other classifications

	Australia		Canada	FitzPatrick	France	Germany	Kubiëna	South Africa
	Handbook	Northcote						
Rendzinas	Rendzinas		Calcareous rego black soils	Rendzinas	Rendzinas	Rendzinas	Rendzinas	
		Uf6.1		[Muc-4CN]				Mayo Milkwood

small, dome-shaped areas with polygonal outlines. In other cases the upper horizon may be massive, coarse platy, puffy or crusty.

In thin sections the middle horizon is seen to contain varying amounts of carbonate concretions while all the ped and pore surface are outlined by fine calcite crystals which usually form encrustations around small objects such as root fragments.

Analytical data (see Fig. 6.49)
The most important property of these soils is their high content of salts which are usually highest near to or at the surface, decreasing with depth. The most common ions are chloride, sulphate, carbonate, bicarbonate, sodium, calcium, magnesium and small amounts of potassium but the variability in the proportions of these ions is extremely wide, however, the data given in Fig. 6.49 are representative of large areas of these soils.

As stated on page 114 the principal criteria used in the classification of Solonchaks are pH values and the electrical conductivity of the soil or saturation extract. The distribution of the individual ions is also taken into account and further subdivisions of the class are often made on the basis of ratios of the various ions present. This is useful in some places where there are regional variations in the proportions of ions as found in central USSR.

The structure throughout the pedo-unit is usually massive except on very sandy materials and it is interesting to compare this type of structure with the granular structure of a Kastanozem which may develop in the same environment on similar parent material.

Genesis
These soils commonly develop in an arid or semiarid environment on flat situations, or in depressions

Table 6.25 Subdivisions of Solonchaks and their equivalents in some other classifications

	Australia		Canada	FitzPatrick	France	Germany	Kubiëna
	Handbook	Northcote					
Solonchaks	Solonchaks	Um1.1 Um1.2 Um2.2		Solonchaks		Solonchaks	Solonchaks
Orthic Solonchaks	Solonchaks		Saline subgroups	[TnHa]	Sols salins		
Mollic Solonchaks	Solonchaks		Saline subgroups	[MuHa]	Sols salins		
Takyric Solonchaks	Solonchaks Solonchaks Solonchaks Solonchaks No suitable group	Uf1.4 Uf6.4 Uf6.5 Uf6.6 Ug5.17		Takyr			
Gleyic Solonchaks				[Mu(GlHa)]			

USA		USSR
Old	New	
Rendzinas	Rendolls	Dern-carbonate soils

where the ground-water-table during the dry season is usually less than about 3 m from the surface. During the wet season the water-table rises often coming to the surface causing some reduction of iron and the development of a mottled pattern in the anaerobic environment. At the same time evaporation or evapotranspiration results in the loss of water causing some of the salts dissolved in the ground water to be deposited on the surface and within the upper part of the soil. When the ground-water-table recedes during the dry period of the year the water left in the upper part of the soil is lost by evaporation and any dissolved salts are deposited. Annual repetition of this cycle of wetting and drying causes a considerable amount of salts to accumulate within the zone of moisture fluctuation but there is no accumulation within the zone of permanent saturation. Therefore, the pattern of salt distribution tends to form a maximum at or near to the surface.

South Africa	USA		USSR
	Old	New	
	Solonchaks		Solonchaks
Oakleaf		Salorthids	
		Salorthidic Calciustolls, Salorthidic Haplustolls.	Solonchaks
			Takyr
			Meadow Solonchaks

If the water-table never comes to the surface the maximum content of salt may occur at some depth within the soil. On the other hand the absence of a salt maximum at the surface may be due to secondary leaching. The amount of soluble salts in some Solonchaks can vary from season to season. During the autumn, following the dry summer, the salt concentration at the surface is several times that of the spring period when melting snow or rainfall causes some downward leaching thereby reducing the salt content.

The texture of the soil seems to influence the rate of salt accumulation; fine textured soils have a higher retentivity, therefore they hold more saline water which upon evaporation leaves a higher amount of salt. Fine textured soils also have a lower permeability which decreases as the content of sodium increases because sodium disperses the clays.

Sometimes it is not clear how the salts in the ground water have originated, usually they are derived from the weathering of rocks but it is difficult to account for large amounts of chloride and carbonate. The former is not a normal constituent of rocks except for small amounts in some sediments, it is usually regarded as coming from the sea either as spray or due to a previous inundation. The presence of carbonate particularly sodium carbonate is more difficult to explain particularly when calcium or magnesium carbonate was either absent or present in only small amounts in the original material. The greatest amount of sodium comes from the weathering of rock either by hydrolysis as in the case of the orthoclase feldspars or by solution from sedimentary rocks. It seems that most of the sodium carbonate is formed from the CO_2 of the atmosphere in a number of stages. At first the CO_2 is dissolved in the soil solution to form H_2CO_3 which reacts with sodium to form sodium bicarbonate which is relatively unstable and is readily transformed into sodium carbonate.

Principal variations in the properties of the class
Parent material
Unconsolidated deposits including loess, alluvium and pedi-sediments are the principal parent materials. On the older landscapes colluvium derived from old soils may form the parent material.

Climate
The most common occurrence of Solonchaks is in mid-latitude and tropical arid and semiarid areas. They are found at the margins of humid continental conditions where evapotranspiration is much greater

Fig. 6.48 Salt efflorescence on the surface of the soil and a partly exposed Solonchak profile

than precipitation. Solonchaks have also been reported from the drier polar areas.

Vegetation

The natural plant cover varies from quite dense to absent, depending upon the degree of salinity. Where the salt content is fairly low the species differ little from the adjacent non-saline areas which usually carry a grassy plant community. As the salt content increases to over 0.5 per cent only halophytic species can grow (see Fig. 6.48). The plants growing on Solonchaks have a high content of ash containing larger amounts of sodium, chloride and sulphate.

Topography

The flat or depression areas where these soils develop are often alluvial terraces, beds of old lakes or else

they are basins surrounded by mountains which shed large amounts of moisture during the wet periods of the year so that they are temporarily waterlogged. The water from the mountains brings with it varying amounts of salts which are left behind when the water is lost by evaporation.

In many areas salt efflorescence occurs in shallow isolated depressions which are easily recognised by the absence of vegetation and their characteristic pale grey colour. As salinity increases the isolated areas grow larger and eventually coalesce to form an almost unbroken salt crust.

Age

These soils form rapidly and most groups seem to have developed during the Holocene Period. Since their main property is the presence of salts they can

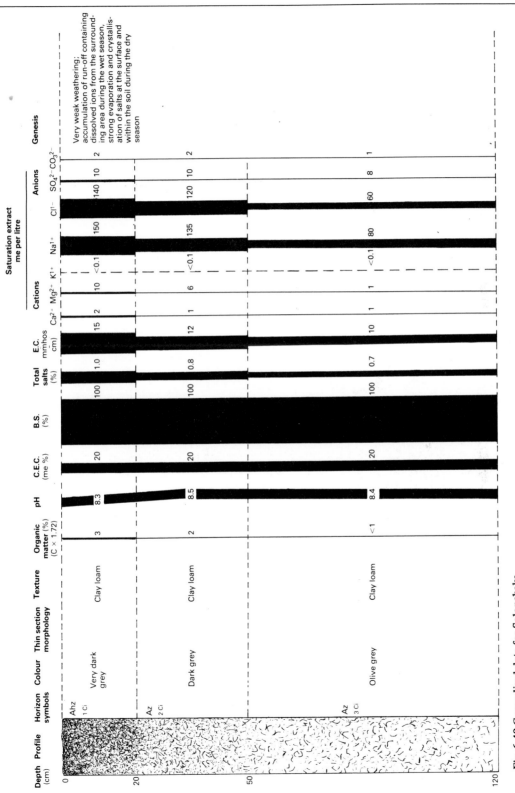

Fig. 6.49 Generalised data for Solonchaks

Table 6.26 Subdivisions of Solonetz and their equivalents in some other classifications

	Australia		Canada	FitzPatrick
	Handbook	Northcote		
Solonetz			**Solonetzic**	**Solonetz**
Orthic Solonetz	Desert loams	Dr1.1, Dr1.3, Dr1.4		[ZoSl]
	Soloths	Dr2.3, Dr2.4, Dr3.3,		[ZoSlCk]
		Dr3.4, Db1.3, Dy2.3,		[ZoSlHl]
		Dy2.4, Dy3.3, Dy3.4,		[TnSl]
		Dd1.3, Dd2.3, Dy3.8,		[SlDu]
		Dy5.4, Dy5.8, Dd1.1,		[Sl(DuCk)]
		Dd1.3, Dd2.3		
	Red Podzolic soils	Dr2.3, Dr2.4, Dr3.3,		
		Dr3.4		
	Red-Brown earths	Dr2.3		
	Solodised Solonetz	Dr3.3, Db1.3, Db2.3,		
		Db2.4, Dy2,3, Dy2.4,		
		Dy3.3, Dy3.8, Dd2.3,		
		Dy5.4, Dy5.8, Dd1.1,		
		Dd1.3, Dd1.4		
	Solodic soils	Dr3.3, Db1.3, Db1.4,		
		Dy2.3, Dy2.4, Dy3.4,		
		Dy3.8, Dy5.4, Dy5.8,		
		Dd1.1, Dd1.3, Dd1.4,		
		Dd2.3		
	Yellow Podzolic soils	Dy2.3, Dy2.4, Dy3.3,		
		Dy3.4, Dy5.4		
	Lateritic Podzolic soils	Dy3.8		
Mollic Solonetz			**Black and Grey Solonetz**	[MuZoSl]
				[MuSl]
				[Mu(MuSl)]
				[MuZoSlCk]
				[MuSlCk]
Gleyic Solonetz	Soloths	Dy5.4, Dy5.8	Gleyed Solonetz	[MiSl]
	Yellow Podzolic soils	Dy5.4, Dy5.8		

be changed rapidly by man's influence or by slight changes in elevation or climate, therefore it is only under somewhat exceptional circumstances that they would be preserved from an earlier period.

Pedo-unit

The principal variations in the properties of these soils lie in the amount and type of ions which can exist in various combinations creating many difficulties with regard to classification and delimiting boundaries because variations can take place rapidly over short distances. This problem can probably be solved by creating four horizons on the basis of the ternary diagram given in Fig. 4.36. The amount of mottling also varies from place to place but probably the greatest variations are found in the upper horizon which can be saline intergrades, since Solonchaks intergrade into a number of directions principally to

Gleysols, Chernozems, Kastanozems, Xerosols and Yermosols.

Distribution (see Fig. 6.50)

Every major land mass has its area of Solonchaks which occur mainly in the drier central parts of continents. However, they are common in many coastal situations where the main wind currents blow from off the land. This is marked on the NW coasts of Africa and South America and the west coast of Australia.

Utilisation

This class of soils probably offers the greatest amelioration problems primarily because of the difficulties encountered when trying to remove the salts. Since water conveyed the salts into the soil it is necessary to use water to remove them, but in an

France	Germany	Kubiëna	South Africa	USA		USSR
				Old	New	
	Solonetz	Solonetz		Solonetz		Solonetz
Sols sodiques à horizon B Solonetz solodisés			Sterkspruit Estcourt		Natrustalfs Natrixeralfs Natrargids Nadurargids	Solonetz
Sols sodiques à horizon B, Solonetz solodisés					Natralbolls Natriborolls Natrustolls Natrixerolls	Solonetz
					Natraqualfs	Meadow Solonetz

arid or semiarid environment there is usually a short-age of water or the available supply may have such a high content of salts as to make it unsuitable for leaching. There are a few notable exceptions such as the waters of the Nile and Colorado rivers. Even when there is an adequate supply of water many problems still exist. The soils usually have a poor structure which makes them slowly permeable so that much of the water that is applied may be lost by evaporation or run-off. Secondly, drains have to be installed so as to remove the saline leachate as well as to reduce the height of the ground-water-table, but since these soils occur in flat or depression situations there are difficulties with regard to removing the drainage water. Even when the drainage water can be removed there remains the problem of disposal for it cannot be put into the natural drainage of the area such as rivers because it would increase their salinity.

Where the area is very low-lying the ground-water-table can be lowered by digging wells and pumping out the water.

In addition to adding water it is normal to add gypsum (calcium sulphate), this dissolves gradually and the calcium is slowly absorbed on to the exchange site thereby replacing sodium which is lost in the drainage. As the clays become more saturated with calcium they flocculate and gradually the structure improves causing an increase in permeability. In some soils the initial content of calcium sulphate is very high so that it is necessary only to irrigate. After these operations have been completed these soils can then be adapted to the system of land use of their surroundings providing there are no other hazards. Sometimes, however, microelements such as boron may be present in toxic proportions.

In a number of arid areas the saline water-table

may be several metres from the surface and without any capillary rise to the surface the soils are not saline. Often when these soils are given an excess of irrigation water, it percolates down through the soil and makes contact with the underground water. This causes capillary rise of moisture to take place and the saline waters are drawn to the surface. This has been a major factor in causing salination of many soils such as in the Sind Valley of Pakistan. Therefore, great caution must be exercised when irrigating a soil with a deep saline ground-water-table.

SOLONETZ

Derivation of name: from the Russian word *sol* = salt; connotative of soils containing salt.

General characteristics

Soils having a natric B horizon; lacking an albic E horizon which shows hydromorphic properties in at least a part of the horizon and an abrupt textural change.

There are three subdivisions of Solonetz:

Orthic Solonetz. These have an ochric A horizon and without hydromorphic properties within 50 cm of the surface.

Mollic Solonetz. These have a mollic A horizon but without hydromorphic properties within 50 cm of the surface.

Gleyic Solonetz. These show hydromorphic properties within 50 cm of the surface.

The subdivisions and their equivalents in some other systems of classification are given in Table 6.26.

Orthic Solonetz

Morphology

At the surface there may be a thin loose leafy litter resting on black humified material about 2 to 3 cm thick. This rests in turn on a brown granular ochric A horizon (ochric horizon; tannon − Tn) up to about 15 cm in thickness. In thin section the matrix is isotropic, forming coatings around the grains or occurring as small granules. Next follows a sharp change into the mottled greyish-brown and brown natric B horizon (natric horizon; solon − Sl) with its prismatic or columnar structure and higher content of clay (Fig. 4.19). In thin section the matrix is predominant-

ly anisotropic with abundant medium and large domains and thin zones of domains some with oblique orientation. Clay coatings occur on some of the surfaces of the peds as well as around pores within peds but their frequency seldom exceeds 2 per cent of the soil. The natric B horizon grades with depth into the somewhat more mottled and massive saline horizon or there may be an intervening gypsic horizon (gypsic horizon; gypson − Gy) with its characteristic clusters of gypsum crystals which are clearly seen in thin section (Fig. 6.51). In many places Solonetz are in a fairly advanced stage of development and an albic E horizon (albic horizon; zolon − Zo) is beginning to form.

Analytical data (see Fig. 6.52)

Perhaps the most conspicuous property of Solonetz is the abrupt and large increase in clay in passing from the upper horizon into the natric B horizon. The increases may be up to threefold with the greatest increase being in the fine clay ($<0.2 \mu m$). The amount of organic matter in the surface mineral horizon varies but is usually less than 10 per cent with a C/N less than 12, indicating a high degree of humification. The pH values are usually between 6.0 and 7.5 at the surface increasing to over 8.5 in the lowest horizons.

The cation-exchange capacity varies with the texture and clay mineralogy but is usually between 15 and 35 per cent and apart from the upper horizons the entire soil is saturated with basic cations. In the upper horizons calcium may be the principal exchangeable cation but in the natric B horizon sodium and magnesium predominate, and together exceed the Ca and H content; alternatively, sodium occupies at least 15 per cent of the exchange complex. With depth calcium may again be the principal exchangeable cation.

Usually the upper horizons are non-saline but salinity increases with depth. The conductivity of the natric B horizon often attains 2.0 mmhos cm^{-1} but in the underlying horizons it may be as much as 15 mmhos cm^{-1} where there is >0.5 per cent salts and sometimes much carbonate. As with Solonchaks the amount and type of salts vary from place to place, but ions of calcium, magnesium, sodium, carbonate, bicarbonate, chloride and sulphate predominate.

The types of clay minerals are mainly inherited from the parent material and generally they are dominated by micas, but kaolinite and montmorillonite can be present in high proportions. The principal minerals produced by hydrolysis are mica, vermiculite and small amounts of montmorillonite.

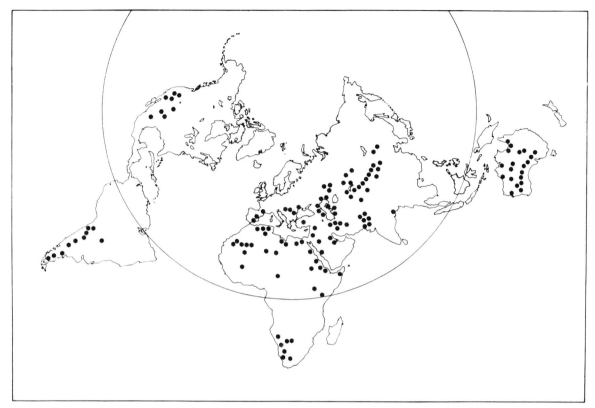

Fig. 6.50 Distribution of Solonchaks

The upper horizons usually contain quartz in the clay fraction.

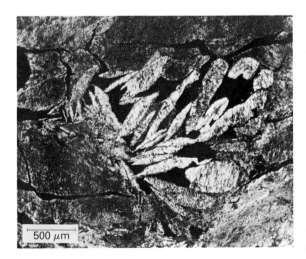

500 μm

Fig. 6.51 A cluster of gypsum crystals in crossed polarised light

Genesis

It is generally accepted that Solonetz formed by the progressive leaching of Solonchaks which are deficient in calcium but have a large amount of sodium ions. This leads to the migration of clay from the upper to the middle position of the soil to form the natric B horizon. When the data for these soils are examined carefully it is difficult for this theory to be sustained. The increase in clay over that in the parent material cannot be attributed to the small amount of clay coatings. Similarly, the amount of clay coatings cannot account for the considerable amount of clay that has been removed from the upper horizon. Thus, there appear to be three processes taking place in these soils, the most important seems to be the destruction and removal of clay from the upper mineral horizons, evidence for which is afforded by the presence of quartz in the clay fraction. A small proportion of clay translocation is taking place hence the presence of clay coatings in the natric B horizon, but often the greatest amount is in its lower part and does not coincide with the maximum amount of clay.

The increase in the clay content in the natric B

horizon over that of the parent material seems to result mainly from weathering *in situ*. Thus the textural contrast is due to clay destruction in the upper horizon and clay formation in the middle horizon together with a small amount of clay translocation.

Although it is probable that Solonetz may have formed by the progressive leaching of Solonchaks, it seems that many can form where sodium rich waters percolate through the soils.

Principal variations in the properties of the class

Parent material
Unconsolidated deposits including loess, alluvium and glacial deposits are the principal parent materials. In areas of old landforms Solonetz may form in the old soils or colluvial material derived from them, this is particularly common in parts of Africa and Australia. Solonetz will not form in material containing calcium carbonate because the calcium displaces the sodium and magnesium but they will form following decalcification.

Climate
These soils are most widespread in the semiarid parts of the world where there is sufficient precipitation to cause leaching of the upper horizons. On the other hand the amount of moisture passing through the soil is insufficient to reduce the alkalinity in the lower horizons.

Vegetation
Various succulents and xerophytically modified species form the normal plant communities on these soils. Generally the combination of low rainfall, high pH and poor structure restricts the development of a complete plant cover.

Topography
Solonetz are confined to flat or gently sloping situations, they tend to be absent from depressions where the water-table comes near to the surface, then Solonchaks form. The topographic relationships are given more fully on page 297.

Age
Although Solonetz may be found associated with old land surfaces their development seems to have taken place during the Holocene Period and may have extended over most of that period since a considerable length of time is required for the destruction and removal of clay from the upper horizons.

Pedo-unit
A common variation is in the nature of the principal upper horizon which depends upon the character of the climate and the other soils with which Solonetz are associated. Therefore, the upper horizons may have similarities with Chernozems, Kastanozems and other upper horizons. The individual properties of Solonetz can vary widely from one soil to the other. For example, there may be a tenfold increase in clay between an albic E horizon and the natric B horizon which in some soils is very dark in colour and contains up to 3 per cent organic matter. At present there is still a measure of disagreement about what constitutes a natric B horizon, some workers use chemical criteria and others use morphological evidence while others require both sets of criteria. The situation is complicated further because the chemical criteria can vary from one worker to the other. Originally, it was suggested that a natric B horizon should have >15 per cent of the exchange complex saturated with sodium but it was discovered that magnesium can be important. Then it was suggested that the exchangeable sodium plus magnesium should be greater than the exchangeable calcium plus hydrogen. At present both chemical and morphological criteria are required, therefore those horizons that have one and not the other are considered to be intergrades. When a horizon has only the chemical characteristics it usually occurs in the lower part of the soil and indicates some influence of saline underground waters and is fairly common in areas of Chernozems, Kastanozems, and Vertisols. In contrast, there are a number of soils that have certain morphological features of Solonetz but do not have the chemical characteristics, these soils are considered to have been Solonetz which have lost their chemical characteristics by progressive leaching through irrigation or after a long period of leaching by rainfall. A somewhat unique set of conditions is found in cold semiarid areas such as in Alberta in Canada, where intergrades between Podzols and Solonetz develop. In these soils the surface horizons are those of Podzols with a natric B horizon in the middle or lower position.

There is a continuous gradational sequence between Solonetz and Solodic Planosols, morphologically expressed through a gradual development of a thick albic E horizon and decreasing thickness of the natric B horizon.

Distribution (see Fig. 6.53)
These soils are of common occurrence in the arid and semiarid regions of the world. They are widespread in Australia, west Pakistan, south-western USSR,

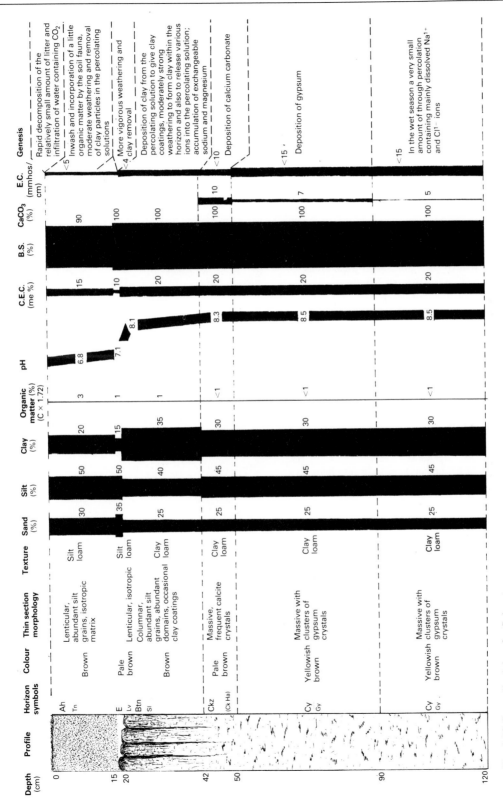

Fig. 6.52 Generalised data for Solonetz

northern and southern Africa. In North America they occur spasmodically in a belt that starts in northern Alberta and continues to western Texas and Mexico. In South America they occur in the narrow coastal belt of northern Chile and also in certain parts of Argentina.

Solonetz may form large contiguous areas but they are found more usually as part of a complex with other soils, especially Chernozems and Kastanozems.

Utilisation

The value of these soils for crop production varies with the content of organic matter and climate. They are regarded as being fairly good for crop growth when they occur in a cool climatic area associated with Chernozems or Kastanozems. On the other hand when they occur in tropical and subtropical areas conditions are usually too dry for crop growth but they may produce forage for grazing.

Where irrigation is possible these soils are highly productive but precautions must be adopted to prevent the salts in the lower part of the soil from coming to the surface by capillary movement. Therefore, adequate drainage has to be provided to remove the excess irrigation water and where necessary to lower the water-table.

The fine texture of the natric B horizon may cause impeded drainage but this can often be alleviated by deep ploughing.

In all cases these soils are deficient in plant nutrient which have to be supplied and often liming materials have to be applied.

VERTISOLS

Derivation of name: from the Latin word *verto* = turn; connotative of turnover of surface soil.

General characteristics

Soils having, after the upper 20 cm have been mixed, 30 per cent or more clay in all horizons to a depth of at least 50 cm; developing cracks from the soil surface downward which at some period in most years (unless the soil is irrigated) are at least 1 cm wide to a depth of 50 cm; having one or more of the following: gilgai microrelief, intersecting slickensides, or wedge-shaped or parallelepiped structural aggregates at some depth between 25 and 100 cm from the surface. These are dark coloured soils having uniform fine or very fine texture and a low content of organic matter; but perhaps their most important property is the dominance in the clay fraction of expanding lattice clay usually montmorillonite which causes these soils to shrink and crack upon drying. Typically, they occur in arid and semiarid areas beneath tall grass or thorn forest.

There are two subdivisions of Vertisols:

Pellic Vertisols. These have moist chromas of less than 1.5 dominant in the soil matrix throughout the upper 30 cm.

Chromic Vertisols. These have moist chromas of 1.5 or more dominant in the soil matrix throughout the upper 30 cm.

The subdivisions and their equivalents in some other systems of classification are given in Table 6.27.

Table 6.27   Subdivisions of Vertisols and their equivalents in some other classifications

	Australia		Brazil	FitzPatrick	France	Germany	South Africa
	Handbook	Northcote					
Vertisols			Grumusols	Vertisols	Vertisols	Pelosols	

Pellic Vertisols	Rendzinas
	Black earths
	Wiesenboden
	No suitable
	group
		Ug5.1
		Ug5.1
		Ug5.1, Ug5.4
		Ug5.4
				[GrVe]
				[DeVe]
					Vertisols à
					   drainage externe
					   nul ou réduit
							Rensburg
							Arcadia

Chromic Vertisols	Grey clays
	Brown clays
	Red clays
	No suitable
	group
		Ug5.2, Ug5.5
		Ug5.2, Ug5.3
		Ug5.3
		Ug5.5
				[Gr(VeRo)]
				[Gr(VeFv)]
					Vertisols à
					   drainage externe
					   possible
							Rensburg
							Arcadia

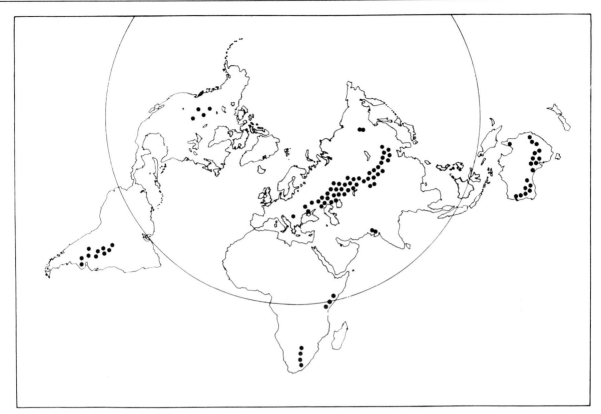

Fig. 6.53 Distribution of Solonetz

Morphology (see Plate IVd)

The surface may have a sparse litter but usually the bare, very dark coloured, fine texture soil is exposed. The uppermost mineral horizon is often thin with a marked granular structure. This is the so-called self-mulching surface (ochric epipedon; granulon — Gr)

USA		USSR
Old	New	
Grumosols	**Vertisols**	
	Pelluderts	Red and Grey Tropical soils
	Pellusterts	Regurs
	Pelloxererts	Vertisols
		Compact soils
	Chromuderts	
	Chromusterts	
	Chromoxererts	
	Torrerts	

that develops naturally due to wetting and drying leading to expansion and contraction. Beneath is the middle similarly dark coloured horizon with compound prismatic and wedge structure and may extend to >1 m (verton — Ve). Near the surface the structure is small wedge but with depth the wedge units become progressively larger with markedly slickensided surfaces (Fig. 4.31).

The colour of these horizons usually has hues of 2.5Y or 10YR with low values of 2 and 3, similarly the chromas seldom exceed 2. The consistence of these three horizons varies widely with moisture content, generally they are hard when dry, firm when moist, plastic and sticky when wet.

In thin section the middle horizon has a dominant anisotropic clayey matrix composed of abundant random fine domains, and frequently there are zones of domains some of which have an oblique orientation. Many of the peds have very thin anisotropic surfaces which are interpreted as slickensides with pressure orientated domains (Fig. 4.14). A characteristic feature of the matrix is the abundance of opaque grains of fine sand and silt size. These are presumed to be

275

composed of manganese dioxide.

With depth there is a gradual change into the underlying material or there may be an irregular mixed pattern or commonly there may be interdigitations. The underlying material is often a relatively unaltered sediment but it may be weathered rock and often it is extremely difficult to determine where the unaltered material begins. Concretions of calcium carbonate are often distributed through the soil and may form a thin layer on the surface. The type of concretion varies from horizon to horizon, usually near the surface they are small (2 to 5 mm) and black due to a coating of manganese dioxide (Fig. 6.54). At depth they are often soft and powdery.

500 μm

Fig. 6.54 Concretion of calcium carbonate in a Vertisol. It is composed mainly of small crystals of calcite and has an outer coating of manganese dioxide which also occurs as dendritic growths within the concretion

Analytical data
The content of clay in the soils is usually uniform throughout the pedo-unit being >35 per cent but in many cases it exceeds 80 per cent. Although the mineralogy of the clay fraction is somewhat variable, it is dominantly montmorillonite or mixed layered minerals which have a high capacity for expansion and contraction following wetting and drying giving

changes in volume of 25 to 50 per cent. In a number of situations mica and kaolinite may be present but the amounts of these two minerals are fairly small.

These soils have density of 1.8 to 2.0 in the middle horizon and therefore are more dense than most soils, probably resulting from repeated expansion and contraction which causes closer and closer packing.

The content of organic matter can be as much as 5 per cent at the surface but usually it is not greater than 1 to 2 per cent with a C/N ratio that is sometimes wide, but usually is 10 to 14.

As expected from the clay content and composition, the cation-exchange capacity is high and varies from 25 to 80 me per cent with a high degree of base saturation which is seldom less than 50 per cent, increasing with depth. A distinctive feature which is shared with some Phaeozems (Ustalfs) of the central USA is that for a given percentage of clay the cation-exchange capacity decreases as the content of organic matter increases. Most Vertisols contain free calcium carbonate in the form of powdery deposits or as concretions but many do not have this property; the content may be as high as 60 per cent, but usually varies from about 5 to 10 per cent. Exchangeable sodium is normally in the range of 5 to 10 per cent and therefore higher than for soils of humid areas but is much less than in saline or alkaline soils.

Generally, the salinity is low since salts seldom appear to accumulate in Vertisols and when they do it is usually below about 30 cm. Vertisols seem to have a self-flushing mechanism, any salts which accumulate on the surfaces of the peds in one season are washed into the lower part of the soil by the rains of the next season. These chemical properties combine to give Vertisols pH values that range from 6.0 to 8.5. However, pH values increase as the exchange complex becomes more saturated with sodium and therefore the soils are intergrading to Solonchaks or Solonetz.

Genesis
Some Vertisols have formed by progressive hydrolysis of the underlying rock, others have formed in fine textured sediments which either contain large amounts of expanding lattice clay or else montmorillonite has formed in these sediments.

The principal process taking place in these soils is the constant churning of the upper horizons. When the soil dries and cracks some of the surface horizon falls into the crack; consequently, when the soil becomes wet and expands, high pressures develop which are released by upward movement of the material. Annual repetition of this cycle results in churning of the soil down to the depth of cracking which is usually about 1 m, hence the relatively deep, uniform pedo-unit.

An additional result of the release of pressure is the differential displacement of the material, causing the formation of the wedge structure and slickensides, as a result of one part of the soil moving and slipping over another part. It appears that two of the important requirements for the formation of these soils are a period of complete saturation with water, and secondly a marked dry season. The period of complete saturation causes anaerobism and reducing conditions. On the other hand the marked dry season causes many of the basic cations to remain in the system thus producing suitable conditions for the formation of montmorillonite. In view of the small amount of organic matter it is difficult to determine the origin of the dark colour of these soils. Singh (1956) has suggested that it might be due to a dark coloured complex of organic matter and montmorillonite which forms in the wet environment when these soils are flooded during the wet period of the year. Thin sections of some Vertisols from several places show the matrix contains a large amount of finely divided opaque material which may be a secondary dark coloured mineral of iron or manganese which might form under anaerobic conditions. Therefore, the dark colour may be due to a combination of properties.

Hydrology

The water content in these soils varies from complete saturation to less than wilting point. With the onset of the dry season the soil dries, shrinks and cracks to give large prisms. Further drying takes place from the surface of the cracks and if there are any soluble salts in the soil they form an effloresence on the surface of the prisms. With the coming of the rainy season, the water from the first rains flows into the cracks and down the surface of the prisms. This dissolves and washes off any salt efflorescence and wets the soil from the bottom upwards. Ultimately, the whole soil becomes saturated with water and there is often free water on the surface.

Principal variations in the properties of the class

Parent material

Many Vertisols have developed in superficial deposits of fine or very fine texture; these are usually alluvium or lacustrine deposits, however, in some cases it is not possible to be certain about their origin. Some deposits appear to be colluvium formed by the insidious erosion

of soils and the accumulation of material in a depression on a flat site. Other Vertisols have developed through progressive weathering of the underlying rock which may be basalt, shale, limestone, or volcanic ash.

Their development is encouraged by a high content of plagioclase feldspars, ferromagnesian minerals and carbonates. Deficiencies of certain minerals in the parent material can sometimes be made good by seepage.

In a number of instances the presence of Vertisols is determined by the occurrence of a particular type of parent material. Usually they tend to become confined more and more to basic or carbonate parent material as the climate becomes more humid, the reason being that both of these materials produce large amounts of cations which maintain suitable conditions for the formation of montmorillonite. In some humid environments they are confined to sedimentary rock containing montmorillonite and carbonates.

Climate

Vertisols develop within the climatic range that includes Marine West Coast, Humid Continental, Tropical Wet-Dry, and Tropical Steppe. Their greatest extent seems to be in Tropical and Middle Latitude Desert and Steppe areas where leaching is at a minimum so that basic cations accumulate in the soil providing conditions for the formation of montmorillonite. Under these conditions precipitation varies from 250 to 750 mm and there is a marked dry season of 4 to 8 months.

Vegetation

Since the greatest extent of these soils occurs in tropical semiarid areas the predominant plant communities are dominated by tall grass, or acacia thorn forest. In some cases the grass communities are considered to be secondary after forest or are a fire climax. In humid environment, as in parts of Java, teak forests occur on these soils.

Topography

Vertisols develop mainly on flat or gently sloping sites, usually on terraces, plains and valley floors; occasionally they occur on low smooth crests but they never occur on slopes $>8°$. These soils are more common at elevations below 300 m but extensive areas in India and Kenya are above this altitude. Commonly, Vertisols have gilgai phenomena which form small but characteristic topographic features whose frequency varies from country to country. They are common in Australia and the USA (Texas) but are less common in India and South Africa (Figs. 3.3 to 3.6).

Age

Vertisols vary in age from Holocene to Pleistocene as indicated by the age of the material on which they are formed. However, where they are formed from the underlying rock they may date from the mid-Pleistocene or earlier but where they are formed in transported old soil material or other sediments they are likely to be of mid-Pleistocene age or later.

Pedo-unit

Some of the principal variations in this class are induced by climatic differences. In a humid environment there is a tendency for large amounts of moisture to pass through the soils so that the content of soluble salts and exchangeable sodium is low. Also, the uppermost horizon may have a platy or massive structure. Often the platy structures develop naturally but the massive structures are usually due to poor cultivation.

With progressive aridity, Vertisols tend to become browner and show a gradual increase in the amount of exchangeable cations and more particularly calcium carbonate which may form a calcic horizon (calcic horizon; calcon — Ck) or it may be fairly uniformly distributed throughout the soil in the form of concretions or as a fine powder. Gypsum may also occur in the lower part of Vertisols. These two horizons are often best developed in situations where the original material in the lower horizons was of a slightly coarser texture and therefore is not disrupted by churning. Some of the best examples are found where Vertisols have developed from rock so that there is a lower horizon of chemically weathered rock with little clay. It is in this layer that carbonate is deposited, thus indicating that some Vertisols may be polygenetic. With an increase in the amount of sodium, the expansion and contraction of these soils upon wetting and drying increase and the development of slickensides often increases in size and may also cause an increased frequency of gilgai.

The stone content of Vertisols is usually low but when the soils are shallow and formed from rock there may be an occasional stone in the pedo-unit but it is more usual for the stones to be carried to the surface by expansion and contraction where they may form a complete layer.

There are also a number of intergrading situations chiefly to Cambisols, Xerosols and Yermosols.

Distribution (see Fig. 6.55)

Vertisols occur mainly in arid, semiarid and Tropical Wet-Dry areas, but they also occur elsewhere making a total area of about 257 million hectares which is equivalent to the size of Western Europe with the largest areas occurring in Australia, India and the Sudan.

Utilisation

The extent to which these soils are utilised depends very much upon the development of local technology. Generally, in tropical areas, the natural grassy vegetation is grazed but this is often a tenuous practice because of water shortages, so that a measure of shifting husbandry has to be practised particularly if there are periods of drought.

Where arable agriculture is carried out moisture conservation is essential through the improvement of infiltration and reduction of losses by excessive evaporation and transpiration. If there is an adequate supply of good water, irrigation is normally practised; perhaps the classical example of this is the Gezira in the Sudan. On the other hand, where moisture is a serious limiting factor, then dry farming is practised.

The high content of clay in Vertisols can impose severe limitations on their utilisation because the moisture range for cultivation is narrow. If cultivation is attempted when they are not at the optimum moisture level, either puddling will result if they are too wet or they will be very intractable if they are too dry.

Generally, however, the level of utilisation is fairly primitive being usually at subsistence level, employing hand tools without the addition of fertilisers or irrigation. Therefore, it is not possible to be certain about the full potential of these soils but in some places such as Australia where a high level of technology is applied crop yields are high and a high standard of living results. Apart from grazing the main crop grown is cotton followed in importance by sugar cane and other grains such as sorghum, millet, rice, wheat and vegetables.

Vertisols are usually deficient in many of the major and micro plant nutrients; the amounts of nitrogen, phosphorus and potassium are low and have to be supplemented by the addition of fertilisers if high yields are required but the response is sometimes disappointing. However, the release of nutrients is usually sufficient to maintain a subsistence level of agriculture for crops like cassava and ground nuts.

Vertisols are highly susceptible to all forms of erosion, even slopes of 5% or less may develop deep gullies in a very short period. A characteristic form

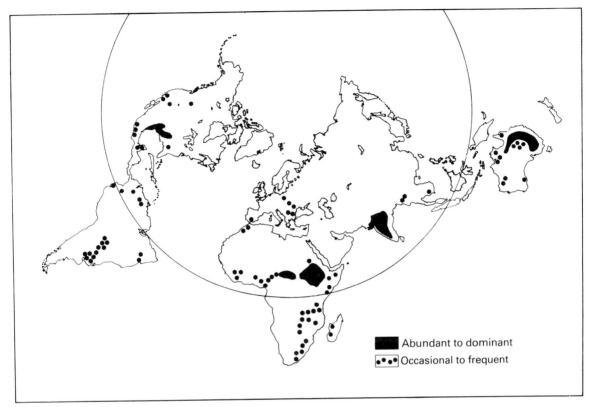

Fig. 6.55 Distribution of Vertisols

of erosion is landslides during which a large area will move as a unit down the slope as the upper horizons slide over the lower layers.

Finally, it can be said that this class of soil has the greatest need for improved utilisation since a tenfold increase in production can result by the application of modern technology.

XEROSOLS

Derivation of name: from the Greek *xeros* = dry; connotative of soils of dry areas.

General characteristics
Soils occurring under an aridic moisture regime; having a weak ochric A horizon and one or more of the following: a cambic B horizon, an argillic B horizon, a calcic horizon, a gypsic horizon; lacking other diagnostic horizons; lacking the characteristics which are diagnostic for Vertisols; lacking high salinity; lacking permafrost within 200 cm of the surface.

There are four subdivisions of Xerosols:

Haplic Xerosols. These have no diagnostic horizons other than a weak ochric A horizon and a cambic B horizon.

Calcic Xerosols. These have a calcic horizon within 125 cm of the surface.

Gypsic Xerosols. These have a gypsic horizon within 125 cm of the surface.

Luvic Xerosols. These have an argillic B horizon; a calcic or a gypsic horizon may be present.

The subdivisions and their equivalents in some other systems of classification are given in Table 6.28.

Morphology
The upper horizon is often a loamy calcareous weak ochric A horizon which varies in colour from yellowish-red to greyish-brown and is about 15 to 20 cm in thickness but can be much thinner. It has a thin crust at the surface and is usually firm or friable with a platy or massive structure and grades quickly into a calcic horizon up to a metre in thickness followed **279**

Table 6.28 Subdivisions of Xerosols and their equivalents in some other classifications

	Australia		FitzPatrick	France	Kubiëna
	Handbook	Northcote			
Xerosols			**Serozems**	**Sierozems**	**Sierozems**
Haplic Xerosols	No suitable group, Terra Rossa soils	Um5.2, Um6.2 Um6.2	[Sn-Azl]	Sols bruns subarides	
Calcic Xerosols	Grey-brown and red calcareous soils, No suitable group, Terra Rossa soils	Um5.1, Um5.6 Um6.2 Um6.2	[SnCk]		
Gypsic Xerosol Luvic Xerosol			[SnGy] [SnAr]	Sols gypseux	

by a gypsic or salic horizon which may be over 3 m in thickness. In many cases there is a slight increase in clay beneath the ochric A horizon to give an argillic B horizon.

In thin section the ochric A horizon has frequent small and medium circular pores and there are usually numerous faunal passages and faecal pellets. The thin section morphology of the calcic or gypsic horizon is similar to that described on pages 197 and 270.

Analytical data

The ochric A horizon usually contains a smaller amount of clay than the underlying horizon and may often contain less silt and fine sand but greater amounts of coarse sand and gravel. The content of organic matter varies from 1 to 2 per cent; the C/N is distinctive being less than 10 and often less than 8. When the parent material contains carbonates the content may exceed 10 per cent increasing to a maximum of over 12 to 15 per cent in the calcic

Table 6.29 Subdivisions of Yermosols and their equivalents in some other classifications

	Australia		FitzPatrick	France	Kubiëna
	Handbook	Northcote			
Yermosols			**Yermosols**		**Serozems**
Haplic Yermosols	Red and brown harapan soils Lithosols Red earths No suitable group	Um5.3 Um5.5 Um5.5 Um5.5, Uf6.7	[Sn-Azl]	Sols bruts xériques inorganisés Sols peu évolués xériques	
Calcic Yermosols			[SnCk]		
Gypsic Yermosols			[SnGy] [SnAr]	Sols gypseux	
Luvic Yermosols	Red earths Calcareous red earths Yellow earths Yellow Podzolic soils	Gn2.1 Gn2.1 Gn2.2, Gn2.3, Gn2.6, Gn2.7 Gn2.3, Gn2.7			
Takyric Yermosols					

South Africa	USA		USSR
	Old	New	
	Desert soils	Typic Aridisols	Semidesert soils, Sierozems
Clovelly Hutton Oakleaf Glenrosa		Camborthids Durorthids	
Oakleaf Mispah		Calciorthids	
		Calciorthids Haplargids Durargids	

Genesis

Most of the processes taking place in these soils are operating very slowly because there is only a small amount of water passing through the system. However, the small amount of organic matter contributed by the sparse vegetation is quickly humified. The presence of many faunal passages and faecal pellets does not indicate a high level of biological activity for features once formed tend to become preserved in the hot, dry environment.

The easily soluble salts are removed from the upper part of the soil and deposited between 1 to 5 m from the surface and small amounts of carbonate are removed from the ochric A horizon and deposited below to form the calcic horizon.

Because of the often thin vegetative cover the surface of the soil is subject to deflation and rapid run-off, both processes causing a removal of the fine material resulting in concentration of gravel at the surface. However, some of the textural difference may be enhanced by weathering *in situ*.

horizon then decreasing in the parent material. Consequently, the pH values vary from 7.0 to 8.0 at the surface to a maximum of about 8.5 in the calcic horizon.

The cation-exchange capacity varies between 10 and 25 me per cent in the ochric A horizon decreasing with depth to about 8 to 10 me per cent and there is complete saturation of the exchange complex with calcium being the dominant ion.

Hydrology

Since these soils occur under semiarid conditions only a small amount of water enters the soil so that little leaching takes place. The most soluble salts are removed to the lower horizons but some tend to come to the surface during the dry part of the year due to the very high rate of evapotranspiration but the absolute content is always low to very low.

Principal variations in the properties of the class

Parent material

The most common parent materials are sediments of Pleistocene or Holocene age. In Europe and the Americas it is often loess or alluvium but in old landscapes it may be a pedi-sediment.

In some cases Xerosols evolve from a previous soil, the most dramatic examples of this occur in the semi-arid areas of Africa and Australia where they are formed in Tertiary soils.

Climate

These soils are restricted to the arid areas of the world occurring in both mid-latitude and tropical areas.

Vegetation

Although most of the plant species growing in these soils have adaptations which permit them to grow in arid areas, the character of the communities is very different. In Eurasia, annuals including grasses and

South Africa	USA		USSR
	Old	New	
	Desert soils, Red desert soils, Grey desert soils	Typic Aridosols	Desert soils
Swartland Valsriver Clovelly Hutton Oakleaf		Camborthids Durorthids	
		Calciorthids	
Oakleaf		Gypsiorthids	
		Argids	
			Takyr

succulents are the principal species but in the Americas various cacti are usually dominant. In tropical and subtropical areas of Africa and Australia there is scrub woodland dominated by species of *Acacia.*

Topography
These soils tend to be restricted to flat or gently undulating situations. On moderate or steep slopes erosion reduces the development of soils so that Lithosols, Regosols and Rankers are the main formations.

Age
Many of these soils started their development at the end of the Pleistocene Period but those of the tropical and subtropical areas may have started to form at a considerably earlier phase of the Pleistocene or even in the Tertiary Period.

Pedo-unit
There is a considerable variation in the properties of this class as determined by differences in parent material, climate and by polygenesis. Where these soils are formed in late Pleistocene sediment, variations in the pedo-units are determined largely by the presence or absence of carbonates, for when carbonates were present in large amounts the soils are calcareous throughout and a calcic horizon has formed. In the absence of carbonate the soils are usually mildly to moderately acid at the surface. This is most marked in areas of old soils.

The colour of Xerosols can vary from brown to red, the former colour occurs mainly in the mid-latitude arid areas whereas in tropical and subtropical areas they are red and may have hues of 2.5YR or 10R. Although goethite is the principal iron oxide in these redder soils, haematite is usually present in small amounts. In some cases the red coloration is inherited from a previous soil cover but reddening seems to be a characteristic feature of these warmer areas. In spite of the dry conditions some weathering does take place in Xerosols and usually results in the formation of a cambic B horizon. Xerosols have continuous gradational sequences to other soils including Luvisols, Kastanozems, Solonetz and Solonchaks.

Where deflation is a vigorous process the gravel and stones increase in frequency to form a hamada which varies very widely in character depending upon local conditions. It may be composed of stones and boulders if the parent material was a sediment or there may be fragments of weathered rock when the soil is developed in an older soil. However, in many tropical and subtropical areas fragments of hardened laterite frequently form the hamada.

Distribution
Xerosols are confined to the arid parts of the world. They are common in southern USSR, in a belt on either side of the Sahara Desert, in parts of Iran, west Pakistan, Afghanistan and Australia.

Utilisation
Under natural conditions rough grazing for cattle is the only form of land use that is practised and even this is a somewhat tenuous system because of the uncertainty of an adequate supply of water for the animals. These soils usually prove to be highly fertile if irrigated but this can be made difficult or impossible through the lack of water. Usually there are very few rivers in areas of Xerosols and the artesian water often has a high content of salts making it unsuitable for irrigation or domestic purposes. However, there are a number of places where rivers flowing through these areas supply suitable water. In some other areas very large installations are used to convey water over long distances. Perhaps the most complex is found in the state of Colorado in the USA where water is pumped over the mountains to irrigate the soils in the valleys on the other side.

YERMOSOLS

Derivation of name: from the Spanish word *yermo* = desert; connotative of very dry areas.

General characteristics
Soils occurring under an aridic moisture regime; having a very weak ochric A horizon and one or more of the following: a cambic B horizon, an argillic B horizon, a calcic horizon, a gypsic horizon; lacking other diagnostic horizons; lacking the characteristics which are diagnostic for Vertisols; lacking high salinity; lacking permafrost within 200 cm of the surface. Where they occur in old landscapes the present Yermosol stage has often been superimposed upon an extremely wide range of previous soils which vary from weakly to extremely strongly weathered. Probably this is best seen in the arid areas of Western Australia and South Africa where many of the red Yermosols started their development under much more humid conditions.

Perhaps it should be pointed out that Yermosols are differentiated from Xerosols by the nature of the ochric A horizon. In the Xerosols it is *weak* whereas in the Yermosols it is *very weak*.

In the future designating of Yermosols, more attention should be given to show those that are polygenetic.

These soils occupy extremely large areas but are of little value for agriculture except when irrigation is possible. The sparse vegetation that grows on them is grazed during the rainy season.

There are five subdivisions of Yermosols:

Haplic Yermosols. These have no diagnostic horizons other than a very weak ochric A horizon and a cambic B horizon.

Calcic Yermosols. These have a calcic horizon within 125 cm of the surface.

Gypsic Yermosols. These have a gypsic horizon within 125 cm of the surface.

Luvic Yermosols. These have an argillic B horizon, a calcic or a gypsic horizon may be present.

Takyric Yermosols. These have takyric features.

The subdivisions and their equivalents in some other systems of classification are given in Table 6.29.

7

Soil relationships

The various relationships can conveniently be considered under the following headings:
 Spatial relationships
 Relationships with parent material
 Relationships with climate
 Relationships with organisms
 Relationships with topography
 Relationships with time
 Horizon relationships
 Property relationships

SPATIAL RELATIONSHIPS

The distribution of soils on the surface of the earth and their relationships one to another are some of the chief interests of soil surveyors whose major concern is to present their distribution pattern in the form of maps. Since soils usually grade gradually from one to another, the ideal method of investigation is to examine exposed transects. Usually this is not possible because of the immense amount of labour and time involved in such operations but some investigations of this type have been conducted when large sections are exposed by road works, pipe laying or similar operations. In a few cases individual investigators have produced large continuous sections but this is exceptional (Mattson and Lönnemark, 1939). Normally, pedo-units are studied at selected points in the landscape, chosen subjectively or at random. The subjective choice is the usual method but some investigators are now planning their operations on an objective basis choosing their sites either randomly or by using a fixed grid. Both of these methods can yield useful information which can be analysed by computer techniques. They suffer from serious disadvantages which are induced by the limitations of the time at the disposal of the operator. If one adheres strictly to statistical methods the number of samples required to give valid results about small areas of a particular soil are much greater than the time at the disposal of the surveyor. Therefore, the error for such areas may be very large and since some of these small areas may be important intergrading situations they may be missed during this type of investigation. On the other hand with a free survey the choice of sites is left to the discretion of the surveyor and the frequency of inspections is varied according to the variability of the soil.

In most cases the spatial relationships are extrapolation based on these points; therefore they are open to error. Thus sites must be sufficiently near to reduce the error to a minimum and at the same time they must not be so close that valuable time is wasted. Probably the most expedient method of investigation is by means of detailed surveys along random line transects.

For practical purposes the areas of relative uniformity are shown on maps where they are delimited by lines which are placed where the rate of change is most rapid. In reality, however, one area gradually merges into another and it is common to find one area of soil merging into two or three others which may themselves merge into other soils or back into a similar soil at some other point forming a repeating pattern over the landscape. When this takes place the pattern is often related to the parent material or topography. Further, it is common for the areas of individual soil to have similar shapes which prompted Hole (1953) to suggest a terminology for describing soils as three-dimensional bodies.

Finally, it should be pointed out that the areas of apparent uniformity shown on maps may have a wide range of variability and in many cases may have small areas or inclusions of different soils which cannot be shown on the scale of the map in use. Thus, it is very difficult to portray accurately the spatial distribution of soils but three examples are given below. Figure 7.1 is an extract from a much larger soil map. It shows the distribution of the soils and the spatial relationship between the various soils. The distribution pattern is determined largely by slope and elevation. The Podzols (Orthods, Podzols) occur above 500 m on a plateau surface. On the slopes leading down from the plateau there are Cambisols (Ochrepts, Altosols),

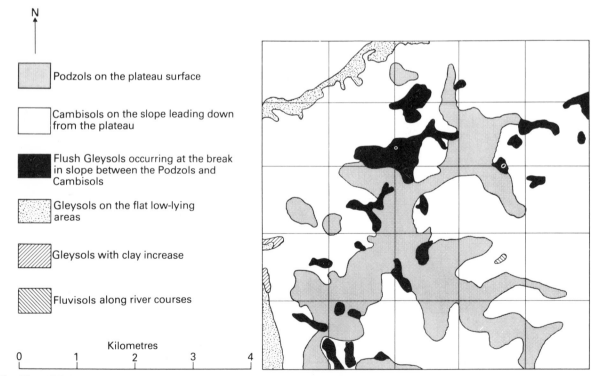

N

Podzols on the plateau surface

Cambisols on the slope leading down from the plateau

Flush Gleysols occurring at the break in slope between the Podzols and Cambisols

Gleysols on the flat low-lying areas

Gleysols with clay increase

Fluvisols along river courses

Kilometres

0   1   2   3   4

284   **Fig. 7.1** Soil map of the area west of Church Stretton in Central England

while flush Gleysols (Aquepts, Subgleysols) occur mainly at the break of slope between the Podzols and Cambisols. In the flat low-lying areas there are Gleyic Luvisols (Aqualfs, Supragleysols). Finally, Fluvisols (Fluvents, Fluvisols) occur along the river courses at the lowest positions in the landscape.

The aerial photograph given in Fig. 7.2 is of a salt marsh and reclaimed land in Norfolk, England, where different types of vegetation are strongly correlated with different soils. Salt marshes are generally areas of silt accumulation which can be reclaimed by constructing a bank and a drain on the landward side of the bank, the material dug out of the drain being used to make the bank. The enclosed creek system is

then rationalised, a few larger creeks are maintained as major drainage channels and the small ones filled in. The distinctive pattern on the marsh is due to differences in the distribution of the natural vegetation, creek pattern and reclamation. Figure 7.3

Fig. 7.2 Salt marsh and reclaimed land in Norfolk
  A) Dominated by the light coloured cord grass with sea poa as the dark areas
  B) Dominantly sea poa
  C) Clearly defined well-drained areas adjacent to the creeks and dominated by sea-couch
  D) The bank
  E) The drain
  F) The pattern of infilled creeks where crop-growth is poor

illustrates the relationship between tone pattern and soil distribution pattern. The underlying rock is chalk with a thin cover of drift. The light areas have Rendzinas and Calcaric Cambisols less than 40 cm deep. Some Calcaric Cambisols and Eutric Cambisols, generally deeper than 50 cm, occur where the drift is deeper and has the darker tones, mainly on the valley floors.

RELATIONSHIPS WITH PARENT MATERIALS

The influence of parent material is expressed through its composition, permeability and surface area of the particles, and is usually most strongly manifest in the initial stages of soil formation. Ultimately, soil forming processes tend to minimise and may even

Fig. 7.3 Relationship between tone pattern and soil distribution pattern

nullify the influence of parent material. For example, in areas with intense and prolonged hydrolysis and solution, most rocks are transformed into clay minerals, oxides and resistant residues. Thus, the chemical composition of the soil in these cases bears little resemblance to that of the original rock. Often this is important in tropical areas where nutrient rich rocks such as basalts and limestone are transformed to almost sterile soils. However, there can be important differences among such soils particularly those that develop from rocks which vary widely in their original composition. This is illustrated in the first example given below. Thus, we find that nearly every type of parent material gives different soils as determined by the variability in the other factors particularly climate. In cool humid areas basalt and more particularly drift deposits derived from basalt often give rise to Cambisols (Eutrochrepts; Altosols), whereas under hot tropical conditions Ferralsols (Orthox; Krasnozems) will form. In a semiarid environment it is normal to find Vertisols developed from basalt. Under similar variations in climate granite will give rise respectively to Podzols (Ortox; Podzols), Ferralsols (Orthods; Krasnozems) and Yermosols (Calciorthids; Yermosols).

Limestone often gives rise to very distinctive soils because of its unique chemical composition. In hot humid areas the early stages in the development of soils on limestone are often Rendzinas (Rendolls; Rendzinas). With time Ferric Cambisols (Drystrochrepts; Flavosols) may form and ultimately Ferralsols (Orthox; Krasnozems) will develop. Under Mediterranean conditions the early stages are the same but the final stages are usually Eutric Nitosols (Rhodudults; Rossosols). Perhaps some of the most interesting parent material–climate relationships occur in the loess belts of the USSR and North America as given on pages 295 and 296).

In parts of Queensland in Australia basalt and fine grained sandstone lie close to each other and both have been deeply weathered. The basalt soils have been inhabited by large earthworms that have formed a marked vermicular structure to a depth of several metres. On the other hand the sandstone has been avoided by the worms so that the original rock structure is present to within 25 cm of the surface but it is very soft and easily dug with a spade to a depth of 2 m.

Probably it is in areas of Pleistocene sediments that the chemical composition of the parent material has its strongest influence on soils. In a present-day Marine West Coast or Humid Continental environment the occurrence of Cambisols or Podzols (Ochrepts or Orthods, Altosols or Podzols) is determined in large measure by the composition of the sediments which are mainly till, solifluction deposits and loess. Such a relationship is described below.

In a Marine West Coast or Humid Continental environment, the permeability of the parent material can be an important factor; on fine textured glacial drift or lacustrine deposits the vertical movement of moisture is restricted so that Podzoluvisols (Glossaqualfs; Supragleysols) develop. This relationship is the third example described below. One other relationship demonstrating an influence of limestone is given and finally the relationship between geological structure, topography, and soil is illustrated in the fifth example.

1. Ferralsol — Acrisol — Podzol relationship

(Orthox — Ustult — Humod; Krasnozem — Luvosol — Podzol)

In a rainy tropical environment as in certain parts of South-East Asia, fine grained basic rocks including basalt and andesite normally give rise to fine textured Rhodic Ferralsols (Orthox; Krasnozems) intermediate and acid rocks may result in medium textured Ferric Acrisols (Paleustults; Luvosols) while Humic Podzols (Humods; Podzols) develop on Pleistocene deposits which are mainly deep quartzose sands of raised beaches. This is shown diagrammatically in Fig. 7.4.

2. Drystic Cambisol — Orthic Podzol — Placic Podzol — Histosol relationship

(Dystrochrept — Orthod — Placorthod — Hemist; Altosol — Podzol — Placosol — Peat)

These soils are common in many areas with a Marine West Coast climate, where their distribution is often determined by the nature of the parent material. Figure 7.5 shows the hypothetical relationship in which the character of the drift (parent material) changes gradually from basic on the left to acid on the right and where there is an accompanying change in the soils. A Dystric Cambisol (Dystrochrept; Altosol) occurs on the basic drift while an Orthic Podzol (Orthod; Podzol) occurs on the acid drift. As the Cambisol grades towards the Podzol the surface organic horizons increase in thickness and gradually become pronounced. The Ah becomes darker and has a higher content of organic matter. Gradually a bleached layer develops, first as small lenses but it is quite prominent in the Podzol. The most significant

287

Rhodic Ferralsol
Orthox
Krasnozem

Ferric Acrisol
Ustult
Luvosol

Humic Podzol
Humod
Podzol

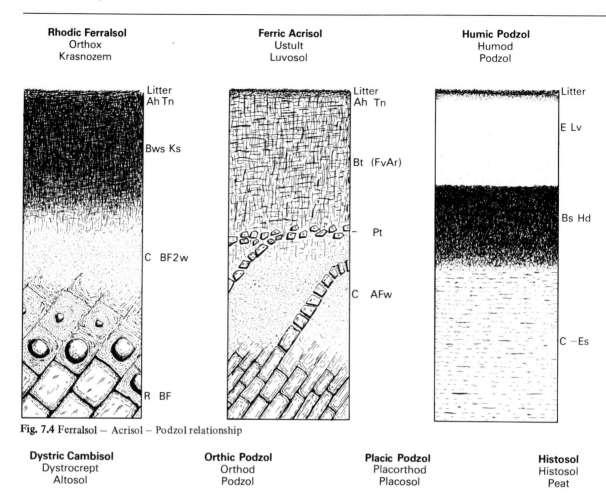

Litter
Ah Tn

Bws Ks

C BF2w

R BF

Litter
Ah Tn

Bt (FvAr)

Pt

C AFw

Litter

E Lv

Bs Hd

C —Es

Fig. 7.4 Ferralsol — Acrisol — Podzol relationship

Dystric Cambisol
Dystrocrept
Altosol

Orthic Podzol
Orthod
Podzol

Placic Podzol
Placorthod
Placosol

Histosol
Histosol
Peat

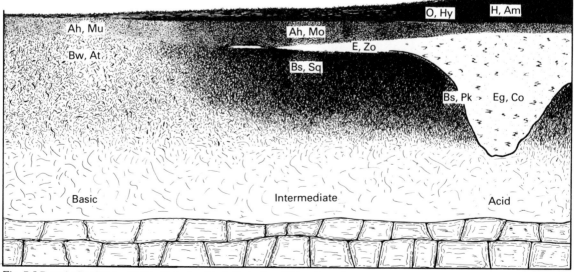

O, Hy

H, Am

Ah, Mu

Ah, Mo

E, Zo

Bw, At

Bs, Sq

Bs, Pk →   Eg, Co

Basic

Intermediate

Acid

288   **Fig. 7.5** Dystric Cambisol — Orthic Podzol — Placic Podzol — Histosol relationship

change takes place in the middle horizon where the Bw (cambic horizion; alton — At) gradually changes into a Bs (spodic horizon; sesquon — Sq) but in some respects the further change to a Placic Podzol (Placorthod; Placosol) is more dramatic. The thin iron pan (placic horizon; placon — Pk) develops at some position in the soil and is quickly followed by the formation of an overlying Eg (albic horizon; candon — Co) because of increased wetness caused by the low permeability of the thin iron pan. Where rainfall is sufficiently high, usually over 750 mm, the organic layers develop into a histic H horizon.

These changes in morphology are accompanied by a number of changes in the physical and chemical properties. The natural vegetation associated with these soils varies from deciduous forest on the Cambisol through coniferous forest on the Podzol to heath on the Placic Podzol and Histosol.

There is a certain amount of variation in this relationship particularly due to the wide differences in precipitation found in a Marine West Coast climate. Where the precipitation is fairly low the relationship can shift towards the right with Cambisols forming on acid parent material. Where the rainfall is high there is a shift towards the left, so that Podzols may develop on intermediate or basic parent material.

A similar relationship can develop through time for there are many situations where Cambisols changed into Podzols through progressive leaching. Sometimes this process is accelerated by a change of vegetation from deciduous to coniferous species or by an increase in precipitation.

3. Orthic Luvisol — Dystric Podzoluvisol — Gleyic Podzoluvisol — Histosol relationship

(Hapludalf — Glossudalf — Glossaqualf — Hemist; Argillosol — Supragleysol — Peat)

Orthic Luvisols usually develop in a moist environment and can be changed quickly into Podzoluvisols either because of an increase in rainfall or due to a slight change in topography. Figure 7.6 shows the sequence of soil that develops when there is an increase in the surface-moisture status of Luvisols due to a change in topography. At the left of the diagram is an Orthic Luvisol in which the first change to the Dystric Podzoluvisol is marked by the occurrence of mottling to form an Eg (albic horizon; candon — Co) immediately beneath the upper mineral horizon which is usually an umbric A horizon (umbric horizon; mullon — Mu). This is often accompanied by the formation of manganiferous concretions in the upper

part of the argillic B horizon (argillic horizon; argillon — Ar) as well as in the Eg itself.

This stage is a Podzoluvisol which has tongues extending into the underlying horizons to form the Btg (argillic horizon; glosson — Gs) as shown in the diagram. With an increase in wetness the Eg increases in thickness and gradually changes into a Bg (cambic horizon; gleyson — Gl). This, in turn changes into a Cr (cambic horizon; cerulon — Cu) which is usually underlain by a relatively unaltered material. In cases of extreme wetness, the main upper horizon is a histic H horizon (histic epipedon; peat).

In addition to being found in the same locality due to changes in topography, the various stages of this sequence can be found at points distant from each other owing to differences in precipitation. Podzoluvisols gradually develop into Planosols (Albaqualfs; Planosols) as hydrolysis increases the clay content of the middle horizons.

4. Luvisol — Cambisol — Rendzina relationship

(Alfisol — Inceptisol — Rendoll; Argillosol — Altosol — Rendzina)

Avery *et al.* (1959), have described a sequence of soils developed on chalk in the Chiltern Hills of southern England. The chalk was strongly weathered during the Tertiary Period to produce a landscape with flat plateaux and steep sided valleys (Fig. 7.7). These soils were possibly Ferralic Cambisols (Tropepts; Flavosols) and are now referred to as clay with flints. During the Pleistocene this area lay outside the limits of glaciation, but within the area subjected to repeated periglacial processes, which caused solifluction of the Tertiary soil down the slopes and in many cases completely removed it to expose the underlying chalk. The resulting solifluction deposits which accumulated in the valleys contain a mixture of the old weathered material and chalk. Superimposed upon the whole landscape is a varying thickness of loess which must have been deposited during or prior to the solifluction since much of it is incorporated into the solifluction deposits. Thus, at the end of the Pleistocene Period, the flat or gently sloping plateaux and valleys were covered with the truncated remnants of the Tertiary soils and loess. The plateau edges and upper slopes had exposed chalk and in the valleys there was a solifluction mixture of the old weathered material, loess or chalk. During the Holocene Period soil formation under a Marine West Coast climate produced Orthic Luvisols (Hapludalfs; Argillosols) in the loess overlying

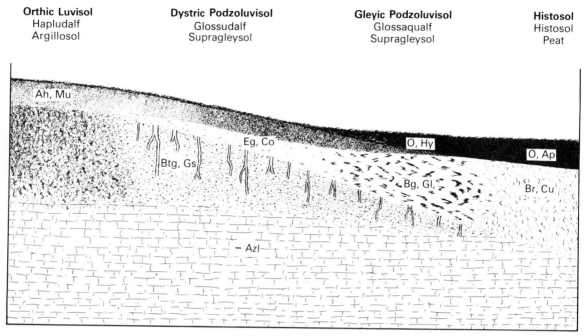

Orthic Luvisol	**Dystric Podzoluvisol**	**Gleyic Podzoluvisol**	**Histosol**
Hapludalf	Glossudalf	Glossaqualf	Histosol
Argillosol	Supragleysol	Supragleysol	Peat

Fig. 7.6 Orthic Luvisol – Dystric Podzoluvisol – Gleyic Podzoluvisol – Histosol relationship

the weathered material, Rendzinas (Rendolls; Rendzinas) on the Chalk and Eutric Cambisols, (Eutrochrepts; Altosols) on the calcareous solifluction deposits. Locally among the Orthic Luvisols where the material is more sandy, Podzols (Humods; Podzols) have developed. The Luvisols are possibly the most interesting class of soils in this area since their horizon sequence is influenced by the nature of the material. Where the loess is thick the horizon sequence is fairly normal, but where it is thin, part of the sequence is formed in the loess and part in the underlying clay with flints. Similar relationships also occur in south-western Germany.

5. Relationships between geological structure and soil pattern

An interesting sequence of soils has been described by Mackney and Burnham (1966) for a part of central England where moderately dipping interbedded strata of limestones, shales and silt stones have produced a scarp landscape with a succession of soils that is related to topography as well as to rock type. This relationship is shown in Fig. 7.8. The softer shales were weathered and differentially eroded during the Tertiary Period leaving the harder limestones and silt-stones which due to their dip have produced two escarpments and long, continuous dip slopes. During

the Pleistocene Period the area was not strongly moulded by erosion or deposition. The valley floors have a covering of till and solifluction deposits but the latter are more widespread on the scarp and dip slopes where they are derived largely from the under-lying material. Therefore, the parent materials of the soils are strongly influenced by the underlying rock in the upland situation but, in the valleys mainly glacial drift occurs. Dystric Cambisols (Dystrochrepts; Altosols) are developed on the dip slopes and colluvial slopes of the relatively coarse siltstones. Orthic Luvisols (Hapludalfs; Argillosols) are found mainly on the dip slope materials and in colluvium over limestones and shales. Towards the bottom of the slope and on flatter sites the Orthic Luvisols inter-grade to Gleyic Luvisols which occupy the lower slope position and the floors of the valleys. Thus, the resulting soil pattern is determined by the parent material as well as by the slope. A further character-istic of this relationship is the spasmodic occurrence of Rendzinas (Rendolls; Rendzinas) on the limestone escarpment.

RELATIONSHIPS WITH CLIMATE

Perhaps the most obvious and striking relationships are those that exist between climate and soils.

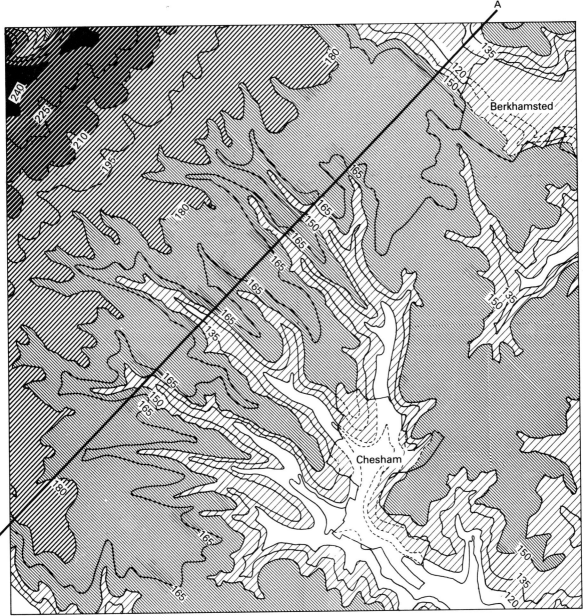

Fig. 7.7(A) Topographic relationship between Luvisols, Cambisols and Rendzinas:

   A) topographic map;
   B) soil map;
   C) transect from A to B showing the soils in relation to
       the topography

Frequently, the influence of climate is so strong that many authors consider it as the dominant factor of soil formation, but in any one climatic area there are many soils that owe their principal characteristics to other factors, therefore this principle can be applied only in a very general way on the macro or continental scale. Nevertheless, climate can be used as a suitable basis for showing geographical distribution patterns and certain relationships between soils, bearing in mind the wide climatic changes that have occurred during the last two million years. Set out in Table 7.1 are the climatic regions according to Strahler (1970) and a frequency estimation of the principal soils that occur in each of these areas. The estimates should not be treated as absolute because of the general dearth of survey information.

291

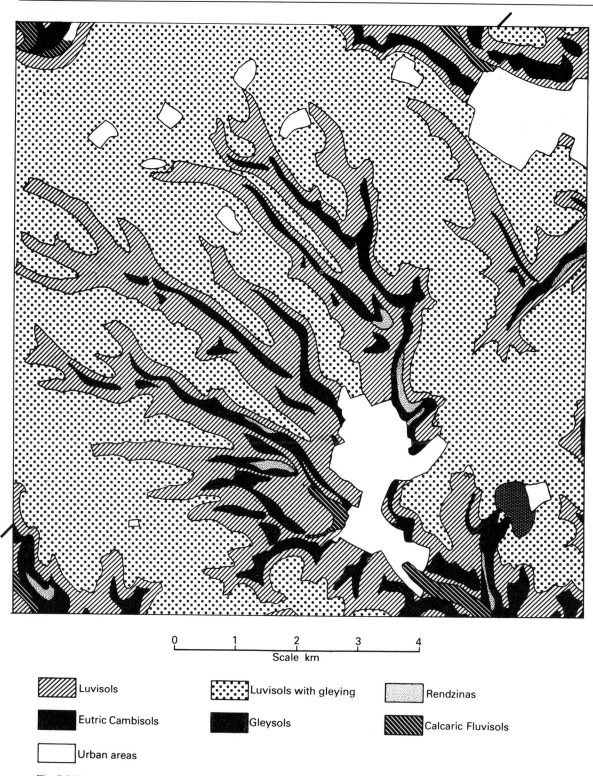

Scale km

Luvisols

Luvisols with gleying

Rendzinas

Eutric Cambisols

Gleysols

Calcaric Fluvisols

Urban areas

Fig. 7.7(B)

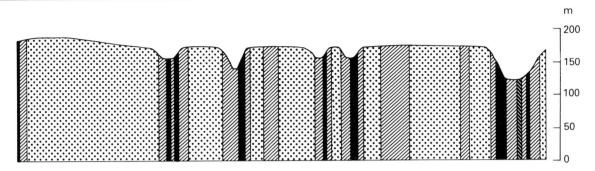

Fig. 7.7(C)

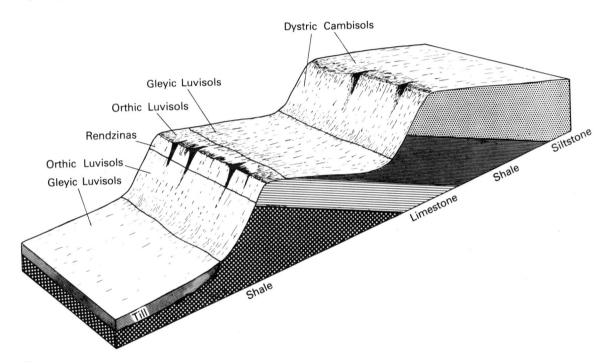

Dystric Cambisols

Gleyic Luvisols

Orthic Luvisols

Rendzinas

Orthic Luvisols

Gleyic Luvisols

Siltstone

Shale

Limestone

Shale

Till

Fig. 7.8 A relationship between geological structure and soil pattern

The second soil–climate relationship that is given is an example of the latitudinal changes in soils and climate through continental land masses.

Latitudinal zonation of soils

Possibly the best example of latitudinal change of soil with climate is seen in the great continental transects that stretch from Novaya Zemlya in the north of the USSR to the Caspian Sea in the south and from the centre of the Sahara Desert to Zaire. These sequences of soils together with their characteristic types of vegetation and climate are shown in Fig. 7.9.

At the left of the diagram there are Gelic Gleysols

(Pergelic Cryaquepts; Cryosols) of the northern latitudes with tundra vegetation, tundra climate, low rainfall and prolonged periods of low temperatures. This zone grades into a moister belt with Humic Gleysols (Cryaquepts; Subgleysols) and Histosols (Histosols; Peat) where larches predominate and there is an increase in rainfall and mean annual temperature. The latter is sufficient to increase the annual growth of vegetation but does not augment evapotranspiration sufficiently to cope with the extra precipitation; hence the wetter conditions. Next comes the area dominated by Podzols (Spodosols; Podzols) carrying a cover of coniferous vegetation, where there is a further increase in precipitation and mean annual temperature. Here

293

Table 7.1 The frequency distribution of the major soils in the main climatic areas. Soil names according to FAO, USDA and FitzPatrick. +++ Abundant > 25%, ++ Frequent 5 to 25%, + Rare < 5%

FAO	Wet Equatorial	Trade Wind Littoral	Tropical Steppe	Tropical Desert	West Coast Desert	Tropical Wet-Dry	Humid Subtropical	Marine West Coast	Mediterranean	Middle Latitude Steppe	Middle Latitude Desert	Humid Continental	High Latitude	Tundra	Highland	USDA	FitzPatrick
Acrisols	++	++				++	+++									Ultisols	Luvosols
Andosols	+	+				+	+	+	+			+	+		+	Andepts	Andosols
Arenosols	+	+	++	+		+	+									Psamments	Arenosols
Cambisols	+	+				+	+	++	+			+	+	+	+	Ochrepts	Altosols
Ferric C.	+	+				+	++		+						+	Tropepts	Rossosols
Chernozems										++						Borolls	Chernozems
Luvisols	++	++	+			+++	++		+							Alfisols	Cumulosols
Ferralsols	+++	+++				++	+									Orthox	Krasnozems
Fluvisols	+	+	+			+	+	+	+	+		+	+	+	+	Fluvents	Fluvisols
Thionic F.	+	+				+	+	+	+							Sulfaquepts	Thiosols
Gleysols	+	+				+	+	++	+			++	++		+	Aquents	Subgleysols
Gelic G.														+++		Pergelic Cryaquepts	Cryosols
Greyzems																Argiborolls	Greyzems
Histosols	+	+				+	++	+				++	+++	++	+	Histosols	Peat
Ironstone	+	+	++	+		++	+		+							Ironstone	Vesosols
Kastanozems										++						Ustolls	Kastanozems
Lithosols	+	+	++	++	++	+	+	+	+	+		+	++	++	+++	Lithic Subgroups	Lithosols
Luvisols	+	+				++	+	+	++	+		++				Alfisols	Argillosols
Nitosols	++	++				++	+		+							Udalfs	Rossosols
Phaeozems										++		+			+	Udolls	Brunizems
Planosols			+			+	+		++	+		+				Aqualfs	Planosols
Solodic P.			++							+						Natraqualfs	Solods
Podzols	+	+				+	+	+++	+			++	+++		++	Spodosols	Podzols
Placic P.								++					++		+	Placorthods	Placosols
Podzoluvisols							++	+				+	+			Glossudalfs	Supragleysols
Rankers	+	+				+	+	+	+	+		+	+	+	++	Lithic Haplumbrepts	Rankers
Regosols	+	+	+	+	+	+	+	+	+	+	+	+	+	+		Orthents	Regosols
Rendzinas	+	+				+	+	+				+	+		+	Rendolls	Rendzinas
Solonchaks			++	+						++	+		+			Salorthids	Solonchaks
Solonetz			++							++						Natrustalfs	Solonetz
Spherosols			+			+	++									Oxisols	Spherosols
Vertisols			+	+++		+	+	+	+	+	+	+				Vertisols	Vertisols
Yermosols			+++	+++	+++					++	+++					Typic Aridisols	Yermosols
Xerosols			+++	+	+					++	+					Mollic Camborthids	Serozems

294

evapotranspiration is relatively high so that accumulations of moisture take place only in depressions where Gleysols (Aquepts; Subgleysols) develop as part of a hydrologic sequence (see pp. 29 and 298).

Continuing the transect, precipitation is maintained but there is an increase in temperature so that Podzols give way to a complex of Luvisols (Alfisols; Argillosols) and Podzoluvisols (Aqualfs; Supragleysols) under deciduous forest. Then comes a narrow belt of Phaeozems (Hapludolls; Brunizems) in the tension zone between deciduous forest and the steppes. Further to the south there are Chernozems (Vermudolls; Chernozems) and with increasing aridity are found Haplic Kastanozems (Haplustolls; Kastanozems) and Luvic Kastanozems (Ustalfs; Burozems) and Xerosols (Mollic Aridisols; Serozems) and finally Yermosols (Typic Aridisols; Yermosols). Similarly, the vegetation changes from tall grass to short grass then to bunchy grass and finally to xerophytically modified succulents on the Yermosols. The vegetation changes result from a steady decrease in precipitation accompanied by an equally steady increase in mean annual temperature and increased transmission of radiant energy through the atmosphere, causing an increase in evapotranspiration. The characteristic feature of this part of the transect is the steady decline in the content of organic matter in the soils and the equally steady increase in the amounts of soluble salts and carbonates. These reach very high proportions in the areas of Luvic Kastanozems (Ustalfs; Burozems) where Solonchaks (Salorthids; Solonchaks) Solonetz (Natrustalfs; Solonetz) and Solidic Planosols (Albic Natrustalfs; Solods) are found as parts of soil complexes. In a number of depressions large amounts of salts may accumulate and the soil surface develops small characteristic puffs or takyrs.

To the south of the desert both the precipitation and the mean annual temperature increase rapidly but because of interception the amount of radiant energy reaching the soil surface is much less than in the desert. Here the beginning of the sequence starts in the desert but thereafter there are marked differences induced by climate and age, for the soils are much older and more highly weathered than those to the north. Rhodic Cambisols (Eutrochrepts; Rossosols), and Vertisols form the next part of the sequence which then continues with Rhodic Ferralsols (Oxisols; Krasnozems) and eroded soils (see pp. 308–9) so that hardened plinthite (ironstone, laterite) occurs at the surface over large stretches of landscape (see pp. 306–309). Finally, Xanthic Ferralsols (Oxisols, Zheltozems) dominated by tropical rain forest occupy the areas with the highest rainfall.

RELATIONSHIPS WITH ORGANISMS

The influence of organisms is usually linked with that of climate which strongly influences the character of the world's flora and fauna. It is in transition zones that the nature of the plant community has a strong influence on the character of the soil. Perhaps one of the best illustrations of this is seen in the marine west coast of north-western Europe where Dystric Cambisols (Dystrochrepts; Altosols) may be maintained under oak forest but if plant communities dominated by heather (*Calluna vulgaris*) or Scots pine (*Pinus sylvestris*) become established the soils may be changed to Orthic Podzols (Orthods; Podzols) by the increased hydrolysis and downward translocation of material caused by the acid litter from these species. Another interesting transition zone lies between the Orthic Luvisols (Ustalfs; Argillosols) and Calcic Chernozems (Udolls; Chernozems) in central USSR. In this zone Phaeozems (Hapludolls; Brunizems) develop beneath deciduous forest but there is a delicate balance and intricate interfingering of these three soils, for if trees invade the grassland Phaeozems develop. The balance can be pushed in the other direction if grasses replace trees.

A conspicuous effect of a change in soil fauna is found in Canada where earthworms were introduced into an area of Luvisols with well developed bleached horizons. The earthworms found favourable conditions for their growth in these soils and multiplied rapidly so that their normal churning activity soon mixed the upper and middle horizons.

The redistribution of material by termites to form Ferralsols is another example which has already been discussed on page 44 *et seq.* Thus, it would appear that quite different soils will develop in the presence or absence of mixing organisms. This is a very important aspect of soil development and may help to account for the presence of well differentiated horizons in soils that have no worms or other churning organisms.

A dramatic example of the influence of vegetation is seen in New Zealand where the Kauri pine (*Agathis australis*) produces a very acid litter which causes Podzol formation to proceed more vigorously near the tree and forms the so-called 'egg cup' Podzol. A similar phenomenon develops in some Acrisols (Ultisols; Luvosols) of South-East Asia where a thick bleached horizon is associated with a species of casuarina (*Gymnostoma nobile*).

A unique example of the relationship between vegetation and the soil pattern occurs in Australia where for most of the humid areas, species of

295

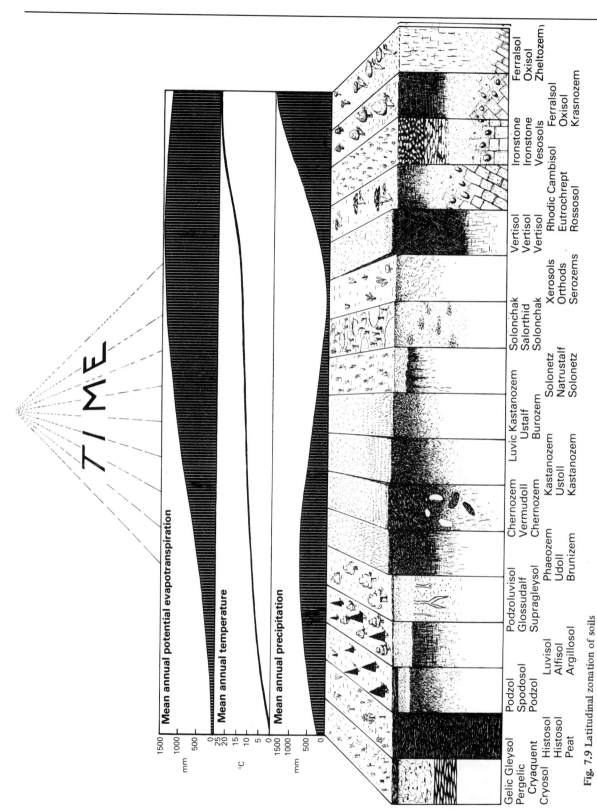

Fig. 7.9 Latitudinal zonation of soils

Eucalyptus are the dominant tree present. It is estimated that there are about 365 species and each seems to have a special habitat and soil. Thus we find the jarrah (*Eucalyptus marginata*) growing on the strongly weathered highly concretionary soils of Western Australia while the snow gums (*Eucalyptus pauciflora*) grow at the higher elevations in the east where the soils are Dystric Cambisols (Dystrochrepts; Altosols). Other species in Australia with special soil requirements include the brigalow (*Acacia harpophylla*) which occurs mainly on Vertisols.

RELATIONSHIPS WITH TOPOGRAPHY

Differences in topography influence the soil in many ways. In areas of moderate to strongly undulating topography and uniform parent material it is common to find a sequence of soils that becomes progressively wetter on descending the slopes. This is known as a hydrologic sequence and is particularly common in areas of hummocky glacial deposits and stabilised sand dunes. The precise types of soils that occur in a hydrologic sequence depends upon a number of factors particularly the nature of the parent material and the climate. Below are given the details of two common sequences.

When there are great differences in elevation between the valley floor and the highest points in the landscape there are differences in climate which can produce altitudinal sequences in the soils and vegetation such as that given below in the third example which also illustrates the influence of aspect.

In polar areas there is often a close relationship between the angle of slope and the nature of the surface phenomena, this is given in the fourth example.

Nearly all landscapes show evidence of erosion and deposition as well as areas of relative stability, thus within any one area there may be soils of widely differing ages and degree of development. This situation is very evident in tropical and subtropical areas such as Nigeria and Australia which supply the fifth and sixth examples that are described.

1. Orthic Podzol — Gleyic Podzol — Dystric Gleysol — Histosol hydrologic sequence

(Orthod — Aquod — Aquept — Histosol; Podzol — Subgleysols — Peat)

A sequence of this type developed on hummocky moraines is illustrated diagrammatically in Fig. 7.10.

It shows the morphological changes from a Podzol (Orthod; Podzol) at the top of a moraine through a Gleyic Podzol and Dystric Gleysol (Aquept; Subgleysol) on the slopes to a Histosol (Histosol; Peat) in the depression. The increase in wetness is caused partly by run-off down the slope but more especially by the ground-water-table which gradually comes closer to the surface on the lower parts of the slope. Finally, the water-table comes to the surface in the depression between the moraines where as a consequence the Histosol forms (Mattson and Lönnemark, 1939).

In addition to the morphological changes there are also significant changes in the chemical, physical and biological properties of the soils. With increasing wetness down the slope the content of cations in the soils increases steadily. This is due to their lateral movement in the drainage water but also due to increased hydrolysis as a result of the increased amount of water. If there are many ferromagnesian minerals present then the content of exchangeable magnesium can exceed that of calcium in the lower wet horizons. The total amount of phosphorus is at a maximum in the middle and lower horizons of the Orthic Podzol and decreases rapidly with increased wetness due to the great solution of apatite in the wetter environment. On the other hand the acetic acid soluble phosphorus is greater in the wet soils.

The increased amount of cations in the system causes a steady increase in the pH of the mineral horizons down the slope. In most cases the pH values of the organic horizons are acid throughout but if the mineral soil contains much calcium or magnesium the Histosol may have pH values up to neutrality.

Perhaps the most marked change is in the content of carbon and C/N ratio. With increasing wetness the content of organic matter at the surface increases while the C/N ratio also increases due to decreased humification.

2. Solonetz — Solodic Planosol relationship

(Natrustalf — Natraqualf; Solonetz — Solod)

Usually there is a simple topographic relationship in which a Solonetz (Natrustalf; Solonetz) occupies the upper slope position and a Solodic Planosol (Natraqualf; Solod) the adjoining depression where moisture accumulates and leaching is greatest. In some cases the reverse relationship is found with the Solodic Planosol on the drier knolls and the Solonetz in the depression. This occurs where there is seepage, a high water-table and reduced leaching in the depression.

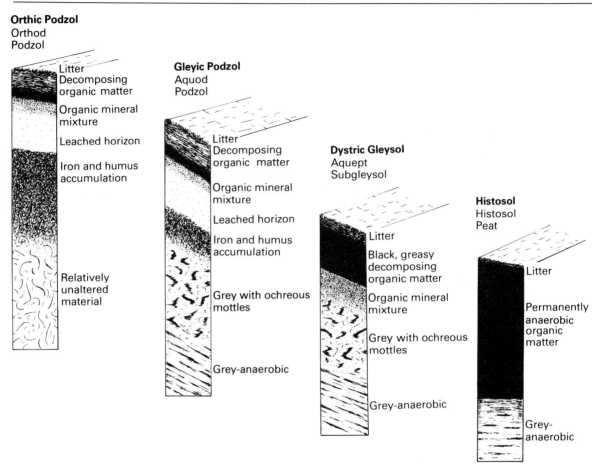

Fig. 7.10 Orthic Podzol – Gleyic Podzol – Histosol hydrologic sequence

The normal relationship is shown in Fig. 7.11.

The most conspicuous feature is the formation of the marked bleached horizon which starts with the development of bleached sand grains on the top of the columnar peds in the Solonetz. Then gradually the bleached horizon becomes thicker and the natric B horizon (natric horizon; solon – Sl) occurs deeper in the soil. This is followed by a stage during which the upper part of the natric B horizon is destroyed to form the bleached horizon which retains vague outlines of the original columnar structure. Finally, in the depression there is a well developed Solodic Planosol which may have evolved from a Solonetz but the evidence for this varies. Alternatively, there is direct development of the Solodic Planosol and only in transition stages does the bleached horizon expand at the expense of the natric B horizon.

3. Altitudinal zonation of soils

Significant changes in soil development take place with increasing elevation but since mountains occur in widely differing environments it is not possible to make many generalisations about the vertical zonation of soils. The example given below is described by Zakharov (1931) who demonstrated for the Caucasus Mountains (Fig. 7.12) a marked zonation of soils and vegetation which parallels the vertical zonation of climate. To some extent the vertical change in soils is similar to the horizontal latitudinal changes described on pages 293, 295–6, but there are variations induced by aspect and differences in day length. Also, mountain soils are shallower and have coarser textures. On the warmer south side of the mountains Luvisols (Alfisols; Argillosols) occupy the lowest position

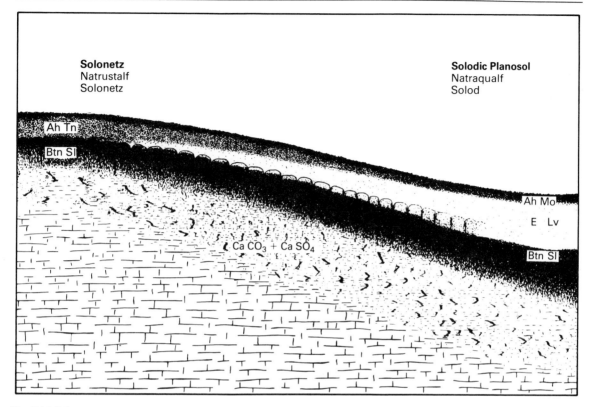

Fig. 7.11 Solonetz – Solodic Planosol relationships

above which are Cambisols (Inceptisols; Altosols) and Phaeozems (Udolls; Brunizems). Xerosols (Aridisols; Serozems) occur at the base of the mountains on the north facing slopes and are continuous with those of the plains to the north. With an increase in elevation on the north slopes there is a change to Kastanozems (Ustolls; Kastanozems), Chernozems (Borolls; Chernozems) and Phaeozems (Hapludolls; Brunizems) which form the first continuous belt from north to south. With increasing altitude one finds Podzols (Orthods; Podzols) and Rankers (Entisols; Rankers) and finally bare rock outcrop. It should be noted that the zonation runs diagonally across the mountain range because the mean annual temperature increases towards the south. Further, the soils on the upper part of the mountains form complete belts whereas the lower ones are found only on one side of the mountain range and in large measure reflect the present differences in climate between the north and south facing slopes — that to the north being generally drier and cooler. The differences in the soils at the lower elevations are due also to the contrasting Tertiary and Pleistocene development that took place

on opposite sides of the mountains. The northern side and upper southern slopes were strongly affected by Pleistocene erosion which removed most of the Tertiary soils whereas in the area of the Black Sea there are extensive remnants of the Tertiary soils so

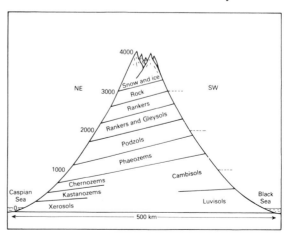

Fig. 7.12 Altitudinal zonation of soils between the Black Sea and the Caspian Sea in southern USSR

that Luvisols have evolved from Ferralsols (Oxisols; Krasnozems and Zheltozems). Another altitudinal relationship has been described by Thorp (1931).

4. Tundra sequence

Figure 7.13 shows a sequence from Svalbard illustrating the relationships that exist between the various surface phenomena, slope and altitude. At the highest elevations the surface is covered with bare angular stones produced by frost action. On the steeper slopes the angular fragments accumulate as scree particularly at the bottoms of the slopes, while differential frost weathering leaves promentories or tors jutting out from the sides of the hills or mountains. This is particularly marked in areas of horizontally bedded sedimentary rocks. Alternating stripes of vegetation and bare soil occur on moderate slopes while at lower elevations on flat or gently sloping situations there is usually abundant vegetation dominated by mosses, grasses and lichens. Such sites tend to be more stable and allow Gelic Gleysols (Pergelic Cryaquents; Cryosols) to form. These soils occur as part of a complex which includes mud polygons (Fig. 6.29), tundra polygons (Fig. 6.26) and ice wedges (Fig. 6.25). Occasionally pingos (Fig. 6.27) are found in this situation and at the lowest position peat may develop if conditions are sufficiently wet. Stone polygons (Fig. 6.30) also form part of the tundra catena developing at all elevations on many flat stony sites.

Permafrost (permafrost; cryons – Cy) occurs throughout the sequence. At the highest elevations it is 1 to 2 m from the surface. On the steeper slopes with scree or stone stripes it occurs at about 1 m but comes progressively closer to the surface as the ground becomes more level and the cover of vegetation increases. Under a thick mat of vegetation and in peat the upper surface of the permafrost may be within a few centimetres of the surface. Where there are mud polygons the upper surface of the permafrost usually has a very irregular outline being greater from the surface beneath the mud polygon which absorbs more heat.

5. Regosol – Ranker – Luvisol – Arenosol relationship

(Lithic soil – Alfisol – Psamment; Regosol – Ranker – Cumulosol – Arenosol)

This relationship is of frequent occurrence in certain parts of West Africa and is illustrated by the Iwo association from western Nigeria (Smyth and Montgomery, 1962). In this area there was profound deep chemical weathering of the basement granitic gneiss in the Tertiary Period followed by a considerable amount of erosion leading to the formation of numerous inselberge. It is possible that weathering and erosion proceeded simultaneously but the latter seems to have been fairly vigorous during the Pleistocene Period causing much disturbance and redistribution of the surface soils. In fact most of the middle and

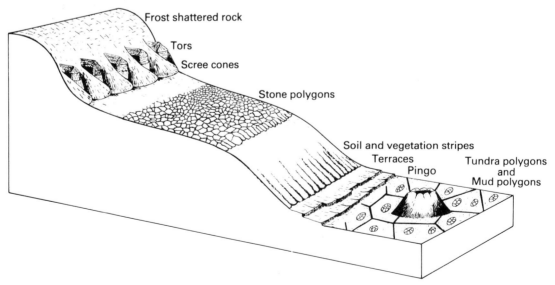

Fig. 7.13 A tundra sequence showing the relationship between slope and surface phenomena

upper horizons in the soils of this and other associations in this area have derived many of their
characteristic features from the Pleistocene Period.
This relationship includes the soils developed on
inselberge and those of the associated pediment
slopes. Also included are the soils of the valley
terraces (Fig. 7.14). The highest elevations are
inselberge which have Rankers (Lithic soils; Rankers)
and many rock outcrops on their upper surfaces and
sides. On the gently sloping pediments surrounding
the inselberge there are Luvisols (Alfisols; Cumulosols)
formed in part by the differential erosion and the
concentration of gravel and concretions in the upper
part of the soil. Usually at the lower end of the
pediment slope there has been the greatest amount of
erosion which may have extended down to the bed
rock followed by a deposition of alluvium in the form
of terraces in which Arenosols (Psamments; Arenosols)
have formed. Commonly in this situation lateral
movement of water has produced a seepage iron pan
(gluton) which may occur in each of the terraces

forming a conspicuous break in slope. On the lowest
terrace Dystric Gleysols (Aquepts, Subgleysols) are
dominant.

6. K-cycles in Australia

In south-eastern Australia there is a succession of
buried soils that is attributed to periods of erosion
and deposition followed by periods of surface
stability and accompanying soil formation. This
periodicity is ascribed to changes in climate from
humid — when soils formed — to arid — when erosion
and deposition were active. Each buried surface has a
specific soil which differs from those of all the other
surfaces, but in some cases the differences are not at
the highest level; nevertheless they are sufficiently
marked to be used for purposes of differentiation and
correlation.

Around Canberra, each cycle commences with
fairly vigorous erosion as indicated by channels
containing current bedded material. Later, material

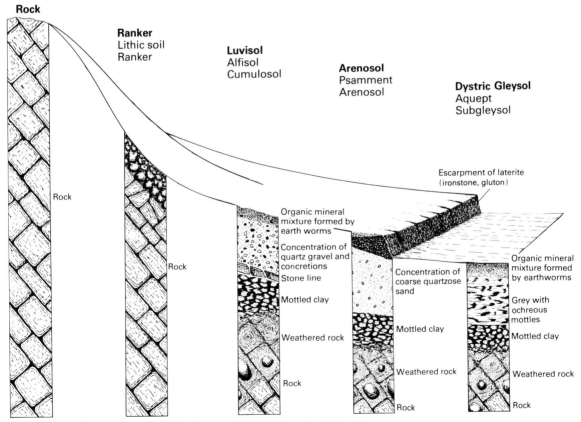

Fig. 7.14 Regosol – Ranker – Luvisol – Arenosol-Dystric Gleysol relationship

sloughed off the upper slopes, filled the channels and buried considerable portions of the land surface, but some surfaces remain relatively unaffected. Thus, within any one period of erosion and deposition, some of the land surfaces were affected while the rest remained relatively unchanged. However, an area that was unaffected during one cycle might be affected during the next. The material of which these deposits are composed is poorly sorted, unbedded and un-orientated and since erosion and deposition are never uniform, the thickness of the mantle varies from slope to slope but on any one hillside it is relatively uniform. Together, these processes have produced three zones on the hillsides; there is the upper part from which material moved *en masse* and which therefore carries no buried soils. The second zone or middle slope position gained material during deposition but also tended to lose it during erosion, therefore there are not many buried soils in this position. The lower zone is the region of accumulation where many full sequences of buried soil occur. In some cases the whole soil was removed while in others the existing soil was buried by new material. Surfaces that were not affected by either erosion or deposition experience changes in climate and processes of soil formation. These latter areas show the effects of several super-imposed sets of soil-forming processes. The result of all of these widely varying processes is a high degree of pedological and geomorphological complexity.

Generally, there are five soils including the present phase, but they are not clearly displayed in all cases. Where remnants of previous soils or superimposed soil formation occur, the true character can be discovered only by means of a lateral traverse to sites where the full soil is preserved or where the two soils become separate due to the deposition of material. These changes from stable periods to erosional phases are termed K-cycles by Butler (1959), cycle K_1 being the youngest. Set out below are the freely drained soils associated with the various cycles in the Canberra area.

	FAO	USDA	FITZPATRICK
K_1	Phaeozems	Udolls	Brunizems
	Kastanozems	Ustolls	Kastanozems
K_2	Nitosols	Udults	Rubosols
	Vertisols	Vertisols	Vertisols
K_3	Acrisols	Ustults	Luvisols
	Nitosols	Udalfs	Rubosols
	Solodic Planosols	Natraqualfs	Solods
K_4	Planosols	Albaqualfs	Planosols
	Solods	Natraqualfs	Solods
K_5	Old weathered remnants		

Cycle K_1 occupies the upper parts of the slopes and is of relatively restricted distribution, while K_2 and K_3 each occupy about one-third of the surface and are more extensive as buried soils. K_4 is also present at the surface but is more often buried. K_5 is red and cream mottled weathered rock and most probably represents highly weathered Tertiary material while the four upper soils are probably Pleistocene in age. Figure 7.15 is an idealised cross-section showing the surfaces in relation to each other and to slope.

To the west of Canberra where loess (or parna as it is known in Australia) and sand dunes are important elements in the landscape, the buried soils are not confined to lower valley slopes, but also occur on the hill tops.

It is clear that these landscapes are comprised of many different soils in an intricate mosaic with soils of different ages lying side by side. This contrasts sharply with other land surfaces where differences in the soil pattern may be due only to moisture as in the case of the hydrologic sequences or to parent material. Landscapes with soils of widely differing ages are fairly common in many tropical and subtropical areas. Other instances are found in areas of repeated mass movement as in certain parts of southern and western Scandinavia where thixotropic marine clays frequently cause catastrophic landslides which expose fresh material as well as burying the older soils with sediment.

RELATIONSHIPS WITH TIME

Undoubtedly, the most fascinating relationships are those that are related to the age of soils. These vary from relatively simple developmental sequences to the very complex evolution involving extremely long periods of time accompanied by changes in climate, vegetation and topography. The first example illustrates a fairly simple development sequence from relatively unaltered material to Podzols. The second example illustrates an evolutionary sequence in Scotland from the Tertiary Period through the Pleistocene to the present. The third example shows a somewhat similar sequence for the central USA. The fourth example attempts to show some of the complex evolution-ary developments encountered in tropical areas. Finally, a hypothetical example is given of the diver-sification and changes in the nature of the soil in an area of landscape in a Marine West Coast environment.

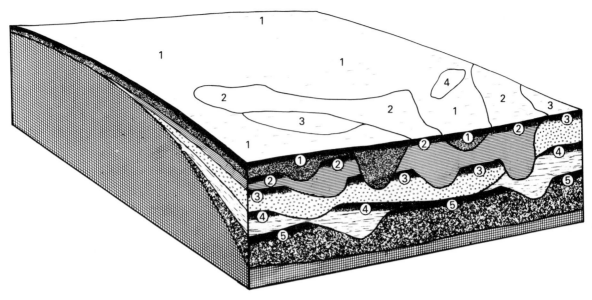

Fig. 7.15 K-cycles in Australia

1. Regosol — Ranker — Podzol relationship

(Entisol — Inceptisol — Spodosol; Primosol — Ranker — Podzol)

Perhaps one of the simplest age relationships is the development of Podzols from unconsolidated siliceous sands as shown diagrammatically in Fig. 7.16. Initially there is the Regosol (Entisol; Regosol) stage which is quickly followed by an accumulation of litter at the surface and the formation of humus horizons. With progressive hydrolysis and leaching, an E horizon (albic horizon; zolon — Zo) forms and at the same time humus, iron and aluminium begin to accumulate in the middle position and initiate the formation of a spodic B horizon (spodic horizon; sesquon). This is the Ranker stage. Further development leads to the differentiation of a spodic B horizon and is sometimes accompanied by an increase in the thickness of the E horizon at the expense of the upper part of the spodic B horizon which is destroyed. The soil may remain in this stage, which is

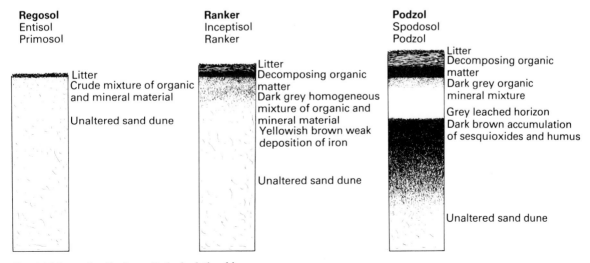

Regosol
Entisol
Primosol

Litter
Crude mixture of organic and mineral material

Unaltered sand dune

Ranker
Inceptisol
Ranker

Litter
Decomposing organic matter
Dark grey homogeneous mixture of organic and mineral material
Yellowish brown weak deposition of iron

Unaltered sand dune

Podzol
Spodosol
Podzol

Litter
Decomposing organic matter
Dark grey organic mineral mixture

Grey leached horizon
Dark brown accumulation of sesquioxides and humus

Unaltered sand dune

Fig. 7.16 Regosol — Ranker — Podzol relationship

303

of frequent occurrence at present, or there may be a gradual hardening of the spodic B horizon as more material is washed in. This latter stage is common in older soils particularly those of cool, moist, oceanic areas.

2. Tertiary soil — Gelic Gleysol — Orthic Podzol — Placic Podzol relationship

(Tertiary soil — Pergelic Cryaquent — Orthod — Placorthod; Tertiary soil — Cryosol — Podzol — Placosol)

It is now generally recognised that within many areas that were glaciated during the Pleistocene Period, there are many places that were not intensively glaciated due either to a thin cover of ice or probably to a single glaciation. In these situations it is possible to trace the evolution of the soil from the Tertiary Period through the Pleistocene Period into the Holocene Period and up to the present. The evolutionary sequence of a Placic Podzol (Placorthod; Placosol), illustrated in Fig. 7.17, probably exemplifies this sequence better than any other soil (FitzPatrick, 1969). Stage 1 shows the reconstructed Tertiary soil at the surface, passing down into chemically weathered rock which has core stones at its base and finally into granite. Stage 2 shows the Tertiary soil truncated by glacial erosion and a deposit of till rests on the remnants of weathered rock. In addition a Gelic Gleysol (Pergelic Cryaquent; Cryosol) has developed, transforming the lower part of the till and the underlying material into permafrost (permafrost; cryon — Cy). It should be noted that the new total thickness is less than at Stage 1 indicating a net reduction in the level of the surface. During a single or successive tundra period solifluction and plastic flow removed the glacial deposits causing the weathered rock to come close to the surface so that the Gelic Gleysol developed entirely in the old Tertiary material as shown in Stage 3 which also shows oriented stones and solifluction arcs. With the amelioration of climate a Podzol (Orthod; Podzol) with a fragipan Cx (fragipan; ison — In) developed, and below it there are the inherited arcs and oriented stones. With a change to cooler and moister conditions of the Atlantic Period a thin iron pan (placic horizon; placon — Pk) formed within the Podzol. This new horizon caused the accumulation of moisture at the surface leading to the formation of an Eg (albic horizon; candon — Co) and histic H horizon. Somewhat similar sequences have taken place on a

variety of other rock types; a common type occurs on basic or intermediate rocks but this usually terminates at Stage 4 as a Cambisol (Ochrept; Altosol)

3. Pleistocene succession in the central USA

Repeated oscillations from cold glacial periods to warm interglacial conditions during the Pleistocene have produced a most fascinating and complex sequence of soil development in the central part of the United States of America, where there is evidence of four major glaciations. During each glaciation erosion and deposition were the dominant processes, whereas during the warm interglacial periods intense weathering prevailed and deep soils developed. In the poorly drained positions, weathering produced thick clay soils which have been called gumbotil by Kay (1916); a term derived from the African word 'gumbo' which refers to the mucilaginous content of okra pods. Also there was profound sheet erosion during the interglacial periods to produce pediment surfaces (Ruhe *et al.*, 1967) and for the accumulation in depressions of thick deposits of clay sediments which were at one time thought to be gumbotil but their sedimentary characteristics have been demonstrated by Frye *et al.* (1960a, 1960b and 1965), who have called them accretion gleys. The freely drained soils of the interglacial periods are not dissimilar to those at the present time and in Illinois they are Luvisols but they are relatively infrequent as compared with gumbotil and accretion gleys.

The simplified sequence with approximate dates shown in Fig. 7.18 starts at the bottom with the Nebraskan glaciation which left a deposit of till, outwash deposits and silts some of which are presumed to be loess. This was followed by an interglacial period and the development of Afton soils which are mainly gumbotil and peat. This initial part of the succession is fairly widespread but it is usually buried deeply beneath deposits of the succeeding glaciations.

The Kansan glaciation was the next, which also left till, outwash deposits and loess followed by the Yarmouth interglacial period and the development of the Yarmouth soils including Luvisols, gumbotil and accretion gleys. Interesting features of this period are the absence of loess overlying the till and the presence of a layer of volcanic ash which has become very important as a stratigraphic marker for the Kansan Period. The third glaciation is known as the Illinoian because of its great extent into the state of Illinois where it covered surfaces that were not glaciated

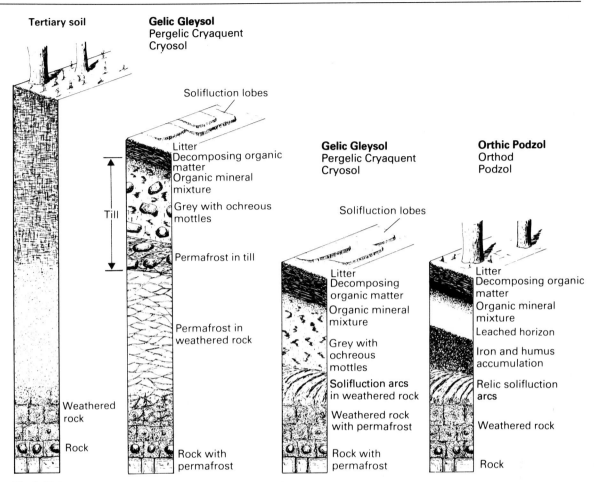

Fig. 7.17 Tertiary soil – Gelic Gleysol – Orthic Podzol – Placic Podzol relationship

previously. Great thicknesses of loess, known as Loveland silts were formed during this period. During the succeeding Sangamonian interglacial period similar soils to those of previous interglacial periods developed but there seems to be a greater extent of freely drained soils from this period as evidenced by the distinctive red colour of the Sangamon soil.

The final or Wisconsinan glaciation produced great thicknesses of till and loess particularly the latter which blankets almost the complete surface of the central states and buries the Wisconsinan till and Sangamon soil. However, Wisconsinan till is exposed at the surface in the states to the north.

The Wisconsinan period was also characterised by a number of interstadial periods some of which were relatively warm and stable so that soils developed; the Brady soil, a Chernozem, is probably the best known and most fully developed example. This

period terminated about 10 000 years B.P. and during the postglacial time, Luvisols, Phaeozems, Gleysols and Planosols have developed on the present land surface.

The extent and distribution of the ice was not always the same so that many areas do not have the full sequence and some have only one layer of till. A significant feature of the glaciations in central USA is the somewhat ineffective erosion, which has caused relatively little removal of the interglacial soils. This is due to the weak erosive power on the flat land surface as well as to the fact that these areas were at the margins of the ice sheet. However, lumps of till and soil of one glaciation are often incorporated in the deposits of succeeding glaciations. To the north, the erosive power of the glaciers was greater so that pre-Wisconsinan tills and soils become less frequent and are completely absent from the northern parts of

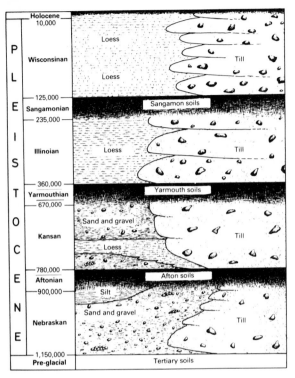

Fig. 7.18 Simplified sequence of Pleistocene events in north-central USA

4. Soil and landscape evolution in the humid tropics from rock to a Ferralsol (Oxisol; Krasnozem) and then to Ironstone (Ironstone; Laterite) at the surface

It is not possible to state with certainty the various stages through which Ferralsols (Oxisols; Krasnozems) pass because the land surfaces on which they occur are very old and there may have been many contrasting phases of pedogenesis, accompanied by changes in topography as weathering and erosion proceeded. Therefore, the various stages given in Fig. 7.19 should be regarded as schematic. At the outset there is a fresh rock surface produced by erosion, uplift or volcanic activity as shown at Stage 1. Rankers (Entisols; Rankers) develop quickly in small depressions and pockets in the rock surface where fine particles accumulate as shown at Stage 2. Thereafter, development involves the progressive hydrolysis of the rock but this is very slow so that there is a gradual change to a Cambisol (Ochrept; Altosol) at Stage 3 in which there are many primary minerals. With further development an oxic B horizon (oxic horizon; krasnon-Ks) forms followed by its thickening and also the thickening of the underlying layer of weathered rock as shown at Stage 4. Accompanying the development of the soil there is natural erosion and landscape development resulting in a gradual lowering of the surface but weathering and soil formation proceed faster than erosion. At each stage the original land surface and volume of material removed is shown by dotted lines. In some situations it seems that there is an intermediate stage between Stages 3 and 4 when the soil has a fairly high content of 2 : 1 clay minerals which are more readily translocated leading to the formation of Nitosols (Alfisols; Rossosols) that have a clay maximum in the middle horizon, but the evidence for this is not unequivocal since many such soils that are regarded as intermediate stages occur on slopes where differential erosion seems to be a major factor in producing the clay differential.

As the topography becomes flatter there is less erosion so that the soil gets deeper. The moisture regime of the soil also changes, the deeper horizons being much moister leading to the formation of the characteristic red and cream mottled layer (plinthite; flambon — Fb) as shown at Stage 5 and in some cases a pallid zone (pallon). In many cases there is an accumulation of iron in the form of concretions or vesicular laterite (plinthite) within the mottled horizon as shown at Stage 6.

The stages that follow are extremely varied and

Michigan and Wisconsin. Although the tills and their interglacial soils must have been frozen during the maximum of each succeeding glaciation, periglacial phenomena and signs of solifluction of these materials are rare. Only during the final phase of the Wisconsinan is there evidence for periglacial processes and this is extensive only in the northern states such as Wisconsin (Black, 1964).

It should be pointed out that the interglacial soils were not confined to the deposits of the preceding glacial period but formed on any exposed surface. In some places strongly developed Luvisols that occur in the southern states can be traced northward beneath Wisconsinan loess indicating that soil formation in the south has proceeded almost without interruption for a long period and also that the interglacial periods were relatively warmer.

In addition, there are places where soil formation of one interglacial period was superimposed upon the soils of the previous interglacial, the most common is the development of a Sangamon soil in a Yarmouth soil.

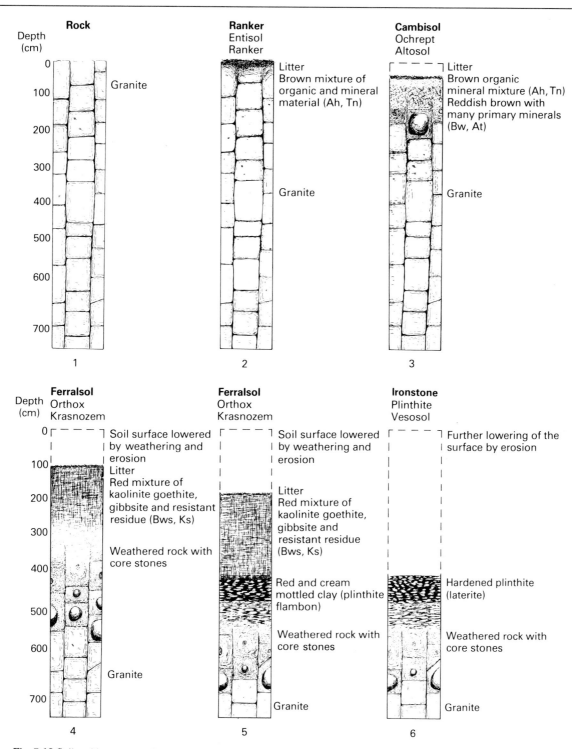

Fig. 7.19 Soil and lanscape evolution in the humid tropics from rock to a Ferralsol and then to ironstone at the surface

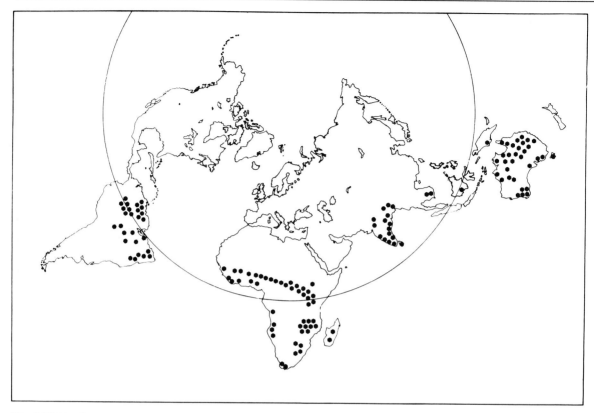

Fig. 7.20 Distribution of ironstone occurring at the surface

display one of the most complex and fascinating evolutions to be found in soils. The simplest development that can follow is for erosion to remove the upper horizons and to lay bare the plinthite (plinthite; laterite) which hardens upon exposure to the atmosphere and forms a protective capping which prevents or reduces the rate of further erosion. This is shown at Stage 6. Such erosion is usually attributed to a change of climate from wet to dry conditions hence the occurrence of many laterite cappings in tropical semiarid areas (Figs. 6.15 and 7.20).

In many instances the climatic change has not been very marked so that the type of soil formation has remained unchanged; nevertheless erosion has proceeded. An example of this is shown in Fig. 7.21 which is taken from the work of Brammer (1962). This sequence starts on the gently sloping upland area where there is a Ferralsol (Oxisol; Krasnozem) containing ironstone. In this case it appears that even this stage may be polygenetic with the upper horizons formed in a sediment and the middle and lower horizons developed from the underlying rock. The change from the upper surface to the valley slopes is

delimited by a small escarpment formed by the hardened and more resistant ironstone (laterite) which is exposed. Below this on the pediment the soils are typical for large parts of the tropical world including Africa. At the surface there is a loamy organic mineral mixture produced by earthworms bringing material to the surface; below is a very distinctive horizon containing abundant fragments of ironstone (laterite) and quartz gravel. This horizon is formed by both differential erosion of the pediment surface — the fine material being moved down slope and the addition of laterite fragments broken from the escarpment and added to the pedi-sediment. There is a sharp boundary to the underlying material which may be mottled clay as shown at Postion 4 or it may be an oxic B horizon (oxic horizon; krasnon — Ks) as shown at Position 5. Further down the slope the fine material accumulates to form clayey soils and finally at the bottom of the slope there are various soils formed in alluvium. These sequences help to explain some of the complexity of the landscape in tropical and subtropical areas.

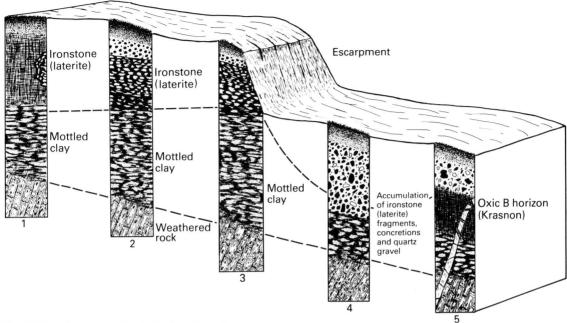

Fig. 7.21 Landscape evolution in the humid tropics

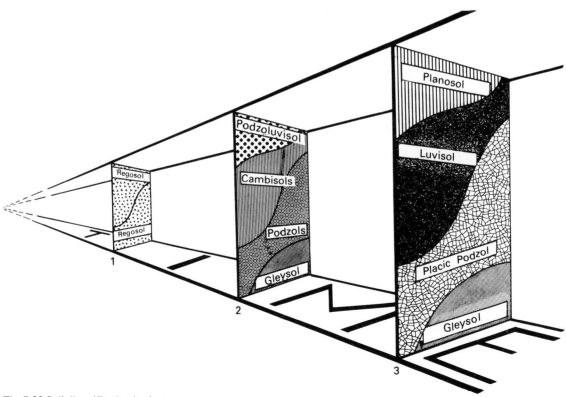

Fig. 7.22 Soil diversification in time

5. Soil diversification in time

The various soils which comprise any particular area of landscape usually evolve in different directions as determined by differences in the factors of soil formation and variations in the intensities of pedogenic processes. Figure 7.22 illustrates three stages in the evolution of a small area of undulating landscape. Stage 1 shows the area as having two different types of parent material, one is a basic silt loam while the other is an acid sand. Also, soil formation has just begun so that Regosols (Entisols; Regosols) occur over the whole area. As soil development proceeds Stage 2 is reached where a Cambisol (Ochrept; Altosol) has formed on the higher and drier area of basic silt loam and a Podzoluvisol (Aqualf; Supragleysol) has formed in the wet depression. Correspondingly, a Podzol (Orthod; Podzol) and a Gleysol (Aquept; Subgleysol) have formed in the acid sand. At Stage 3 which can be regarded as the present time the soils show further development. The Cambisol has changed into a Luvisol (Alfisol; Argillosol) because progressive leaching has caused acidification with clay destruction or translocation downwards. The Podzoluvisol (Aqualf; Supragleysol) has become a Planosol (Aqualf; Planosol) because progressive hydrolysis has led to increased clay formation in the gleyed horizon. On the acid sand the Podzol has changed into a Placic Podzol, (Placorthod; Placosol) while a histic H horizon has accumulated at the surface of the Gleysol.

HORIZON RELATIONSHIPS

It is estimated that within any one area of about $10\,000$ km^2 there may occur about 35 to 40 different horizons together with numerous intergrades. Set out in Fig. 7.23 are the principal intergrading pathways among the horizons in the soils of Scotland. Since nearly every horizon can intergrade to parent material this aspect is not shown.

In tropical and subtropical environments where the time factor has been greater than elsewhere, horizons intergrade in time through weathering stages as well as in space through different climates so that the end products of weathering vary from place to place. Some of these relationships are presented in Figs. 7.24 and 7.25. Figure 7.24 shows the transformation of acid rock such as granite under conditions that range from hot humid to hot subhumid conditions. During its transformation under hot, humid, freely draining conditions the feldspars and any ferromagnesian minerals are decomposed, their places being taken by

kaolinite, goethite and gibbsite. The amount of quartz and accessories is reduced, thus the end product is a typical oxic B horizon (oxic horizon; krasnon). As climatic conditions become drier the gibbsite content diminishes and mica becomes increasingly important. Thus the diagram shows a variety of end products of weathering and a number of development stages which contain varying proportions of weatherable minerals. By varying the initial composition of the rock the content of quartz and weatherable minerals in the developmental stages is also varied with consequent changes in the end products.

While Fig. 7.24 shows only one set of end products for the areas of higher precipitations, Fig. 7.25 attempts to show the variability that is possible on account of variations in local conditions. The amounts of kaolinite, gibbsite and goethite are to be regarded as ratios and not absolute amounts since there are usually variable amounts of quartz and other resistant minerals. This is a theoretical diagram, therefore it is likely that the end members do not exist but most of the other combinations appear to be possible with area A, including the most common combinations. The frequency of horizons with small amounts of kaolinite is low but they do exist, therefore it may be necessary to create segments to accommodate these situations. Alternatively, one can construct a similar diagram to show the continuous variability that exists between quartz, weatherable minerals and weathering products.

SOIL PROPERTY RELATIONSHIPS

Clay relationships

The amount, type and distribution patterns of clay are probably three of the most important properties of soils for they are used more generally than any of the others as distinguishing and differentiating criteria. In Fig. 7.26 the quantitative distribution patterns for the clay fraction (<2 μm) are given for a selected number of soils. These clay patterns fall into the following six groups:

1. Gradual decrease with depth — Cambisol (Ochrept; Altosol)
2. Sharp increase followed by a gradual decrease — Luvisol (Alfisol; Argillosol)
3. Gradual increase followed by a gradual decrease — Ferralsol (Oxisol; Krasnozem)

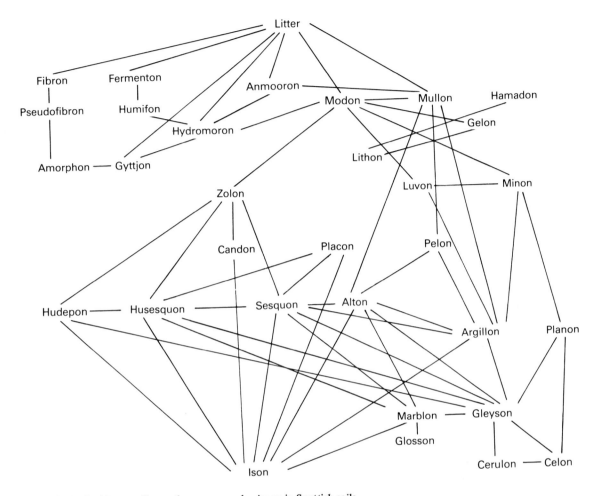

Fig. 7.23 Principal intergrading pathways among horizons in Scottish soils

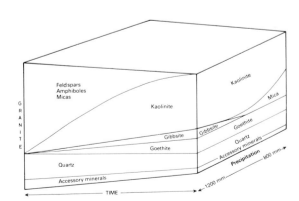

Fig. 7.24 The relationships between products of weathering, time and precipitation in humid tropical areas

4. Sharp increase, then uniform, and then a decrease — Acrisol (Ultisol; Luvosol)
5. Decrease followed by a sharp increase, then a gradual decrease — Solonetz (Natrustalf; Solonetz)
6. Uniformly distributed — Vertisol (Vertisol; Vertisol)

Northcote (1971) has suggested that three types of clay patterns should be recognised viz. uniform, gradational, and duplex which correspond to types 6, 3 and 2 above. But this suggestion is an over-simplification since there are at least six patterns of clay distribution. Of particular interest is the fact that as certain soils develop the clay content of the upper horizons becomes less than that of the middle

311

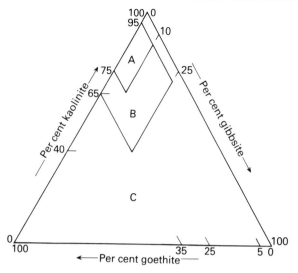

Fig. 7.25 The relative amounts of kaolinite, goethite and gibbsite in the clay fraction of Ferralsols. The frequence of soils falling within the various areas are (A) abundant > 25 per cent, (B) occasional to frequent 5–25 per cent, (C) rare to locally abundant <5 per cent

horizons. This is found even in Chernozems and Kastanozems in spite of vigorous churning, particularly in the latter. This increase in clay with depth has been interpreted by most workers as resulting from translocation but this is not supported by detailed laboratory analyses which indicate that at least two other processes are involved. They are the destruction of clay in the upper horizon and *in situ* weathering in the middle horizon. These three processes may proceed singly, simultaneously or successively; therefore the significance attached to an increase in clay with depth will vary from soil to soil.

There is a need for more precise criteria for the recognition of translocated clay. Some soils may show a greater amount of clay in the middle horizon than above and below, yet show little or no evidence for illuvial clay in the form of coatings. Therefore, those soils that contain less than about 5 per cent of clay coatings in their middle horizons are not regarded as having an argillic B horizon (argillic horizon; argillon).

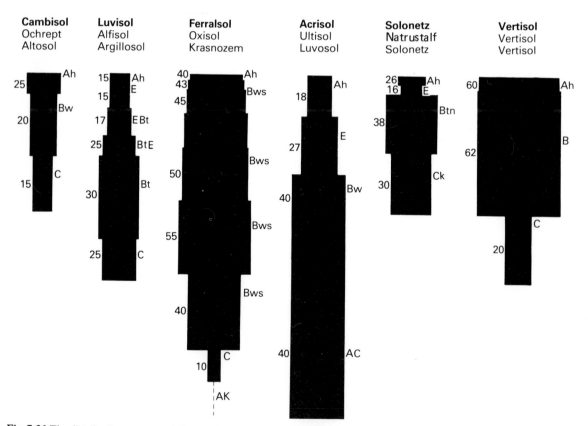

Fig. 7.26 The distribution patterns of the clay fraction for a number of selected soils

Appendix I

SECONDARY MINERAL MATERIALS IN SOILS

Allophane

Composition. $Al_2O_3 . xSiO_2 . nH_2O$

Colour and pleochroism.   Colourless but usually stained brown with ferric hydroxide, non-pleochroic.

Form and occurrence.   Allophane occurs as shapeless masses, granules and coatings around minerals, it is non-crystalline to microcrystalline but cannot be distinguished with the optical microscope. *In rocks* it is of very restricted distribution being associated mainly with certain clay deposits and altered volcanic rocks. *In soils* allophane is the initial product of weathering of a variety of silicates especially volcanic ash. It usually occurs as a brown isotropic mixture with ferric hydroxide and is the principal component of the matrix of the middle horizon in Andosols.

Relief.   Very low

Birefringence.   Nil

Interference colours.   Isotropic

Alteration products.   With time allophane gradually changes from its amorphous state to crystalline halloysite or kaolinite. In many cases there is a well recognised sequence: allophane → halloysite → kaolinite.

Identification.   The recognition of allophane in thin sections is difficult particularly when it is mixed with ferric hydroxide. Then it is very similar to well humified organic matter. Positive identification requires techniques such as D.T.A.

Calcite

Composition. $CaCO_3$

Colour and pleochroism.   Colourless but has an oatmeal appearance when present in fine grain clusters or containing impurities, non-pleochroic.

Form and occurrence.   *In rocks* calcite usually occurs as fine grained aggregates or clusters of rhombohedral crystals. It is the principal constituent of sedimentary and metamorphic limestones and occurs also in some igneous rocks. *In soils* calcite may be derived from the parent material or it may be formed *in situ*. Parent materials containing calcite vary from massive rocks to a variety of sediments including most loesses and parna. Since calcite is readily soluble in carbonic acid it is easily removed from the soils of humid areas but it is common in many soils of the arid and semiarid areas. Apart from ice, calcite is probably the most common and abundant crystalline substance to form in soils where it has a variety of forms. It may be present as aggregates of acicular crystals like those that form the pseudomycelium in soils such as Chernozems. However, rhombohedral grains of various sizes are more frequent. These may form coatings on the surfaces of peds or they may grow on the sides of pore spaces or around objects such as rock fragments. Subspherical concretions or aggregates of calcite grains are common in some soils.

Cleavage.   Perfect rhombohedral

Relief.   Very low to moderate causing a twinkling effect when rotated in plane polarised light.

Birefringence.   Extreme

Interference colours.   The maximum is a pale colour of the high orders. Bright colours of the lower orders may show on the edges of the grains.

Extinction.   Symmetrical to the cleavage

Alteration products.   After being dissolved it can be lost in the drainage water, taken up by plants or

313

translocated within the soil to reform various types of calcite crystals.

Identification.   The variable relief and high order interference colours are distinctive. Therefore, it is not easy to confuse calcite with other minerals except in fine grain sediments such as loess when identification of all minerals in thin sections is difficult because of their small size.

Chalcedony

Composition. SiO_2

Colour and pleochroism.   Colourless to pale yellow or brown, non-pleochroic

Form and occurrence.   *In rocks* chalcedony is a secondary mineral occurring in the cavities of igneous rocks, as fibres and in bands with fibres normal to the bands. It may be present as aggregates sometimes radially spherulitic or as nodules in sedimentary limestone, flint and chert and as a replacement of calcareous fossils. *In soils* chalcedony occurs in a variety of situations. It is found in some strongly weathered horizons where it is present as pseudomorphs or more characteristically as coatings which are sometimes stained brown due to ferric hydroxide. In plane polarised light the coatings show the typical laminar structure but in crossed polarised light there is aggregate polarisation and transverse banding.

Relief.   Low

Birefringence.   Weak

Interference colours.   The maximum interference colour is white of the first order, may show aggregate structure.

Extinction.   Parallel to the length of the fibres

Alteration products.   Not known

Identification.   Very readily identified due to its form, low relief and first order interference colours.

Chlorite

Composition. $(Mg, Al, Fe)_{12} [(SiAl)_8 O_{20}] (OH)_{16}$

Colour and pleochroism.   Green, dark green, greenish-yellow with weak to moderate pleochroism.

Form and occurrence.   *In rocks* chlorite occurs as laminar grains similar to the micas. It is widely distributed in low grade metamorphic rocks and occurs in some igneous rocks as a product of the hydrothermal alteration of ferromagnesian minerals. *In soils* chlorite occurs as single laminar grains or small aggregates. It may be derived from the parent material or it may be formed *in situ* as in certain sediments. When formed *in situ* chlorite may pseudomorph other minerals such as biotite. However, the greatest amount of authigenic chlorite is submicroscopic and usually occurs interstratified with vermiculite. Therefore, identification has to be made by X-ray or D.T.A. techniques.

Cleavage.   Perfect in one direction

Relief.   Low

Birefringence.   Weak

Interference colours.   The maximum colour is grey to white of the first order, often ultra blue

Extinction. $0-10°$ almost straight in most sections

Alteration products.   Chlorite alters readily to vermiculite and montmorillonite.

Identification.   Large grains are fairly easily identified as detrital grains or in thin sections by their green colour, low relief and first order colours but distinguishing chlorite from weakly weathered biotite is difficult sometimes.

Ferric hydroxide

Composition. $Fe(OH)_3$

Colour and pleochroism.   Yellowish-brown, non-pleochroic

Form and occurrence.   *In rocks* it is present as an alteration product of many iron-bearing minerals forming shapeless masses, granules and coatings. *In soils* ferric hydroxide is an extremely common substance particularly in the early stages of weathering and soil formation. It is one of the principal substances forming coating on mineral grains and a major constituent of the matrix of the middle horizon of many Podzols. It is also the chief component of bog iron ore.

Relief. Very high

Birefringence. Nil

Interference colours. Isotropic

Alteration products. With time ferric hydroxide crystallises to form goethite or haematite.

Identification. Since ferric hydroxide usually occurs associated with other substances in the matrix of some soils, positive identification and frequency estimation by optical methods are difficult. In addition, it is easily confused with well humified organic matter which has similar optical properties. In a number of cases material that appears to be ferric hydroxide has been shown to contain microcrystalline goethite or haematite.

Gibbsite

Composition. $Al(OH)_3$

Colour and pleochroism. Colourless, non-pleochroic

Form and occurrence. In rocks gibbsite occurs as aggregates of small or very small hexagonal crystals mainly as a weathered product of many igneous and metamorphic rocks and is one of the principal minerals in bauxite where it is associated with cliachite — amorphous aluminium hydroxide. *In soils* it is found mainly in strongly weathered situations of tropical and subtropical areas where the frequency varies from absent to dominant. In some cases it forms the principal component of the matrix occurring as randomly orientated submicroscopic crystals causing the matrix to be isotropic. In the lower horizons where minerals are undergoing vigorous hydrolysis, gibbsite occurs as pseudohexagonal euhedral crystals filling cavities, forming pseudomorphs after feldspars and outlining cleavages. It also occurs as vermiforms and concretions in many soils.

Cleavage. Perfect in one direction but difficult to see in small crystals.

Relief. Low

Birefringence. Moderate to strong

Interference colours. The maximum colour is middle second order. These colours are seldom seen because the crystals are usually small.

Extinction. Oblique up to 25° but difficult to measure because of the small size of the crystals.

Alteration products. Many soils that contain large amounts of gibbsite in their lower horizons have only a small amount in their upper and middle horizons. From this it is concluded that gibbsite is altered to other minerals particularly kaolinite. Conversely, kaolinite can be weathered to gibbsite.

Identification: Gibbsite often occurs as submicroscopic crystals which can be identified only by X-ray or D.T.A. analyses. When it occurs as larger crystals it can be confused with kaolinite and muscovite. It has a higher birefringence than kaolinite but a lower birefringence and smaller extinction angle than muscovite. In many cases positive identification can be established only by X-ray or D.T.A. analysis.

Goethite

Composition. $FeO \cdot OH$

Colour and pleochroism. Yellow, brown or orange-red, weakly to strongly pleochroic.

Form and occurrence. Since goethite is mainly a product of weathering and recrystallisation, its form in rocks and soils is very similar. It occurs usually as submicroscopic particles forming irregular masses, concretions or uniformly distributed throughout the mass. *In rocks* it is of restricted distribution occurring mainly in sedimentary iron ore deposits, as thin coatings on grains in other sediments and as a product of weathering. *In soils* goethite is one of the principal colouring substances usually occurring as submicroscopic grains evenly distributed throughout the matrix. Although it is conspicuous through its colour it is often of very low frequency. Goethite is usually formed by the slow crystallisation of ferric hydroxide within the matrix or it may grow as crystals on various surfaces from percolating waters containing dissolved compounds of iron. In many cases its colour can be masked by the presence of haematite or more particularly by manganese dioxide. In certain horizons, coatings of goethite composed of euhedral crystals may occur outlining pores.

Cleavage. Perfect in one direction but difficult to distinguish in small crystals.

Relief. Very high

Birefringence. Extreme

Interference colours. The maximum is a pale colour of the high orders but usually masked by the colour of the mineral.

Alteration products. Goethite can be altered to ferric hydroxide or haematite depending upon environmental conditions.

Identification. Because goethite often occurs as submicroscopic crystals positive identification can be made only by X-ray or D.T.A. When it occurs as large crystals it is easily recognised by its colour, prismatic form and high interference colours.

Gypsum

Composition. $CaSO_4 \cdot 2H_2O$

Colour and pleochroism. Colourless, non-pleochroic

Form and occurrence. *In rocks* gypsum usually occurs as anhedral and subhedral aggregates, acicular or fibrous. It is of restricted distribution, occurring in evaporites and certain massive rocks. *In soils* authigenic gypsum occurs as euhedral gains, with characteristic diamond shape. It may be derived from the parent material but more commonly it forms *in situ* as single crystals or clusters. Since gypsum is fairly soluble in carbonic acid it is easily removed from soils of humid areas but it is a common mineral in many soils of arid and semiarid areas where it is formed mainly *in situ*.

Cleavage. Perfect in one direction, imperfect in two other directions.

Relief. Low

Birefringence. Weak

Interference colours. The maximum is pale yellow of the first order.

Extinction. Parallel to the cleavage

Alteration products. Gypsum dissolves readily in the soil solution then the two ions may be re-distributed separately or together.

Identification. The form, cleavage and interference colours are distinctive.

Halite

Composition. NaCl

Colour and pleochrism. Colourless but may contain inclusions, non-pleochroic.

Form and occurrence. Halite occurs as anhedral or euhedral crystals, the latter have square or triangular outlines. *In rocks* it occurs principally as beds of massive rock salt often accompanied by gypsum. *In soils* it may be derived from the parent material but it usually crystallises *in situ* from a saline soil solution particularly in arid and semiarid areas. The crystals usually accumulate in pore spaces and outline ped surfaces.

Cleavage. Perfect cube

Relief. Low

Birefringence. Nil

Interference colours. Isotropic

Alteration products. Halite dissolves readily, then the two ions may be redistributed separately or together.

Identification. Because halite is colourless, isotropic and has a similar refractive index to impregnating resins, its identification in thin sections is very difficult. When halite is expected the slide should be examined very carefully. On the other hand its presence in soils can easily be determined in an aqueous extract.

Halloysite and Metahalloysite

Composition. $Al_2O_3 2SiO_2 4H_2O$ or $Al_4Si_4(OH)_8O_{10} \cdot 8H_2O$

Colour and pleochroism. Colourless to white

Form and occurrence. In both soils and rocks they occur as submicroscopic particles forming irregular masses; very rarely as concretions. *In rocks* these two minerals are of restricted distribution occurring associated with allophane, kaolinite and gibbsite and other substances as a result of hydrothermal alteration.

In soils halloysite and metahalloysite usually occur as submicroscopic alteration products, particularly of feldspars and may make up the greater part of the matrix of some soils. They often crystallise from allophane and therefore are associated with many soils developed from volcanic ash.

Cleavage.   None

Relief.   Low

Birefringence.   Very weak

Extinction.   Straight

Alteration products.   With time these two minerals often change into kaolinite and form part of the sequence allophane – halloysite – kaolinite.

Identification.   Because of their small particle size they are difficult to distinguish optically, positive identification is usually made by X-ray, D.T.A. or other techniques.

Haematite

Composition. Fe_2O_3

Colour and pleochroism.   Usually opaque in sections of normal thickness; deep red (blood red) in very thin sections, moderately pleochroic.

Form and occurrence.   *In rocks* haematite occurs as very small flakes or grain masses and as inclusions in other minerals. In igneous rocks it is rare but is common as a secondary mineral in many meta-morphic and sedimentary rocks. It is a common constituent of some massive iron ores in which it has a botryoidal or compact splinter form. *In soils* haematite occurs as very small submicroscopic grains and is usually a secondary mineral forming as a result of alteration of other minerals containing iron. It is a common constituent of the matrix of many tropical and subtropical soils and imparts a bright red colour to the soils. Larger microscopic particles also occur in some soils. Soils formed from certain old sediments such as those of the Devonian or Permo–Triassic age may inherit haematite from the sediments.

Relief.   Very high

Birefringence.   Extreme

Interference colours.   The maximum is a pale colour of the high orders but usually masked by the colour of the mineral.

Extinction.   Straight

Identification.   When haematite is present as sub-microscopic particles, positive identification can be achieved only by X-ray, D.T.A. or other techniques. Microscopic particles are easily identified by their deep red colour.

Ice

Composition. H_2O

Colour and pleochroism.   Colourless, non-pleochroic

Form and occurrence.   In both soils and rocks as anhedral crystal masses forming wedges, subhedral crystals in lenses and veins and as acicular crystals in masses. In soils of the polar and subpolar areas, particularly the arctic and subarctic, ice is a very common constituent of the lower horizons. It is probably the most abundant secondary crystalline material in soil, particularly as vertically orientated subhedral crystals in veins. In many soils of temperate areas ice forms in the upper and middle horizons during the winter period, either as masses of acicular prismatic crystals, the so called needle-ice, or as veins.

Cleavage.   Poor

Relief.   Moderate

Birefringence.   Very weak

Interference colours.   Thin sections of ice are seldom less than 0.5 mm thick which give colours of the second order.

Extinction.   Straight

Identification.   Ice is easily identified because of its melting point at 0 °C.

Iddingsite

Composition.   Mg-Fe-silicate. This mineral is generally regarded as a mixture of Mg-silicate and haematite or goethite, hence its strong colour.

317

Colour and pleochroism. Shades of reddish-brown and red with slight to distinct pleochroism.

Form and occurrence. *In rocks* it occurs as rims around olivine or as complete pseudomorphs of olivines and pyroxenes. It is polycrystalline but the grains show optical properties of a single crystal because the minute crystals of both constituents are strongly oriented after the original mineral. *In soils* it is present as irregular grains and is abundant in the humid tropics in soils derived from basic rocks.

Cleavage. Good

Relief. Moderate

Birefringence. Extreme

Interference colours. The maximum colour is a pale colour of a high order but masked by the colour of the mineral.

Extinction. Straight

Alteration products. Not studied

Identification. Iddingsite is recognised in soils by its colour, cleavage and pleochroism. In rocks its pseudomorphic form is an additional diagnostic characteristic.

Jarosite

Composition. $KFe_3(SO_4)_2(OH)_6$

Colour and pleochroism. Shades of pale yellow to deep yellow or pale brown with distinct pleochroism.

Form and occurrence. *In rocks* jarosite occurs as a weathering product of sulphides or forms from hydro-thermal solutions but it is relatively rare. *In soils* it occurs as individual or clusters of fine euhedral cubes with square diamond or rectangular cross-sections. It occurs principally as an oxidation product of pyrite mainly in estuarine and other tidal materials which have been reclaimed from the sea and allowed to oxidise. It may occur in some marine sediments that have been uplifted and oxidised naturally.

Cleavage. Perfect in one direction

Relief. High

Birefringence. Extreme

Interference colours. The maximum is a pale colour of the high orders.

Extinction. Straight

Alteration products. Not identified

Identification. The form, high relief and pale high order colours are distinctive.

Kaolinite

Composition. $Al_4[Si_4O_{10}](OH)_8$

Colour and pleochroism. Colourless to pale yellow may be cloudy or grey, readily absorbs dyes and becomes markedly pleochroic.

Form and occurrence. *In rocks* kaolinite occurs mainly as submicroscopic platy flakes in clayey sediments or may form pseudomorphs of minerals, particularly feldspars, as a result of hydrothermal alteration. *In soils* and weathered rocks kaolinite is probably the most common and widespread clay mineral. In soils it usually occurs as submicroscopic plates that occur singly with random orientation thus causing the matrix to be isotropic. It may be aggregated to form domains or deposited on surfaces to form clay coatings. Kaolinite may be derived from the parent material or it may be formed *in situ*. In many lower strongly weathered horizons kaolinite commonly forms vermicular masses within pores spaces or as pseudomorphs after many minerals particularly feldspars and biotite. Certain concretionary horizons in tropical areas may have kaolinite forming sinuous ramifications between the concretions. Perhaps the purest and highest concentrations occur in the pallid zones (pallons).

Cleavage. Perfect in one direction similar to the micas

Relief. Low

Birefringence. Weak

Interference colours. The maximum is grey and white of the first order.

Alteration products. Kaolinite is extremely resistant

to weathering but will decompose to give dilute solutions of silica and alumina with or without the formation of gibbsite.

Identification. Its low relief and first order colours are characteristic features but it can easily be confused with gibbsite and some other minerals, therefore positive identification when the distinctive crystal form is absent requires X-ray analysis or other techniques such as the adsorption of dyes.

Lepidocrocite

Composition. $FeO \cdot OH$

Colour and pleochroism. Yellow to orange with strong pleochroism.

Form and occurrence. *In rocks* it forms submicroscopic masses and is a weathering product of iron bearing minerals, it occurs in various iron ores associated with goethite. *In soils* it occurs also as submicroscopic masses as a weathering product and is one of the principal iron oxides in moist soils where it often causes the characteristic ochreous coloured mottles.

Cleavage. Perfect in one direction, moderate in two others but is seldom seen because the crystals are too small.

Relief. Very high

Birefringence. Strong

Interference colours. The maximum is middle second order but masked by the colour of the mineral.

Extinction. Straight

Alteration products. Not studied

Identification. Lepidocrocite is usually associated with goethite from which it cannot be distinguished optically, X-ray analysis is required.

Manganese dioxide

Composition. MnO_2

Colour and pleochroism. Opaque – black in reflected light

Form and occurrence. *In rocks* manganese dioxide occurs as an amorphous alteration product but it is very rare. *In soils* manganese dioxide is also amorphous and most abundant in moist situations and where seepage is common. It occurs as irregular or dendritic growth on the surface of peds around pores and within the matrix, or it may dominate the whole horizon imparting an almost completely black colour. It may also be associated with ferric hydroxide in concretions and as a cementing material in some horizons. In many Vertisols manganese dioxide occurs as coating and dendritic growths in calcium carbonate concretions and may also occur in a finely particulate form embedded in the matrix.

Alteration products. Not identified

Identification. The combination of colour and form are distinctive.

Montmorillonite

Composition. $(1/2\ Ca, Na)_{0.7}(Al, Mg, Fe)_4$
$$[(Si, Al)_8O_{20}]\ (OH)_4 \cdot nH_2O.$$

Colour and pleochroism. Colourless or brown, may be pale pink, or green, non-pleochroic.

Form and occurrence. *In rocks* it is the chief mineral of bentonite formed by the alteration of volcanic glass. It is also a constituent of various sediments and usually occurs as microcrystalline aggregates in the form of shards. *In soils* montmorillonite may be derived from sedimentary parent material or it may be formed *in situ* either from consolidated or unconsolidated material. These are normally basic or else there is a source of magnesium. Montmorillonite forms where leaching is not a dominant process, thus is a major constituent of the soils of arid and semiarid areas. It usually occurs as submicroscopic particles which may coalesce to form domains that are visible in crossed polarised light, only rarely are large microscopic aggregates present.

Cleavage. Perfect in one direction hence the laminar form.

Relief. Low to very low R.I., less than resin

Birefringence. Strong

Interference colours. The thin crystals seldom give colours above first order.

Extinction. Straight

Alteration products. Under neutral or alkaline conditions montmorillonite is fairly stable but it will alter to mica or kaolinite if conditions change and the soil becomes acid.

Identification. Because of the small size of the particles, optical identification is difficult, positive identification is usually made by X-ray, D.T.A. or other techniques.

Opal

Composition. SiO_2 $(H_2O) x$

Colour and pleochroism. Colourless to milky, non-pleochroic

Form and occurrence. Opal is generally regarded as randomly orientated submicroscopic grains of cristobalite. *In rocks* it occurs as encrustations, veinlets and cavity fillings usually without any structure. It is present in many volcanic and intrusive rocks. *In soils* opal occurs as phytoliths. These are cellular residues which formed originally in plants and accumulate when the plants decompose. These are some of the commonest forms of opal in soils and can in some instances become a major component. Opal also forms the skeletons of diatoms and accumulations in pores.

Cleavage. Absent

Relief. High R.I., less than resin.

Birefringence. Nil

Interference colours. Isotropic, some varieties show pale first order colours due to strain.

Alteration product. Not determined

Identification. The cellular form, high relief (low R.I.) and isotropism are distinctive.

Pyrite

Composition. FeS_2

Colour and pleochroism. Opaque – brassy with metallic lustre in reflected light

Form and occurrence. *In rocks* pyrite is a common mineral as a secondary formation in igneous rocks, shales and slates where it occurs as euhedral cubic crystals with square, rectangular or triangular sections. It may occur also as irregular single grains or clusters. *In soils* pyrite may be derived from the parent material but more usually it is formed *in situ.* When derived from the parent material it usually occurs as isolated grains which may have partially corroded edges and a thin isotropic coating of ferric hydroxide. Such occurrences are extremely rare and may pass unnoticed. Pyrite usually forms in many tidal or marshy soils through the interaction of iron in the soil and sulphate in the sea water. The grains so formed occur in clusters or singly within the matrix usually associated with organic matter. It may also be concretionary or botryoidal.

Alteration product. Pyrite in most freely drained soils weathers to give brown ferric hydroxide but this is a relatively unimportant transformation because of the small amount of pyrite found in most soils. On the other hand the weathering of pyrite in tidal soils is very important for there it is usually present in fairly large quantities. In such soils weathering takes place only when the soil is drained causing the pyrite to be oxidised to crystals of jarosite and sulphuric acid.

Identification. The brassy metallic lustre in reflected light is distinctive.

Quartz

Composition. SiO_2

Colour and pleochroism. Colourless, inclusions often present

Form and occurrence. *In rocks* the form varies widely depending upon the composition and formation of the rock. The grains are anhedral in igneous rocks while they are elongate with interlocking margins and have undulose extinction in quartzose rocks. In vein quartz the grains may be platy, radial or feathery. Quartz occurs in nearly all rocks except the very basic igneous and metamorphic rocks and very pure limestone. *In soils* quartz is probably the most common mineral and occurs mainly as single irregular grains. It is usually derived from the parent material but under certain specialised conditions euhedral or subhedral quartz may occur as a secondary mineral.

Cleavage. Rhombohedral or prismatic but this is seen only in very thin sections or strongly weathered grains.

Relief. Low

Birefringence. Weak

Interference colours. The maximum interference colour is white to pale yellow of the first order. This is an extremely constant property which makes it suitable for measuring the thickness of thin sections, particularly during their preparation.

Extinction. Straight but may be undulose due to strain.

Alteration products. Quartz is extremely resistant to weathering, therefore it tends to become more and more concentrated in soil as weathering proceeds thus causing soils to contain more quartz than their parent materials. In a strongly weathering environment particles of quartz <50 μm are weathered and gradually disappear hence the low frequency of this size range in many tropical soils. Some quartz grains may disintegrate partially along cleavages which may become outlined by depositions of ferric oxides. The alteration products of quartz have not been identified, presumably they are some form of soluble silica.

Identification. The combination of form, general absence of cleavage, low relief, low order interference colours are usually distinctive. Quartz may sometimes be confused with certain feldspars but in soils feldspars are usually cloudy and slightly weathered whereas quartz is usually clear.

Siderite

Composition. $FeCO_3$

Colour and pleochroism. Colourless to grey but usually has brown spots due to alteration, non-pleochroic.

Form and occurrence. *In rocks* siderite is associated with certain iron-bearing sediments of which certain oolitic iron stones and some bauxites are the principal examples. *In soils* siderite usually occurs as clusters of acicular crystals, euhedral rhombohedra or sub-hedral crystals. It is found as a replacement substance in peat and as an *in situ* formation in certain soils of semiarid areas.

Cleavage. Perfect rhombohedral as in calcite

Relief. Low to high causing a twinkling effect when rotated in plane polarised light, both refractive indices are greater than Canada balsam.

Birefringence. Extreme

Interference colours. The maximum interference colour is white of a high order.

Extinction. Parallel to cleavage traces

Alteration products. The principal substance produced on weathering seems to be amorphous or poorly crystalline ferric hydroxide hence the brown staining on the outsides of the crystals.

Identification. The very variable relief and high order interference colours are distinctive but there may be some confusion with calcite, dolomite and magnesite but all of these other minerals have one index of refraction less than balsam and most impregnating resins.

Vermiculite

Composition. $(Mg, Ca)_{0.7}(Mg, Fe^{3+}Al)_{6.0}$
$[(Al, Si)_8 O_{20}] (OH)_4 \cdot 8H_2O$

Colour and pleochroism. Pale brown to yellow, weakly to moderately pleochroic

Form and occurrence. *In rocks* vermiculite is of restricted distribution occurring as hydrothermal alteration product. It has a laminar form similar to biotite. *In soils* it may be derived from the parent material but it usually forms *in situ* by weathering. Generally, the individual particles are submicroscopic but there are often larger grains forming partial or complete pseudomorphs of biotite, amphiboles and pyroxenes.

Cleavage. Perfect in one direction

Relief. Low

Birefringence. Moderate to strong

Interference colours. The maximum colour is upper second order as seen in coatings. In weathered rocks it tends to be disordered and shows lower colours.

Extinction. Straight

Alteration products. Vermiculite is common in many soils that are not highly weathered. With time it may change to amorphous materials, mica or kaolinite depending upon the environment. Therefore, it can be regarded as an intermediate weathering product.

Identification. Because of the small size of the particles optical identification is difficult. Positive identification is usually made by X-ray, D.T.A. or other techniques. Large particles resemble biotite but they have weaker pleochroism, less perfect cleavage and lower interference colours.

Appendix II

SUMMARY DEFINITIONS OF SOIL HORIZONS ACCORDING TO FITZPATRICK

The properties used are defined in Chapter 4.
The abbreviations used are as follows:

OM	=	organic matter, expressed as a percentage by weight of the <2 mm fraction
C/N	=	carbon : nitrogen ratio
CEC	=	cation-exchange capacity, expressed in me per cent of the <2 mm fraction
BS	=	base saturation expressed as a percentage of the CEC
MM	=	micromorphology of the horizon
Ge	=	genesis of the horizon
VP	=	Von Post scale of peat decomposition
C	=	Celtic
E	=	English
F	=	French
G	=	German
Gk	=	Greek
I	=	Italian
J	=	Japanese
L	=	Latin
OE	=	Old English
R	=	Russian
S	=	Swedish

Qualifying suffixes:

c	=	containing calcium and/or magnesium carbonate from the parent material
h	=	high base status
d	=	hardened

The equivalent FAO symbols and names are given in parenthesis

Alkalon Ak (Az, Bz, Cz, saline horizon)

Upper, middle or lower; 10–100 cm; yellowish-brown brownish-grey, mottled; variable texture; massive, prismatic; firm, friable, hard, pH 8.5–10.5; CEC variable; BS 100 per cent; OM < 2 per cent; C/N 12–20; $(Ca + Mg)CO_3 > 1$ per cent; $> 30\% CO_3^{2-}$ and HCl_3^{1-} in the saturation extract, EC > 4 mmhos; Ge formation of carbonates and bicarbonates mainly of sodium.
From alkaline

Alton Al (Bw, cambic B horizon)

Middle; < 50 cm to > 1 m; usually brown to dark brown but may inherit other colours from the parent material; loam to silt loam to clay loam, usually has similar clay content to the horizon above; crumb, subangular or angular blocky, may have shiny ped faces but coatings few to absent; firm to friable; pH 5.5–7.5; CEC 10–30 me per cent, BS 30–100 per cent; OM < 2 per cent; C/N 6–12; MM isotropic to weakly anisotropic matrix; Ge *in situ* weathering and deposition of oxides with little addition from other horizons.
From L. *alter* = alter

Amorphon Ap (H, histic H horizon)

Upper, middle or lower; 5–30 cm; very dark brown to black; usually saturated with water, sticky and plastic when wet, hard when dry; pH 3–5.0 but may be higher if the saturating waters are derived from base rich area; OM > 60 per cent; C/N 20–35; MM isotropic with < 30 per cent plant fragments, some diatoms and phytoliths may be present also fungal hyphae and sclerotia; VP 7–10; Ge accumulation and slow decomposition of organic matter under anaerobic conditions.
From Gk. *Amorphus* = shapeless

Andon An (Bw, cambic B horizon)

Middle; 10–50 cm; brown, yellowish-brown, reddish-brown; loam, silt loam, silt, usually has similar clay content to the horizon above; granular, blocky, massive; friable or fluffy and usually thixotropic; bulk density < 1.0; pH 5.0–7.0; CEC 15–80 Me per varying with technique used; OM 3–8 per cent. C/N 8–12; dominant allophane; MM yellowish-brown isotropic matrix; Ge weathering and *in situ* formation of ferric hydroxide and allophane.
From J. *An* = soil and *do* = dark

Anmooron Am (Ah, umbric A horizon)

Upper; 10–20 cm; very dark grey, very dark brown to black with faint to strong mottling especially along root passages; loam; massive to blocky, some show evidence of earthworm activity; firm, friable, plastic; wet for long periods of the year; pH 4.5–7.5; OM 10–30 per cent; C/N 12–20; $(Ca + Mg)CO_3$ variable from parent material; MM alveolar, granular, vermicular with many faecal pellets and plant fragments; Ge incorporation and decomposition of organic matter in a wet but not permanently saturated environment.
From Kubiëna

Arenon Ae (Bw, cambic B horizon)

Middle or lower; 25–> 200 cm; yellow to red; loamy sand, sand, < 10 per cent silt plus clay; single grain, massive; loose to firm, often hard; pH 4.0–7.0; CEC < 5 me per cent; OM < 2 per cent; MM fine material coating grains and sometimes forming bridges; Ge preferential removal of fine material on slopes, development from coarse textured material.
From L. *arena* = sand

Argillon Ar (Bt, argillic B horizon)

Middle or lower; 1–> 30 cm; yellow, brown, red, sometimes with weak mottling; variable texture usually loams, clay loams, silt loams, often but not always has more clay than the horizon above; blocky, prismatic; friable, firm, hard; pH 4.5–7 often slightly lower than horizon above; MM ped surfaces have conspicuous clay coatings which should exceed 5 per cent of the total soil area, some coatings may be incorporated within the soil mass; Ge the accumulation and formation of clay coatings on surfaces of peds, this usually takes place best in the lower horizons.
From L. *argilla* = clay

Buron Bu (Ah, mollic A horizon)

Upper; 25–35 cm; dark brown; loam to silt; granular or blocky, prismatic with depth; firm to friable rarely hard; pH 7–8.5; CEC 12–20 me per cent; BS 80–100 per cent; OM 2–3 per cent; C/N 8–12; $(Ca + Mg)CO_3$ 1–3 per cent, may be higher from parent material, pseudomycelium and concretions often in the lower part; MM isotropic to weakly anisotropic matrix; Ge incorporation of organic matter in a semiarid environment.
From R. *buri* = brown

Calcon Ck (Bk, Bkm, Bckm, Ck, Ckm, Cckm, calcic horizon)

Middle or lower, may be upper due to erosion; 5–100 cm; pale brown, greyish-brown; blocky, prismatic, massive; firm, friable, hard, when hard Ckd; pH 7.5–8.5; CEC variable; BS 100 per cent;

323

OM <2 per cent; C/N 8–10; $(Ca + Mg)CO_3$ 2–25 per cent as powdery fillings, concretions, pseudomycelium and pendants below stones and as sheets; MM needles of calcite in pores and comprising pseudomycelia, rhombohedral and prismatic crystals of calcite occurring singly, in clusters, lining pores and surrounding rock fragments; Ge accumulation of calcium carbonate.
From L. *calx* = lime

Candon Co (Eg, albic E horizon)
Upper or middle; 10–>50 cm; pale brown or olive, often with ochreous mottling; loam, sandy loam, loam sand, similar or more clay than the horizon above; stones absent to dominant; massive; firm, friable; pH 3.5–5.5; CEC 5–20 me per cent; BS 10–40 per cent; OM <10 per cent; C/N 8–15; MM massive with occasional rounded pore, very pale and transparent anisotropic matrix with small domains, occasional clay coating; Ge leaching and reduction near the surface.
From L. *candela* = candle

Cerulon Cu (Br, Cr, cambic B horizon)
Middle or lower; 20–200 cm; olive, grey or blue, variable texture; massive, permanently saturated, may be prismatic if dried; pH variable usually >5.0; CEC variable; BS >40 per cent; OM <1 per cent; C/N 8–15; $(Ca + Mg)CO_3$ variable from parent material; MM massive with large domains; Ge strongly reduced.
From L. *caeruleus* = blue

Chernon Ch (Ah, mollic A horizon)
Upper, extending into the middle; 50–150 cm; black, very dark grey, dark greyish-brown; silt loam to silt; crumb, granular, subangular blocky, vermicular; firm, friable; pH 6.5–8.0; CEC 15–40 me per cent; BS 50–100 per cent; OM 3–15 per cent >20 kg m^{-2}; C/N 8–15; $(Ca + Mg)CO_3$ 0–60 per cent from parent material, pseudomycelium and concretions in the lower part; vigorous worm activity; MM dominant earthworm and/or enchytralied worm vermiforms; Ge rapid decomposition and incorporation of organic matter in a cool dry environment.
From R. *cherni* = black

Chloron Ci (Az, Bz, Cz, saline horizon)
Upper middle or lower 0–100 cm; yellowish-brown, brownish-grey, mottled; variable texture; massive, prismatic; firm, friable, hard; pH 7.5–8.5; CEC variable; BS 100 per cent; OM <3 per cent; C/N <12;

$(Ca + Mg)CO_3$ variable from the parent material; >80% Cl^{1-} in the saturation extract, EC >4 mmhos; Ge accumulation of chloride ions in an arid or semiarid environment.
From chloride.

Cryon Cy (C, permafrost)
Middle or lower; >100 cm; dark olive, greyish-brown, olive brown; variable texture; variable stones with a coating of ice, thickest on the underside; massive in sands, ice veins in silts and loams with the veins of ice enclosing lenticular mineral soil peds, subcuboidal in clays; very hard; pH 5.5–8.5; CEC 10–50 me per cent; BS 20–100 per cent; OM <3 per cent; C/N 10–40; $(Ca + Mg)CO_3$ variable; MM dense isotropic matrix; ice in veins is crystalline; Ge gradual freezing of a wet mass of soil.
From Gk. *kruos* = frost

Cumulon Cm (Bos, cambic B horizon)
Upper but may extend into the middle position; 10–100 cm; variable colour, often red or reddish-brown; loam to sand, >30 per cent gravel stones and sesquioxide concretions, contains less clay than horizon below; single grain, alveolar, crumb; loose to firm; pH 4.5–6.0; CEC 5–20 me per cent; BS 10–80 per cent; OM <3 per cent; C/N 8–12; $(Ca + Mg)CO_3$ absent; MM abundant quartz gravel and concretions with thin coatings of fine material containing abundant domains; Ge concentration of quartz gravel and concretions by differential erosion, mass movement and termite activity.
From L. *cumulus* = heap

Dermon De (Ah, umbric A horizon)
At the surface; 1–15 cm; black, very dark grey; dark brown; clay loam to clay; massive to coarse platy; hard when dry, plastic when wet; pH 5.5–7.0; CEC 40–70 me per cent; BS 60–100 per cent; OM <2 per cent; C/N 8–12; $(Ca + Mg)CO_3$ 0–50 per cent mainly from parent material, small concretions sometimes present; MM massive or platy with brown isotropic matrix; Ge puddling of the surface, sometimes by rain drop impact.
From Gk. *derma* = skin

Duron Du (Cmq, duripan)
Middle or lower; 10–50 cm; brown, reddish-brown, red; variable texture; massive; hard and cemented; OM <2 per cent; MM deposits of opal in pore spaces and as pendants under stones; sometimes clay coatings; Ge cementation through the deposition of opal.
From L. *durus* = hard

Fermenton Fm (O)
Upper; 1—20 cm; very dark brown to black; laminated with easily recognisable plant fragments; fibrous; pH 3.5—5.0; CEC 80—120 me per cent; BS 20—40 per cent; OM >75 per cent; C/N 25—30; MM dominant partially decomposed plant fragments, faecal pellets and fungal hyphae, sometimes mycorrhizae; Ge accumulation and decomposition or organic matter under aerobic conditions.
From L. *fevere* = to boil

Ferron Fr (Bs, spodic B horizon)
Middle or lower; 20—60 cm; brown, yellowish-brown; sand, loam; stones absent to abundant; fine granular; loose, friable, firm; pH 4.5—5.5; CEC 10—20 me per cent; BS 10—20 per cent; OM <3 per cent; C/N 12—20; MM small granules, singly or in clusters with yellowish-brown isotropic matrix, yellowish brown isotropic coatings on sand grains; Ge accumulation of oxides and hydroxides of iron and aluminium translocated from the horizon above and released by *in situ* weathering.
From L. *ferrum* = iron

Fibron Fi (H, histic H horizon)
Upper or middle; 50—300 cm; very dark brown, black; spongy, massive; fibrous when wet, hard when dry; pH 3.5—4.0 higher when base rich water is present; CEC 80—100 me per cent; BS 10—30 per cent, higher when base rich water is present; OM >75 per cent; C/N 25—40; MM brown laminated >60 per cent recognisable plant remains, some fungal hyphae and sclerotia; Ge accumulation and slow decomposition of organic matter under anaerobic conditions.
From L. *fibra* = fibre

Flambon Fb (C)
Middle or lower; 100—500 cm; mottled red, reddish-brown and cream; sandy clay loam to clay; massive, partially preserved rock structure; firm, plastic, may harden on drying; pH 5.0—7.0; CEC <15 me per cent; BS 10—100 per cent; OM <1 per cent; C/N 6—10; MM partially preserved rock structure, abundant clay coatings, occasional to rare concretions and macro-crystalline kaolinite and gibbsite; Ge hydrolysis and clay mineral formation in moist rock in a humid tropical climate.
From F. *flambe* = flame

Flavon Fv (Bsw, cambic B horizon)
Middle or lower; 50—200 cm; yellowish-brown, brown, yellowish-red, reddish-brown; clay loam to clay >30 per cent silt plus clay; subangular blocky usually with smooth shiny ped surfaces, firm to friable; pH 4.5—5.5; CEC >20 me per cent clay; BS <20 per cent; OM <5 per cent; C/N 8—12; MM massive, incomplete subangular blocky, labyrinthine, abundant random domains, clay coatings absent to rare; Ge progressive hydrolysis through a long period of time.
From L. *flavus* = yellow

Fragon Fg (Cx, fragipan)
Middle or lower with sharp to abrupt upper boundary; 10—100 cm; yellowish-brown, olive or similar colour to the parent material; loam, loamy sand, sandy loam, sand; massive, platy, lenticular; firm to hard, explosive rupture when pressed between thumb and forefinger, compacted and weakly cemented; pH 5.5—7.0; CEC 3—15 me per cent; BS 30—100 per cent; OM <2 per cent; C/N 8—12; $(Ca + Mg)CO_3$ absent; MM massive with rare to occasional pores, domains absent, rare to occasional clay coatings; Ge unknown.
From L. *fragilis* = fragile

Gelon Gn (Ah, some ochric and umbric A horizons)
Upper; 10—100 cm; brown, greyish-brown, yellowish-brown; loams, silts; lenticular; loose friable when thawed, hard when frozen; pH 4.5—7.5; CEC variable; BS variable; OM <10 per cent much as recognisable plant fragments; C/N 12—25; MM lenticular, the lenses are dense and non-porous with irregular underside and smooth upper surfaces coated with fine materials. Ge freezing and ice lense formation during the winter.
From L. *gelare* = to freeze

Gleyson Gl (Bg, cambic B horizon)
Middle or lower; 20—200 cm; grey, olive with yellow or brown mottling >5 per cent prominent mottles; variable texture; massive, prismatic; soft, firm, plastic; pH 4.5—7.0; CEC 10—40 me per cent; BS 20—100 per cent; OM <4 per cent; C/N 8—15; $(Ca + Mg)CO_3$ 0—50 per cent from parent material; MM massive, occasional pores, abundant domains, some clay coatings on pores; Ge alternate oxidation and reduction.
From R. *glei* = grey sticky loam

Glosson Gs (Bg, Cg, cambic B horizon)
Middle or lower; 30—150 cm; brown to red, mottled with pale vertical streaks or tongues; medium to fine texture often silty, finer than horizon above;

massive, prismatic; firm when wet, hard when dry; pH 5.5–7.5; CEC 15–30 me per cent; BS 40–100 per cent; OM <2 per cent, few to many black manganese dioxide concretions; MM the tongues have thick coatings of clay, silt and fine sand, between the tongues is massive with abundant domains, few pores with clay coatings; Ge the tongues may be relic ice lenses or formed by shrinking and cracking then translocation of material down the cracks and removal of iron from the sides of the tongues.
From Gk. *glossa* = tongue

Gluton Gt (Bms, ironstone)
Upper or middle; 10–50 cm; dark brown, reddish-brown, black; massive; hard to very hard; MM sand grains coated and cemented by goethite and manganese dioxide forming a black isotropic matrix; Ge cementation of sandy material by goethite and manganese dioxide brought in by lateral seepage.
From L. *gluten* = glue

Granulon Gr (Al, mull, self-mulching surface)
At the surface; 2–15 cm; black, very dark grey, dark brown; clay loam to clay; distinctive granular structure; firm or hard when dry, plastic when wet; pH 5.5–7.5; CEC 40–70 me per cent; BS 60–100 per cent; OM <2 per cent; C/N 8–12; $(Ca + Mg)CO_3$ 0–50 per cent mainly from parent material, small concretions sometimes present; MM granular structure with brown isotropic matrix; Ge repeated wetting and drying in a dry environment.
From E. = granular

Gypson Gy (By, Bym, Bcym, Cy, Cym, Ccym, gypsic horizon)
Middle or lower; 5–50 cm; pale brown, brown, greyish-brown; massive, blocky, prismatic; firm, hard; pH 7.0–9.0; CEC variable; BS 100 per cent; OM <2 per cent; C/N 8–12; $(Ca + Mg)CO_3$ <5 per cent; $CaSO_4$ >5 per cent; frequent to abundant concretionary clusters of gypsum crystals; MM clusters of well formed gypsum crystals and occasionally individual crystals; Ge gradual growth of gypsum crystals.
From Gk. *gypsus* = gypsum

Gyttjon Gj (H, histic H horizon)
Upper, middle; 20–200 cm; black, very dark brown; massive; pH 3.5–7.5; CEC 50–100 me per cent; BS 10–30 per cent higher when developed in base rich water; OM 30–60 per cent; MM dark brown, isotropic, <30 per cent recognisable plant fragments, some diatoms, phytoliths and fungal hyphae; Ge

accumulation of organic matter under very wet conditions often with free water at the surface.
From S. *gyttja*

Halon Hl (Az, Bz, Cz, saline horizon)
Upper, middle or lower; 10–100 cm; yellowish-brown, brownish-grey; massive, prismatic; firm, friable, hard; pH 7.5–8.5; CEC variable; BS 100 per cent; OM <5 per cent; C/N <12; $(Ca + Mg)CO_3$ 0–50 per cent from the parent material; <80 per cent Cl and 60 per cent SO_4^{2-} in the saturation extract, E.C. >4 mmhos; Ge accumulation of soluble salts in an arid or semiarid environment.
From Gk. *hals* = salt

Hamadon Ha
At the surface; 5–15 cm; black, dark grey, brown; >30 per cent of the surface covered by stones and boulders; Ge accumulation of stones and boulders at the surface following removal of fine material by deflation or surface wash, frost heaving of stones and boulders to the surface.
From Kubiëna hamada

Hudepon Hd (Bh, spodic B horizon)
Middle or lower; 5–60 cm; black, very dark brown; medium to coarse; massive; loose, firm, hard; pH 4.5–5.5; CEC 15–40 me per cent; BS 10–20 per cent; OM 3–25 per cent; C/N 18–22; MM single grain or bridge with very dark brown to black isotropic coatings on the grains; Ge accumulation of translocated humus with only very small amounts of iron and aluminium.
Hu from humus and *dep* from deposit

Humifon Hf (O)
Upper; 1–10 cm; black, very dark brown; plastic; pH 3.5–4.5; CEC 80–100 me per cent; BS 20–50 per cent; OM >75 per cent; C/N 15–20; MM dark brown, fine organic residues; Ge decomposition of organic matter by arthropods, bacteria and fungi.
From E. humification

Husesquon Hs (Bhs, spodic B horizon)
Middle or lower; 10–50 cm; black, very dark brown; loamy sand to loam; weak crumb, blocky, massive; friable, firm, hard; pH 4.5–5.5; CEC 10–30 me per cent; BS 10–30 per cent; OM <20 per cent; C/N 8–22; MM small granules, singly or in clusters, very dark brown to black isotropic coatings or sand grains and rock fragments; Ge accumulation of humus and sesquioxides translocated from the horizon above

also some *in situ* weathering.
Hu from humus and *sesq* from sesquioxide

Hydromoron Hy (O)

Upper; 5–25 cm; very dark brown, black; massive when wet, dries with vertical cracks; plastic; pH 3.0–5.0; CEC 50–100 me per cent; BS 20–40 per cent; OM >50 per cent; C/N 15–25; MM massive, dark brown, isotropic with occasional faecal pellets; Ge accumulation of organic matter under wet but not waterlogged conditions.
From Duchaufour hydromor

Ison In (Cx, fragipan)

Lower with sharp to abrupt upper boundary; 10–200 cm; greyish-brown, olive; loam to sand; massive, lenticular, sometimes with vertical cracks; firm, hard with explosive rupture when pressed between thumb and forefinger; pH 5.5–7.0; CEC 3–15 me per cent; BS 30–100 per cent; OM <2 per cent; C/N 8–12; (Ca + Mg)CO_3 absent to dominant; MM isotropic matrix rare to occasional pore, prominent silt coating on upper surfaces of stones, boulders and sometimes on pore surfaces; Ge disappearance of ice from preceding cryon (permafrost) and the formation of silt coatings.
From OE *is* = ice

Jaron Ja (Br, sulphuric horizon)

Middle or lower; 10–50 cm; mottled grey, bluish-grey, olive; medium to fine texture; massive, prismatic; firm, plastic; pH 3–4.5; CEC 20–40 me per cent; BS 30–50 per cent; OM <3 per cent; MM abundant domains and clusters of jarosite crystals; Ge drainage of a thion causing oxidation of pyrite to form straw coloured jarosite.
From jarosite

Kastanon Kt (Ah, mollic A horizon)

Upper; 35–50 cm; dark brown, very dark grey; loam, silt; crumb or granular sometimes prismatic in the lower part; firm, friable; pH 7.0–8.5; CEC 10–40 me per cent; BS 100 per cent; OM 2–5 per cent, 15–20 kg m^{-2}; C/N 8–12; (Ca + Mg)CO_3 variable from parent material, pseudomycelium and concretions in the lower part; MM abundant faecal material and worm casts, isotropic, needles of calcite in lower part; Ge rapid decomposition and incorporation of organic matter in a cool semiarid environment.
From R. *kastano* = chestnut

Krasnon Ks (Bsw, oxic B horizon)

Middle or lower; 50–200 cm; red; clay; blocky, weak crumb, massive; pH 4.5–6.0; CEC <15 me/100 g clay; BS usually <20 per cent; OM <2 per cent; C/N 8–12; <30 per cent concretions; MM incomplete structure, very small domains seen only in sections 15 μm thick; Ge very advanced weathering in humid tropical environment.
From R. *krasni* = red

Kuron Ku (Ah, umbric A horizon)

Upper; 10–25 cm; black, very dark brown; loam to silt; granular, crumb, subangular blocky; friable, greasy; dominant allophane; pH 4.5–6.5; CEC 20–50 me per cent, variable with technique; BS 10–30 per cent; OM 5–20 per cent; C/N 12–20; MM isotropic matrix; Ge decomposition and blending of organic and mineral material.
From J. *kurobou* = dark

Limon Lm

Upper, middle; 3–100 cm; grey to white; firm, plastic; pH 7.5–9.0; OM <5 per cent; (Ca + Mg)CO_3 >30 per cent; Ge accumulation of precipitated calcium carbonate and calcareous skeletons of aquatic animals on the bottom of a lake.
From L. *limus* = slime

Lithon Lh

Upper, middle or lower; 20–100 cm; brown; <30 per cent fine earth; dominantly angular rock fragments; pH variable; OM <2 per cent; Ge physical shattering of rock *in situ* or moved only a short distance down the slope.
From Gk. *lithos* = stone

Luton Ln (Ahg, some ochric and umbric A horizons)

Upper; 10–30 cm; black, dark brown, brown with faint mottling particularly along root passages; medium to fine texture; massive, blocky; firm, hard, plastic; pH 5.0–7.5; CEC 15–100 me per cent; BS 30–100 per cent; OM 1–10 per cent; C/N 8–15; (Ca + Mg)CO_3 variable from parent material; MM alveolar, isotropic matrix, some vermicular areas; brown staining around some pores; Ge decomposition and incorporation of organic matter in a moist soil, degradation of structure due to poor utilisation.
From L. *lutum* = mud

Luvon Lv (E, albic A horizon)

Upper; 3–80 cm; very pale; medium to coarse texture; single grain, weak crumb, lenticular; loose,

327

firm; pH 3.5–5.5; CEC 5–20; BS 10–20 per cent; OM <5 per cent; C/N 10–20; MM single grain, bridge, alveolar, isotropic; Ge leaching and/or the destruction of clay.

From L. *luo* = to wash

Marblon Mb (Bg, cambic B horizon)

Middle or lower; 10–50 cm; yellow to brown with marbled pattern, ped surfaces often grey; massive prismatic, blocky; firm, friable, hard; pH 5.0–7.5; CEC 15–30 me per cent; BS 40–100 per cent; OM <3 per cent; C/N 8–12; MM alveolar with frequent to abundant domains, occasional clay coatings; Ge weak reducing conditions for a short period during each year.

From Gk. *marmaros* = sparkle

Minon Mi (Eg, albic E horizon)

Upper; 3–80 cm; pale to very pale with yellow and brown mottling; medium to coarse texture with significantly less clay than the horizon below; single grain, massive, weak crumb; loose, firm, may be very hard when dry; pH 3.5–5.5; CEC 5–20 me per cent; BS 10–20 per cent; OM <5 per cent; C/N 8–20; occasional to abundant black concretions; MM alveolar with occasional to abundant domains, occasional to abundant ochreous mottles; Ge removal or destruction of clay in a partially anaerobic environment, formation of mottles and concretions.

From L. *minus* = less

Modon Mo (Ah, umbric A horizon)

Upper; 3–20 cm; dark to very dark; loams to loamy sand; massive or poorly developed blocky; firm, friable, slightly plastic; pH 3.5–6.5; CEC 10–40 me per cent; BS 10–30 per cent; OM 5–20 per cent; C/N 15–20; $(Ca + Mg)CO_3$ variable from parent material; MM varies from massive black and isotropic with abundant sand grains to mixed single grains and black isotropic granules. Ge weathering and incorporation of organic matter at low pH values.

From Kubiëna's moder

Mullon Mu (Ah, mollic A horizon)

Upper; 10–50 cm; dark to very dark; medium to fine texture; very well developed crumb, granular, vermicular; pH 5.5–7.5; CEC 20–80 me per cent; BS 40–100 per cent; OM 3–10 per cent; C/N 8–12; $(Ca + Mg)CO_3$ variable from parent material; MM well developed alveolar, granular or vermicular, isotropic with variable sand and silt; Ge vigorous mixing of organic and mineral material by soil fauna particularly earthworms.

From E, mull

Nekron Nk (Ah, umbric A horizon)

Upper; 5–20 cm; very dark when wet but becomes significantly lighter when dry with change of 2–4 units of value; humose loam to loamy sand; massive when wet, vertical cracks form upon drying; firm, friable, slightly plastic; pH 3.0–5.0; CEC 20–80 me per cent; BS 10–40 per cent; OM 10–20 per cent many decomposing roots; C/N 15–20; MM massive black isotropic humose matrix with abundant sand grains, many of the feldspars and other weatherable minerals show signs of weathering, occasional decomposing root; Ge weathering of minerals and decomposition of dead roots under wet conditions.

From Gk. *nekros* = corpse

Oron Or (Bms, ironstone)

Middle or lower; 5–30 cm; black, very dark brown; medium to coarse texture; single grain, massive, nodular; loose, firm, hard; Ge accumulation of hydrous oxides of iron and manganese at the top of the water-table.

From OE. *ar* = brass

Pallon Pl

Middle or lower; 100 to more than 300 cm; white, very pale grey; clay loam to clay; massive or prismatic; plastic but hardens on drying; pH 4.5–5.5; CEC 10–20 me per cent; BS <50 per cent; OM <1 per cent; some pallons have up to 0.5 per cent sodium chloride which seems to be a secondary accumulation; MM massive, partially preserved rock structure, variable amounts of vermiform kaolinite; Ge progressive hydrolysis and clay mineral formation in a lower horizon of complete saturation.

From L. *paller* = pale

Pelon Pe (Bw, cambic B horizon)

Middle or lower; 20–50 cm; brown, reddish-brown, red, black, grey; sandy clay to clay from parent material; blocky prismatic, massive, with some slickensides; firm, hard, plastic; pH 5.0–7.5; CEC 20–40 per cent; BS 30–100; OM <2 per cent; C/N 4–12; $(Ca + Mg)CO_3$ <30 per cent from parent material; MM wedge, incomplete blocky frequent small domains occasional anisotropic lines; Ge weak expansion and contraction, weak weathering.

From Gk. *pelos* = mud

Pesson Ps (Bcs, cambic B horizon)

Middle or lower; 20–200 cm; reddish-brown, red, very dark brown; sandy clay to clay; massive with abundant concretions; firm, hard, cemented;

pH 4.5–5.5; OM <1 per cent; MM abundant
concretions in a brown matrix containing macrocrystalline kaolinite.
From Gk. *pessos* = oval stone

Petron Pt

Middle or lower; <20 cm; abundant to dominant
fragments of rock or laterite.
From L. *petra* = oval stone

Placon Pk (Bsm, thin iron pan)

Middle or lower and may have a very irregular
boundary and usually continuous through stones;
<2 cm; very dark brown, black; massive, hard to very
hard; OM <1–20 per cent; MM black, dark brown,
commonly red, reddish-yellow, dominantly isotropic,
uniform or granular; Ge not fully understood, it
seems that various organic substances in the soil
solution chelate iron as they percolate through the
soil and at a critical concentration precipitation takes
place.
From Gk. *plax* = plate

Plaggon Pg (Ap, plaggen A horizon)

Upper; sometimes middle; 50–100 cm; dark brown,
brown; medium texture; crumb, granular, blocky;
firm, friable, pH 5.5–7.0; CEC 20–40 me per cent;
BS 60–100 per cent; OM 3–15 per cent; C/N 8–15;
Ge addition of organic and mineral material to the
surface by man to maintain fertility.
From G. *plaggen* = meadow

Planon Pn (Btg, argillic B horizon)

Middle or lower usually with sharp upper boundary;
30–160 cm; brown with distinct mottling; medium
to fine texture with significantly more clay than the
horizon above; prismatic, angular blocky, massive;
firm, hard, plastic; pH 5.5–7.0; CEC 20–50 me per
cent; BS 20–100 per cent; OM <2 per cent;
C/N 8–15; concretions few to abundant; MM massive,
incomplete angular blocky, frequent anisotropic
aureoles, occasional to rare clay coatings; Ge
weathering and clay mineral formation in a wet
environment may contain some clay translocated
from the horizon above.
From L. *planus* = level

Primon Pr (Ah)

Upper; 5–20 cm; brown, greyish-brown. grey;
OM 1–5 per cent; Ge mixing of organic and mineral
material in the initial stages of soil formation.
From L. *primus* = first

Pseudofibron Pd (H, histic H horizon)

Upper, middle or lower; 10–20 cm; black, very dark
brown; massive, appears fibrous but is easily broken
down between the fingers; spongy to plastic;
pH 3.0–5.0 but may be higher; CEC 80–120 me per
cent; BS 10–30 per cent but may be higher;
OM >60 per cent; C/N 25–40; MM dark brown,
mainly isotropic with <30 per cent recognisable
plant fragments, some diatoms, phytoliths and fungal
hyphae may be present; Ge accumulation of partly
decomposed organic matter under anaerobic
conditions.
Gk. *pseudo* = false L. *fibra* = fibre

Rosson Ro (Bsw, cambic B horizon)

Middle or lower; 20–100 cm; red, reddish-brown,
2.5YR or redder; >30 per cent clay; granular, blocky;
firm, friable, plastic; pH 4.5–7.0; CEC >20 me per
cent clay; BS >30 per cent; OM <5 per cent; C/N
8–12; MM incomplete blocky, abundant small
domains, absent to rare clay coatings; Ge hydrolysis
and clay mineral formation in a warm or hot humid
climate.

Rubon Ru (Bsw, cambic B horizon)

Middle or lower; 20–100 cm; red, reddish-brown,
2.5YR or redder; loam to clay; alveolar, granular,
blocky; firm, friable, plastic; pH 4.5–5.5; CEC <20 me
cent; OM <2 per cent; C/N 8–12; 5–20 per cent
weatherable minerals; MM incomplete blocky,
abundant small and medium domains; absent to rare
clay coatings; Ge hydrolysis and clay mineral
formation in hot wet and dry conditions.
From L. *rufus* = brownish-red

Rufon Rf (Bsw, cambic B horizon)

Middle or lower; 20–100 cm; red, reddish-brown,
2.5YR or redder; loam to clay; alveolar, granular,
blocky; firm, friable, plastic; pH 4.5–5.5; CEC <20 me
per 100 g clay; BS >30 per cent; OM <5 per cent;
C/N 8–12; 5–20 per cent weatherable minerals; MM
incomplete blocky, abundant small and medium
domains; absent to rare clay coatings; Ge hydrolysis
and clay mineral formation in a warm or hot climate.
From *rufus* = brownish red.

Sapron Sa (C, weathered rock)

Middle or lower; 20–100 plus cm; dark brown,
yellowish-brown, reddish-brown, brownish-yellow;
loam to clay; massive with partially preserved rock
structure; loose, firm; Ph 4–7; CEC 5–20 me per
cent; BS 10–50 per cent; OM <1 per cent;
C/N 8–12; MM partially preserved rock structure,

329

Table 11.1 Soil horizons and symbols

FitzPatrick		World		USDA	
Name	Symbol	Name	Symbol	Name	Symbol
Alkalon	Ak	Saline horizon	Az, Bz, Cz	Salic horizon	Asa, Bsa, Csa
Alton	At	Cambic B horizon	Bw	Cambic horizon	B2
Amorphon	Ap	Histic H horizon	Ha	Histic epipedon	O2
Andon	An	Cambic B horizon	Bw	Cambic horizon	B2
Anmooron	Am	Umbric A horizon	Ah	Umbric epipedon	A1
Arenon	Ae		Bsw		B2
Argillon	Ar	Argillic B horizon	Bt	Argillic horizon	B2t
Buron	Bu	Mollic A horizon	Ah	Mollic epipedon	A1
Calcon	Ck	Calcic horizon	Bk, Bkm, Bckm, Ck, Ckm, Cckm	Calcic horizon	Cca
Candon	Co	Albic E horizon	Eg	Albic horizon	A2g
Cerulon	Cu	Cambic B horizon	Br, Cr	Cambic horizon	B2g, Cg
Chernon	Ch	Mollic A horizon	Ah	Mollic epipedon	A1
Chloron	Ci	Saline horizon	Az, Bz, Cz	Salic horizon	Asa, Bsa, Csa
Cryon	Cy	Permafrost	C	Permafrost	Cf
Cumulon	Cm	Cambic B horizon	Bcs	Cambic horizon	B2
Dermon	De	Umbric A horizon	Ah	Umbric epipedon	A1
Duron	Du	Duripan	Cmq	Duripan	Csi
Fermenton	Fm	–	O	–	O1
Ferron	Fr	Spodic B horizon	Bs	Spodic horizon	Bir
Fibron	Fi	Histic H horizon	Hi	Histic epipedon	O1
Flambon	Fb	Plinthite (some)	C	Plinthite (some)	C
Flavon	Fv	Cambic B horizon	Bsw	Cambic horizon	B2
Fragon	Fg	Fragipan	Cx	Fragipan	Cx
Gelon	Gn	Umbric or ochric A horizon	Ah	Umbric or ochric epipedon	A1
Gleyson	G1	Cambic B horizon	Bg, Cg	Cambic horizon	Bg
Glosson	Gs	Cambic B horizon	Bg, Cg	Cambic horizon	B & A
Gluton	Gt	Ironstone	Bms		Cm
Granulon	Gr	Ochric A horizon	Ah	Ochric epipedon	A1
Gypson	Gy	Gypsic horizon	By, Bym, Bcym, Cy, Cym, Ccym	Gypsic horizon	Ccs
Gyttjon	Gj	Histic H horizon	Ha	Histic epipedon	O2
Halon	Hl	Saline	Az, Bz, Cz	Saline horizon	Asa, Bsa, Csa
Hamadon	Ha	–	A	–	–
Hudepon	Hd	Spodic B horizon	Bh	Spodic horizon	Bh
Humifon	Hf	–	O	–	O2
Husesquon	Hs	Spodic B horizon	Bhs	Spodic horizon	Bhir
Hydromoron	Hy	–	O	–	O2
Ison	In	Fragipan	Cx	Fragipan	Cx
Jaron	Ja	Sulphuric horizon	Br	Sulphuric horizon	B2g, Cg

pseudomorphs of oxides of clay minerals; Ge profound weathering under humic tropical and subtropical conditions.
Gk. *sapros* = rotten.

Seron Sn (Ah, ochric A horizon)
Upper; 5–20 cm; often thin due to erosion; yellowish-brown, reddish-brown, red; variable texture often contains less clay than the horizon below; weak laminar, alveolar, massive with many pores; pH 7.0–8.0; CEC 5–30 me per cent; BS 100 per cent; OM 1–2 per

cent; C/N 8–12; $(Ca + Mg)CO_3$ 1–5 per cent may be higher from parent material then Sec; MM isotropic matrix, thin coatings of calcite on ped surface; Ge wetting and drying with little leaching in a hot dry environment, some deflation.
From R. *seri* = grey

Sesquon Sq (Bs, spodic B horizon)
Middle or lower; <30 cm—>1 m; brown to dark brown; loamy sand to loam; weak crumb; massive, blocky; friable to firm to hard; pH 4.5–5.5;

Table 11.1 (continued)

FitzPatrick		World		USDA	
Name	Symbol	Name	Symbol	Name	Symbol
Kastanon	Kt	Mollic A horizon	Ah	Mollic epipedon	A1
Krasnon	Ks	Oxic B horizon	Bsw	Oxic horizon	B2
Kuron	Ku	Umbric A horizon	Ah	Umbric epipedon	A1
Limon	Lm	–	–	–	–
Lithon	Lh	–	–	–	–
Litter	Lt	Litter	O	Litter	O1
Luton	Ln	Umbric A horizon	Ahg	Umbric epipedon	A1
Luvon	Lv	Albic E horizon	E	Albic horizon	A2
Marblon	Mb	Cambic B horizon	Bg	Cambic horizon	B2
Minon	Mi	Albic E horizon	Eg	Albic horizon	A2g
Modon	Mo	Umbric A horizon	Ah	Umbric epipedon	A1
Mullon	Mu	Mollic A horizon	Ah	Mollic, umbric epipedon	A1
Nekron	Nk	Umbric A horizon	Ah	Umbric epipedon	A1
Oron	Or	Ironstone	Bms	–	Birm
Pallon	Pl	–	C	–	C
Pelon	Pe	Cambic B horizon	Bw	Cambic horizon	B2
Pesson	Ps	Cambic B horizon	Bcs	Cambic horizon	Bcn
Petron	Pt	–	B	–	–
Placon	Pk	Thin iron pan	Bms	Placic horizon	Birm
Plaggon	Pg	Plaggen A horizon	Ap	Plaggen epipedon	Ap
Planon	Pn	Argillic B horizon	Btg	Argillic horizon	B2tg
Primon	Pr	Ochric A horizon	Ah	Ochric epipedon	A1
Pseudofibron	Pd	Histic H horizon	He	Histic epipedon	O2
Rosson	Ro	Cambic B horizon	Bsw	Cambic horizon	B2
Rubon	Ru	Cambic B horizon	Bsw	Cambic horizon	B2
Rufon	Rf	Cambic B horizon	Bsw	Cambic horizon	B2
Sapron	Sa	–	C	–	C
Seron	Sn	Ochric A horizon	Ah	Ochric epipedon	A1
Sesquon	Sq	Spodic B horizon	Bs	Spodic horizon	Bir
Sideron	Sd	–	–	–	O2
Solon	Sl	Natric B	Btn	Natric horizon	Bt
Sombron	So	–	Bh	Sombric horizon	Bth
Sulphon	Su	Saline horizon	Az, Bz, Cz	Salic horizon	Asa, Bsa, Csa
Tannon	Tn	Ochric A horizon	Ah	Ochric epipedon	A1
Thion	To	Sulphidic material	Br, Cr	Sulphidic material	B2g, Cg
Verton	Ve	Vertic properties	Bu	–	A1
Veson	Vs	Plinthite (some) Ironstone (some)	Bms	Plinthite (some)	C
Zhelton	Zh	Oxic B horizon	Bsw	Oxic horizon	B2
Zolon	Zo	Albic E horizon	E	Albic horizon	A2

CEC 10–30 me per cent; OM 3–10 per cent; C/N 12–20; MM small brown granules, singly or in clusters, brown sesquioxide coatings on sand grains and rock fragments; Ge accumulation of sesquioxides and humus translocated from the horizon above also some *in situ* weathering.
From L. *sesqui* = one and a half

Sideron Sd
Middle or lower; 20–100 cm; grey to dark brown becomes dark rusty brown when dry; massive; plastic;

pH 4.5–6.5; OM <50 per cent; C/N 15–40; MM massive, laminar, some plant fragments, small concretions, crystals and needles of siderite; Ge formation of siderite in an anaerobic environment usually in peat.
From siderite

Solon Sl (Btn, natric B horizon)
Middle or lower; 20 to 100 cm; grey, greyish-brown, brown with fine yellow mottles and sometimes with manganese dioxide concretions; loam to clay with

significantly more clay that the horizon above; columnar or prismatic; firm, hard, plastic; pH >8.5 in some part; CEC 15–30 me per cent; BS 100 per cent; >15 per cent exchangeable sodium; OM <3 per cent; soluble salts <4 mmhos; MM massive within columns or prisms variable domains, clay cloating may be present; Ge formation of clay *in situ* and some translocated from above.
From R. *sol* = salt

Sombron So (Bh, sombric horizon)

Middle; 20–50 cm; dark brown to black; loam to clay; blocky, prismatic; pH 5.0–6.0; CEC 15–30 me per cent; BS <50 per cent; OM 2–10 per cent; MM coating of organic matter on ped and pores; Ge deposition of translocated humus.

Sulphon Su (Az, Bz, Cz, saline horizon)

Upper, middle or lower; >20 cm; greyish-brown, mottled, olive brown; variable texture; massive, prismatic; firm, hard; pH 8.0–9.0; CEC variable; BS 100 per cent; OM <2 per cent; C/N 12–20; $(Ca + Mg)CO_3$ variable; soluble salts >0.5 per cent; >60% SO_4^{2-} in the saturation extract, EC >4 mmhos; Ge accumulation of sulphate ions in an arid or semiarid environment.
From sulphur.

Tannon Tn (Ah, ochric A horizon)

Upper; 5–25 cm; brown, yellowish-brown, reddish-brown; variable texture; crumb, granular, blocky; firm, friable, plastic; pH 4.5–7.0; CEC 5–20 me per cent; BS 10–100 per cent; OM <2 per cent; C/N 8–12; $(Ca + Mg)CO_3$ variable from the parent material; MM crumb, granular, vermicular, isotropic to weakly anisotropic; Ge mixing of organic and mineral material by soil fauna.
From C. *tann* = oak

Thion To (Br, Cr sulphidic material)

Middle or lower; 50–100 cm; grey, olive, bluish-grey with black pyrite grains; medium to fine texture; massive, prismatic when dry; soft, firm, hard, plastic; pH about neutral when wet, very acid when dry; CEC 20–40 me per cent; BS 30–50 per cent; OM 0–30 per cent; C/N 20–35; MM massive, abundant domains random pyrite associated with organic matter; Ge reduction of iron and formation of pyrite in an anaerobic environment.
From Gk. *theion* = sulphur

Verton Ve (A1, vertic properties)

Middle or lower; 20–200 cm; black, very dark grey, very dark brown; >30 per cent expanding lattice clay; wedge or blocky; hard, firm plastic; pH 5.5–7.0; CEC 50–100 me per cent; BS 50–100 per cent; OM <2 per cent; C/N 8–12; sometimes small calcite concretions; MM characteristic wedge structure, moderately anisotropic with abundant small domains; Ge clay mineral formation in a basic environment and churning due to expansion and contraction in response to wetting and drying.
From L. *vertere* = to turn

Veson Vs (Bms, some ironstone)

Upper, middle or lower; 50–300 cm; mottled red, brown and yellowish-brown; vesicular; plastic, hard.
From L. *vesica* = cyst

Zhelton Zh (Bsw, oxic B horizon)

Middle or lower; 25–200 cm; yellowish-brown, brown, reddish-yellow; >30 per cent clay; blocky; firm, friable, plastic; pH 5.5–7.5; CEC <15 me per cent clay; BS 10–50 per cent; OM <5 per cent; MM incomplete blocky, labyrinthine, moderately to strongly anisotropic matrix with large domains; Ge hydrolysis and clay mineral formation in moist hot environment.
From R. *zhelti* = yellow

Zolon Zo (E, albic E horizon)

Upper or middle; 1 cm to > 1 m; grey, pale brown; loam to sand; loose, friable, rarely hard; pH 3.5–5.0; CEC 5–20 me per cent; BS <20 per cent; OM 0.5–5 per cent; C/N 16–20; MM single grain, alveolar, isotropic matrix; Ge strong weathering and removal by leaching of iron, aluminium, clay and other cations.
From R. *zola* = ash

Appendix III

Table 11.2 South African Soil Forms and their horizon sequences

Arcadia:	vertic A
Avalon:	orthic A, yellow-brown apedal B, soft plinthic B
Bainsvlei:	orthic A, red apedal B, soft plinthic B
Bonheim:	melanic A, pedocutanic B
Cartref:	orthic A, E horizon, lithocutanic B
Champagne:	organic 0
Clovelly:	orthic A, yellow-brown apedal B
Constantia:	orthic A, E horizon, yellow-brown apedal B
Dundee:	orthic A, stratified alluvium
Estcourt:	orthic A, E horizon, prismacutanic B
Fernwood:	orthic A, regic sand
Glencoe:	orthic A, yellow-brown apedal B, hard plinthic B
Glenrosa:	orthic A, lithocutanic B
Griffin:	orthic A, yellow-brown apedal B, red apedal B
Houwhoek:	orthic A, E horizon, ferrihumic B, saprolite
Hutton:	orthic A, red apedal B
Inanda:	humic A, red apedal B
Inhoek:	malanic A, stratified alluvium or neocutanic B
Katspruit:	orthic A, G horizon
Kranskop:	humic A, yellow-brown apedal B, red apedal B
Kroonstad:	orthic A, E horizon, gley cutanic B
Lamotte:	orthic A, E horizon, ferrihumic B, unconsolidated material
Longlands:	orthic A, E horizon, soft plinthic B
Magwa:	humic A, yellow-brown apedal B
Mayo:	melanic A, lithocutanic B
Milkwood:	melanic A, hard rock or ferricrete or calcrete or silcrete or dorbank
Mispah:	orthic A, hard rock or ferricrete or calcrete or silcrete or dorbank
Nomanci:	humic A, lithocutanic B
Oakleaf:	orthic A, neocutanic B
Plinedene:	orthic A, yellow brown apedal B, gleycutanic B
Rensburg:	vertic A, G horizon
Shepstone:	orthic A, E horizon, red apedal B
Shortlands:	orthic A, red structured B
Sterkspruit:	orthic A, prismacutanic B
Swartland:	orthic A pedocutanic B, saprolite
Tambankulu:	melanic A, soft plinthic B
Valsriver:	orthic A, pedocutanic B, unconsolidated material
Vilafontes:	orthic A, E horizon, neocutanic B
Wasbank:	orthic A, E horizon, hard plinthic B
Westleigh:	orthic A, soft plinthic B
Willowbrook:	melanic A, G horizon

Glossary

Abrasion: The physical weathering of a rock surface by running water, glaciers or wind laden with fine particles. See **Roches moutonnées, Ventifact**.

Absorption: The physical uptake of water and/or ions by a substance. For example, soils absorb water.

Accelerated erosion: An increased rate of erosion usually caused by man's improper use of the land.

Accessory minerals: Minerals occurring in small quantities in a rock whose presence or absence does not affect the true nature of the rock.

Accretion: The gradual addition of soil or sediment to the surface, may result from movement down-slope or flooding by a river.

Accumulation: An increase formed by transportation and deposition of one, or more of the existing constituents in the system. Often refers to the residuum left after the removal of one or more constituents.

Acidity: A measure of the activity of the hydrogen and aluminium ions in wet soil; usually expressed as pH value.

Acid rock: An igneous rock that contains more than 65 per cent silica.

Acid soil: Specifically a soil with pH value < 7.0 caused by the presence of active hydrogen and/or aluminium ions. The pH value decreases as the activity of these two ions increases.

Actinomycetes: A group of organisms intermediate between the bacteria and the true fungi, mainly resembling the latter because they usually produce branched mycelium.

Adhesion: The molecular attraction between two substances causing their surfaces to remain in contact for example the present of water on rock fragments.

Adsorption: The attachment of a particle, ion or molecule to a surface. Calcium is adsorbed on to the surface of clay or humus.

Adsorption complex: The various substances in the soil that are capable of adsorption, these are mainly clay and humus.

Aeolian: Pertaining to or formed by wind action.

Aeolian deposits: Fine sediments transported and deposited by wind, they include loess, dunes, desert sand and some volcanic ash.

Aeration: The process by which atmospheric air enters the soil. The rate and amount of aeration depends upon the size and continuity of the pore spaces and the degree of waterlogging. The atmosphere in well aerated soils differs only slightly from the atmosphere above the soil surface; poorly aerated soils usually have a higher content of carbon dioxide.

Aerial photo: See **Aerial photograph**.

Aerial photograph: A photograph of the earth's surface taken from an aeroplane or some other type of airborne equipment.

Aerobic: (*a*) Conditions having a continuous supply of molecular oxygen.
(*b*) A horizon that is usually moist or dry with a sufficiently moist period to support plant growth.

Aerobic decomposition: The breakdown of a substance in the presence of molecular oxygen.

Aerobic organism: Organisms living or becoming active in the presence of molecular oxygen.

Afforestation: The planting of a forest crop on land that has not previously or not recently carried a forest crop.

Aggregate: A cluster of soil particles forming a ped (cf. **Fragment**).

Aggregation: The process during which particles coalesce to form aggregates.

Agronomy: That part of agriculture devoted to the production of crops and soil management – the scientific utilisation of agricultural land.

Air-dry: The dryness of a soil when at equilibrium with the moisture content of the surrounding atmosphere. The atmosphere is usually that of the laboratory.

Alkali: A substance capable of furnishing hydroxyl-OH ions to its solution. The two most important alkali metals are potassium and sodium.

Alkaline soil: Specifically a soil with pH value > 7.0 caused by the presence of carbonates of calcium, magnesium, potassium and more especially sodium. Usually refers to soils with pH values > 8.5.

Allochtonous: A term applied to features not formed *in situ*.

Alluvial fan or alluvial cone: Sediments deposited in a characteristic fan or cone shape by a mountain stream as it flows on to a plain or flat open valley.

Alluvial plain: A flat area built up of alluvium.

Alluvial soil: A general term for those soils developed on fairly recent alluvium, usually they show no horizon development.

Alluvium: A sediment deposited by streams and varying widely in particle size. The stones and boulders when present are usually rounded or subrounded. Some of the most fertile soils are derived from alluvium of medium or fine texture.

Amino acid: An organic compound containing both the amino (NH_2) and carboxyl (COOH) groups. Amino acid molecules combine to form proteins, therefore they are a fundamental constituent of living matter. They are synthesised by autotrophic organisms principally green plants.

Ammonia fixation: Adsorption of ammonium ions by clay minerals, rendering them insoluble and non-exchangeable.

Ammonification: The production of ammonia by micro-organisms through the decomposition of organic matter.

Amorphous humus: Strongly decomposed organic material of colloidal size.

Anaerobic: Conditions that are free of molecular oxygen. In soils this is usually caused by excess wetness.

Anaerobic organism: One that lives in an environment without molecular oxygen.

Anastomosing: Branching and rejoining of branches.

Anhedral: Form of minerals which are not bounded by distinct (own) crystal faces.

Anion: An ion having a negative electrical charge.

Anion exchange capacity: The total amount of anions that a soil can absorb, usually expressed as milliequivalents per 100 g of soil.

Anisotropy: The quality of being anisotropic. A substance is said to be anisotropic if the magnitude of one or more of its properties is dependent on the direction in which it is measured in the system. When the measured magnitudes are independent of the direction considered, the substance is isotropic for the properties in question. (1) Soil: The presence of different superposed horizons in soil makes vertical anisotropy an essential characteristic of soil. (2) Optically: all-crystalline materials, not belonging to the cubic crystal system are optically anisotropic as is expressed by their birefringence.

Annelid: Red-blooded worm such as an earthworm.

Annual plant: A plant that completes its life cycle within one year.

Apparent specific gravity: See **Bulk density.**

Aréte: A sharp, jagged mountain ridge.

Arid: A term applied to a region or climate in which pre-cipitation is too low to support crop production.

Arthropod: A member of the phylum Arthropoda which is the largest in the animal kingdom. It includes, insects, spiders, centipedes, crabs, etc.

Aspect: The compass direction of a slope.

Auger: See **Soil auger.**

Autochthonous: A term applied to features formed *in situ.*

Autotrophic organism: Organisms that utilise carbon dioxide as a source of carbon and obtain their energy from the sun or by oxidising inorganic substances such as sulphur, hydrogen, ammonium, and nitrite salts. The former include the higher plants and algae and the latter various bacteria (cf. **Heterotrophic**).

Available elements: The elements in the soil solution that can readily be taken up by plant roots.

Available nutrients: See **Available elements.**

Available water: That part of the water in the soil that can be taken up by plant roots.

Available water capacity: The weight percentage of water which a soil can store in a form available to plants. It is approximately equal to the moisture content at field capacity minus that at the wilting point.

Bacteria: Unicellular or multicellular microscopic organisms. They occur everywhere and in very large numbers in favourable habitats such as sour milk and soil where they number many millions per gram.

Badlands: An arid region with innumerable deep gullies caused by occasional torrential rain. The distribution and total precipitation are insufficient to support a protective vegetative cover.

Bar: 10^5 Pascal or 10^5 (Nm^{-2})

Basalt: A fine grained igneous rock forming lava flows or minor intrusions. It is composed of plagioclase, augite and magnetite; olivine may be present.

Base level: The level to which a river can lower its bed. Sea level is the permanent base level but there may be many temporary base levels provided by lakes, other rivers or resistant rock strata.

Base saturation: The extent to which the exchange sites of a material are occupied by exchangeable basic cations; expressed as a percentage of the cation-exchange capacity.

Basic rock: An igneous rock that contains less than 55 per cent silica.

Bedrock: The solid rock at the surface of the earth or at some depth beneath the soil and superficial deposits.

Bentonite: A very plastic clay that swells extensively when wet. It consists of montmorillonite and beidellite.

Biennial: A plant that completes its life cycle in two years.

Biomass: (*a*) The weight of a given organism in a volume of soil that is 1 m^2 at the surface and extending down to the lower limit of the organism's penetration.
(*b*) The weight of an organism or number of organisms in a given area or volume.

Biota: The flora and fauna of a region.

Birefringence: The numerical difference in value between the highest and lowest refractive index of a mineral.

Birefringent: Having birefringence.

Bleicherde: The light coloured bleached horizon in Podzols and other soils.

Bog iron ore: A ferruginous deposit in bogs and swamps formed by oxidising algae, bacteria or the atmosphere on iron in solution.

Boulder clay: See **Till.**

BS: An abbreviation for **Base saturation.**

Buffer: A substance that prevents a rapid change in pH when acids or alkalis are added to the soil, these include clay, humus and carbonates.

Bulk density: Mass per unit volume of undisturbed soil, dried to constant weight at 105 °C. Usually expressed as g cm^{-3}.

Calcareous soil: A soil that contains enough calcium carbonate so that it effervesces when treated with hydrochloric acid.

Calcification: Used by some to refer to the processes of calcium carbonate accumulation. This term is not specific and should not be used.

Calcite: Crystalline calcium carbonate, $CaCO_3$. Crystallises in the hexagonal system, the main types of crystals in soils being prismatic, needle shaped, nodular, granular and compact.

Caliche: A layer or horizon cemented by the deposition of calcium carbonate. It usually occurs within the soil but may be at the surface due to erosion.

Capillarity: The process by which moisture moves in any direction through the fine pores and as films around particles.

Capillary fringe: The zone just above the water-table that remains practically saturated with water.

Capillary moisture: That amount of water that is capable of movement after the soil has drained. It is held by adhesion and surface tension as films around particles and in the finer pore spaces.

Catena: A sequence of soils developed from similar parent material under similar climatic conditions but whose characteristics differ because of variations in relief and drainage.

Cation: An ion having a positive electrical charge.

Cation exchange: The exchange between cations in solution and another cation held on the exchange sites of minerals and organic matter.

Cation-exchange capacity: The total potential of soils for adsorbing cations, expressed in milligram equivalents per 100 g of soil. Determined values depend somewhat upon the method employed.

CEC: An abbreviation of Cation-Exchange Capacity.

Cemented: Massive and either hard or brittle depending upon the degree of cementation by substances such as calcium carbonate, silica, oxides of iron, aluminium and manganese, or humus.

Chalk: The term refers to either (*a*) soft white limestone which consists of pure calcium carbonate and leaves little residue when treated with hydrochloric acid, sometimes consists largely of the remains of foraminifers, echinoderms, molluscs and other marine organisms, or (*b*) the upper or final member of the Cretaceous System.

Chroma: The relative purity of a colour directly related to the dominance of the determining wavelength. One of the three variables of colour.

Chronosequence: A sequence of soils that change gradually from one to the other with time.

Cirque: A large semiamphitheatre or armchair excavation in the mountains formed by ice erosion.

Clay: Either (1) Mineral material $< 2 \mu m$; (2) A class of texture; (3) Silicate clay minerals.

Clay mineral: Crystalline or amorphous mineral material, $< 2 \mu m$ in diameter.

Claypan: A middle or lower horizon containing significantly more clay than the horizon above. It is usually very dense and has a sharp upper boundary. Claypans generally impede drainage, are usually plastic and sticky when wet and hard when dry.

Cleavage: The ability of a mineral or rock to split along predetermined planes.

Climate, continental: see Continental climate

Climate, oceanic; See Oceanic climate.

Climax vegetation: A fully developed plant community that is in equilibrium with its environment.

Clod: A mass of soil produced by disturbance.

Coating: A coating or deposit of material on the surface of a ped, stone, etc. A common type is the clay coating caused by translocation and deposition of clay particles on a ped surface.

Coefficient of linear extensibility: The ratio of the difference between the moist and dry lengths of a clod to its dry length, $(Lm - Ld)/Ld$ when Lm is the moist length (at $\frac{1}{3}$ atmospheres) and Ld is the air-dry length. The measure correlates with the volume change of a soil upon wetting and drying.

COLE: An abbreviation of coefficient of linear extensibility.

Colloid: The inorganic and organic material with very fine particle size and therefore high surface area which usually exhibits exchange properties.

Colluvium: Soil materials with or without rock fragments that accumulate at the base of steep slopes by gravitational action.

Compaction: Increase in bulk density due to mechanical forces such as tractor wheels.

Compost: Plant and animal residues that are arranged into piles and allowed to decompose, sometimes soil or mineral fertilisers may be added.

Concept: General notion or idea.

Concretion: Small hard local concentrations of material such as calcite, gypsum, iron oxide or aluminium oxide. Usually spherical or subspherical but may be irregular in shape.

Conglomerate: A sedimentary rock composed mainly of rounded boulders.

Coniferous forest: A forest consisting predominantly of cone-bearing trees with needle-shaped leaves: usually evergreen but some are deciduous, for example the larch forests (*Larix dehurica*) of central Siberia. Their greatest extent is in the wide belt across northern Canada and northern Eurasia. Coniferous forests produce soft wood which has a large number of industrial applications including paper making.

Consistence: The resistance of the soil to deformation or rupture as determined by the degree of cohesion or adhesion of the soil particles for each other.

Consolidated: A term that usually refers to compacted or cemented rocks or other materials.

Continental climate: A climate with wide diurnal and seasonal variations in temperature. Such conditions usually occur in the interior of continents protected by mountains from the moderating influence of the oceans.

Continuously anaerobic (very poorly drained): A horizon that is saturated with water throughout the year, it is blue, olive or grey.

Creep: Slow movement of masses of soil down slopes that are usually steep. The process takes place in response to gravity facilitated by saturation with water.

Crotovina (Krotovina): A large animal burrow which has been filled with material from another horizon.

Croute calcaire: A synonym for caliche.

Crust: A surface layer of soils that becomes harder than the underlying horizon when dry.

Crystal morphology: See Euhedral, Subhedral, Anhedral.

Cuesta: A ridge, or belt of hilly land which has a gentle dip slope on one side, and a relatively steep escarpment slope on the other.

Cutan: Alteration of material at a surface, includes material deposited on a surface and material altered by stress.

Deciduous forest: A forest composed of trees that shed their leaves at some season of the year. In tropical areas the trees lose their leaves during the hot season in order to conserve moisture. Deciduous forests of the cool areas shed their leaves during the autumn to protect themselves against the cold and frost of winter. Deciduous forests produce valuable hardwood timber such as teak and mahogany from the tropics, oak and beech from the n the cooler areas.

Deflation: Preferential removal of fine soil particles from the surface soil by wind. See **Desert pavement.**

Deflocculate: To separate or dispense particles of clay dimensions from a flocculated condition.

Delta: A roughly triangular area at the mouth of a river composed of river transported sediment.

Dendroid: Branching after the manner of a tree with no rejoining of branches.

Denitrification: The biological reduction of nitrate to ammonia, molecular nitrogen or the oxides of nitrogen, resulting in the loss of nitrogen into the atmosphere and therefore undesirable in agriculture.

Denudation: Sculpturing of the surface of the land by weathering and erosion; levelling mountains and hills to flat or gently undulating plains.

Deposit: Material placed in a new position by the activity of man or natural processes such as water, wind, ice or gravity.

Desert crust: A hard surface layer in desert regions containing calcium carbonate, gypsum, or other cementing materials.

Desert pavement: A layer of gravel or stones remaining on the surface of the ground in deserts after the removal of the fine material by wind. See **Deflation** and **Hamada**.

Desert varnish: A glossy sheen or coating on gravel and stones in arid regions.

Devonian: A period of geological time extending from 320–280 million years B.P.

Diatoms: Algae that possess a siliceous cell wall which remains preserved after the death of the organisms. They are abundant in both fresh and salt water and in a variety of soils.

Disperse: See **Dispersion.**

Dispersion: The process whereby the structure or aggregation of the soil is destroyed so that each particle is separate and behaves as a unit.

Doline or dolina: A closed depression in a karst region, often rounded or elliptical in shape, formed by the solution and subsidence of the limestone near the surface. Sometimes at the bottom there is a sink hole into which surface water flows and disappears underground.

Domain: A bundle of clay particles that are only visible in crossed polarised light.

Drift: A generic term for superficial deposits including till (boulder clay), outwash gravel and sand, alluvium, solifluction deposits and loess.

Drumlin: A small hill, composed of glacial drift with hog-back outline, oval plan, and long axis oriented in the direction of ice moved. Drumlins usually occur in groups, forming what is known as basket of eggs topography.

Dry farming: A method of farming in arid and semiarid areas without using irrigation, the land being treated so as to conserve moisture. The technique consists of cultivating a given area in alternate years, allowing moisture to be stored in the fallow year. Moisture losses are reduced by producing a mulch and removal of weeds. In Siberia, where melting snow provides much of the moisture for spring crops, the soil is ploughed in the autumn providing furrows in which snow can collect, preventing it from being blown away and evaporated by strong winds. Usually alternate narrow strips are cultivated in an attempt to reduce erosion in the fallow year. Dry farming methods are employed in the drier regions of India, USSR, Canada and Australia.

Dunes, sand dunes: Ridges or small hills of sand which have been piled up by wind action on sea coasts, in deserts and elsewhere. Barkhans are dunes with characteristic crescentic forms, they may occur singly or in groups.

Ecology: The study of the interrelationships between individual organisms and between organisms and their environment.

Ecosystem: A group of organisms interacting among themselves and with their environment.

Edaphic: (1) Of or pertaining to the soil. (2) Influenced by soil factors.

Edaphology: The study of the relationships between soil and organisms including the use of the land by man.

Eluvial horizon: An horizon from which material has been removed either in solution or suspension.

Eluviation: Removal of material from the upper horizon in solution or suspension.

Equatorial forest or tropical rain forest: A dense, luxuriant, evergreen forest of hot, wet, equatorial regions containing many trees of tremendous heights, largely covered with lianas and epiphytes. Individual species of trees are infrequent but they include such valuable tropical hardwoods as mahogany, ebony and rubber. Typical equatorial forests occur in the Congo and Amazon basins and South-East Asia.

Erosion: The removal of material from the surface of the land by weathering, running water, moving ice, wind and mass movement.

Erosion pavement: A layer of gravel or stones left on the surface of the ground after the removal of the fine particles by erosion.

Esker: A long narrow ridge, chiefly of gravel and sand, formed by a melting glacier or ice sheet.

Euhedral: A crystal with good to perfect crystallographic form. See **Anhedral** and **Subhedral.**

Eutrophic: Containing an optimum concentration of plant nutrients.

Evapotranspiration: The combined processes of evaporation and transpiration.

Excessively aerobic: A horizon which is usually too dry to support adequate plant growth.

Excessively drained: See **Excessively aerobic.**

Exfoliation: A weathering process during which thin layers of rock peel off from the surface. This is caused by heating of the rock surface during the day and cooling at night leading to alternate expansion and contraction. This process is sometimes termed 'onion skin weathering'.

Extinction: The position at which a crystal goes black in crossed polarised light.

Extinction angle: The angle at which a crystal goes black in crossed polarised light.

Fallow: Leaving the land uncropped for a period of time. This may be to accumulate moisture, improve structure or induce mineralisation of nutrients.

Family: One of the categories in soil classification intermediate between the great soil group and the soil series.

Fen peat: Peat that is neutral to alkaline due to the presence of calcium carbonate.

Ferralitisation: Used by some to refer to the processes of formation of ferralitic soils. This term is not specific and should not be used.

Fertiliser: A material that is added to the soil to supply one or more plant nutrients in a readily available form.

Field capacity or field moisture capacity: The total amount of water remaining in a freely drained soil after the excess has flowed into the underlying unsaturated soil. It is expressed as a percentage of the oven-dry soil.

Fine texture: Containing > 35 per cent clay.

Fiord: A long, narrow coastal inlet, usually having steep sides. They have been formed by glaciers over-deepening valleys which were previously cut by steams.

Flood plain: The land adjacent to a stream built of alluvium and subject to repeated flooding.

Fluvio-glacial: See **Glacio-fluvial deposits.**

Fragment: A small mass of soil produced by disturbance.

Freely drained: A soil that allows water to percolate freely and is continuously well aerated.

Friable: A term applied to soils that when either wet or dry crumble easily between the fingers.

Fulvic acid: The mixture of organic substances remaining in solution upon acidification of a dilute alkali extract of soil.

Fungi: Simple plants that lack chlorophyll and are composed of cellular filamentous growth known as hyphae. Many fungi are microscopic but their fruiting bodies, viz. mushrooms and puffballs are quite large.

Gastropod: A member of the Gastropoda class of molluscs which includes snails and slugs.

Geological erosion: See **Natural erosion.**

Geomorphology: The study of the origin of physical features of the earth, as they are related to geological structure and denudation.

Gilgai: A distinctive microrelief of knolls and basins that develops on clay soils that exhibit a considerable amount of expansion and contraction in response to wetting and drying.

Glacial drift: Material transported by glaciers and deposited, directly from the ice or from the meltwater.

Glacier: A large mass of ice that moves slowly over the surface of the ground or down a valley. They originate in snow fields and terminate at lower elevations in a warmer environment where they melt.

Glacio-fluvial deposits: Material deposited by meltwaters coming from a glacier. These deposits are variously stratified and may form outwash plains, deltas, kames, eskers, and kame terraces. See **Glacial drift** and **Till.**

Gleisation: See **Gleying.**

Gleying: The reduction of iron in an anaerobic environment leading to the formation of grey or blue colours. Not a very good term because it includes a very large number of individual processes.

Granite: An igneous rock that contains quartz, feldspars and varying amounts of biotite and muscovite.

Gravitational water: The water that flows freely through soils in response to gravity.

Great soil group: One of the categories in soil classification.

Ground-water-table: The upper limit of the ground water.

Gully: A shallow steep-sided valley that may occur naturally or be formed by erosion.

Gully erosion: A form of catastrophic erosion that forms gullies.

Gyttja: Peat consisting of faecal material, strongly decomposed plant remains, shells of diatoms, phytoliths, and fine mineral particles. Usually forms in standing water.

Halomorphic soil: A soil containing a significant proportion of soluble salts.

Halophyte: A plant capable of growing in salty soil; i.e. a salt tolerant plant.

Halophytic vegetation: Vegetation that requires or tolerates saline conditions.

Hamada: An accumulation of stones at the surface of deserts, formed by the washing or blowing away of the finer material.

Hardpan: An horizon cemented with organic matter, silica, sesquioxides or calcium carbonate. Hardness or rigidity is maintained when wet or dry and samples do not slake in water.

Heavy soil: (Obsolete) A soil that has a high content of clay and is difficult to cultivate.

Heterotrophic organisms: Those that derive their energy by decomposing organic compounds (cf. **Autotrophic**).

Holocene Period: The period of time extending from 10 000–0 B.P.

Horizon: Relatively uniform material that extends laterally, continuously or discontinuously throughout the pedo-unit; runs approximately parallel to the surface of the ground and differs from the related horizons in many chemical, physical and biological properties.

Hue: The dominant spectral colour and one of the three colour variables.

Humic Acid: Usually refers to the mixture of ill-defined dark organic substances precipitated upon acidification of a dilute alkali extract of soil. Some workers use it to include only the alcohol-insoluble portion of the precipitate.

Humification: The decomposition of organic matter leading to the formation of humus.

Humin: Usually applied to that part of the organic matter that remains after extraction with dilute alkali.

Humus: The well-decomposed, relatively stable part of the organic matter found in aerobic soils.

Hydration: The process whereby a substance takes up water.

Hydraulic conductivity: The ratio of the flow velocity to the driving force for the viscous flow under saturated conditions of a specific liquid in a porous medium.

Hydrologic cycle: Disposal of precipitation from the time it reaches the soil surface until it re-enters the atmosphere by evapotranspiration to serve again as a source of precipitation.

Hydrolysis: In soils it is the process whereby hydrogen ions from water are exchanged for cations such as sodium, potassium, calcium and magnesium, and the hydroxyl ions combine with the cations to give hydroxides.

Hydromorphic soil: Soils developed in the presence of excess water.

Hygroscopic water: Water that is adsorbed on to a surface from the atmosphere.

Igneous rock: A rock formed by the cooling of molten magma including basalt and granite.

Illuvial horizon: An horizon that receives material in solution or suspension from some other part of the soil.

Illuviation: The process of movement of material from one horizon and its deposition in another horizon of the same soil; usually from an upper horizon to a middle or lower horizon in the pedo-unit. Movement can take place also laterally.

Immature soil: Lacking a well-developed pedo-unit.

Impeded drainage: Restriction of the downward movement of water by gravity. May result in the development of anaerobic conditions due to waterlogging.

Imperfectly drained: See **Weakly anaerobic.**

Impervious: Not easily penetrated by roots or water.

Indicator plants: Plants that are indicative of specific site or soil conditions.

Infiltration: The process whereby water enters the soil through the surface.

Inselberg: (pl. Inselberge) A steep-sided hill composed predominantly of hard rock and rising abruptly above a plain; found mainly in tropical and subtropical areas.

Interglacial period: A relatively mild period occuring between two glacial periods.

Intergrade: A soil which contains the properties of two or more distinctive and genetically different soils or soil horizons.

338

Interpedal: Between peds.

Interstadial period: A slightly warmer phase during a glacial period.

Intrapedal: Within peds.

Intrazonal soils: One of the three orders of the zonal system of soil classification. They have well developed characteristics resulting from the dominant influence of a dominant local factor such as topography and parent material.

Isomorphous replacement: The replacement of one ion by another in the crystal lattice without changing the structure of the mineral.

Isotropic: Not visible in crossed polarised light. See **Anisotropy.**

Kame: A small hill of stratified gravel or sand formed by a melting glacier or ice sheet.

Karst topography: An irregular land surface in a limestone region. The principal features are depressions (e.g. dolines) which sometimes contain thick soils which have been washed off the rest of the surfaces leaving them bare and rocky. Drainage is usually by underground streams.

Kettle hole or kettle: A hollow or depression in a melting glacier area, probably formed by a block of ice which was covered by gravel and subsequently melted, allowing the debris to settle.

Krotovina: See **Crotovina.**

Lacustrine: Pertaining to lakes.

Lacustrine deposit: Materials deposited by lake waters.

Landslide or landslip: The movement down the slope of a large mass of soil or rocks from a mountain or cliff. Often occurs after torrential rain which soaks into the soil making it heavier and more mobile. Earthquakes and the undermining action of the sea are also causative agents.

Laterisation: Used by some to refer to the processes of formation of laterite or red and yellow tropical soils. This term is not specific and should not be used.

Lattice structure: The orderly arrangement of atoms in crystalline material.

Leaching: The washing out of material from the soil, both in solution and suspension.

Light soil: (Obsolete) A soil which has a coarse texture and is easily cultivated.

Lime: Compounds of calcium used to correct the acidity in soils.

Litter: The freshly fallen plant material occurring on the surface of the ground.

Lodging: The collapse of top-heavy plants, particularly grain crops because of excessive growth or beating by rain.

Loess: An aeolian deposit composed mainly of silt which originated in arid regions, from glacial outwash or from alluvium. It is usually of yellowish-brown colour and has a widely varying calcium carbonate content. In the USSR, loess is regarded as having been deposited by water.

Lysimeter: Apparatus installed in the soil for measuring percolation and leaching.

Macroelement: Elements such as nitrogen that are needed in large amounts as nutrients for plant growth.

Macronutrient: See **Macroelement.**

Mangrove swamp: A dense jungle of mangrove trees which have the special adaptation of extending from their branches long arching roots which act as anchors and form an almost impenetrable tangle. They occur in tropical and subtropical areas, particularly near to the mouths of rivers.

Manure: Animal excreta with or without a mixture of bedding or litter.

Matrix: Something within which something else originates or is enclosed or embedded.

Mature soil: A well developed soil usually with clearly defined horizonation.

Meristem: The region of active cell-division in plants, it is the tips of stems and roots in most plants. The cells so formed then become modified to form the various tissues such as the epidermis and cortex.

Mesofauna: Small organisms such as worms and insects.

Metamorphic rock: A rock that has been derived from other rocks by heat and pressure. The original rock may have been igneous, sedimentary or another metamorphic rock.

Microclimate: The climate of a very small region.

Microelements: Those elements that are essential for plant growth but are required only in very small amounts.

Microfauna: The small animals that can only be seen with a microscope, they include protozoa, nematodes, etc.

Microflora: The small plants that can only be seen with a microscope, they include algae, fungi, bacteria, etc.

Micronutrient: See **Microelement.**

Microorganism: The members of the microflora and microfauna that can only be seen with a microscope.

Microrelief: Small differences in relief, including mounds, or pits that are a few metres across and have differences in elevation not greater than about two metres.

Milliequivalent: A thousandth of an equivalent weight.

Mineralisation: The change of an element in an organic form to an inorganic form by microorganisms.

Mineral soil: A soil containing less than 20 per cent organic matter or having a surface organic layer less than 30 cm thick (cf. **Organic soil**).

Mites: Very small members of the Arachnida which include spiders; they occur in large numbers in many organic surface soils.

Moder: A kind of decomposition and humus formation which produces an advanced but incomplete humification of the remains of organisms due to good aeration.

Mor: An accumulation of acid organic matter at the soil surface beneath forest.

Moraine: Any type of constructional topographic form consisting of till and resulting from glacial deposition.

Mottling: Patches or spots of different colours usually used for the colour pattern developed due to partial anaerobism.

Mulch: A loose surface horizon that forms naturally or may be produced by cultivation and consists of either inorganic or organic materials.

Mull: A crumbly intimate mixture of organic and mineral material formed mainly by worms particularly earthworms.

Mull-like moder: A term to indicate humus forms which are characterised by: (1) the external appearance of mull; (2) a relatively weak binding together of organic matter and the mineral soil particles; (3) little or no clay and a relatively high percentage of silt-sized and sand-sized mineral grains.

Neutral soil: A soil with pH value around 7.

Nitrification: The oxidation of ammonia to nitrite and nitrite to nitrate by microorganisms.

Nitrogen fixation: The transformation of elemental nitrogen to an organic form by microorganisms.

Non-silicate: Rock forming minerals that do not contain silicon.

Onion skin weathering: See **Exfoliation.**

Organic soil: A soil that is composed predominantly of organic matter, usually refers to peat.

Ortstein: A brown to black cemented middle horizon in Podzols.

Pans: Soil horizons that are strongly compacted, cemented or have a high content of clay. See **Hardpan.**

Patent material: The original state of the soil. The relatively unaltered lower material in soils is often similar to the material in which the horizons above have formed.

Peat: An accumulation of dead plant material often forming a layer many metres deep. It shows various stages of decomposition and is completely waterlogged.

Ped: A single individual naturally occurring soil aggregate such as a granule or prism (cf. **Clod** or **Fragment**).

Pedalfer: (Obsolete) The large group of soils in which sesquioxides accumulate relative to silica.

Pedocal: (Obsolete) The large group of soils in which calcium accumulates during soil formation.

Pedogenesis: The natural process of soil formation.

Pedology: The study of soils as naturally occurring phenomena taking into account their composition, distribution and method of formation.

Pedoturbation: All mixing of soil components that is not caused by illuviation.

Pedo-unit: A selected column of soil containing sufficient material in each horizon for adequate field and laboratory characterisation.

Peneplain: A large flat or gently undulating area. Its formation is attributed to progressive erosion by rivers and rain, which continues until almost all the elevated portions of the land surface are worn down. When a peneplain is elevated, it may become a plateau which then forms the initial stage in the development of a second peneplain.

Perched water: An accumulation of water within the soil due to an impermeable layer such as a pan or a high content of clay.

Perched water-table: The upper limit of perched water. See **Perched water.**

Percolation: (soil water) The downward or lateral movement of water through soil.

Perennial: A plant that continues to grow from year to year.

Permafrost: Permanently frozen subsoil.

Permeability: The ease with which air, water, or plant roots penetrate into or pass through a specific horizon.

pH: The negative logarithm of the hydrogen-ion concentration of a solution. It is the quantitative expression of the acidity and alkalinity of a solution and has a scale that ranges from 0 to 14. pH 7 is neutral <7 is acid and >7 is alkaline.

pH soil: The negative logarithm of the hydrogen-ion activity of a soil. The degree of acidity (or alkalinity) of a soil expressed in terms of the pH scale, from 2 to 10.

Physical weathering: The comminution of rocks into smaller fragments by physical forces such as by frost action and exfoliation.

Physiological drought: A temporary daytime state of drought in plants due to the losses of water by transpiration being more rapid than uptake by roots even though the soil may have an adequate supply. Such plants usually recover during the night.

Phytolith: Opaline formation in plant tissue that remains

in the soil after the softer plant tissue has decomposed.

Plagioclimax: A plant community which is maintained by continuous human activity of a specific nature such as burning or grazing.

Plastic: A moist or wet soil that can be moulded without rupture.

Platy: Soil aggregates that are horizontally elongated.

Pleistocene Period: The period following the Pliocene Period, extending from 2 000 000–10 000 years B.P. In Europe and North America, there is evidence of four or five periods of intense cold during the Pleistocene Period when large areas of the land surface were covered by ice – glacial periods. During the interglacial periods the climate ameliorated and the glaciers retreated.

Pluvial Period: A period of hundreds of thousands of years of heavy rainfall.

Podzolisation: Used by some to refer to the process of formation of a Podzol. This term is not specific and should not be used.

Polder: A term used in Holland for an area of land reclaimed from the sea or a lake. A dyke is constructed around the area which is then drained by pumping the water out. Polders form valuable agricultural land or pasture land for cattle.

Polygenetic soil: A soil that has been formed by two or more different and contrasting processes so that all the horizons are not genetically related.

Poorly drained: See **Strongly anaerobic.**

Pore: A discrete volume of soil atmosphere completely surrounded by soil (cf. **Pore space**).

Pore space: The continuous and interconnecting spaces in soils.

Porosity: The volume of the soil mass occupied by pores and pore space.

Primary mineral: 1. A mineral such as feldspar or a mica which occurs or occurred originally in an igneous rock. 2. Any mineral which occurred in the parent material of the soil.

Profile: A vertical section through a soil from the surface into the relatively unaltered material.

Puddle: To destroy the structure of the surface soil by physical methods such as the impact of rain drops, poor cultivation with implements and trampling by animals.

Rainfall interception: The interception and accumulation of rainfall by the foliage and branches of vegetation.

Rain splash: The redistribution of soil particles on the surface by the impact of rain drops. On slopes this can cause a large amount of erosion.

Rain splash erosion: See **Rain splash.**

Raised beach: A beach raised by earth movement thus forming a narrow coastal plain. There may be raised beaches at different levels resulting from repeated earth movement.

Raw humus: A humus form consisting predominantly of well preserved, though often fragmented plant remains with few faecal pellets.

Regolith: The unconsolidated mantle of weathered rock, soil and superficial deposits overlying solid rock.

Rhizosphere: The soil close to plant roots where there is usually an abundant and specific microbiological population.

Rill: A small intermittent water course with steep sides, usually only a few centimetres deep.

Rill erosion: The formation of rills as a consequence of poor cultivation.

Roches moutonnées: Small hills of rock smoothed and striated by glacier on the upstream side and roughened on the downstream side, from which fragments have been plucked during ice movement.

Roundness: Relates to the sharpness of the corners of a phenomenon irrespective of its shape and relationship to a sphere (cf. **Sphericity**).

Rubifaction: The development of a red colour in soil – reddening.

Saline soil: A soil containing enough soluble salts to reduce its fertility.

Salinisation: The process of accumulation of salts in soil.

Sand: Mineral or rock fragments that range in diameter from 2–0.02 mm in the international system or 2–0.05 mm in the USDA system.

Saturated flow: The movement of water in a soil that is completely filled with water.

Secondary mineral: Those minerals that form from the material released by weathering. The main secondary minerals are the clays and oxides.

Sedimentary rock: A rock composed of sediments with varying degrees of consolidation. The main sedimentary rocks include sandstones, shales, conglomerates and some limestones.

Self-mulching soil: A soil with a naturally formed well aggregated surface which does not crust and seal under the impact of rain drops.

Sesquioxides: Usually refers to the combined amorphous oxides of iron and aluminium.

Sheet erosion: The gradual and uniform removal of the soil surface by water without forming any rills or gullies.

Silicates: Rock forming minerals that contain silicon.

Silt: Mineral particles that range in diameter from 0.02–0.002 in the international system or 0.05–0.002 mm in the USDA system.

Slickenside: The polished surface that forms when two peds rub against each other as the soil expands in response to wetting.

Slickspot: Small areas of surface soil that are smooth and slippery when wet because of alkalinity or high exchangeable sodium.

Soil: (1) The natural space–time continuum occurring at the surface of the earth and supporting plant life. (2) Anything so called by a competent authority.

Soil auger: A tool used for boring into the soil and withdrawing small samples for field or laboratory examination.

Soil fabric: The nature and arrangement of the constituents of a soil and their relation to each other.

Soil horizon: See Horizon.

Soil monolith: A vertical section through the soil preserved with resin and mounted for display.

Soil piping or tunnelling. Erosion causing the formation of subterranean passages and tunnels.

Soil profile: A section of *two dimensions* extending vertically from the earth's surface so as to expose all the soil horizons and a part of the relatively unaltered underlying material.

Soil survey: The systematic examination and mapping of soil.

Solifluction: Slow flowage of material on sloping ground, characteristic, though not confined to regions subjected to alternate periods of freezing and thawing.

Solum: The part of the soil above the relatively unaltered material.

Sphericity: Relates to the overall shape of a phenomenon irrespective of the sharpness of its edges and is a measure of the degree of its conformity to a sphere (cf. **Roundness**).

Springtails: Very small insects that live in the surface soil.

Strip cropping: The practice of growing crops in strips along the contour in an attempt to reduce run-off, thereby preventing erosion or conserving moisture.

Strongly anaerobic: (Poorly drained) A soil that remains very wet or waterlogged for long periods of the year and as a result develops a mottled pattern of greys and browns.

Structure: The spatial distribution and total organisation of the soil system as expressed by the degree and type of aggregation and the nature and distribution of pores and pore space.

Subhedral: Minerals with partly developed crystallographic form.

Symbiosis: Two organisms that live together for their mutual benefit. Fungus and alga that form a lichen or nitrogen fixing bacteria living in roots are examples of symbiosis. The individual organisms are called symbionts.

Talus: Angular rock fragments that accumulate by gravity at the foot of steep slopes and cliffs.

Tectonic: Rock structures produced by movements in the earth's crust.

Terrace: A broad surface running along the contour. It can be a natural phenomenon or specially constructed to intercept run-off thereby preventing erosion or conserving moisture. Sometimes they are built to provide adequate rooting depth for plants.

Tertiary Period: The period of time extending from 75 000 000–2 000 000 years B.P.

Thermophilic bacteria: Bacteria which have optimum activity between about 45 and 55 °C.

Thorn forest: A deciduous forest of small, thorny trees, developed in a tropical semiarid climate.

Tile drain: Short lengths of concrete or pottery pipes placed end to end at a suitable depth and spacing in the soil to collect water from the soil and lead it to an outlet.

Till: An unstratified or crudely stratified glacial deposit consisting of a stiff matrix of fine rock fragments and old soil containing subangular stones of various sizes and composition many of which may be striated (scratched). It forms a mantle from less than 1 m to over 100 m in thickness covering areas which carried an ice-sheet or glaciers during the Pleistocene and Holocene Periods.

Till plain: A level or undulating land surface covered by glacial till.

Tilth: The physical state of the soil that determines its suitability for plant growth taking into account texture, structure, consistence and pore space. It is a subjective estimation and is judged by experience.

Toposequence: A sequence of soils whose properties are determined by their particular topographic situation.

Toxic substance: A substance that is present in the soil or in the above ground atmosphere that inhibits the growth of plants and ultimately may cause their death.

Translocation: The movement of material in solution, suspension or by organisms from one horizon to another.

Transpedal: Traversing peds.

Trellised: Branching and rejoining forming a regular network of long horizontal channels interconnected by shorter vertical channels.

Triassic: A period of geological time extending from 190 000 000 to 150 000 000 years B.P.

Tropical rain forest: See **Equatorial forest.**

Unavailable nutrients: Plant nutrients that are present in the soil but cannot be taken up by the roots because they have not been released from the rock by weathering or from organic matter by decomposition.

Unavailable water: Water that is present in the soil but cannot be taken up by plant roots because it is strongly adsorbed on to the surface of particles.

Unconsolidated: Sediments that are loose and not hardened.

Unsaturated flow: The movement of water in a soil that is not completely filled with water.

Varnish (desert): A dark shiny coating on stones in deserts, probably composed of compounds of iron and manganese (cf. **Desert varnish**).

Varve: A layer representing the annual deposit of a sediment, it usually consists of a lighter and darker portion due to the change in rate of deposition during the year. The material may be of any origin but the term is most often used in connection with glacial lake sediments.

Ventifact: A faceted or moulded pebble caused by wind action, usually forms in polar and desert areas. The flat facets meet at sharp angles.

Very poorly drained: See **Continuously anaerobic.**

Volcanish ash (volcanic dust): Fine particles of lava ejected during a volcanic eruption. Sometimes the particles are shot high into the atmosphere and carried long distances by the wind.

Waterlogged: Saturated with water.

Water-table (ground): The upper limit in the soil or under-lying material permanently saturated with water.

Water-table perched: See **Perched water-table.**

Weakly anaerobic: A horizon that is anaerobic for short periods and moist for long periods. The colours are less bright than aerobic horizons and they are usually marbled or weakly mottled.

Weathering: All the physical, chemical and biological processes that cause the disintegration of rocks at or near the surface.

Well drained: See **aerobic.**

Wilting point: The percentage of water remaining in the soil when the plants wilt permanently.

Xerophytes: Plants that grow in extremely dry areas.

References

Afanaseva, E. A., 1927. In Vilensky, D. G., 1957, *Soil Science,* pp. 488. Moscow.

Afanasiev, J. N., 1927. The classification problem in Russian soil science, *Russ. Pedol.,* **5,** 51.

Arkley, R., 1976. Statistical methods in soil classification research, *Adv. Agon.,* **28,** 37–70.

Aubert, G. and Duchaufour, Ph., 1956. Projet de classification des sols., *Trans. 6th Int. Cong. Soil Sci.,* Paris, Vol. E, 597–604.

Avery, B. W., 1973. Soil classification in the soil survey of England and Wales, *J. Soil Sci.,* **24,** 324–38.

Avery, B. W., Stephen, I., Brown, G. and Yaalon, D. H., 1959. The origin and development of Brown Earths on clay-with-flints and Coombe deposits, *J. Soil Sci.,* **10,** 177–95.

Babbington, B., 1821. Remarks on the geology of the country between Tellicherry and Madra, *Trans. Geol. Soc. Lond.,* **5,** 328–9.

Babel, U., 1975. Micromorphology of soil organic matter, ch. 7 in *Soil Components,* Vol. 1, *Organic Components.* Ed. J. E. Gieseking, pp. 369–471.

Baldwin, M., Kellogg, C. E. and Thorp, J., 1938. Soil classification, in *Soils and Men (Yearbook of Agriculture,* 1938). Washington, USDA, pp. 1232.

Barley, K. P., 1959. Earthworms and soil fertility, *Aust. J. Agric. Res.,* **10,** 171–85.

Basinski, J. J., 1959. The Russian approach to soil classification and its recent development, *J. Soil Sci.,* **10,** 14–26.

Bidwell, O. W. and Hole, F. D., 1964(a). Numerical taxonomy and soil classification, *Soil Sci.,* **97,** 58–62.

Bidwell, O. W. and Hole, F. D., 1964(b). An experiment in the numerical classification of some Kansas soils, *Soil Sci. Soc. Amer. Proc.,* **28,** 263–8.

Black, R. F., 1952. Growth of ice wedge polygons in perma-frost near Barrow, Alaska, *Bulletin, Geological Society of America,* **63,** 1235–6.

Black, R. F., 1960. Ice Wedges in Northern Alaska (Abstract of Paper) 19th International Geographical Congress, Stockholm, p. 26.

Black, R. F., 1964. Periglacial studies in the United States 1959–1963, *Biuletyn Peryglacjalny,* **14,** 5–29.

Black, R. F., 1973. Periglacial phenomena of Wisconsin, north central United States, Vol. IV, *Report of VI INQUA Congress,* Warsaw, 1961; Lodz, 1964, 21–8.

Bloomfield, C., 1964. Mobilization and immobilization phenomena in soils, *Paleoclimatology.* Ed. A. E. M. Nairn, Interscience, New York.

Bonnevie-Svendsen, C. and Gjems, O., 1957. Amount and chemical composition of the litter from larch, beech, Norway spruce and Scots pine stands and its effect on the soil, *Medd. Norske Skogsforsoksv,* 48, 111–74.

Brammer, H., 1962. *Soils,* chapter 6, Agriculture and land use in Ghana. Oxford Univ. Press, 88–126.

Brewer, R., 1964. *Fabric and Mineral Analysis of Soils,* John Wiley & Sons Inc., pp. 470.

Brewer, R. and Haldane, A. D., 1956. Preliminary experiments in the development of clay orientation in soils, *Soil Sci.,* 84, pp. 301–416.

Brown, G., 1961. The X-ray Identification and Crystal Structure of Clay Minerals. Ed., Mineralogical Society (Clay Minerals Group), London.

Buchanan, F., 1807. *A Journey from Madras Through the Countries of Mysore, Canara and Malabar,* Lond., 3 vols., see Vol. 2, pp. 436–7, 440–1, 460, 559; Vol. 3, pp. 66, 89, 251, 258.

Buol, S. W., Hole, F. D. and McCracken, R. J., 1973. *Soil Genesis and Classification,* Ames, Iowa, pp. 360.

Butler, B. E., 1959. *Periodic Phenomena in Landscape as a Basis for Soil Studies,* CSIRO., Australia Soil Pub. No. 14, pp. 20.

Cheshire, M. V., Cranwell, P. A., Franshaw, C. P., Floyd, A. J. and Howarth, R. D., 1967. Humic acid – II, structure of humic acid, *Tetrahedron,* 23, 1668–82.

Clark, F. W., 1924. The data of geochemistry, *U.S. Geol. Survey Bull.,* 770.

Clements, F. E. and Weaver, J. E. Experimental vegetation, *Carnegie Institution,* 192.

Coffey, G. N., 1912. A study of the soils of the United States, *U.S. Dep. Agr. Bureau of Soils Bull.,* 85, 114. *Commission de pédologie et de Cartographie des sols 1967 Classification des sols,* Lab. Geol. Pedol., Ecole Nat Super. Agron (Grignon, France), pp. 87.

Crowther, E. M., 1953. The sceptical soil chemist, *J. Soil Sci.,* 4, 107–22.

Dalrymple, J. R., Blong, R. J. and Conacher, A. J., 1968. A hypothetical nine unit landsurface model, *Zeitschrift für Geomorphologie,* 12, 60–76.

Danckelmann, B., 1887. Streuertragstafel für Buchen- und Fichtenhochwaldunger *Z. Forst- u. Jagdw.,* 19, 577–87.

Davis, W. M., 1954. *Geographical essays,* Dover Publ., pp. 777.

De Coninck, F. and Laruelle, J., 1964. Soil development in sandy materials of the Belgian Campine, *Soil Micromorphology,* Ed. A. Jongerius, 169–88.

Deer, W. A., Howie, R. A. and Zussman, J., 1966. *An Introduction to the Rock-forming Minerals,* Longman, pp. 528.

Del Villar, E., 1937. *Los Suelos de la Peninsula Luso-iberica.* English translation by G. W. Robinson. Madrid, 1937; Murby, London, 1937.

Despande, T. L., Greenland, D. J. and Quirk, J. P., 1964. Role of iron oxide in the bonding of soil particles, *Nature,* London, 201, 107–8.

Dokuchaev, V. V., 1883. *Russian Chernozem,* St. Petersburg.

Douglas, L. A. and Tedrow, J. C. F., 1960. Tundra soils of Arctic Alaska, *Trans. 7th Inter. Cong. Soil Sci.,* IV, 291–304.

Dudal, R., 1965. *Dark Clay soils of Tropical and Subtropical Regions,* FAO, Rome, pp. 161.

Eswaran, H., 1967. Micromorphology study of a 'cat-clay' soil, *Pedologie,* 17, 259–65.

Everette, D. H., 1961. The thermodynamics of frost damage to porous solids, *Trans. Faraday Soc.,* 57, 1541–51.

Eyre, S. R., 1968. *Vegetation and Soils, A World Picture,* 2nd Edition, Edward Arnold, pp. 328.

FAO-UNESCO, 1974. *Soil Map of the World,* Vol. 1, Legend, Paris, pp. 59.

Fieldes, M. and Swindale, L. D., 1954. Chemical weathering of silicates in soil formation, *N.Z. J. Sci. Tech.,* 36B, 140–54.

FitzPatrick, E. A., 1956. An indurated soil horizon formed by permafrost, *J. Soil Sci.,* 7, 248–54.

FitzPatrick, E. A., 1967. Soil nomenclature and classification, *Geoderma,* 1, 91–105.

FitzPatrick, E. A., 1969. Some aspects of soil evolution in north-east Scotland, *Soil Sci.,* 107, 403–8.

FitzPatrick, E. A., 1971. *Pedology, a Systematic Approach to Soil Science,* Oliver and Boyd, Edinburgh, pp. 306.

FitzPatrick, E. A., 1976. Soil horizons and homology, *Class. Soc. Bull.,* 3, 68–89.

FitzPatrick, E. A., 1977. *Soil Description,* Univ. of Aberdeen, pp. 66.

Flaig, W., 1960. Comparative chemical investigations of natural humic compounds and their model substances, *Sci. Proc. Roy. Dublin Soc.,* 1, 149–62.

Frye, J. C., Schaffer, P. R., Willman, H. B. and Ekblaw, G. E., 1960a. Accretion-gley and the gumbotil dilemma, *Amer. J. Sci.,* 258, 185–90.

Frye, J. C., Willman, H. B. and Glass, H. D., 1960b. Gumbotil, accretion-gley, and the weathering profile. III. *State Geol. Survey, Circular,* 295, pp. 39.

Frye, J. C., Willman, H. B. and Black, R. F., 1965. *Outline of Glacial Geology of Illinois and Wisconsin. The Quaternary of the United States.* Edited by H. E. Wright and D. G. Frey.

Gedroiz, K. K., 1929. Der absorbierende Bodenkomplex und die adsorbierten Bodenkationen als Grundlage der genetischen Boden-Klassification Kolloidchem. Beihaft, 1–112.

Glinka, K. D., 1914. *Die Typen der Bodenbildung,* Berlin.

Grim, R. E., 1969. *Clay Mineralogy,* 2nd Edition, New York, McGraw-Hill, pp. 596.

Guppy, E. M. and Sabine, P. A., 1956. Chemical analysis of igneous rocks, metamorphic rocks and minerals, *Mem. Geol. Surv. Great Britain,* pp. 78, HMSO.

Hallsworth, E. G., Robertson, G. K. and Gibbons, F. R., 1955. Studies in pedogenesis in New South Wales, VII. The Gilgai soils, *J. Soil Sci.,* 6, 1–31.

Handley, W. R. C., 1954. Mull and Mor formation in relation to forest soils, *For. Comm. Bull.,* 23, pp. 115, HMSO.

Hardon, H. F., 1936. Factoren, die het organische staf-en het stikstofgehalts van tropische gronden beheerschen, Medeleelingen alg. *Proefst Landbouw,* 18, Buitenzorg.

Harrison, Sir J. B., 1933. *The Katamorphism of Igneous Rocks Under Humid Tropical Conditions,* Imp. Bureau of soils Science, Rothamsted Experimental Station, Harpenden, pp. 79.

Hesse, P. R., 1971. *A Textbook of Soil Chemical Analysis,* J. Murray, London, pp. 520.

343

Hilgard, E. W., 1906. *Soils,* MacMillan Company, New York.

Hodgson, J. M., 1974. *Soil Survey Field Handbook,* Tech. Mono. No. 5, Soil Surv. of England and Wales, Harpenden, pp. 99.

Hole, F. D., 1953. Suggested terminology for describing soils as three dimensional bodies, *Soil Sci. Soc. Amer. Proc.,* 17, 131–5.

Jackson, M. L., 1958. *Soil Chemical Analysis,* Prentice-Hall Inc., Englewood Cliffs, N.J., pp. 498.

Jackson, M. L., 1964. Chemical composition of soils, in *Chemistry of the Soil,* Ed. F. E. Bear, pp. 71–141.

Jackson, M. L., 1968. Weathering of primary and secondary minerals in soils, *Trans. 9th Int. Cong. Soil Sci.,* 4, 281–92.

Jackson, M. L. and Sherman, G. D., 1953. Chemical weathering of minerals in soils, *Advan. Agron.,* 5, 219–318.

Jenny, H., 1941. *Factors of Soil Formation,* McGraw-Hill Book Co., Inc., pp. 281.

Jones, T. A., 1959. Soil classification – destructive criticism, *J. Soil Sci.,* 10, 196–200.

Kay, G. F., 1916. Gumbotil, a new term in Pleistocene geology, *Science,* 44, 637–8.

Kendrick, W. B. and Burges, A., 1962. Biological aspects of the decay of *Pinus sylvestris* leaf litter, *Nova Hedwigia,* 4, 313–42.

King, L. C., 1967. *Morphology of the Earth,* Oliver and Boyd, Edinburgh, pp. 726.

Kononova, M. M., 1961. *Soil Organic Matter,* 2nd Edn., Pergamon Press, pp. 544.

Koppi, A. J., Private communication.

Kovda, V. A., Lobova, Ye. V. and Rozanov, V. V., 1967. Classification of the world's soils, *Soviet Soil Sci.,* 851–63.

Kubiëna, W. L., 1953. *The Soils of Europe,* Murby, London, pp. 317.

Kubiëna, W. L., 1958. The classification of soils, *J. Soil Sci.,* 9, 9–19.

Lang, R., 1915. Versuch einer exakten klassifikaten der böden in klimatischen und geologischen hinsicht, *Int. Mitt. Bodenk.,* 5, 312–46.

Leeper, G. W., 1956. The classification of soils, *J. Soil Sci.,* 7, 59–64.

McCaleb, S. B., 1959. The genesis of red-yellow podzolic soils, *Soil Sci. Soc. Amer. Proc.,* 23, 164–8.

Mackney, D. and Burnham, C. P., 1966. *The Soils of the Church Stretton District of Shropshire,* Mem. Soil Surv. Great Britain. England and Wales, Rothamsted Experimental Station, Harpenden, HMSO, pp. 247.

McRae, S. G. and Burnham, C. P., 1976. Soil Classification, *Class. Soc. Bull.,* 3, 56–64.

MacVicar, C. N., Loxton, R. F., Lambrechts, J. J. N., Le Roux, J., DeVilliers, J. M., Verster, E., Merryweather, F. R., Van Rooyen, T. H. and Harmse, H. J. von M., 1977. *Soil Classification – A binomial system for South Africa,* Dept. Agric. Tech. Serv., pp. 150.

Marbut, C. F., 1928. A scheme for soil classification, *Proc. 1st. Int. Congr. Soil Sci.,* 4, 1–31.

Mattson, S. and Lönnemark, H., 1939. The pedology of hydrologic podsol series. I. *Ann. Agr. Coll. Sweden,* 7, 185–227.

Meyer, A., 1926. Uber einige Zusammenhänge Zwichen Klima und Böden in Europe, *Chem. der Erde,* 290–347.

Miljkovic, N., 1965. General review of the salt-affected ('Slatina') soils of Yugoslavia and their classification, *Agrokemia es Talajtan,* 14 Supplement, 235–42.

Milner, H. B., 1962. *Sedimentary Petrography.* Vol. I, *Methods in Sedimentary Petrography,* pp. 643. Vol. II, *Principles and Application,* pp. 715. Allen and Unwin.

Moormann, F. R., 1963. Acid sulfate soils (Cat-clays) of the tropics, *Soil Sci.,* 95, 271–5.

Mückenhausen, E., 1962. The soil classification system of the Federal Republic of Germany, *Trans. Int. Soil Conf., N.Z.,* 377–87.

Mulcahy, M. J. and Hingston, F. J., 1961. The development and distribution of the soil of the York-Quairading area, Western Australia, in relation to landscape evolution, *CSIRO, Aust. Soil Publ,* 17, pp. 43.

Müller, P. E., 1879. Studies over Skovjord, *Tidsskr. Skovbrug.,* 3, 1–124.

Müller, P. E., 1884. Studies over Skovjord, *Tidsskr. Skovbrug.,* 7, 1–232.

Müller, P. E., 1887. *Studien üder die natürlichen Humus formen,* Springer, Berlin, pp. 324.

Munsell Soil Colour Charts, 1954. Munsell Color Co., U.S.A.

Neustruev, S. S., 1926. Bull. Geogr. Inst. Russian, in J. N. Afanasiev, 1927. The Classification Problem in Soil Science, *Suss. Pedol.,* 5, pp. 51.

Northcote, K. H., 1971. *A Factual Key for the Recognition of Australian Soils,* CSIRO. Aust. Soils Div. Rep. 3rd Edition, pp. 123.

Nye, P. H., 1954. Some soil-forming processes in the humid tropics. 1. A field study of a catena in the West African forest, *J. Soil Sci.,* 4, 7–21.

Oertel, A. C., 1961. Pedogenesis of some red brown soils based on trace element profiles, *J. Soil Sci.,* 12, 242–58.

Paton, A. M., Private communication.

Peltier, L., 1950. The geographic cycle in periglacial regions as it is related to climatic geomorphology, *Ann. Assoc. Amer. Geog.,* 40, 214–36.

Penck, W., 1953. *Morphological Analysis of Landforms,* Translated by H. Czech and K. C. Boswell. Macmillan & Co., pp. 429.

Penman, H. L., 1956. Evaporation: an introduction survey, *Neth J. Agric, Sci.,* 4, 9–29.

Péwé, J. L., 1959. Sand-wedge polygons (Tesselations) in the McMurdo Sound Region, Antartica – a progress report, *Amer. J. Sci.,* 257, 545–52.

Piper, C. S., 1947. *Soil and Plant Analysis,* Adelaide Univ. Press, pp. 368.

Pons, L. J. and Zonneveld. I. S., 1965. *Soil Ripening and Soil Classification,* Int. Inst. for Land Reclamation and Improvement. Pub. 13, pp. 128.

Ramann, E., 1911. *Bodenkunde.* Berlin.

Richards, B. N., 1974. *Introduction to the Soil Ecosystem.* Longman, pp. 266.

Richtofen, F. Von, 1886. Führer fur Forschungsreisende, in Glinka, 1914, pp. 23–24.

Riley, D. and Young, A., 1966. *World Vegetation,* Cambridge University Press, pp. 96.

Riquier, J., 1960. Les Phytolithes de Certain Sols Tropicaux et des Podzols, *Trans. 7th Inter. Cong. Soil Sci.,* IV, 425–31.

Rozov, N. N. and Ivanova, E. N., 1967. Classification of the soils of the USSR, *Soviet Soil Sci.,* 2, 147–56, Rosov, N. N. and Ivanova, E. N. In World Soil Resources Report 32, FAO, Rome.

Ruhe, R. V., Daniels, R. B. and Cady, J. G., 1967. *Landscape Evolution and Soil Formation in South Western Iowa,* USDA, Tech. Bull., 1349, pp. 242.

Saitô, T., 1957. Chemical changes in beech litter under microbiological decomposition, *Ecol. Rev. Japan,* **14,** 209–16.

Saitô, T., 1965. Microbiological decomposition of beech litter, *Ecol. Rev. Japan,* **14,** 141–7.

Saussure, Théodore De., 1804. *Recherches Chimiques sur la végétation,* Paris.

Sharpe, C. F. S., 1938. *Landslides and Related Phenomena,* a study of mass-movement of soil and rock. Columbia University Press, New York, pp. 137.

Sibirtsev, N. M., in Glinka 1914.

De'Sigmond, A. A. J., 1933. Principles and scheme of a general soil system, *Soil Res.,* **3,** 103–26.

Simonson, R. W., 1949. Genesis and classification of red-yellow podzolic soils. *Soil Sci. Soc. Amer. Proc.,* **14,** 316–319.

Simonson, R. W., 1968. Concept of soils, *Adv. Agron.,* **20,** 1–47.

Singh, S., 1956. The formation of dark-coloured clay-organic complexes in black soils, *J. Soil Sci.,* **7,** 43–58.

Smyth, A. J. and Montgomery, R. F., 1962. *Soils and Land Use in Central Western Nigeria,* Ibadan, pp. 265.

Sneath, P. H. A. and Sokal, R. R., 1973. *Numerical Taxonomy – The Principles and Practise of Numerical Classification,* W. H. Freeman and Co., San Fransisco, pp. 573.

Sokolovsky, A. N., 1930. The nomenclature of the genetic horizons of the soil. *Proc. 2nd. Inter. Soc. Soil Sci.,* Leningrad – Moscow July 20–31, **5,** 153–4.

Stace, H. T. C., Hubble, G. D., Brewer, R., Northcote, K. H., Sleeman, J. R., Mulcahy, M. J. and Hallsworth, E. G., 1968. *A Handbook of Australian Soils,* Rellim Tech. Press, Adelaide, pp. 435.

Stanek, W. and Silc, T., 1977. Comparison of four methods. for determination of degree of peat humification (decomposition) with emphasis on the Von Post method, *Can. J. Soil Sci.,* **57,** 109–17.

Stebutt, A., 1930. *Lehrbuch der allgemeinen Bodenkunde,* Berlin, pp. 293.

Stewart, V. I. and Adams, W. A., 1968. The quantitative description of soil moisture states in natural habitats with special reference to moist soils, in *The Measurement of Environmental Factors in Terrestrial Ecology,* edited by R. M. Wadsworth. Blackwell Scientific Publications, Oxford and Edinburgh, 161–73.

Stoopes, G. and Jongerius, A., 1975. Proposal for a micro-morphological classification of soil materials. I. A classification of the related distributions of fine and coarse particles, *Geoderma,* **13,** 189–99.

Strahler, A. N., 1970. *Introduction to Physical Geography,* 2nd Edition, Wiley, pp. 457.

Swederski, W., 1931. Untersuchungen über die Gebirgsböden in den Ostkarpaten, I, Mémoires de L'Institut National Polonais d'Economie Rurale a Pulawy, **12,** 115–54.

Taylor, S. A. and Ashcroft, G. L., 1972. *Physical Edaphology,* W. H. Freeman & Co., San Fransisco, pp. 533.

Thorp, J., 1931. The effects of vegetation and climate upon soil profiles in northern and northwestern Wyoming, *Soil Sci.,* **32,** 283–301.

Thorp, J., Johnson, W. M. and Reed, E. C., 1951. Some post-Pliocene buried soils of central United States, *J. Soil Sci.,* **2,** 1–19.

Transeau, E. N., 1905. Forest centers of eastern America, *Am. Naturalist,* **39,** 875–89.

Ugolini, F. C., Bockheim, J. G. and Anderson, D. M., 1973. Soil development and patterned ground evolution in Becon Valley, Antarctica, in permafrost, North American Contribution, Second International Permafrost Conference, Yakutsk, USSR, National Academy of Science Publication 2115, pp. 246–54.

USDA, 1951. *Soil Survey Manual,* Handbook No. 18, pp. 503.

USDA, 1975. *Soil Taxonomy,* Agricultural Handbook No. 436, pp. 754.

Vilensky, D. G., 1927. Concerning the principles of a genetic soils' classification. Contributions to the study of the soils of Ukrania, *1st. Inter. Cong. Soil Sci.,* **6,** 129–51.

Von Post, L., 1922. Sveriges geologiska undersöknings trovinventering och nâgra av dess hittills vunna resultat, *Sv. Mosskulturförening, Tidskr.,* **1,** 1–27.

Waksman, S. A. and Iyer, K. R. N., 1932. Contribution to our knowledge of the chemical nature and origin of humus, *Soil Sci.,* **33,** 43–69.

Walker, G. F., 1949. The decomposition of biotite in the soil, *Mineral Mag.,* **28,** 693–703.

Walscher, H. L., Alexander, J. D., Ray, B. W., Beavers, A. H. and Odell, R. T., 1960. Characteristics of soils associated with glacial tills in north eastern Illinois, *Bull.* 665, Univ. of Ill., pp. 155.

Warcup, J. H., 1967. Fungi in soil, in *Soil Biology,* Ed. A. Burges and H. Raw, Academic Press, pp. 51–100.

Watson, J. P., 1964. A soil catena on granite in Southern Rhodesia, *J. Soil Sci.,* **16,** 31–43.

Webb, I. S., 1973. Pedological studies of some soils of the Solomon Islands, Ph.D. Thesis Univ. Abdn., pp. 154.

Webster, R., 1965. A catena of soils on the northern Rhodesia plateau, *J. Soil Sci.,* **16,** 31–43.

Webster, R., 1968. Fundamental objections to the 7th Approximation, *J. Soil Sci.,* **19,** 354–66.

Webster, R., 1976. The nature of soil variation, *Class. Soc. Bul.,* **3,** 43–55.

West, R. G., 1961. Late and postglacial vegetational history in Wisconsin, particularly changes associated with the Valders readvance, *Amer. J. Sci.,* **259,** 766–83.

Whiteside, E. P., 1953. Some relationships between the classification of rocks by geologists and the classification of soil by soil scientists, *Soil Sci. Soc. Amer. Proc.,* **17,** 138–43.

Whitney, M., 1909. *Soils of the United States,* U.S. Dept. Agr. Bur. Soils Bull. 55. U.S. Govt. Printing Office, Washington.

Williamson, W. D., 1947. The fabric, water distribution, drying – shrinkage and porosity of some shaped discs of clay, *Am. Jour., Sci.,* **245,** 645–62.

Witkamp, M. and van der Drift, J., 1961. Breakdown of forest litter in relation to environmental factors, *Plant and Soil,* **15,** 295–311.

Yaalon, D. H., 1959. Weathering reactions, *J. Chem. Education,* **36,** 73–6.

Yaalon, D. H., 1960. Some implications of fundamental concepts of pedology in soil classification, *Trans. 7th, Inter. Cong. Soil Sci.,* **IV,** 119–23.

Zakharov, S. A., 1930. On the nomenclature of soil horizons, *Proc. 2nd Inter. Soc. Soil Sci.,* Leningrad-Moscow, July 20–31, **5,** 150–2.

Zakharov, S. A., 1931. *A Course in Pedology,* 2nd edition, Moscow.

Index

imperfectly drained soil, 112
improved drainage, 112
Inanda form, 195, 205
Inceptisols, 141–2, 146, 157, 195,
   289, 299, 303
incomplete structure, 98
index
   climatic, 33
   elements, 67
   minerals, 67
Inhoek form, 198, 233, 243
interfingering, diagnostic property, 180
interglacial
   periods, 52
   soils, 51, 304
intergrade
   horizons, 164
   segments, 164
intergrading of horizons, 5, 121, 164,
   311
Intrazonal soils, 124
iron
   alteration compound, 64
   manganese concretions, 116
   microcrystalline oxides, 115
   oxidation of, 68
   Podzolic, 247
   Podzols, 247
   reduction of, 68
ironstone, 295, 306, 307, 308, 309
irregular blocky structure, 100
isomorphous replacement, 9
ison, 248, 252, 253, 254, 304, 327

jaron, 327
jarosite, 212, 318

Kalk Braunerde, 195
kaolinite, 13, 17, 19, 20, 64, 66, 67,
   310, 312, 318–19
kastanon, 231, 327
Kastanozems, 40, 157, 230–4, 295,
   299, 302, 312
Kauri Podzol, 66
K-cycles, in Australia, 301
key for soils, 166
Kranskop form, 185, 195, 205
krasnon, 202, 306, 308, 310, 327
Krasnozems, 189, 190, 204, 242, 287,
   295, 306, 308, 310
Kuroboku, 188
kuron, 186, **327**

labyrinthine structure, 99
laminar structure, 100
landscape models, 46
larger separates, 85–7
laterite, 207–10, 295, 306, 308
Laterite soils, 146
Lateritic Podzolic soils, 190, 236, 268
latitudinal zonation of soils, 293–5
Latosolic brown forest soils, 146
Latosols, 146, 204, 205
350 Lenticular structure, 100

Leptic Podzols, 248
limon, 327
lines, anisotropic, 96
lithon, 327
Lithosols, 146, 190, 232, 233, 234,
   262, 280
litter supply, 39, 40
Lophodermium pinastri, in needle
   decomposition, 73
Low Humic Gley soils, 146
Low Humic Latosols, 146
low latitude climates, 35
luton, 327
Luvisolic soils, 236
Luvisols, 157, **234–9**, 285, 289, 290,
   295, 298, 300, 305, 306, 310
luvon, 244, 327
Luvosols, 184, 242, 287, 295, 302, 311

magnesium hydroxide sheet, structure
   of, 11
Magwa form, 185, 195, 205
mammals, 40–1
manganese dioxide, 66, 110, 167,
   188, 276, 289, 319
Mangrove soils, 215
marblon, 328
marine subarctic climate, 37
marine west coast climate, 38
mass movement, 49
massive structure, 100
master horizons, 170
matrix, 92, 93, 94, 96
Mayo form, 198, 233, 243, 264
Meadow
   Grey Forest soils, 255, 258
   Soils, 217
   Solonchaks, 265, 269
   Solonetz, 269
Mediterranean climate, 36
Medifibrist, 157
Medihemist, 157
mesic temperature regime, 38
Mesiosol, 157
mesofauna, 43–5, 54
metahalloysite, 316–17
mica, 19, 67
micelle, of clay, 17
microcrystalline oxides, 115
microorganism, 41–3, 55
middle-latitude climates, 36
middle latitude desert and steppe
   climates, 36
Milkwood form, 233, 243, 264
minerals
   composition of, 20
   hydrolysis of, 59–62
   in rocks, 9–12
   in soils, 18–19
   resistance to transformation, 64
   solution of, 62
   structure of, 9
minon, 244, 257, 328
Mispah form, 232, 281

mites, 43–4, 74
Moder gley, 217
modon, 184, 215, 244, 248, 328
moisture, 28–35
moles, 40
molecular ratios, 115
mollic A horizon, **173,** 186, 195, 197,
   215, 223, 230, 240, 243, 261,
   263, 270
Mollisols, 142–3, 146, 199, 225, 233
montmorillonite, 14, 17, 19, 20, 58,
   64, 66
Moorböden, 230
morphogenetic regions, 46
mottled clay in Ferralsols, 203–6, *passim*
mottling, 91
movement, *en masse,* 58
moving ice, 46–7
mud polygons, 223, 300
Mull gleys, 217
mullon, 192, 244, 261, 289, 328
muscovite, 12, 17, 18, 19, 20, 64
mutualism, 43
mycorrhiza, 43

Nadurargids, 269
Natralbolls, 269
Natraqualfs, 247, 269, 297, 302
Natrargids, 269
Natriborolls, 157, 269
natric B horizon, **175,** 243, 244, 270,
   298
Natrixeralfs, 269
Natrixerolls, 269
Natrustalfs, 269, 295, 297, 311
Natrustolls, 269
needle decomposition of pine, 73–4
needle ice, 56
nekron, 328
nematodes, 43
Nitosols, **239–40,** 287, 302
nitrification, 74–6
Nitrobacter, 42–74
Nitrococcus, 74
nitrogen
   cycle, 74
   fixation of, 76
   total in soils, 119
Nitrosomonas, 42, 74
Nomanci form, 185, 195
nomenclature and classification, 120–82
Non-calcareous brown soils, 236
Non-Calcic Brown soil, 146
non-silicates, 9, 19–20
n-value, 212

O soils, 158, 214, 230
Oakleaf form, 195, 265, 281
Ochrepts, 142, 195, 284, 290, 304,
   306, 310
ochric A horizon, **174,** 183, 186, 189,
   191, 201, 210, 214, 235, 243, 248,
   260, 263, 270, 282, 283